D1749192

Computational Modeling for Homogeneous and Enzymatic Catalysis

Edited by
Keiji Morokuma and
Djamaladdin G. Musaev

Further Reading

van Santen, R. A., Neurock, M.

Molecular Heterogeneous Catalysis
A Conceptual and Computational Approach

2006
ISBN 978-3-527-29662-0

Dronskowski, R.

Computational Chemistry of Solid State Materials
A Guide for Materials Scientists, Chemists, Physicists and others

2006
ISBN 978-3-527-31410-2

Computational Modeling for Homogeneous and Enzymatic Catalysis

A Knowledge-Base for Designing Efficient Catalysts

Edited by
Keiji Morokuma and Djamaladdin G. Musaev

WILEY-VCH

WILEY-VCH Verlag GmbH & Co. KGaA

The Editors

Prof. Dr. Keiji Morokuma
Emory University
Department of Chemistry
1515, Dickey Drive
Atlanta, GA 30322
USA

Dr. Djamaladdin G. Musaev
Emory University
C.L. Emerson Center
1521, Dickey Drive
Atlanta, GA 30322
USA

■ All books published by Wiley-VCH are carefully produced. Nevertheless, authors, editors, and publisher do not warrant the information contained in these books, including this book, to be free of errors. Readers are advised to keep in mind that statements, data, illustrations, procedural details or other items may inadvertently be inaccurate.

Library of Congress Card No.: applied for

British Library Cataloguing-in-Publication Data
A catalogue record for this book is available from the British Library.

Bibliographic information published by the Deutsche Nationalbibliothek
The Deutsche Nationalbibliothek lists this publication in the Deutsche Nationalbibliografie; detailed bibliographic data are available in the Internet at ⟨http://dnb.d-nb.de⟩.

© 2008 WILEY-VCH Verlag GmbH & Co. KGaA, Weinheim

All rights reserved (including those of translation into other languages). No part of this book may be reproduced in any form – by photoprinting, microfilm, or any other means – nor transmitted or translated into a machine language without written permission from the publishers. Registered names, trademarks, etc. used in this book, even when not specifically marked as such, are not to be considered unprotected by law.

Printed in the Federal Republic of Germany
Printed on acid-free paper

Composition Asco Typesetters, Hong Kong
Printing betz-druck GmbH, Darmstadt
Binding Litges & Dopf GmbH, Heppenheim
Cover Design Adam Design, Weinheim

ISBN 978-3-527-31843-8

Dedicated to Eiko, Matanat, Miki and Aiten

Contents

Preface *XV*

List of Contributors *XVII*

1 Computational Insights into the Structural Properties and Catalytic Functions of Selenoprotein Glutathione Peroxidase (GPx) *1*
Rajeev Prabhakar, Keiji Morokuma, and Djamaladdin G. Musaev

1.1	Introduction *1*	
1.2	Catalytic Functions *3*	
1.2.1	Peroxidase Activity *3*	
1.2.2	Reductase Activity *3*	
1.3	Computational Details *5*	
1.3.1	Computational Methods *5*	
1.3.2	Computational Models *5*	
1.4	Results and Discussion *6*	
1.4.1	Refinement of the Active Site *6*	
1.4.2	Catalytic Functions: Peroxidase Activity *8*	
1.4.3	Catalytic Functions: Effect of the Surrounding Protein on the Peroxidase Activity *12*	
1.4.3.1	Hydrogen Peroxide Coordination *12*	
1.4.3.2	Formation of Selenenic Acid [E–Se–OH] *14*	
1.4.4	Catalytic Functions: Reductase Activity *16*	
1.4.4.1	Peroxynitrite/Peroxynitrous Acid (ONOO$^-$/ONOOH) Coordination *17*	
1.4.4.2	Oxidation Pathway *17*	
1.4.4.3	Nitration Pathways *20*	
1.5	Summary *23*	
	References *24*	

2 A Comparison of Tetrapyrrole Cofactors in Nature and their Tuning by Axial Ligands 27
Kasper P. Jensen, Patrik Rydberg, Jimmy Heimdal, and Ulf Ryde

2.1 Introduction 27
2.2 Methodology 29
2.3 Comparison of the Intrinsic Chemical Properties of the Tetrapyrroles 31
2.3.1 Introduction 31
2.3.2 Spin States 32
2.3.3 Tetrapyrroles Prefer Their Native Ions 33
2.3.4 Cavity Size and Flexibility of the Tetrapyrroles 33
2.3.5 Cytochrome-like Electron Transfer 35
2.3.6 Stability of a Metal–Carbon Bond 36
2.3.7 Metallation Reaction 37
2.4 Tuning of Tetrapyrrole Structure and Function by Axial Ligands 38
2.4.1 Introduction 38
2.4.2 Importance of the Lower Axial Ligand for B_{12} Chemistry 38
2.4.3 Lower Axial Ligand in Cofactor F430 41
2.4.4 Importance of Axial Ligands for the Globins 42
2.4.5 Role of Axial Ligands for the Cytochromes 42
2.4.6 Role of the Axial Ligand in Heme Enzymes 44
2.4.7 Tuning the His Ligand by Hydrogen Bonds in Heme Proteins 47
2.4.8 Axial Ligand in Chlorophylls 50
2.5 Concluding Remarks 51
References 53

3 Modeling of Mechanisms for Metalloenzymes where Protons and Electrons Enter or Leave 57
Per E. M. Siegbahn and Margareta R. A. Blomberg

3.1 Introduction 57
3.2 Energy Diagrams 59
3.2.1 Photosystem II 59
3.2.2 Cytochrome *c* Oxidase 64
3.2.3 Nitric Oxide Reduction 70
3.2.4 NiFe-hydrogenase 72
3.2.5 Molybdenum CO Dehydrogenase 76
3.3 Conclusions 79
References 80

4 Principles of Dinitrogen Hydrogenation: Computational Insights 83
Djamaladdin G. Musaev, Petia Bobadova-Parvanova, and Keiji Morokuma

4.1 Introduction 83
4.2 Reaction Mechanism of the Coordinated Dinitrogen Molecule in Di-zirconocene-N_2 Complexes with a Hydrogen Molecule 87

4.2.1	Mechanism of the Reaction (3)	87
4.2.2	Mechanisms of the Reactions (4) and (5)	89
4.3	Factors Controlling the N_2 Coordination Modes in the Di-zirconocene-N_2 Complexes	91
4.4	Why the $[(\eta^5\text{-}C_5Me_nH_{5-n})_2Ti]_2(\mu_2,\eta^2,\eta^2\text{-}N_2)$ Complex Cannot Add a H_2 Molecule to the Side-on Coordinated N_2, while its Zr- and Hf-analogs Can	95
4.4.1	Relative Stability of the Lowest Singlet (S) and Triplet (T) Electronic States of the Complexes $[(\eta^5\text{-}C_5Me_nH_{5-n})_2M]_2(\mu_2,\eta^2,\eta^2\text{-}N_2)$, II_M (for M = Ti, Zr, and Hf, and $n = 0$ and 4)	96
4.4.2	Reactivity of the Lowest Singlet and Triplet States of the Complexes $[(\eta^5\text{-}C_5Me_nH_{5-n})_2M]_2(\mu_2,\eta^2,\eta^2\text{-}N_2)$, II_M, (for M = Ti, Zr, and Hf, and $n = 0$ and 4) towards H_2 Molecules	99
4.5	Why Dizirconium-dinitrogen Complexes with bis(Amidophosphine) (P2N2) and Cyclopentadienyl (Cp) Ligands React differently with the Hydrogen Molecule: Role of Ligand Environment of the Zr Centers	101
4.6	Several Necessary Conditions for Successful Hydrogenation of a Coordinated Dinitrogen Molecule	103
	Appendix: Computational Details	105
	References	106

5	**Mechanism of Palladium-catalyzed Cross-coupling Reactions**	**109**
	Ataualpa A. C. Braga, Gregori Ujaque, and Feliu Maseras	
5.1	Introduction	109
5.2	Oxidative Addition	112
5.3	Transmetalation	114
5.3.1	Suzuki–Miyaura Reaction	114
5.3.1.1	Role of the Base	115
5.3.1.2	Cis versus Trans Species in bis(Phosphine) Systems	118
5.3.1.3	Monophosphine Systems	120
5.3.2	Stille Reaction	122
5.4	Reductive Elimination	125
5.5	Isomerization	126
5.6	Concluding Remarks	128
	References	129

6	**Transition Metal Catalyzed Carbon–Carbon Bond Formation: The Key of Homogeneous Catalysis**	**131**
	Valentine P. Ananikov, Djamaladdin G. Musaev, and Keiji Morokuma	
6.1	Introduction	131
6.1.1	Catalytic C–C Bond Formation via the Reductive Elimination Pathway	131
6.1.2	Reductive Elimination of Alkyl Groups (Alkyl–Alkyl Coupling)	133

6.2	C–C Coupling of Unsaturated Ligands 133
6.2.1	Reductive Elimination from the Symmetrical $R_2M(PH_3)_2$ Complexes 134
6.2.2	Reductive Elimination from the Asymmetrical $RR'M\,(PH_3)_2$ Complexes 137
6.2.3	Homocoupling versus Heterocoupling Pathways via C–C Reductive Elimination 140
6.2.4	Metal Effect on C–C Reductive Elimination Reaction 141
6.2.5	Ligand Effect on C–C Reductive Elimination Reaction 142
6.2.6	Dissociative Mechanism of C–C Bond Formation 142
6.2.7	Solvent Effect on C–C Reductive Elimination Reaction 144
6.2.8	Reductive Elimination of Unsaturated Organic Molecules Involving Cyano and Carbonyl Groups 144
6.3	Conclusions 145
	References 145

7 Olefin Polymerization Using Homogeneous Group IV Metallocenes   149
Robert D. J. Froese

7.1	Introduction 149
7.2	Computational Details 152
7.3	Results and Discussion 152
7.3.1	Chain Propagation 152
7.3.2	β-Hydride Elimination 164
7.3.3	Chain-transfer-to-monomer 166
7.3.4	Chain-transfer-to-hydrogen 168
7.3.5	Absolute Rates of Reactions 170
7.3.6	Kinetic Considerations 173
7.4	Conclusions 176
	References 177

8 Group Transfer Polymerization of Acrylates with Mono Nuclear Early d- and f-block Metallocenes. A DFT Study   181
Simone Tomasi and Tom Ziegler

8.1	Introduction 181
8.2	Computational Details 182
8.3	Discussion 184
8.3.1	Polymerization Mechanism 184
8.3.1.1	Initiation: Formation of a Metallocene-enolate Complex 184
8.3.1.2	Conformations of the Metallocene-acrylate-enolate Complexes 185
8.3.1.3	Transition States of the Coupling Reaction 188
8.3.1.4	Ring-opening Reactions 190
8.3.2	Kinetic Scheme for the Prediction of Stereoregularity 195
8.3.3	Side Reactions Involving Metallacycles 198
8.3.3.1	Backbiting Reaction in Zr- and Sm-eight-membered Metallacycles 200

8.3.3.2	Proton Transfer in Zr- and Sm-eight-membered Metallacycles 201
8.3.3.3	Backbiting in Ten-membered Metallocenes 201
8.3.4	Polymer Chain Transfer Reactions 204
8.3.5	Minimization of Side Reactions 207
8.4	Conclusions 211
	References 212

9 Insights into the Mechanism of H_2O_2-based Olefin Epoxidation Catalyzed by the Lacunary $[\gamma\text{-}(SiO_4) W_{10}O_{32}H_4]^{4-}$ and di-V-substituted-γ-Keggin $[\gamma\text{-}1,2\text{-}H_2SiV_2W_{10}O_{40}]^{4-}$ Polyoxometalates. A Computational Study 215

Rajeev Prabhakar, Keiji Morokuma, Yurii V. Geletii, Craig L. Hill, and Djamaladdin G. Musaev

9.1	Introduction 215
9.2	Computational Details 217
9.2.1	Methods 217
9.2.2	Models 217
9.3	Results and Discussion 218
9.3.1	Mechanism of the Olefin Epoxidation Catalyzed by the Lacunary POM, $[\gamma\text{-}(SiO_4)W_{10}O_{32}H_4]^{4-}$ (1) 218
9.3.1.1	Formation of W-hydroperoxy (W—OOH) Species 218
9.3.1.2	Formation of Epoxide 220
9.3.1.3	Counter Cation Effect 221
9.3.2	Mechanism of the Olefin Epoxidation Catalyzed by the di-V-substituted γ-Keggin POM $[\gamma\text{-}1,2\text{-}H_2SiV_2W_{10}O_{40}]^{4-}$ (2) 223
9.3.2.1	Mechanism of Reaction (3a) 224
9.3.2.2	Mechanism of the Epoxidation Reaction (4a) 225
9.3.2.3	Counter Cation Effect 226
9.4	Polyoxometalate-catalyzed Ethylene Epoxidation by Hydrogen Peroxide. Comparison of the Hydroxyl Mechanism involving Complexes 1 and 2 228
	References 229

10 C—H Bond Activation by Transition Metal Oxides 231

Joachim Sauer

10.1	Introduction 231
10.2	Gas Phase and Surface Species Considered 232
10.3	Oxidation of Methanol to Formaldehyde 233
10.4	Oxidative Dehydrogenation of Alkanes on Supported Transition Metal Oxide Catalysts 235
10.5	C—H Activation of Alkanes by Transition Metal Oxide Species in the Gas Phase 238
10.6	Conclusions 241
	References 243

11	**Mechanism of Ru- and Mo-catalyzed Olefin Metathesis** 245
	Andrea Correa, Chiara Costabile, Simona Giudice, and Luigi Cavallo
11.1	Introduction 245
11.2	Models and Computational Details 247
11.2.1	Models 247
11.3	Results and Discussion 250
11.3.1	Nomenclature 250
11.3.2	Ru-based Catalysts 251
11.3.3	Mo-based Catalysts 257
11.4	Conclusions 260
	References 261

12	**Heterolytic σ-Bond Activation by Transition Metal Complexes** 265
	Shigeyoshi Sakaki, Noriaki Ochi, and Yu-ya Ohnishi
12.1	Introduction 265
12.2	Characteristic Features of Heterolytic σ-Bond Activation 266
12.2.1	Theoretical Study of the Shilov Reaction 266
12.2.2	Fujiwara–Moritani Reaction: Driving Force and Orbital Interaction 267
12.3	Heterolytic C–H σ-Bond Activation of Methane 270
12.3.1	Methane C–H Bond Activation by Pt(II) Complexes 270
12.3.2	Methane C–H bond Activation by Late Transition Metal Catalysts 273
12.3.3	Heterolytic Methane C–H Bond Activation by Early Transition Metal Systems 274
12.4	Heterolytic σ-Bond Activation of Dihydrogen and Alcohol Molecules 277
12.4.1	Heterolytic H–H Bond Activation 277
12.4.2	Heterolytic σ-Bond Activation in Hydrogen Transfer Reactions 280
12.5	Summary 282
	References 282

13	**Hydrosilylation Reactions Discovered in the Last Decade: Combined Experimental and Computational Studies on the New Mechanisms** 285
	Yun-Dong Wu, Lung Wa Chung, and Xin-Hao Zhang
13.1	Introduction 285
13.2	General Mechanistic Pathways in the 20th Century 286
13.2.1	Chalk–Harrod and Modified Chalk–Harrod Mechanisms 286
13.2.2	Stereochemistry 289
13.2.3	Carbonyl Compounds 291
13.3	New Mechanistic Pathways Discovered at the Beginning of the 21st Century 292
13.3.1	Main Group Metal Complexes (K, Ca, Sr and B) 292
13.3.2	Early Transition Metal Complexes (Zr and Ta) 294

13.3.2.1	Alkenes *294*	
13.3.2.2	Dinitrogen *296*	
13.3.3	Middle Transition Metal Complexes (Mo, W and Re) *298*	
13.3.4	Late Transition Metal Complexes (Ru and Os) *303*	
13.3.4.1	Alkynes *303*	
13.3.4.2	Alkenes *309*	
13.4	Conclusions *312*	
	References *313*	

14   Methane Hydroxylation by First Row Transition Metal Oxides *317*
Kazunari Yoshizawa

14.1	Introduction *317*
14.2	Reactivity of the MO$^+$ Species *319*
14.3	Energy Profile for Methane Hydroxylation *320*
14.4	Intrinsic Reaction Coordinate (IRC) Analysis *324*
14.5	Spin–Orbit Coupling (SOC) in Methane Hydroxylation *326*
14.6	Kinetic Isotope Effect (KIE) for H-atom Abstraction *329*
14.7	Regioselectivity in Alkane Hydroxylation *331*
14.8	Concluding Remarks *334*
	References *335*

15   Two State Reactivity Paradigm in Catalysis. The Example of X–H (X = O, N, C) and C–C Bonds Activation Mediated by Transition Metal Compounds *337*
Maria del Carmen Michelini, Ivan Rivalta, Nino Russo, and Emilia Sicilia

15.1	Introduction *337*
15.2	General Methods *338*
15.3	Activation of X–H (X = O, N, C) Bonds by First-row Transition Metal Cations *339*
15.3.1	Reaction of the First-row Transition Metal Ions with a Water Molecule *340*
15.3.2	Insertion of First-row Transition Metal Ions into the N–H bond of Ammonia *343*
15.3.3	Bond Activation of CH_4 by bare First-row Transition Metal Cations *347*
15.4	Activation of C–C Bond: Cyclotrimerization of Acetylene by Second-row Transition Metal Atoms *351*
15.5	Use of the Electron Localization Function to Characterize the Bonding Evolution in Reactions involving Transition Metals *355*
15.5.1	Reaction of Mn$^+$ (in its ^{7}S and ^{5}S States) with H_2O, NH_3 and CH_4 Molecules *356*
	References *361*

Subject Index *367*

Preface

The design and realization of catalysts capable of a particular task is a challenge and has long occupied most capable investigators in homogeneous catalysis, heterogeneous catalysis, and biocatalysis. It requires knowledge of the fundamental principles of catalysis, as well as the understanding of mechanisms and factors controlling the catalytic processes. Success in this complex field requires comprehensive and collaborative experimental and theoretical investigations.

The development of new computational methods (like hybrid density functional, integrated molecular orbital and molecular mechanics, QM/MM, local correlation methods, and more) and the increase in computer power have opened up new opportunities for guiding the design of more efficient catalysts for various vital problems (like olefin polymerization, organic oxidation, nitrogen fixation, water oxidation, hydrocarbon functionalization, the design of new and more efficient drugs against numerous diseases, etc.) using computational approaches. Several research groups around the world have seized this opportunity and actively applied modern computational tools to many important catalytic processes. The research covered in the chapters of this book, written by experts in the field, exemplifies the strength and also the present limitations of quantum chemical methods for giving insights into the mechanism of transition metal and enzyme catalyzed reactions.

The first three chapters deal with understandings the mechanism and controlling factors of important biological/enzymatic catalytic processes. Namely, Chapter 1 deals with selenoproteins and their peroxidase and reductase activities with a goal of designing new and more efficient anti-cancer drugs against various diseases. Chapters 2 and 3 deal with computational studies of metalloenzymes that aim to guide the engineering of better biomimetic catalysts for oxidation and hydrogenation processes.

Activation of the N≡N triple bond and its utilization is a global problem that occupies the minds of many scientists. The authors of Chapter 4 describes the fundamental principles of hydrogenation of dinitrogen, i.e., utilization of two inert gases, N_2 and H_2, to produce ammonia. Theoretical predictions presented in this chapter have led to the discovery several new catalysts for dinitrogen hydrogenation.

Another interesting issue of modern organometallic chemistry is the search for

Computational Modeling for Homogeneous and Enzymatic Catalysis.
A Knowledge-Base for Designing Efficient Catalysts. K. Morokuma and D. G. Musaev (Eds.)
Copyright © 2008 WILEY-VCH Verlag GmbH & Co. KGaA, Weinheim
ISBN: 978-3-527-31843-8

new and more efficient tools for organic synthesis. In this respect, understanding transition metal catalyzed cross-coupling, oxidative addition, transmetalation, reductive elimination and metathesis reactions is extremely important. Chapters 5 and 6 beautifully demonstrate the advantages of computation approaches in elucidating the mechanism and controlling factors of these important reactions.

Chapters 7 and 8 deal with the understanding of industrially important olefin polymerization catalyzed by early d- (group IV) and f-block metallocenes. These studies are instrumental for building more efficient catalysts for olefin polymerization, and for elucidating the factors controlling the structure (branched vs. linear) and physicochemical properties of polymers. Chapter 11 subsequently continues the theme transition metal catalyzed olefin polymerization, dealing with the mechanism of Ru- and Mo-catalyzed olefin metathesis.

The use of nano-scaled catalysts has opened up a new era in catalysis science. Size-selectivity, cooperativity, and bifunctionality of nano-structures make this aggregate stage of species very attractive. Chapter 9 addresses some key issues of organic oxidation in depth by metal oxide cluster anions (polyoxometalates or "POMs"). This chapter addresses a topic that has been the focus of considerable experimental and computational research – the selective (non-radical) oxidation of organic substrates by O2 catalyzed by metal centers, and, in particular, by POMs substituted with two adjacent d-electron-containing redox-active metals and by lacunary POMs. Chapter 10 deals with supported, as well as gas-phase, metal oxide catalysts. It elucidate the role of the support in catalysis, which may have a significant effect.

Transition metal catalyzed σ-bond activation (including the C–H, Si–H, H–H, etc.) is another fundamental process of organometallic chemistry, an understanding of which facilitates the utilization of hydrocarbons, dihydrogen, silanes and so for. Chapters 12–15 elucidate the mechanisms of Pd(II)- and Pt(II)-complexes, "naked" first-row transition metal oxides and transition metal atoms and cations catalyzed alkane oxidation, hydrosilylation and, in general, -bond activation reactions.

This book provides a faithful snapshot of the status of computational studies of transition metal- and metalloenzyme-catalyzed reactions. It also provides principles for designing better catalysts for many vital processes. The various chapters are of interest to both experimentalists and theoreticians.

We thank all the contributors for their positive response to our call, their dedicated efforts and timely submission of their chapters. We also thank the people at Wiley-VCH, especially Elke Maase and Rainer Muenz for their support and patience.

Finally, we thank the reviewers for their hard work and timely submission of their very insightful comments and suggestions.

November 31, 2007

Jamal Musaev and Keiji Morokuma
Emory University, Atlanta, GA, USA

List of Contributors

Valentine P. Ananikov
N.D. Zelinsky Institute of Organic
Chemistry
Russian Academy of Sciences
Leninski Prospect 47
Moscow 119991
Russia

Margareta R. A. Blomberg
Stockholm University
Arrhenius Laboratory
Department of Biochemistry and
Biophysics
106 91 Stockholm
Sweden

Petia Bobadova-Parvanova
Department of Chemistry
1515 Dickey Dr.
Atlanta, GA 30322
USA

Ataualpa A. C. Braga
Institute of Chemical Research of
Catalonia (ICIQ)
Av. Països Catalans, 16
43007 Tarragona
Spain

Luigi Cavallo
University of Salerno
Department of Chemistry
Via Salvador Allende
84081 Baronissi (SA)
Italy

Lung Wa Chung
The University of Science and
Technology
Department of Chemistry
Clear Water Bay
Kowloon
Hong Kong

Andrea Correa
University of Salerno
Department of Chemistry
Via Salvador Allende
84081 Baronissi (SA)
Italy

Chiara Costabile
University of Salerno
Department of Chemistry
Via Salvador Allende
84081 Baronissi (SA)
Italy

Robert D. J. Froese
The Dow Chemical Company
Midland, MI 48674
USA

Yurii V. Geletii
Cherry L. Emerson Center for
Scientific Computation
Emory University
Department of Chemistry
1515 Dickey Dr.
Atlanta, GA 30322
USA

Simona Giudice
University of Salerno
Department of Chemistry
Via Salvador Allende
84081 Baronissi (SA)
Italy

Jimmy Heimdal
Department of Theoretical
Chemistry
Lund University
Chemical Centre
P.O. Box 124
221 00 Lund
Sweden

Craig L. Hill
Cherry L. Emerson Center for
Scientific Computation
Emory University
Department of Chemistry
1515 Dickey Dr.
Atlanta, GA 30322
USA

Kasper P. Jensen
Technical University of Denmark
Department of Chemistry
Kemitorvet 207
2800 Kgs. Lyngby
Denmark

Feliu Maseras
Institute of Chemical Research of
Catalonia (ICIQ)
Av. Països Catalans, 16
43007 Tarragona
Spain

Maria del Carmen Michelini
University of Calabria
Department of Chemistry
via P. Bucci
Cubo 14/C Floor 7
87030 Arcavacata di Rende (CS)
Italy

Keiji Morokuma
Cherry L. Emerson Center for Scientific
Computation
Emory University
Department of Chemistry
1515 Dickey Dr.
Atlanta, GA 30322
USA

Djamaladdin G. Musaev
Cherry L. Emerson Center for Scientific
Computation
Emory University
1515 Dickey Dr.
Atlanta, GA 30222
USA

Noriaki Ochi
Kyoto University
Graduate School of Engineering
Department of Molecular Engineering
Nishikyo-ku
Kyoto 615-8510
Japan

Yu-ya Ohnishi
Kyoto University
Graduate School of Engineering
Department of Molecular
Engineering
Nishikyo-ku
Kyoto 615-8510
Japan

Rajeev Prabhakar
University of Miami
Department of Chemistry
1301 Memorial Dr.
Coral Gables, FL 33146
USA

Ivan Rivalta
University of Calabria
Department of Chemistry
via P. Bucci
Cubo 14/C Floor 7
87030 Arcavacata di Rende (CS)
Italy

Nino Russo
University of Calabria
Department of Chemistry
via P. Bucci
Cubo 14/C Floor 7
87030 Arcavacata di Rende (CS)
Italy

Patrik Rydberg
Department of Theoretical
Chemistry
Lund University
Chemical Centre
P.O. Box 124
221 00 Lund
Sweden

Ulf Ryde
Department of Theoretical Chemistry
Lund University
Chemical Centre
P.O. Box 124
221 00 Lund
Sweden

Shigeyoshi Sakaki
Kyoto University
Graduate School of Engineering
Department of Molecular Engineering
Nishikyo-ku
Kyoto 615-8510
Japan

Joachim Sauer
Humboldt Universität
Institut für Chemie
Arbeitsgruppe Quantenchemie
Unter den Linden 6
10099 Berlin
Germany

Emilia Sicilia
University of Calabria
Department of Chemistry
via P. Bucci
Cubo 14/C Floor 7
87030 Arcavacata di Rende (CS)
Italy

Per E. M. Siegbahn
Stockholm University
Arrhenius Laboratory
Department of Biochemistry and
Biophysics
106 91 Stockholm
Sweden

Simone Tomasi
University of Calgary
Department of Chemistry
University Drive 2500
Calgary
Alberta T2N 1N4
Canada

Gregori Ujaque
Unitat de Química Física
Edifici Cn
Universitat Autónoma de
Barcelona
08193 Bellaterra
Spain

Yun-Dong Wu
The University of Science and
Technology
Department of Chemistry
Clear Water Bay
Kowloon
Hong Kong

Kazunari Yoshizawa
Institute for Materials Chemistry and
Engineering
Kyushu University
Fukuoka 819-0395
Japan

Xin-Hao Zhang
The University of Science and
Technology
Department of Chemistry
Clear Water Bay
Kowloon
Hong Kong

Tom Ziegler
University of Calgary
Department of Chemistry
University Drive 2500
Calgary
Alberta T2N 1N4
Canada

1
Computational Insights into the Structural Properties and Catalytic Functions of Selenoprotein Glutathione Peroxidase (GPx)

Rajeev Prabhakar, Keiji Morokuma, and Djamaladdin G. Musaev

1.1
Introduction

Selenium (Se) was discovered in 1817, and in mammals its deficiency has been associated with many fatal diseases, such as cancer, HIV, cardiovascular and Keshan-Back [1, 2]. In most selenoproteins discovered so far, Se is present as a selenocysteine residue that plays an important role in the catalytic activities of these enzymes [3]. One of the first selenoproteins to be discovered was Glutathione Peroxidase (GPx), which demonstrated a strong anti-oxidant activity [4, 5]. This protein reduces numerous reactive oxygen species (ROS), including hydrogen peroxide [4] and peroxynitrite [6], by utilizing various reducing substrates, and protects cell membranes and other cellular components against oxidative damage [4].

Four different classes of Se-dependent GPx have been classified: (1) cytosolic (GPx-1), (2) gastrointestinal tract (GPx-2), (3) extracellular (GPx-3) and (4) phospholipid hydroperoxide (GPx-4) [3]. All four types are easily reduced by glutathione while they also accept some other reducing substrates. For example, GPx-3 can be reduced by glutaredoxin and thioredoxin [7], and GPx-4 utilizes protein thiols as reducing substrates [8, 9]. The catalytic activity of GPx has been mimicked in various organoselenium compounds possessing a direct Se—N bond; the most prominent among them is an anti-inflammatory drug called ebselen [10–12].

The only crystal structures elucidated are of bovine erythrocyte (intracellular enzyme, GPx-1) and human plasma (extracellular enzyme, GPx-3) GPx, resolved at 2.0 and 2.9 Å resolution, respectively [13, 14]. The complete X-ray structure of human plasma GPx (2.9 Å) is available in the literature [4]. These enzymes are tetrameric, with two asymmetric units containing two dimers that exhibit half-site reactivity. The dimeric X-ray structure is shown in Figure 1.1 and each monomer contains a critical selenocysteine residue. The following experimental observations explicitly implicate this residue in the catalytic cycle: (a) Treatment of the oxidized enzyme with cyanide destroys the catalytic activity and releases selenium from the enzyme [15]. (b) Iodoacetate inhibits only the substrate-reduced enzyme and reacts with the selenocysteine residue [16]. (c) A photoelectron spectroscopic study shows that the redox state of selenium in GPx depends on the substrate present [17]. (d) In the crystalline state the selenium sites of the oxidized enzyme can be reduced [18]. (e) The formation of E–Se–S–G complex through selenosulfide linkage has been demonstrated [19]. The overall active site structures of GPx-1 and GPx-3 enzymes are very similar, while the environments around their selenocysteine residue are quite different. Only half of the residues at the active site, within a range of 10 Å, are conserved in both enzymes [14].

As shown in Figure 1.1(C) at the active site the selenocysteine residue of the enzyme exists in the "resting" seleninic acid, E–Se(O)(OH), form. The active site residues Gln83 and Trp157 are located within hydrogen bonding distance of the selenium atom and have been suggested to play a critical role in the catalysis [13]. These two residues are conserved in the entire glutathione peroxidase superfamily and their homologues, which probably accounts for the similarities in

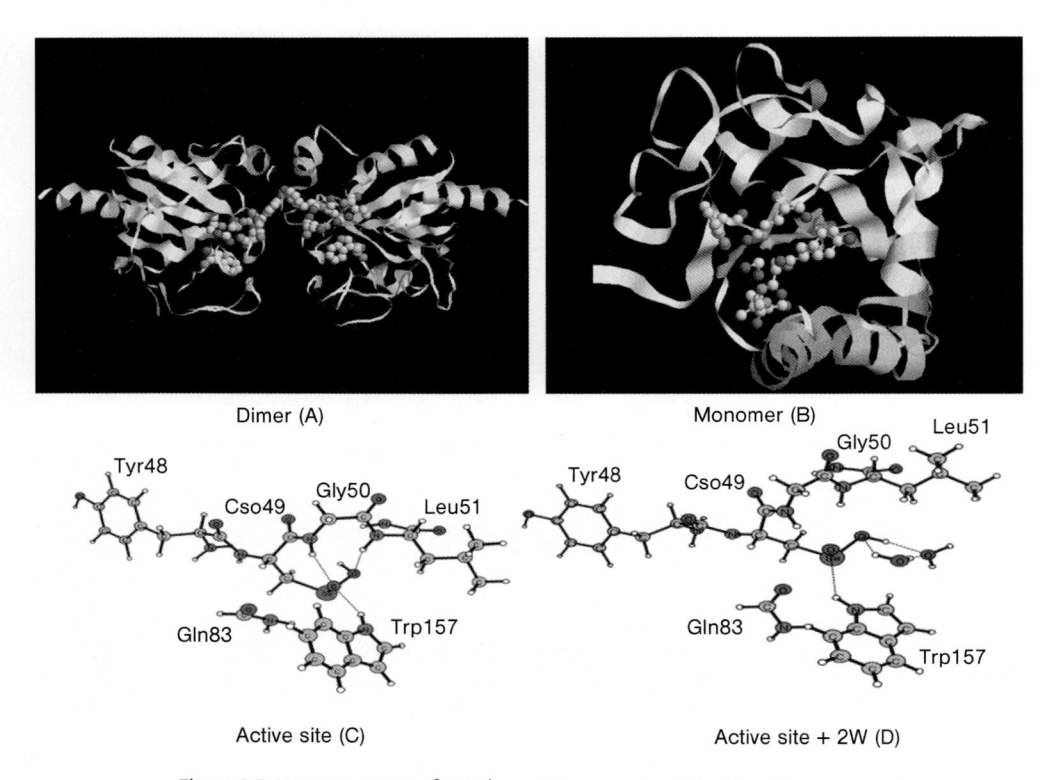

Figure 1.1 X-ray structures of (A) dimer, (B) monomer, (C) active site, and (D) optimized structure of the active site including two water molecules.

their catalytic mechanisms [20]. The active site seleninic acid residue is coordinated to Gly50 and Tyr48 in a tetradic arrangement [14]. The catalytically active form of GPx could either be as selenolate anion (E–Se$^-$) or selenol (E–SeH) [13].

1.2 Catalytic Functions

Based on substrate specificity the catalytic function of GPx enzyme is divided into two parts: (1) Peroxidase activity, and (2) Reductase activity.

1.2.1 Peroxidase Activity

GPx catalyzes the reduction of hydrogen peroxide, utilizing two molecules of glutathione. The proposed mechanism for the peroxidase activity is described by three elementary reactions (Figure 1.2). In the first step, Reaction (1), hydrogen peroxide reduction is accompanied by oxidation of the selenolate anion (or selenol) to selenenic acid. The experimentally measured rate for this reaction of 0.51 s^{-1} corresponds to a barrier of 14.9 kcal mol^{-1} [21]. In the second step, Reaction (2), the formed selenenic acid subsequently reacts with the substrate glutathione (GSH) to produce a seleno-sulfide adduct (E–Se–SG), which has been observed [19]. In the third step, Reaction (3), a second molecule of GSH attacks the seleno-sulfide adduct to regenerate the active form of the enzyme and form the disulfide (GS–SG). This step is suggested to be the rate-determining step of the entire mechanism [22]. Thus, in the full catalytic cycle, two equivalents of GSH are consumed to produce the disulfide and water.

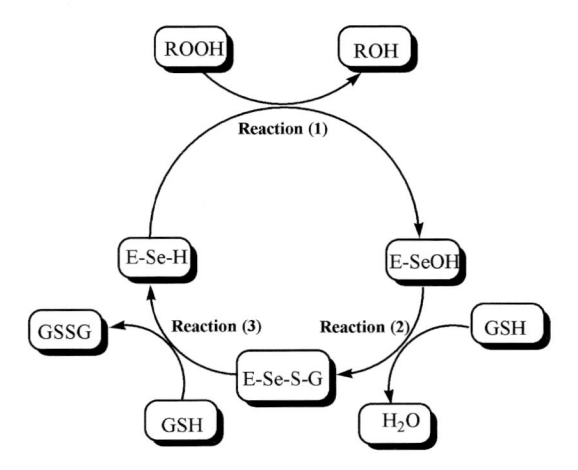

Figure 1.2 Experimentally suggested mechanism for the peroxidase activity of GPx.

1.2.2
Reductase Activity

Based on the extensive experimental [14, 23, 24] and theoretical [25] information for the peroxidase activity of GPx, the proposed catalytic mechanism for peroxynitrite/peroxynitrous acid (ONOO$^-$/ONOOH) reduction by this enzyme is shown in Figure 1.3. The suggested mechanism incorporates the fact that under physiological conditions the peroxynitrite reductase activity of GPx *in vivo* can follow either the "oxidation" or "nitration" of the critical selenocysteine residue [13, 24]. Therefore, these two processes are referred as "oxidation" and "nitration" pathways, respectively. As discussed below, only the first step, Reaction (1), of these pathways is completely different. In the "oxidation" pathway, ONOO$^-$/ONOOH reduction is accompanied by the oxidation of the selenol (E–SeH) to the selenenic acid (E–Se–OH), whereas in the "nitration" pathway the selenocysteine residue is nitrated by ONOO$^-$/ONOOH to generate E–Se–NO$_2$. For substrate ONOO$^-$, the experimentally measured rate for the formation of E–Se–OH expressed per monomer of reduced GPx of $2.0 \pm 0.2 \times 10^6$ M^{-1} s^{-1} corresponds to a barrier of 8.8 kcal mol^{-1} [23]. After the oxidation or nitration of the selenocysteine residue, the subsequent chemistry, Reactions (2) and (3) (Figure 1.2), in both pathways is similar. In the second step, Reaction (2), substrate glutathione (GSH) reacts with E–Se–OH or E–Se–NO$_2$ to produce seleno-sulfide adduct (E–Se–S–G). In Reaction (3), a second molecule of GSH attacks the seleno-sulfide adduct to regenerate the active form of the enzyme and form the disulfide (GS–SG).

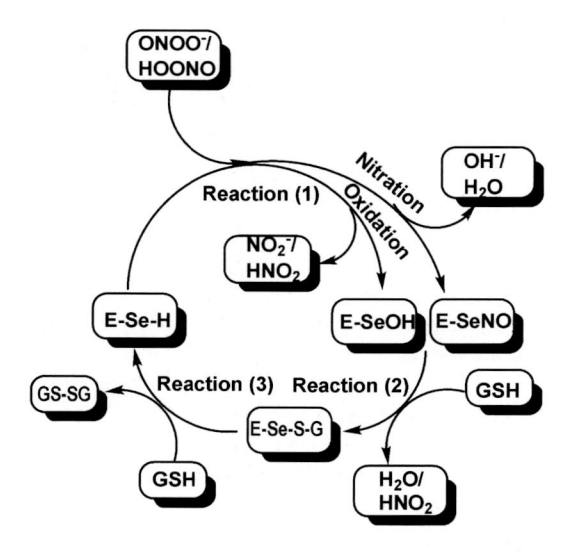

Figure 1.3 Experimentally suggested mechanism for the reductase activity of GPx.

Despite the availability of a great wealth of experimental information, the catalytic mechanism of GPx enzyme and the factors controlling its activity were not known with certainty. We have employed high-level quantum chemical approaches, namely pure quantum mechanics (QM) and hybrid quantum mechanics/molecular mechanics (QM/MM), incorporating all the available experimental information, to elucidate the catalytic mechanism of this interesting enzyme.

1.3
Computational Details

1.3.1
Computational Methods

All calculations were performed using the *Gaussian 03* program [26]. In the pure QM approach, the geometries of reactants, intermediates, transition states and products were optimized without any symmetry constraint using the B3LYP method [27] with the 6-31G(d) basis set. All degrees of freedom of proposed structures were optimized and frequency calculations were performed for all optimized minima and transition states. It was confirmed that the calculated minima have no imaginary frequency, while all transition states have one imaginary frequency corresponding to the reaction coordinate. The final energetics of the optimized structures were improved by performing single point calculations using a triple-zeta quality basis set 6-311+G(d,p). Since it was computationally not feasible to calculate zero-point energy and thermal corrections using the large basis set, these effects were estimated at the B3LYP/6-31G(d) level and added to the final B3LYP/6-311+G(d,p) energetics. This type of correction is an adequate approximation and has commonly been used in quantum chemical studies [28]. Dielectric effects from the surrounding environment were estimated using the self-consistent reaction field IEF-PCM method [29] at the B3LYP/6-31G(d) level. These calculations were performed with a dielectric constant of 4.3 corresponding to diethyl ether, close to 4.0 generally used to describe the protein surrounding. Throughout this chapter the energies obtained at the B3LYP/(6-311+G(d,p)) + zero-point energy (un-scaled) and thermal corrections (at 298.15 K and 1 atm) + solvent effects (the last three terms at B3LYP/6-31G(d) level) are used, while energies without the solvent effects are provided in parentheses.

The hybrid QM/MM approach allows the explicit inclusion of both steric and electrostatic effects from the protein surroundings. In this approach, the entire system (referred to as the "real system") is divided into two subsystems. The QM region (or model system) contains the active site, and is treated by quantum mechanics [B3LYP/6-31G)d level], and the MM region is treated at the MM level using the Amber force Field, and contains the rest of the protein. We used the two layer ONIOM(QM:MM) method [30, 31]. In this method, the interface between the QM region and the MM regions is treated by link atoms, and the interaction between the two layers is included at the classical level (mechanical embedding) [32].

1.3.2
Computational Models

Experimental studies on bovine erythrocyte GPx suggested a half-site reactivity of the enzyme [13], which justifies the use of the active site of only a monomer to investigate the enzyme reactivity. In the "active site only" approach, the first question to be considered is the choice of an appropriate model for the enzyme active site that retains all its basic features. Since the selenocysteine residue is, experimentally, suggested to play a critical role in the catalytic cycle, it is included in the model. The active site Gln83 and Trp157 residues, conserved in all known GPx's and experimentally suggested to be involved in the catalytic mechanism [13], are also included in the model. In addition, in the X-ray structure [13], the Tyr48, Gly50 and Leu51 residues form a part of the cage around the selenocysteine residue; therefore, they are also included in the model. Since the active site of GPx has been suggested to contain water molecules, a water molecule is also included in the active site model. Based on earlier experience, glutamine and tryptophan residues are modeled by formamide and indole, respectively. The substrate GSH (γ-glutamylcysteinylglycine, γ-GluCysGly) is a tri-peptide, which is modeled by ethanethiol (C_2H_5SH).

In the QM:MM approach, the entire monomer of GPx enzyme has been utilized as a model. In these studies, the aforementioned model of the active site forms the QM region, whereas the remaining part of the monomer constitutes the MM region. The overall system consists of 3113 atoms (86 in the QM region and 3027 in the MM region). Hydrogen atoms not included in the PDB-structure were added to the system, containing 196 amino acid residues, using the *GaussianView* program [33].

1.4
Results and Discussion

1.4.1
Refinement of the Active Site

The most notable difference between the active sites of GPx-1 and GPx-3 enzymes is the presence of two water molecules in the former, which, due to the low resolution of the X-ray structure, could not be observed in the latter. First, we explored the presence of water molecules at the active site of GPx-3 using the two-layer ONIOM (QM:MM) method [34]. In this study, the entire monomer (system **I**, Figure 1.1B) is chosen as a "real" system, extracted from the dimeric X-ray structure [4] (Figure 1.1A). The choice of monomeric unit is justified by the fact that in the crystal structure of the mammalian GPx (at 2.9 Å resolution) the active site selenocysteine residues are well separated, with an Se–Se distance of 23.2 Å.

Based on experimental information, the "active" part (Figure 1.1C) of the "real" system **I** includes SeCys, Tyr48, Gly50, Leu51, Gln83, and Trp157 residues. Later, system **I** was extended by adding two water molecules in the "active" part (Figure 1.1D) and called system **I(2W)**. All the structures belonging to the systems **I** and **I(2W)** are fully optimized at the Amber, ONIOM(HF/STO-3G:Amber)_ME and ONIOM(B3LYP/6-31G(d):Amber)_xx levels, where xx = ME or EE represents the mechanical embedding (ME) and the electronic embedding (EE) schemes.

Table 1.1 gives all the results, and comparisons have been made by the means of root-mean-square (RMS) deviations between the X-ray and optimized structures using only non-hydrogen atoms. The RMS deviations are 1.72, 1.71, and 1.71 Å between the X-ray and the Amber, ONIOM(HF/STO-3G:Amber)_ME and ONIOM(B3LYP/6-31G(d):Amber)_ME structures, respectively. Treatment of QM-MM interactions using the electronic embedding [ONIOM(B3LYP/6-31G(d):Amber)_EE] scheme gives an almost similar RMS deviation of 1.73 Å. These results indicate that irrespective of the method used the calculated RMS deviations remain the same and significant. The largest deviations between the calculated and X-ray structures correspond to the residues positioned near the second monomer, which should be improved with the inclusion of the second monomer into calculations. However, such large calculations were, technically, not feasible.

The RMS deviation of the critical active-site atoms is calculated to be 1.04 and 1.22 Å for the Amber and ONIOM(HF/STO-3G:Amber)_ME levels, respectively. Application of a higher-level method, ONIOM(B3LYP/6-31G(d):Amber)_ME, reduces it to 0.97 Å, which is still significant.

One major reason for such a significant deviation between the calculated and X-ray structure could be the low resolution (2.9 Å) of the X-ray structure. To corroborate the existence of water molecules at the active site of GPx-3 (2.9 Å), calculations have been performed that include two water molecules at the active site of system **I** from the X-ray structure of GPx-1 (2.0 Å). However, other possibilities, suggesting the presence of either one or more than two water molecules, can not be ruled out. As shown in Table 1.1, the inclusion of these two water molecules

Table 1.1 Calculated RMS deviations (in Å) for the monomer and the active site.

	Model	Monomer	Active site
X-ray – Amber	I	1.72	1.04
X-ray – ONIOM(HF:Amber)_ME	I	1.71	1.22
X-ray – ONIOM(B3LYP:Amber)_ME	I	1.71	0.97
X-ray – ONIOM(B3LYP:Amber)_EE	I	1.73	1.17
X-ray – ONIOM(B3LYP:Amber)_ME	I(2W)		0.79
X-ray – B3LYP	I, "active-site only"		2.26
X-ray – B3LYP	I(2W), "active-site only"		1.48

[system **I(2W)**] indeed reduces the RMS deviation to 0.79 Å. This result suggests the existence of two bound water molecules at the active site of the mammalian GPx.

This conclusion is also supported by the "active-site only" calculations, which give a very large RMS deviation of 2.26 Å for the active site of system **I**. The inclusion of two water molecules into the "active-site only" system [active site of system **I(2W)**] reduces this deviation to 1.48 Å, which clearly indicates the importance of these two water molecules at the active site. However, notably, even the RMS deviation of 1.48 Å for the "active-site only" calculation of system **I(2W)** is still much larger than 0.79 Å obtained on including the protein environment. Thus this result explicitly demonstrates the significance of the protein–active-site interaction for the refinement of the active-site structure and, consequently, for the enzyme activity. Without excluding other possibilities, this study suggests that the active site of the 2.9 Å resolution X-ray structure needs to be complemented by two water molecules, which could be crucial for the catalytic activity of the enzyme.

1.4.2
Catalytic Functions: Peroxidase Activity

The catalytic cycle of the peroxidase activity of GPx can be described by the following overall reaction:

$$[E-SeH](H_2O) + HOOH + 2RSH \rightarrow [E-SeH](H_2O)_2 + RS-SR + H_2O$$

As discussed above, this cycle consists of three elementary reactions and has been investigated at the B3LYP level [35]. In the first elementary reaction, $[E-SeH] + H_2O_2 \rightarrow [E-SeOH] + H_2O$ (1), the active selenol [E–SeH] form of GPx takes up an hydrogen peroxide molecule and produces selenenic acid [E–SeOH] and a water molecule (Figure 1.4). The first step of Reaction (1) is the coordination of hydrogen peroxide molecule to the active site selenol [E–SeH] state of the enzyme to form a complex **II**. In **II**, the H_2O_2 molecule binds to the active site by forming strong hydrogen bonds with the water molecule and the Gly50 residue. This process is found to be exothermic by 6.3 (13.5) kcal mol^{-1}. From the complex **II**, the reaction is suggested to proceed via a stepwise pathway, which is divided into two parts: (a) proton transfer from [E–SeH] to the Gln83 residue through a hydrogen peroxide and a water molecule, leading to the formation of [E–Se$^-$] and protonated Gln83 (**III**), and (b) O–O bond cleavage via protonation of the terminal oxygen atom of H_2O_2 by the proton earlier transferred to Gln83 residue. As a result, a water molecule and selenenic acid (**IV**) are formed.

In the first part, the Se–H bond of the selenol [R–Se$^-$] is broken and, simultaneously, the proton is transferred through the oxygen atom of hydrogen peroxide and a water molecule to the neighboring Gln83, providing an intermediate product **III** (Figure 1.4). As a result, the selenolate anion [R–Se$^-$] and protonated Gln83 are formed. DFT calculations show that the first part of this stepwise path-

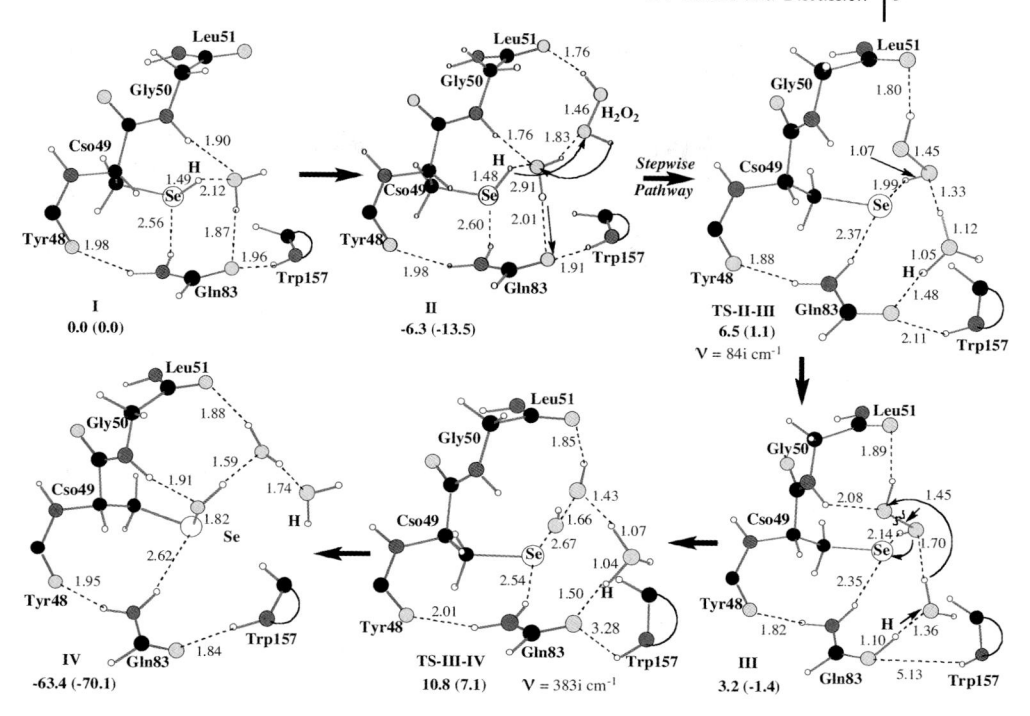

Figure 1.4 Optimized structures (in Å) and energies [with and without (in parenthesis) solvent effects, in kcal mol^{-1}] of reactant, intermediates and transition states for the stepwise mechanism of the reaction [E−SeH] + H$_2$O$_2$ → [E−SeOH] + H$_2$O (Reaction 1).

way occurs through transition state **TS-II-III** with a 12.8 (14.6) kcal mol^{-1} barrier and is endothermic by 9.5 (11.9) kcal mol^{-1}.

In the second part, the O−O bond of H$_2$O$_2$ is cleaved and the hydroxyl of H$_2$O$_2$ is transferred to the selenolate anion [E−Se$^-$]. In this part, the proton previously transferred to Gln83 residue moves to the terminal oxygen atom of H$_2$O$_2$ and cleaves the O−O bond, forming selenenic acid and a water molecule (**IV**). The barrier for this process is 7.6 (8.5) kcal mol^{-1}. Since the O−O bond cleavage is followed by the formation of **III**, which is endothermic by 9.5 (11.9) kcal mol^{-1}, the overall barrier for the formation of the selenenic acid [E−SeOH], Reaction (1), becomes 17.1 (20.6) kcal mol^{-1}. The calculated barrier for the hydrogen peroxide reduction is in excellent agreement with the experimentally measured barrier of 14.9 kcal mol^{-1} [21]. Reaction (1) is calculated to be exothermic by 63.4 (70.1) kcal mol^{-1}.

Based on the results obtained in the first part of this reaction, the Gln83 residue is suggested to play a key role of a proton acceptor, which is consistent with the experimental proposal indicating that the Gln83 residue participates in the catalytic cycle [13].

The second elementary reaction, [E–SeOH] + GSH → [E–Se–SG] + HOH (2), starts with the coordination of the first glutathione molecule to the previously formed selenenic acid [E–SeOH] to give a weak selenenic acid–glutathione complex (**V**) (Figure 1.5). The binding energy of [E–SeOH]–(GSH) is calculated to be 4.2 (7.5) kcal mol^{-1}. The reaction proceeds further through the transition state **TS-V-VI**, climbing a barrier of 17.9 (22.6) kcal mol^{-1}. In this process, synchronously, the S–H bond of the glutathione is broken and a proton is transferred to the hydroxyl group of the selenenic acid accompanied by the formation of a Se–S bond. In the product, the seleno-sulfide adduct (E–Se–SG) and a water molecule (**VI**) are formed. Reaction (2) is calculated to be exothermic by 15.9 (23.4) kcal mol^{-1}. Formation of E–Se–SG complex through selenosulfide linkage has been observed experimentally [19].

In the third and final elementary reaction, [E–Se–SG] + GSH → [E–SeH] + GS–SG (3),the second glutathione molecule (GSH), reacts with the seleno-sulfide adduct (**VI**) to produce the selenol [E–SeH] and oxidized disulfide (GS–SG) form of glutathione (**X**) (Figure 1.6).

We have explored several pathways for this reaction before suggesting the most plausible one. In this step, the S–H bond of the glutathione is broken and the proton is transferred via two water molecules to the amide backbone of Gly50. This process leads to the formation of the oxidized form of glutathione C_2H_5S–SC_2H_5 and protonated Gly50 residue. The amide backbone of Gly50 directly participates in this process and the presence of two water molecules is essential to bridge the proton donor (glutathione) and the acceptor (Gly50) sites. The calculated barrier for this step is 21.5 (25.5) kcal mol^{-1} and corresponds to the transition state **TS-VII-VIII**. This step is proposed to be a rate-determining step of the entire catalytic mechanism, which is in agreement with experiments

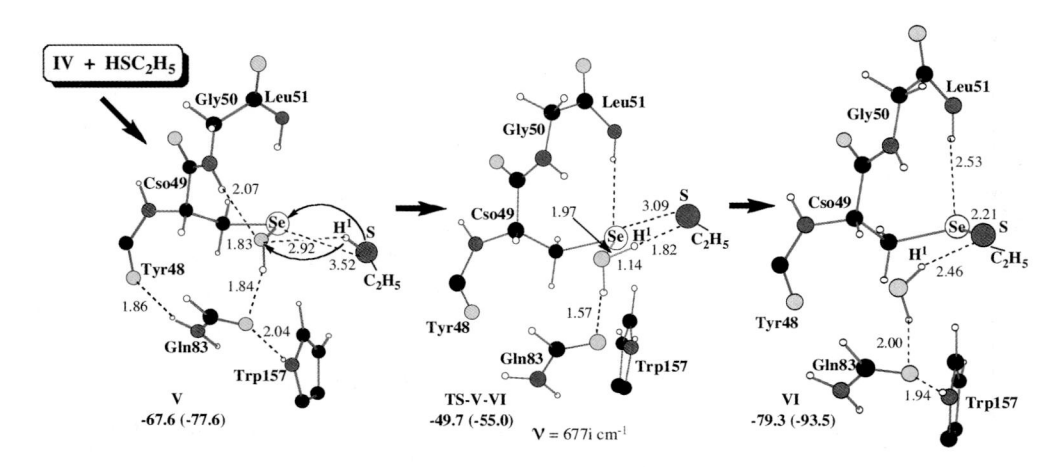

Figure 1.5 Optimized structures (in Å) and energies [with and without (in parenthesis) solvent effects, in kcal mol^{-1}] of reactant, intermediates and transition states for the reaction [E–SeOH] + GSH → [E–Se–SG] + HOH (Reaction 2).

1.4 Results and Discussion | **11**

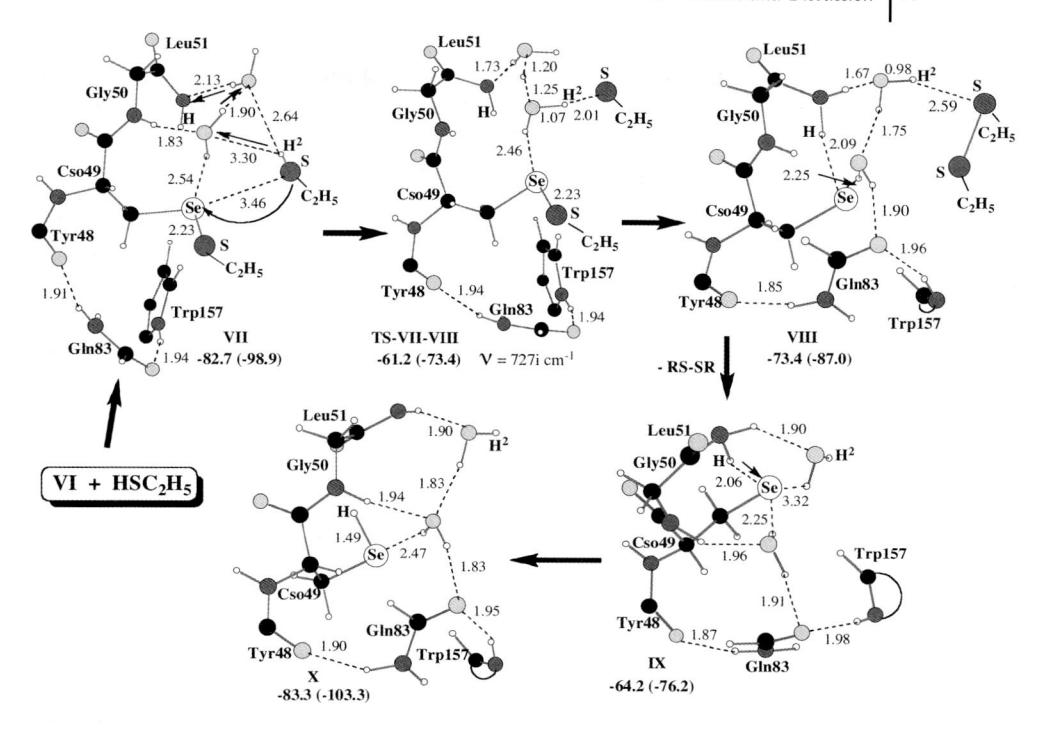

Figure 1.6 Optimized structures (in Å) and energies [with and without (in parenthesis) solvent effects, in kcal mol^{-1}] of reactant, intermediates and transition states for [E−Se−SG] + GSH → [E−SeH] + GS−SG (Reaction 3).

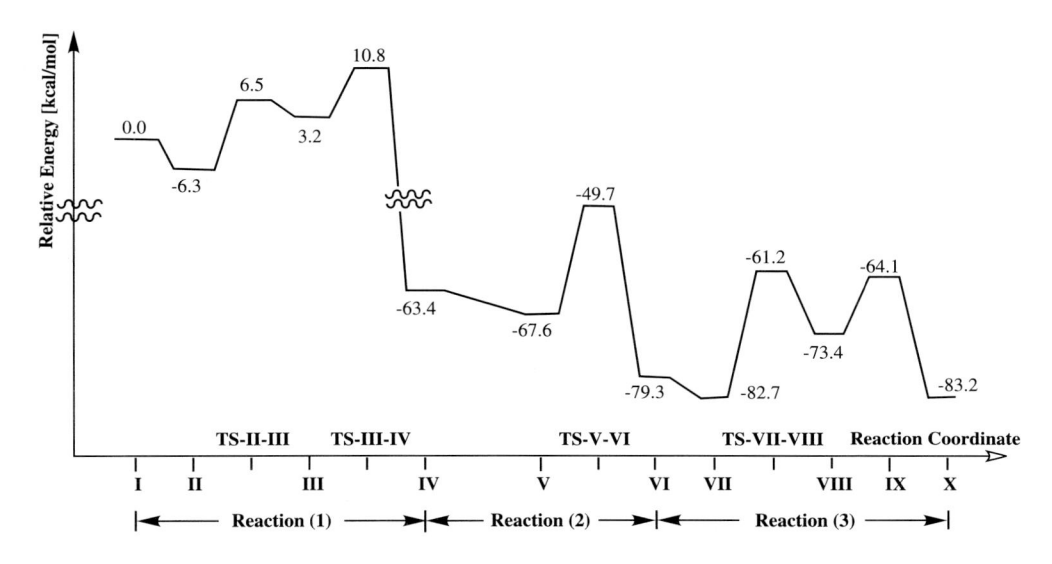

Figure 1.7 Energy diagram for the peroxidase activity of GPx (including solvent effects).

[22]. The product **VIII** formed in this step could be described as a weakly-bound complex where the selenocysteine residue is in its selenolate (E–Se$^-$) form. Later, this product rearranges to the final product of selenol **X**, with almost no barrier. Formation of the selenol (E–SeH) is exothermic by 19.1 (27.1) kcal mol^{-1} and the enzyme returns to its original form.

Figure 1.7 shows the energy diagram of the peroxidase activity of GPx.

1.4.3
Catalytic Functions: Effect of the Surrounding Protein on the Peroxidase Activity

To evaluate the quantitative effect of the protein surroundings on the energy we investigated the catalytic mechanism of hydrogen peroxide reduction by glutathione peroxidase (GPx) by including the complete protein environment in the calculations [35]. The two-layer ONIOM(QM:MM) method was employed to explore the potential energy surface (PES) of Reaction (1), [E–SeH] + H$_2$O$_2$ → [E–SeOH] + H$_2$O (1), in the catalytic cycle. Comparison of this PES with that reported in our previous "active site only" study allows us to assess the effect of the protein environment on the proposed mechanism. The most relevant results of the previous study are:

1. According to experimental data [13], the active state of the selenocysteine residue could be either the selenolate anion [E–Se$^-$] or selenol [E–SeH]. Our calculations showed that E–SeH is the most preferable active form of the enzyme [25].
2. From the X-ray structure [14], the geometry optimization led to two different conformers of Gln83 residue in the "active site only" model (either the oxo or the –NH$_2$ group of Gln83 facing Trp157), while Trp157 deviated significantly from its position in the X-ray structure.
3. In general, selenenic acid [E–SeOH] formation could occur either via a concerted or stepwise mechanism. Our calculations using the "active site only" models show that the barrier for the concerted mechanism is 4.2 kcal mol^{-1} higher than that for the stepwise mechanism [25].

In this ONIOM (QM:MM) study, the aforementioned theoretical conclusions from the previous study are fully utilized to investigate the reaction mechanism of H$_2$O$_2$ reduction catalyzed by GPx; here we only investigate the mechanisms identified as plausible in the "active site only" study.

1.4.3.1 Hydrogen Peroxide Coordination
The starting point of this study is the optimization of the structure of the enzyme. In contrast to the "active site only" study, the ONIOM calculations made in the presence of the surrounding protein yield only a single stable conformation of Gln83 (**I**, Figure 1.8). During optimization both Gln83 and Trp157 re-

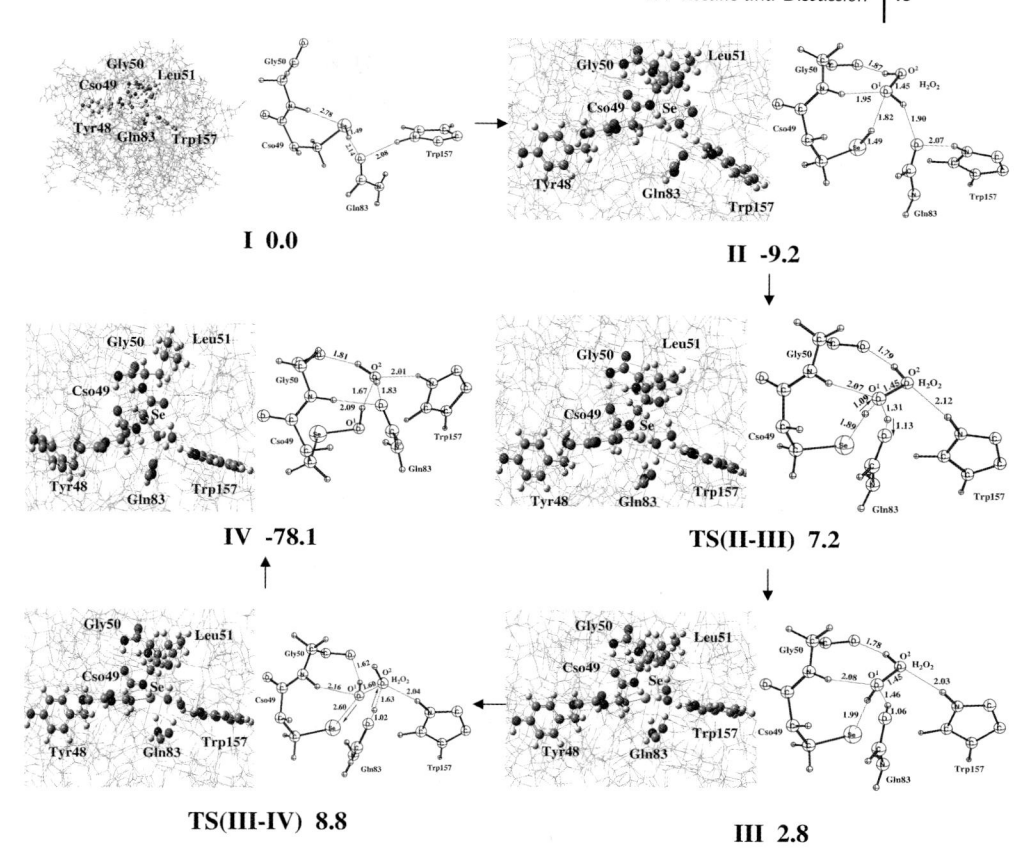

Figure 1.8 Optimized ONIOM structures (C atoms in gray, H atoms in white, and O and N atoms in black) with critical reaction coordinates (separately displayed in the adjoining views, in Å) and energies (kcal mol^{-1}) of reactant, intermediates and transition states and product for the H$_2$O$_2$ reduction mechanism of GPx.

sidues largely retained their positions shown in the X-ray structure. The effect of the protein environment on the structure of the active site is also reflected in the RMS deviations between the optimized and the X-ray structures, which are 1.48 and 0.97 Å for the "active site only" and QM/MM calculations, respectively.

In the first step of Reaction (1), similarly as for the "active site only" system, a hydrogen peroxide molecule coordinates to the active site of the enzyme to produce **II** (Figure 1.8). In **II**, H$_2$O$_2$ forms strong hydrogen bonds with Cso49, Gln83 and Gly50 residues, and has a binding energy of 9.2 (6.3) kcal mol^{-1}. In the presence of the protein environment the SeH–O^1 bond (1.82 Å) is shorter than the corresponding bond (1.91 Å) in the "active site only" system. As discussed above, in the present study we did not include a water molecule in the active site, which in the case of "active site only" investigations was shown to reduce the H$_2$O$_2$

binding energy by 1.0 kcal mol^{-1}. The RMS deviation between the MM parts of structures **I** and **II** is only 0.09 Å, which indicates that hydrogen peroxide binds to the active site without affecting the surrounding protein environment.

1.4.3.2 Formation of Selenenic Acid [E—Se—OH]

In the next stage of the reaction, selenenic acid (E–Se–OH) is formed. As shown previously, the stepwise mechanism consists of two steps: (a) formation of selenolate anion (E–Se$^-$) and (b) O–O bond cleavage.

Formation of Selenolate Anion [E—Se$^-$] In the first step, formation of the selenolate anion (E–Se$^-$) occurs via proton transfer from the Se through the oxygen (O^1) atom of hydrogen peroxide to the neighboring Gln83, leading to intermediate **III** (Figure 1.8). In this process the Gln83 residue plays a role of proton acceptor. This proposal concerning the participation of the Gln83 residue in the reaction is consistent with the available experimental information [20]. The computed barrier for the creation of selenolate anion from structure **II** is 16.4 (12.8) kcal mol^{-1}. However, this value could be slightly overestimated because B3LYP is known to overestimate the activation energy of long-range proton transfer processes [36]. The fully optimized transition state structure [**TS(II-III)**] associated with this barrier is shown in Figure 1.8. As seen from this figure, **TS(II-III)** is stabilized by hydrogen bonds from the Gly50, Gln83, Trp157 and Cso49 residues. In the presence of the surrounding protein, Trp157 forms hydrogen bond with H$_2$O$_2$, whereas in the "active site only" study Trp157 is hydrogen bonded to Gln83. Compared with the "active site only" study, Gln83H$^+$–O^1 and Se–O^{1}H bond distances are 0.05 and 0.03 Å longer, respectively. This step is calculated to be endothermic by 12.0 (9.5) kcal mol^{-1}. The correction introduced for the different atom types between structures **II** and **III** (in the present case from HS to HO type) reduce the endothermicity by 1.7 kcal mol^{-1}. In intermediate **III**, the O–H bond length in the protonated Gln83 is 1.05 Å, which is 0.03 Å shorter than in the "active site only" study. The absence of the active site water molecule slightly increases both barrier and exothermicity by 1.0 and 1.1 kcal mol^{-1}, respectively, in the "active site only" calculations. RMS deviations between the MM parts of **II-TS(II-III)** and **II-III** structures are 0.20 and 0.21 Å, respectively.

O—O Bond Cleavage In the second step of the stepwise mechanism, the O^1–O^2 bond of H$_2$O$_2$ is cleaved. During this process, one hydroxyl fragment (O^{1}H) is transferred to the selenolate anion [R–Se$^-$] to form selenenic acid [R–SeO^{1}H] while, simultaneously, the second hydroxyl fragment (O^{2}H) accepts the previously transferred proton from Gln83 to form a water molecule (**IV**, Figure 1.8).

Figure 1.8 shows the optimized transition state [**TS(III-IV)**] for this process. As seen from this figure, all the corresponding distances indicate that this process is synchronous. The calculated barrier for this process is 6.0 (7.6) kcal mol^{-1}. Compared with the "active site only" study, the Se–O^1 and Trp157–O^2 bond distances are 0.11 and 0.43 Å shorter, respectively. Since this step follows the 12.0 kcal mol^{-1} endothermic selenolate anion formation step (from **II** to **III**), the overall barrier (from **II** to **IV**) for the formation of selenenic acid (E–Se–OH) becomes 18.0 (17.3) kcal mol^{-1}. These results show that the presence of surrounding protein slightly increases the overall barrier by 0.7 kcal mol^{-1}, which is still in a good agreement with the experimentally measured barrier of 14.9 kcal mol^{-1} [21]. Here, it has to be stressed that the active site is not deeply buried inside the enzyme; instead it is located on the interface of two monomers, which is the main reason why the inclusion of protein surroundings does not exert any significant influence on the energetics of the reaction. However, in methane monooxygenase (MMO), cytochrome P450 and triose-phosphate isomerase (TIM) enzymes, where the active sites are deeply buried, the protein surroundings are reported to exhibit considerable effects [37]. The absence of the water molecule at the active site increases the barrier by 2.0 kcal mol^{-1}. This step of the stepwise formation of selenenic acid is calculated to be exothermic by 80.9 (66.6) kcal mol^{-1}. Here, the effect of changing MM atom types (from O2 to OS) is 1.4 kcal mol^{-1}. RMS deviations between the MM parts of **III-TS(III-IV)** and **III-IV** structures are 0.15 and 0.28 Å, respectively, again indicating that there are no major changes in the protein environment in this step of the mechanism.

Figure 1.9 shows the ONIOM (QM:MM) potential energy diagram for the H_2O_2 reduction mechanism of GPx.

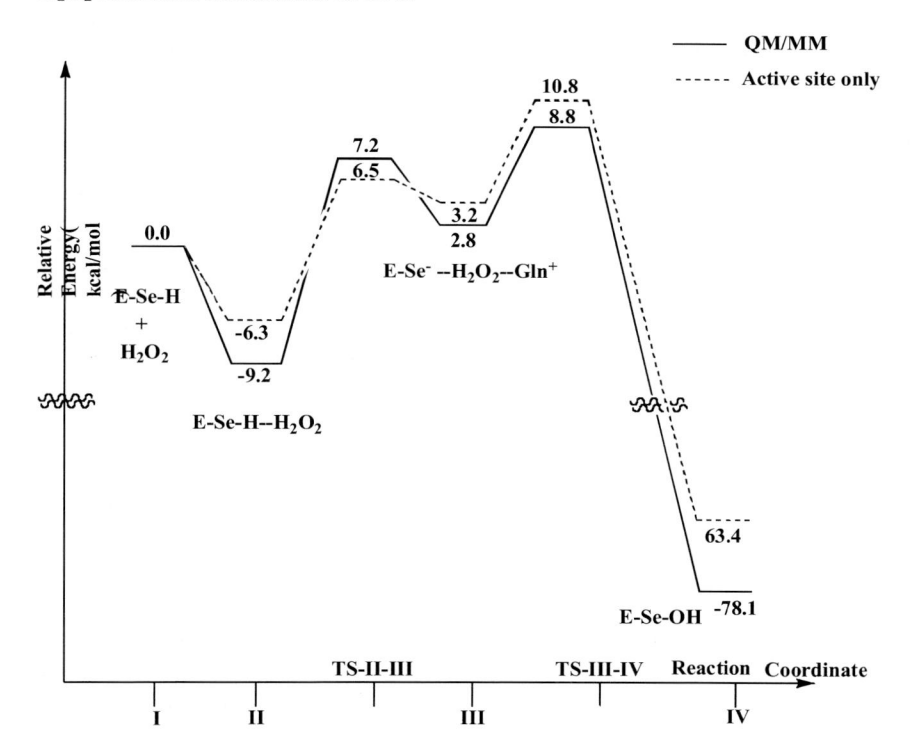

Figure 1.9 Potential energy diagram for the H_2O_2 reduction mechanism of GPx (including solvent effects).

1.4.4
Catalytic Functions: Reductase Activity

The suggested mechanism shown in Figure 1.3 was used as a starting point for this study [38], and for peroxynitrite (ONOO$^-$) and peroxynitrous acid (ONOOH) substrates the overall mechanism is described by the following two reactions, which are calculated to be exothermic by 76.7 (69.9) and 66.4 (62.9) kcal mol^{-1}, respectively:

$$\text{ONOO}^- + 2\text{C}_2\text{H}_5\text{SH} \rightarrow \text{C}_2\text{H}_5\text{S-SC}_2\text{H}_5 + \text{NO}_2^- + \text{H}_2\text{O} \qquad (A)$$

$$\text{ONOOH} + 2\text{C}_2\text{H}_5\text{SH} \rightarrow \text{C}_2\text{H}_5\text{S-SC}_2\text{H}_5 + \text{HNO}_2 + \text{H}_2\text{O} \qquad (B)$$

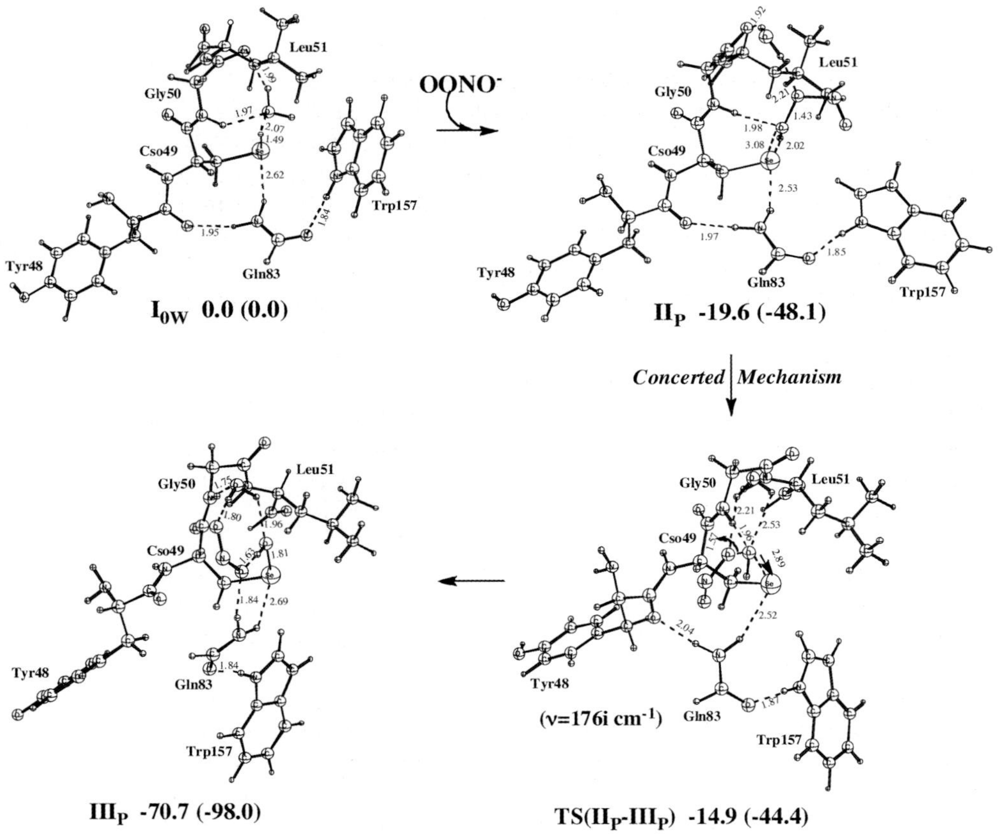

Figure 1.10 Optimized structures (distances in Å) and energies relative to the reactants [with and without (in parenthesis) solvent effects, in kcal mol^{-1}] of intermediates and transition states in Reaction (1) for the concerted "oxidation" mechanism using peroxynitrite (ONOO$^-$) as a substrate.

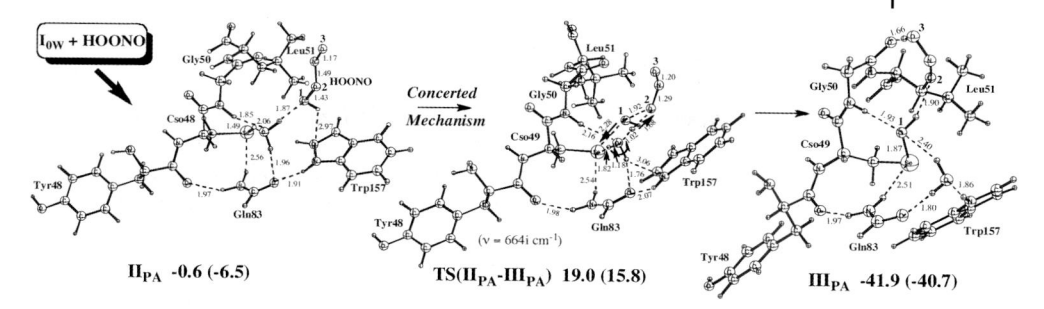

Figure 1.11 Optimized structures (distances in Å) and energies relative to the reactants [with and without (in parenthesis) solvent effects, in kcal mol^{-1}] of intermediates and transition states in Reaction (1) for the concerted "oxidation" mechanism using peroxynitrous acid (ONOOH) as a substrate.

1.4.4.1 Peroxynitrite/Peroxynitrous Acid (ONOO$^-$/ONOOH) Coordination

Coordination of substrate at the active site of GPx enzyme is a first step of both "oxidation" and "nitration" pathways. Coordination of peroxynitrite (ONOO$^-$) to the active site of the enzyme (structure I_{0W}, Figure 1.10) yields structure II_P (E−Se$^-$···HO^{1}O^{2}NO^3···H$_2$O). ONOO$^-$ acts as a strong base, which abstracts a proton from the E−SeH upon binding to I_{0W}. In II_P, peroxynitrite interacts with the active site through hydrogen bonds with water molecule, selenolate and the Gly50 residue. The binding energy of a free peroxynitrite to I_{0W} is calculated to be 19.6 (48.1) kcal mol^{-1}.

The protonated form of ONOO$^-$, peroxynitrous acid (ONOOH), binds at the active site of the enzyme (structure I_{0W}) and forms a weakly interacting complex E−Se−H···H$_2$O···HO^{1}O^{2}NO^3 (structure II_{PA}, Figure 1.11). The computed binding energy of free peroxynitrous acid is only 0.6 (6.5) kcal mol^{-1}. After the coordination of ONOO$^-$/ONOOH, the catalytic cycle proceeds through the "oxidation" and "nitration" pathways (Figure 1.3), which are discussed separately.

1.4.4.2 Oxidation Pathway

The first process occurring in this pathway is the formation of selenenic acid [E−Se−OH]. The mechanism is quite different for the peroxynitrite and peroxynitrous acid substrates.

Concerted "Oxidation" Mechanism for Peroxynitrite Formation of the oxidation product (E−Se−OH) requires a hydroxyl group. For peroxynitrite (ONOO$^-$), the hydroxyl group required could either be donated by peroxynitrite (now in the form of HOONO in II_P) or a water molecule located near the Se-center. Since the source of the hydroxyl group is not known, both possibilities are explored in this study. In the case where the hydroxyl group is donated by peroxynitrite, the O^1−O^2 bond of the previously formed E−Se$^-$···HO^{1}O^{2}NO^3···H$_2$O complex (II_P) is broken and the hydroxyl group (O^{1}H) is concertedly transferred from the substrate to the selenolate (E−Se$^-$) ion to produce ESe−O^{1}H (III_P). The optimized structure of the corresponding transition state **TS(II$_P$-III$_P$)** is shown in Figure 1.10. It is exothermic by 51.1 (49.9) kcal mol^{-1} and proceeds with a barrier of 4.7 (3.7) kcal mol^{-1}. The low barrier for ONOO$^-$ reduction agrees with the experimentally measured barrier of 8.8 kcal mol^{-1} for both GPx and ebselen [23] and the computed value of 7.1 kcal mol^{-1} for ebselen [12]. The removal of a water molecule hydrogen-bonded to ONOO$^-$ from the model makes the peroxynitrite a stronger nucleophile and reduces the barrier by 2.6 (3.3) kcal mol^{-1}. However, notably, this process is most likely to occur in the presence of a water molecule as ONOO$^-$ binding does not remove the water molecule from the active site. The barrier associated with the donation of a hydroxyl group from the water molecule to the E–Se$^-$ is prohibitively high, ca. 40.0 kcal mol^{-1}. These results explicitly indicate that the hydroxyl group required to produce the E–Se–OH is provided by peroxynitrite (ONOO$^-$) that is converted into ONOOH during its coordination. There is no stepwise mechanism corresponding to this step, as the O–O bond cleavage and hydroxyl group transfer are strongly coupled and cannot go through an intermediate.

Concerted "Oxidation" Mechanism for Peroxynitrous Acid For peroxynitrous acid (ONOOH), E–Se–OH formation can take place via either a concerted or a stepwise mechanism. The stepwise mechanism consists of two parts: (a) formation of selenolate anion (E–Se$^-$) and (b) O–O bond cleavage.

In the concerted mechanism for peroxynitrous acid (O^{3}NO^{2}O^{1}H), starting from **II$_{PA}$** complex (Figure 1.11), the Se–H bond of the selenol (E–SeH) is broken and, with the help of a bridging water molecule, a proton is transferred to O^{3}NO^{2}O^{1}H, which in turn facilitates the O^1–O^2 cleavage and, subsequently, the formation of ESe–O^{1}H. Figure 1.11 shows the corresponding transition state **TS(II$_{PA}$-III$_{PA}$)**. The overall process **II$_{PA}$ → III$_{PA}$** is exothermic by 41.3 (34.2) kcal mol^{-1} and proceeds with a 19.6 (22.3) kcal mol^{-1} barrier. In the absence of a water molecule in the model this barrier is further increased by 5.2 kcal mol^{-1} (in the gas phase).

Stepwise "Oxidation" Mechanism for Peroxynitrous Acid In the first step of this mechanism (Figure 1.12) the Se–H bond of the selenol (E–SeH) is broken and, simultaneously, the proton is transferred through the oxygen atom (O^1) of O^{3}NO^{2}O^{1}H and a water molecule to the neighboring Gln83, producing the **II$'_{PA}$** intermediate involving an E–Se$^-$ + Gln83$^+$ ion-pair. The corresponding transition state, **TS(II$_{PA}$-II$'_{PA}$)**, for this process is stabilized by hydrogen bonds with Gly50 and Trp157 residues. The **II$_{PA}$ → II$'_{PA}$** process is endothermic by 5.1 (10.0) kcal mol^{-1} and proceeds with a 6.5 (10.4) kcal mol^{-1} barrier from **II$_{PA}$**. In **II$'_{PA}$** the O–H bond (1.08 Å) in Gln83$^+$ is slightly longer than the normal (0.98 Å) O–H bond. The absence of the water molecule in the model increases the barrier (by 6.4 kcal mol^{-1} in the gas phase) and endothermicity (by 6.2 kcal mol^{-1} in the gas phase) of the reaction.

In the second step of the stepwise pathway of E–Se–OH formation, the O^1–O^2 bond of O^{3}NO^{2}O^{1}H is cleaved and the hydroxyl (O^{1}H) is transferred to the sele-

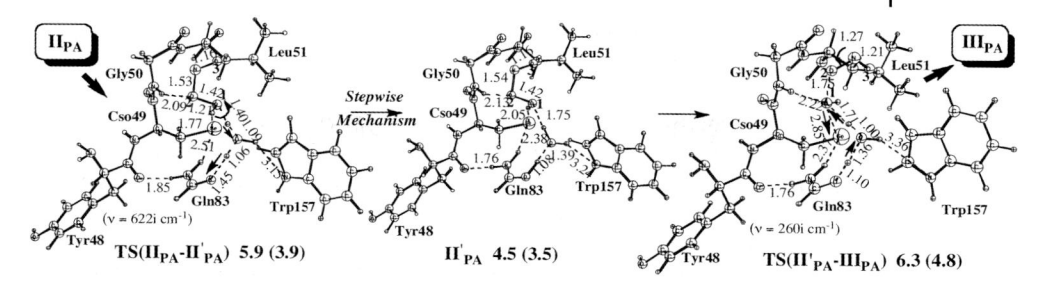

Figure 1.12 Optimized structures (distances in Å) and energies relative to the reactants [with and without (in parenthesis) solvent effects, in kcal mol^{-1}] of intermediates and transition states in Reaction (1) for the stepwise "oxidation" mechanism using peroxynitrous acid (ONOOH) as a substrate.

nolate anion (E–Se$^-$). In this step, the proton previously transferred to the Gln83 residue moves to the oxygen atom (O^2) of O^{3}NO^{2}O^{1}H and initiates the O^1–O^2 bond cleavage. As a result of this process, the E–SeO^{1}H and HNO$_2$ (structure III$_{PA}$) are produced. The transition state **TS(II$'_{PA}$-III$_{PA}$)** indicates that this step is synchronous, i.e., all bond distances change smoothly from the intermediate II$'_{PA}$ to the product III$_{PA}$. The barrier for this process is only 1.8 (1.3) kcal mol^{-1}, which makes the overall barrier (II$_{PA}$ → III$_{PA}$) for the formation of the selenenic acid (E–Se–OH) 6.9 (11.3) kcal mol^{-1}. The presence of the water molecule significantly stabilizes the transition state as its removal from the model increases the barrier by 13.0 kcal mol^{-1} (gas phase). This step of Reaction (1) is calculated to be exothermic by 46.4 (44.2) kcal mol^{-1}. Since the overall barrier for the stepwise mechanism [6.9 (11.3) kcal mol^{-1}] is substantially lower than the barrier [19.6 (22.3) kcal mol^{-1}] for the concerted mechanism of the E–Se–OH formation, the latter mechanism is ruled out.

The aforementioned results explicitly indicate that the Gln83 residue plays a key role as proton acceptor (step 1) and donor (step 2), which is consistent with the available experimental suggestion that the Gln83 residue participates in the catalytic cycle [20]. Moreover, the water molecule located near the Se-center also plays a very important role by directly participating in the reaction and reducing the barriers. Compared with H$_2$O$_2$ reduction by GPx, where the calculated barrier for the enzyme oxidation through the identical stepwise mechanism is reported to be 17.1 (20.6) kcal mol^{-1}, the overall barrier for ONOOH reduction by this enzyme is lower by 10.2 (9.3) kcal mol^{-1}. These results demonstrate that ONOOH is a more efficient substrate for the oxidation of selenocysteine than H$_2$O$_2$. In contrast, compared with ebselen (a mimic of GPx), GPx catalyzes the reduction of ONOOH with a barrier higher by 3.9 kcal mol^{-1} (gas phase) [11].

Remaining Steps of the Oxidation Pathway In this pathway, subsequent to the formation of selenenic acid [E–Se–OH], the remaining two steps (Reactions 2 and 3) follow identical mechanisms, as suggested for H$_2$O$_2$ reduction by GPx in our previous computational study [25]. Therefore, these two steps are not discussed here.

1.4.4.3 Nitration Pathways

This pathway, after the binding of ONOO$^-$/ONOOH to the active site (structure **II$_{PA}$**), leads to the nitration of the selenol [E−SeH] (Figure 1.13). In general, the nitration product [E−Se−NO$_2$] can exist in two different isomeric forms, nitro [E−Se−NOO] and nitrito [E−Se−O−N=O], which can be formed by the reactions of both peroxynitrite (ONOO$^-$) and peroxynitrous acid (ONOOH) with the active site selenocysteine residue. However, the use of ONOO$^-$ creates a highly reactive hydroxyl (OH$^-$) ion that could readily react with the active site amino acid residues. The enzyme may be set-up to control these ions but this unknown regulation process is hard to model. Therefore, to avoid these side reactions, in this study, the "nitration" pathway is investigated only for HNOOH substrate. The B3LYP calculations suggest that the formation of the E−Se−NOO (structure **III$_{NOO}$**) is thermodynamically, 18.8 (19.0) kcal mol^{-1}, more favorable than the E−Se−O−N=O (structure **III$_{ONO}$**), and can occur through the following two

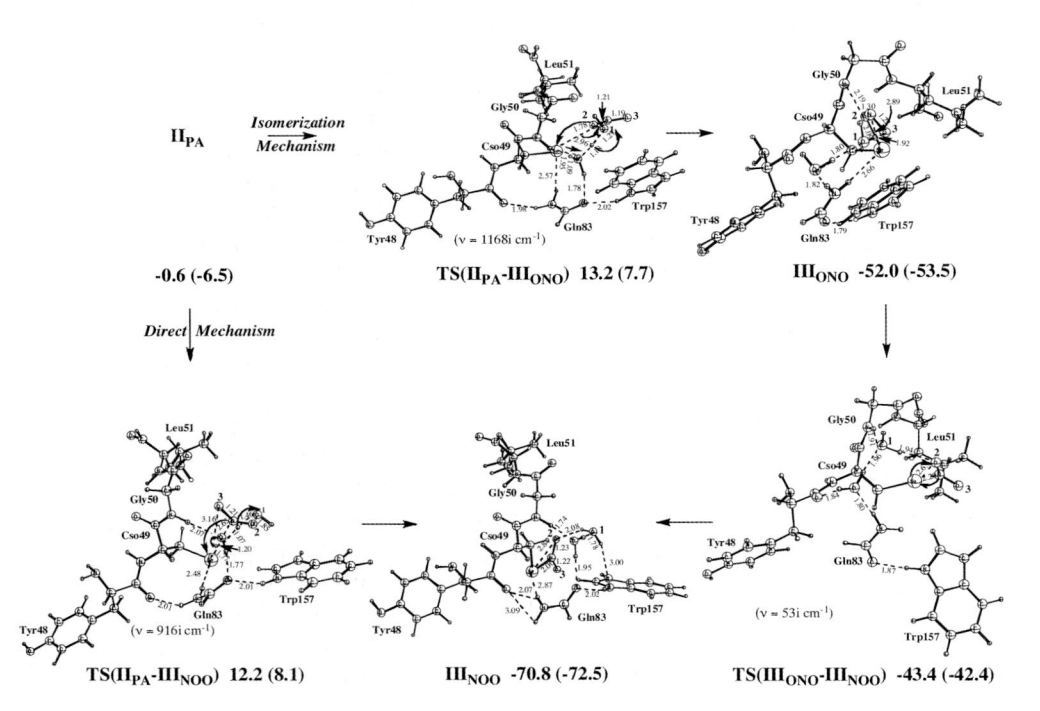

Figure 1.13 Optimized structures (distances in Å) and energies relative to the reactants [with and without (in parenthesis) solvent effects, in kcal mol^{-1}] of intermediates and transition states in Reaction (1) for the isomerization and direct mechanisms in the "nitration" pathway using peroxynitrous acid (ONOOH) as a substrate.

mechanisms: (1) isomerization and (2) direct. In the former, the III_{ONO} species is generated first, which then isomerizes to the energetically more favorable product III_{NOO}, whereas in the latter III_{NOO} is generated directly.

Isomerization Mechanism In the first part of this mechanism, the Se–H bond of the selenol (E–SeH), II_{PA}, is broken and with the help of the bridging water molecule a proton is transferred to the terminal oxygen atom (O^1) of $O^3NO^2O^1H$, which facilitates O^1–O^2 bond cleavage and leads to the formation of E–Se–O^3–N=O^2 (III_{ONO}) and a water molecule. Figure 1.13 shows the optimized transition state for this process, structure **TS(II_{PA}-III_{ONO})**. The computed barrier for the formation of III_{ONO} is 13.8 (14.2) kcal mol^{-1}, which could be slightly overestimated because the B3LYP method overestimates the activation energy of long-range proton transfer [36]. The formation of this intermediate is exothermic by 51.4 (47.0) kcal mol^{-1}. Similar to the "oxidation" pathway, here also the water molecule plays a key role in the mechanism by keeping the barrier low. Removal of the water molecule from the model increases the barrier for this process by 7.2 kcal mol^{-1} (gas phase).

A stepwise mechanism, similar to the one (involving Gln83) investigated for the formation of III_{PA} in the "oxidation" pathway, was also investigated for the generation of III_{ONO}. However, all attempts to locate the transition state for the O^1–O^2 bond splitting of $O^3NO^2O^1H$ failed as its optimization always led to the above discussed concerted mechanism. Based on this result it can be concluded that the stepwise mechanism for the formation of the E–Se–O–N=O product does not exist.

In the second part of this isomerization mechanism, the formed III_{ONO} isomerizes to III_{NOO}. During this isomerization, the Se–O^2 bond of III_{ONO} is broken and the Se–N bond is formed. In the associated transition state, **TS(III_{ONO}-III_{NOO})**, the Se–O^2 and Se–N bond distances of 2.65 and 2.20 Å, respectively, are between the corresponding distances in III_{ONO} (2.09 and 2.89 Å) and III_{NOO} (2.84 and 2.00 Å), which clearly indicates the transformation of the intermediate E–Se–O–N=O into E–Se–NOO. $III_{ONO} \rightarrow III_{NOO}$ isomerization is found to be exothermic by 18.8 (19.0) kcal mol^{-1} and proceeds with a barrier of 8.6 (11.1) kcal mol^{-1}.

Direct Mechanism In this mechanism, in II_{PA} the Se–H bond of the selenol (E–SeH) is broken, and with the help of a bridging water molecule a proton is transferred to the terminal oxygen atom (O^1) of $O^3NO^2O^1H$, which in turn cleaves the O^1–O^2 bond and produces the E–Se–NO^2O^3 (III_{NOO}) species and a water molecule (H_2O^1). In the corresponding transition state **TS(II_{PA}-III_{NOO})** (Figure 1.13) the Se–N and Se–O^2 bond distances of 2.63 and 3.16 Å, respectively, clearly indicate the formation of III_{NOO}. The barrier for this concerted mechanism is 12.8 (14.6) kcal mol^{-1} and this process is exothermic by 70.2 (66.0) kcal mol^{-1}. In this process a long-range proton transfer also takes place, which could be slightly overestimated by the B3LYP method [36]. Since for the generation of III_{ONO} the exclusion of water molecule in the model raises the barrier by 7.2 kcal mol^{-1} (in the gas phase), this possibility is not explored here. As discussed above for the formation of III_{ONO}, the stepwise mechanism involving the Gln83 residue does not exist in the direct generation of III_{NOO}.

The calculated barriers for the nitration of the selenocysteine by peroxynitrous acid through the isomerization and direct mechanisms, 13.8 (14.2) and 12.8 (14.6) kcal mol^{-1}, are very close. While the direct mechanism is slightly preferred, the accuracy of the methods applied in this study does not allow a clear discrimination. Therefore, both these mechanisms for the selenocysteine nitration by ONOOH are plausible.

Reaction of Nitro Product (E–Se–NOO) with Glutathione (GSH) Reaction (2) of the "nitration" pathway of Figure 1.3 starts with the coordination of the first unbound glutathione molecule to III_{NOO} and leads to the formation of a weakly bound (E–SeNO^{2}O^3)–(GSH) complex (**IV**, Figure 1.14) with a binding energy of 1.8 (2.7) kcal mol^{-1}. From **IV** the reaction proceeds through the transition state **TS(IV-V)** and produces the seleno-sulfide (E–Se–S–G) adduct and HNO$_2$ (structure **V**). As shown in Figure 1.14, at the transition state **TS(IV-V)**, synchronously, the S–H bond of glutathione is broken and, through the water molecule, a proton is transferred to O^3 of the E–Se–NO^{3}O^2 accompanied by the formation of a Se–S bond and nitrous acid (HNO$_2$). The barrier for this step is calculated to be 9.5 (9.6) kcal mol^{-1} and it is exothermic by 13.9 (12.9) kcal mol^{-1}.

Also in this reaction, the water molecule plays a critical role by significantly reducing the barrier by 15.7 kcal mol^{-1} (in the gas phase). This large effect could be explained by comparing the TS structures with and without the participation of the water molecule. In **TS(IV-V)**, the Se–S and Se–N distances are considerably shorter (0.14 and 0.62 Å, respectively) than the corresponding distances without the water molecule. Moreover, the direct participation of a water molecule also provides an additional hydrogen bond. In the presence of a water molecule **TS(IV-V)** appears to be optimum for the formation of the Se–S bond.

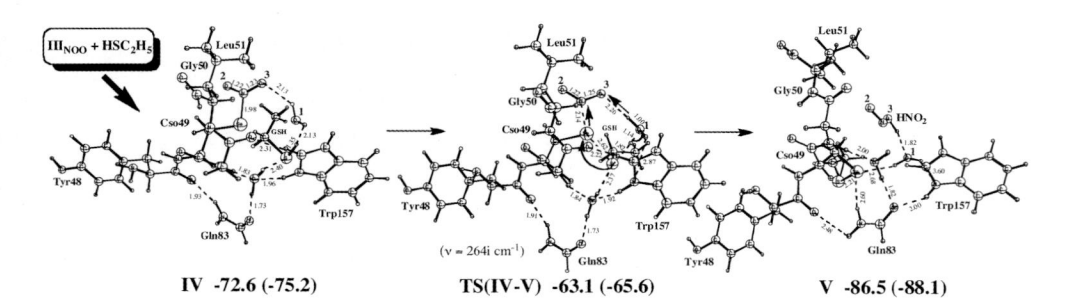

Figure 1.14 Optimized structures (distances in Å) and energies relative to the reactants [with and without (in parenthesis) solvent effects, in kcal mol^{-1}] of intermediates and transition states in Reaction (2) for the "nitration" pathway using peroxynitrous acid (ONOOH) as a substrate.

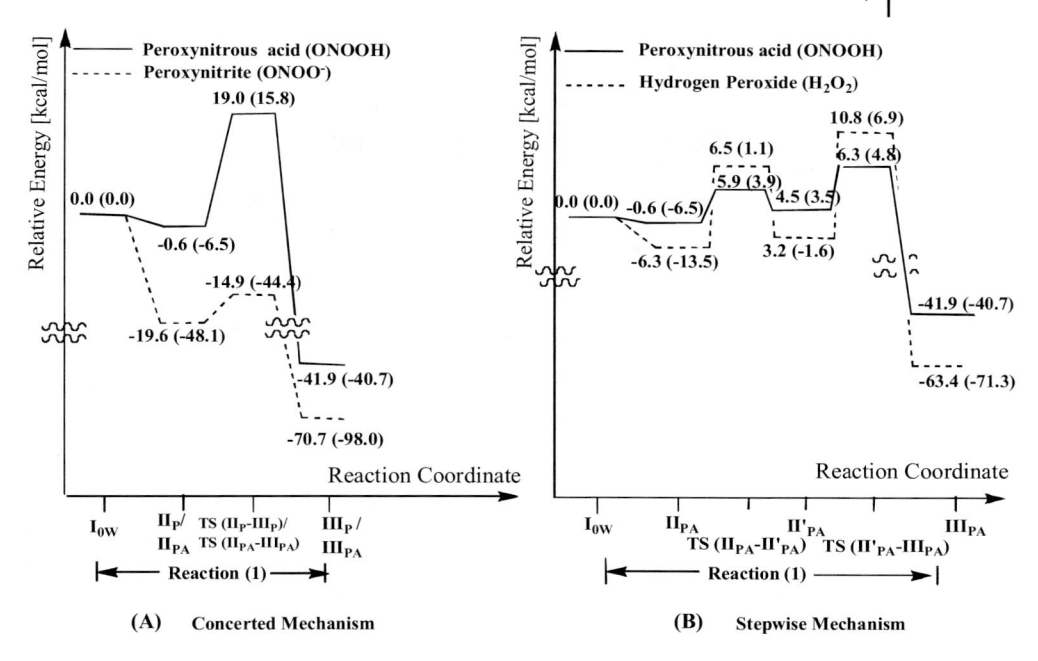

Figure 1.15 (A) Energy diagram for Reaction (1) of the "oxidation" pathway [with and without (in parenthesis) solvent effects, in kcal mol^{-1}]. The energy scale is setup for the values including solvent effects. (B) Energy diagram for reactions (1) and (2) of the "nitration" pathway [with and without (in parenthesis) solvent effects, in kcal mol^{-1}]. The energy scale is setup for the values including solvent effects.

In a similar way, III_{ONO} is also found to react with the substrate GSH to produce the seleno-sulfide (E−Se−S−G) adduct and HNO_2. First, a molecule of GSH binds to III_{ONO} to form a $(E-Se-O^3-N=O^2)-(GSH)$ complex (structure IV_{ONO}) with a binding energy of 1.8 (2.7) kcal mol^{-1}, which then rearranges to V through $TS(IV_{ONO}-V)$ with a small barrier of 1.3 (1.0) kcal mol^{-1}.

After formation of the E−Se−S−G adduct, the third and final step, Reaction (3), of the overall mechanism is identical in both "oxidation" and "nitration" mechanisms (as discussed above). Figure 1.15(A) and (B) shows the overall potential energy diagrams for the concerted and stepwise mechanisms of the "oxidation" pathway.

1.5 Summary

In this chapter we have been explored the structural properties and catalytic functions of the selenoprotein glutathione peroxidase (GPx), utilizing pure quantum mechanics (QM) and hybrid quantum mechanics/molecular mechanics

(QM/MM) approaches. Our discussion is divided into four parts. In the first, the active site of the mammalian GPx is refined using a two-layer ONIOM(QM:MM) method. It was found that the inclusion of two water molecules at the active site provides the lowest root mean square (RMS) deviation from the X-ray structure. Based on these results it was concluded that active site of the enzyme is most likely to have two water molecules. The second part investigated the entire catalytic cycle corresponding to the peroxidase activity of this enzyme, using pure QM ("active site only") approach. This study reveals important catalytic roles played by Gln83, Gly50 residues and two water molecules. In addition, the generation of the oxidized form of glutathione was proposed to occur with a barrier of 21.5 (25.5) kcal mol^{-1} in the rate-determining step, which is in line with experimental findings. The third part explored the role of the protein surroundings in the mechanism of H_2O_2 reduction, using whole monomer (3113 atoms in 196 amino acid residues), with ONIOM(QM:MM) method. The protein surroundings were calculated to exert a net effect of only 0.70 kcal mol^{-1} (in comparison with the "active site only" model) on the overall barrier, which is most likely due to the active site being located at the enzyme surface. The fourth and final part investigated the peroxynitrite reductase activity of GPx at B3LYP level. Our calculations suggest that for peroxynitrite/peroxynitrous acid (ONOO$^-$/ONOOH) substrates the enzyme utilizes two different "oxidation" and "nitration" pathways. In the "oxidation" pathway for ONOO$^-$, the oxidation of GPx and the subsequent formation of the selenenic acid (E–Se–OH) occurs through a concerted mechanism with an energy barrier of 4.7 (3.7) kcal mol^{-1}, which is in good agreement with the computed value of 7.1 kcal mol^{-1} for the drug ebselen and the experimentally measured barrier of 8.8 kcal mol^{-1} for both ebselen and GPx. For ONOOH, formation of the E–Se–OH prefers a stepwise mechanism with an overall barrier of 6.9 (11.3) kcal mol^{-1}, which is 10.2 (11.2) kcal mol^{-1} lower than that for hydrogen peroxide (H_2O_2), indicating that ONOOH is a more efficient substrate for GPx oxidation. The nitration of GPx by ONOOH produces a nitro (E–Se–NO$_2$) product via either of two different mechanisms, isomerization and direct, having almost the same barrier heights. Comparison of the rate-determining barriers of the "oxidation" and "nitration" pathways suggests that the oxidation of GPx by ONOOH is more preferable than its nitration. It was also shown that the rate-determining barriers remain the same, 21.5 (25.5) kcal mol^{-1}, in the peroxynitrite reductase and peroxidase activities of GPx.

References

1 Combs, G. F. Jr., Lü, J., *Selenium Its Molecular Biology and Role in Human Health*, ed. D. L. Hatfield, Kluwer Academic Publishers, Boston, 2001, p. 205.

2 Zhao, L., Cox, A. G., Ruzicka, J. A., Bhat, A. A., Zhang, W., Taylor, E. W. *Proc. Natl. Acad. Sci. U.S.A.* 2000, **97**, 6356.

3 Birringer, M., Pilawa, S., Flohè, L. Trends in selenium Chemistry, *Nat. Prod. Rep.* 2002, **19**, 693–718.

4 Mills, G. C. *J. Biol. Chem.* 1957, **229**, 189–197.
5 Flohè, L., *Glutathione*, eds. Dolphin, D., Avramovic, O., Poulson R., John Wiley & Sons, New York, 1989, pp. 644–731.
6 Sies, H., Sharov, V. S., Klotz, L. O., Briviba, K. *J. Biol. Chem.* 1997, **272**, 27812–27817.
7 Björnstedt, M., Xue, J., Huang, W., Akesson, B., Holmgren, A. *J. Biol. Chem.* 1994, **269**, 29382.
8 Godeas, C., Sandri, G., Panfili, E. *Biochim. Biophys. Acta* 1994, **1191**, 147.
9 Roveri, A., Ursini, F., Flohé, L., Maiorino, M. *Biofactors* 2001, **14**, 213.
10 Sies, H., Masumoto, H. *Adv. Pharmacol.* 1997, **38**, 229–246.
11 Musaev, D. G., Hirao, K. *J. Phys. Chem. A* 2003, **107**, 9984–9990.
12 Musaev, D. G., Geletii, Y. V., Hill, C. L., Hirao, K. *J. Am. Chem. Soc.* 2003, **125**, 3877–3888.
13 Epp, O., Ladenstein, R., Wendel, A. *Eur. J. Biochem.* 1983, **133**, 51–69.
14 Ren, B., Huang, W., Åkesson, B., Ladenstein, R. *J. Mol. Biol.* 1997, **268**, 869–885.
15 Prohaska, J. R., Oh, S. H., Hoekstra, W. G., Ganther, H. E. *Biochem. Biophys. Res. Commun.* 1977, **74**, 64–71.
16 Forstrom, J. W., Zakowski, J. J., Tappel, A. L. *Biochemistry* 1978, **17**, 2639–2644.
17 Wendel, A., Pilz, W., Ladenstein, R., Sawatzki, G., Weser, U. *Biochim. Biophys. Acta* 1975, **377**, 211–215.
18 Ladenstein, R., Epp, O., Bartels, K., Jones, A., Huber, R., Wendel, A. *J. Mol. Biol.* 1979, **134**, 199–218.
19 Kraus, R. J., Ganther, H. E. *Biochem. Biophys. Res. Commun.* 1980, **96**, 1116–1122.
20 Ursini, F., Maiorino, M., Brigelius-Flohé, R., Aumann, K. D., Roveri, A., Schomburg, D., Flohé, L. *Methods Enzymol.* 1995, **252**, 38–53.
21 Roy, G., Nethaji, M., Mugesh, G. *J. Am. Chem. Soc.* 2004, **126**, 2712–2713.
22 Mugesh, G., du Mont, W., Sies, H. *Chem. Rev.* 2001, **101**, 2125–2179.
23 Briviba, K., Kissner, R., Koppenol, W. H., Sies, H. *Chem. Res. Toxicol.* 1998, **11**, 1398–1401.
24 Padmaja, S., Squadrito, G. L., Pryor, W. A. *Arch. Biochem. Biophys.* 1998, **349**, 1–6.
25 Prabhakar, R., Vreven, T., Morokuma, K., Musaev, D. G. *Biochemistry* 2005, **44**, 11864–11871.
26 *Gaussian 03 (Revision C1)*, Frisch, M. J. et al. (2004) Gaussian Inc., Pittsburgh, PA.
27 (a) Becke, A. D. *Phys. Rev. A* 1988, **38**, 3098–3100. (b) Lee, C., Yang, W., Parr, R. G. *Phys. Rev. B* 1988, **37**, 785–789. (c) Becke, A. D. *J. Chem Phys* 1993, **98**, 5648–5652.
28 Siegbahn, P. E. M., Blomberg, M. R. A. *Chem. Rev.* 2000, **100**, 421–437.
29 Cances, E., Mennucci, B., Tomasi, J. *J. Chem. Phys.* 1997, **107**, 3032–3041.
30 Maseras, F., Morokuma, K. *J. Comp. Chem.* 1995, **16**, 1170–1179.
31 Dapprich, S., Komaromi, I., Byun, S., Morokuma, K., Frisch, M. J. *J. Mol. Struct. (Theochem)* 1999, **461**, 1–23.
32 Bakowies, D., Thiel, W. *J. Phys. Chem.* 1996, **100**, 10580.
33 *GaussianView 3.0*, Gaussian Inc., Pittsburgh, PA, 2003.
34 Prabhakar, R., Musaev, D. G., Khavrutskii, I. V., Morokuma, K. *J. Phys. Chem. B* 2004, **108**, 12643–12645.
35 Prabhakar, R., Vreven, T., Morokuma, K., Musaev, D. G. *J. Phys. Chem. B* 2006, **110**, 13608–13613.
36 Prabhakar, R., Blomberg, M. R. A., Siegbahn, P. E. M. *Theor. Chem. Acc.* 2000, **104**, 461–470.
37 Friesner, R. A., Gullar, V. *Annu. Rev. Phys. Chem.* 2005, **56**, 389–427.
38 Prabhakar, R., Vreven, T., Morokuma, K., Musaev, D. G. *Biochemistry* 2006, **45**, 6967–6977.

2
A Comparison of Tetrapyrrole Cofactors in Nature and their Tuning by Axial Ligands

Kasper P. Jensen, Patrik Rydberg, Jimmy Heimdal, and Ulf Ryde

2.1
Introduction

The well-being and every-day life functions of all living creatures are determined by an overwhelming and many-faced work force of efficient catalysts, the enzymes. These molecules have evolved by random trial and error during the 3–4 billion years that life forms have existed on earth to a stage where they are so highly specialized that a small chemical change in such a molecule may cause severe disorder, disease, or even death of the affected organism.

About one-third of these proteins depend either structurally or functionally on metals [1]. Two types of metals can be distinguished by their properties: main-group metals and transition metals. Both occur in biological systems in their ionic forms. The main-group metals, Na^+, Mg^{2+}, K^+, Ca^{2+}, and Zn^{2+}, have closed electronic configurations and occur in only one oxidation state. They have structural roles, roles as second messengers, or function as redox-inactive catalysts, e.g., Lewis acids in general acid–base chemistry. During evolution, it has become apparent that more complex tasks, including transport and storage of oxygen, containment and degradation of oxygen-based radicals, as well as electron transfer and redox reactions, have only been possible by the use of transition metals, particularly from the bio-available first row of the d-block. With these elements, a much more diverse chemistry can be achieved in living organisms. Transition metals occur in several oxidation states, a feature that is important for their function – they are usually redox active.

In many cases, the metal binds directly to the protein through various amino-acid ligands. However, in other cases, some ligands are provided by a pre-arranged non-protein ligand structure, a rigid and ready-to-use catalyst module, to be incorporated into the protein. Several similar near-planar tetradentate ligands exist in nature. They define the equatorial ligand field of an octahedral ion, leaving open two axial coordination sites. They all consist of four five-membered rings with four carbons and a nitrogen atom, similar to pyrrole and are, therefore, usually referred to as tetrapyrroles. The nitrogen atoms coordinate to the central ion in

Computational Modeling for Homogeneous and Enzymatic Catalysis.
A Knowledge-Base for Designing Efficient Catalysts. K. Morokuma and D. G. Musaev (Eds.)
Copyright © 2008 WILEY-VCH Verlag GmbH & Co. KGaA, Weinheim
ISBN: 978-3-527-31843-8

all cases. Figure 2.1 shows the four tetrapyrrole cofactors that are the subject of this chapter, i.e., coenzyme F430, cobalamin, heme, and chlorophyll.

Tetrapyrroles are present in more than 5% of the protein structures in the Protein Data Bank, directly implying their importance. Porphyrin is by far the most abundant (found in over 2000 structures) and diverse, in terms of function. It normally binds iron in the center of the ring, forming the well-known heme group. Heme is used for many different functions, e.g., electron transfer in the cytochromes, binding and transport of small molecules, e.g., O_2 in the globins, and for the catalysis of a great wealth of chemical reactions, e.g., in oxidases, peroxidases, catalases, and cytochromes P450.

Chlorophylls are found in $\sim$100 crystal structures and their prime use is as pigments in photosynthesis. They contain a Mg^{2+} ion in the center of the tetrapyrrole ring, although the same ring system is sometimes used in nature without the metal ion (pheophytin).

The cobalamins, including coenzyme B_{12}, have a Co ion in the center of the ring. This ion forms an organometallic Co^{III}–C bond to a methyl or a 5'-deoxyadenosyl group. The former is used in methyl transfer reactions, whereas the latter is employed in radical-based 1,2-shifts or elimination reactions [2]. Approximately 35 protein structures with cobalamin are available in the Protein Data Bank.

Coenzyme F430, finally, is found in methanogenic archaebacteria, as part of the methylcoenzyme M reductase (MCR) complex. It contains a Ni ion, which

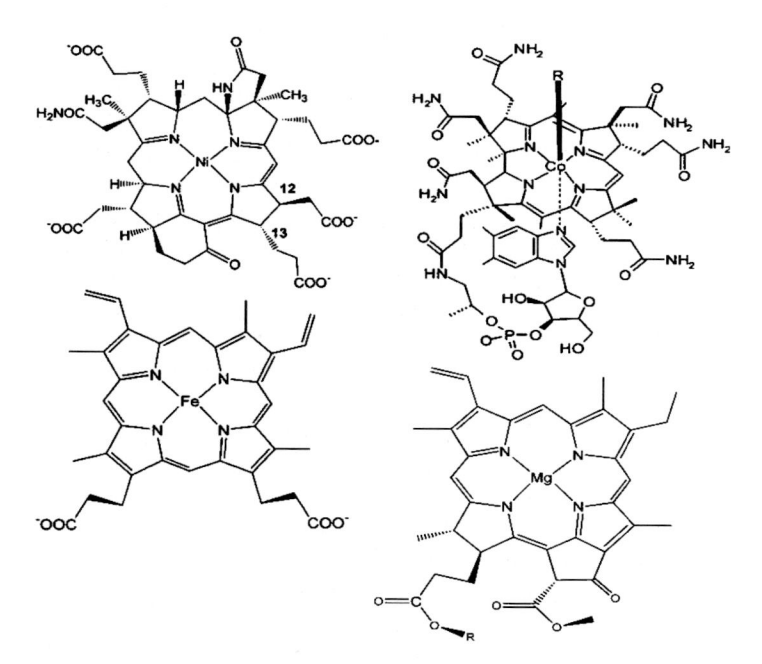

Figure 2.1 Tetrapyrrole cofactors: F430 (upper left), cobalamin (upper right), heme *b* (lower left) and chlorophyll *a* (lower right).

is involved in methyl-transfer reactions [3]. It has been speculated that a Ni^{III}–C bond is formed during catalysis, in analogy with the cobalamins [4].

The apparent similarity of the four cofactors in Figure 2.1 is in contrast to their widely different functions. To understand these differences, systematic studies are needed. Theoretical chemistry may play a key role in this regard, being able to answer questions that are not easily accessed by experimental means. The present chapter reviews recent progress in understanding the similarities and differences between tetrapyrrole cofactors and how they have been designed as catalysts, with an emphasis on our own theoretical work. The first part deals with general differences between tetrapyrroles and aims to elucidate the design principles of the ring systems and the choice of the metal ions. The second part puts further emphasis on the axial ligands and how the function of tetrapyrroles has been optimized by means of particular axial ligands and their interactions with the surroundings.

2.2
Methodology

The method used almost exclusively in the reviewed work is density functional theory (DFT). The three-parameter hybrid functional B3LYP [5] achieves an impressive average absolute error of $\sim$10 kJ mol^{-1} in relative energies of formation and 0.013 Å for bond distances for the main-group G2 test set [6]. This has made B3LYP the most widely used functional today, also within the field of inorganic chemistry [6]. With this state-of-the-art approach, reviews of many important theoretical studies of metalloproteins have been published in recent years, contributing significantly to the understanding of metalloprotein chemistry [6–12].

Experience has shown that most properties, including structures, frequencies, and energies of conversions that preserve orbital occupation, can be modeled accurately with B3LYP for transition metals [13, 14] and, even before the advent of hybrid functionals, successful work was carried out with generalized gradient approximation functionals [15–17] such as BP86, which typically gives better geometries than B3LYP [18]. However, the energy gap between spin states is much harder to estimate and different functionals may give results that differ by over 50 kJ mol^{-1} [19–21]. In general, pure functionals overestimate the stability of low-spin states, whereas hybrid functionals such as B3LYP overestimate the stability of high-spin states, although to a smaller degree [19, 22]. Newer functionals have been suggested with improved performance [23–26]. Moreover, it has been shown that B3LYP underestimates the bond dissociation energy (BDE) of tetrapyrrole models by favoring the open-shell dissociation products, whereas other functionals such as BP86 perform very well [27, 28]. Therefore, much of this work has been performed with the BP86 functional for geometry optimization and BDEs, whereas other energies have been calculated with the B3LYP functional, which is consistent with the strengths of each functional.

DFT methods have successfully been applied to the study of tetrapyrroles, including both porphyrin models [29–47], cobalamin models [27, 48–64], F430 models [65–69] and chlorophyll models [70–73]. This chapter concentrates on work concerned with comparison of the tetrapyrrole cofactors and their tuning by the axial ligands.

Standard procedures have been used, implying optimizing geometries with double-ζ basis sets, including polarization functions on heavy atoms. More accurate energies were usually calculated with a triple-ζ all-polarized basis set. Energies are usually well-converged, and only in rare cases (see below) have competing configurations been a problem during optimization of the Kohn–Sham determinant. Errors in final energies are expected to be $\sim$20 kJ mol^{-1} for barriers, whereas relative isomerization energies are more accurate. Spin-splitting energies have larger errors, but an estimate of the error is not straight-forward. However, most functionals reproduce the correct spin state of small transition metal systems [25], so that, qualitatively, most conclusions in this work are expected to be valid.

We also review perturbations caused by the axial ligands. Effects of side chains are probably small [50, 74] and are not intrinsic to the cofactor in the same way, because they are affected by the surrounding protein. Thus, the molecular models applied have consisted of the bare rings – i.e., corrin (Cor), hydrocorphin (Hcor), porphine (Por), and chlorin (Chl) (Figure 2.2), combined with various axial ligands.

Thus, this chapter is directed primarily towards comparing the *intrinsic* properties for the four metals Fe, Co, Ni, and Mg in the four tetrapyrrole ring systems

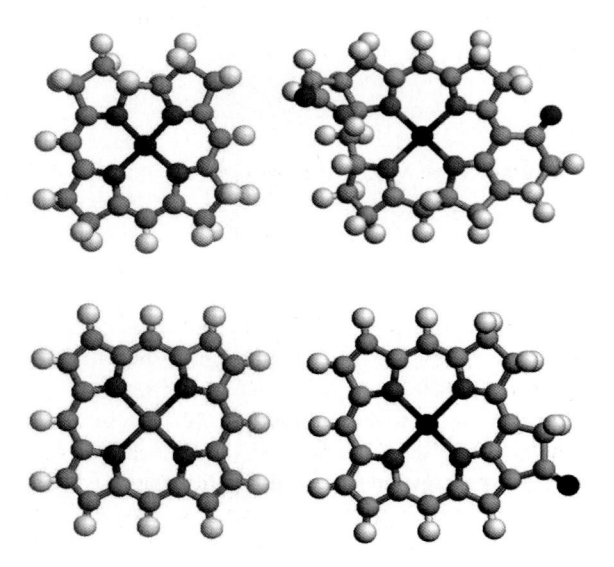

Figure 2.2 Corrin (top left), hydrocorphin (top right), porphine (bottom left), and chlorin (bottom right) rings, with Co, Ni, Fe, and Mg as central ions.

Por, Cor, Hcor, and Chl, and with various sets of axial ligands. For such studies, DFT calculations in vacuum are ideal, because we want to obtain results that are not biased by differences in the surrounding proteins. However, in most cases we check the general effect of the surroundings by repeating the calculations in a continuum solvent with a dielectric constant of 4 and 80. This should illustrate possible solvation effects in any protein, because the effective dielectric constant of a protein is normally assumed to be between 2 and 20 [75]. In most cases, the effect of solvation is small. However, when the reaction involves creation or annihilation of charges, the effect may be large, e.g., for the calculation of reduction potentials and protonation reactions. Only in those cases are solvation effects discussed explicitly in this chapter.

In a few cases, we have gone one step further and studied reactions in specific enzymes. In those cases, we are no longer interested in the intrinsic reactivity of a site or how it compares with similar sites, but rather in the detailed energetics of a certain enzyme reaction. Then, the detailed structure of the surrounding protein is of course crucial and needs to be included. This is normally done by QM/MM methods, in which the active site is treated by quantum mechanical (QM; typically DFT) methods, whereas the surrounding protein is treated by less accurate but much faster molecular mechanics (MM) methods [76, 77]. Such methods normally give excellent structures, but stable and reliable energies are sometimes harder to obtain, because dynamic, entropic, and solvation effects are not properly treated [78]. We have developed recently a method to solve these problems, by performing free energy perturbations, both at the MM level and between the QM/MM and MM level, so that we can still study the interesting reaction at the DFT level. This method is called QM/MM thermodynamic cycle perturbation (QTCP) [79, 80].

2.3
Comparison of the Intrinsic Chemical Properties of the Tetrapyrroles

2.3.1
Introduction

Molecular evolution is a local optimization constrained by the biochemical environment available at any given time. This is why most biomolecules are astonishingly similar and belong to quite a few distinct groups of compounds.

The same situation is true for the tetrapyrrole cofactors: They differ from each other in some modest, but functionally crucial, ways. Por is the most symmetric, with D_{4h} symmetry, all four rings being equivalent and the entire ring fully conjugated. Chl has one tetrapyrrole saturated and a lactone ring is fused with an adjacent pyrrole, making it a five-membered ring system. In Cor, one of the bridging methine groups is missing, and the ring is saturated on all peripheral carbons, rendering the conjugated π-system smaller. Cor has a rotation axis, giving it C_2 symmetry. Finally, the Hcor ring is completely asymmetric, with both a five-membered lactam and a six-membered lactone ring attached. The Hcor ring is even more saturated than Cor, with only five double bonds, of which two pairs are conjugated. The external rings force Hcor to be distorted, with the N–N distances differing substantially. The Por and Chl rings are dianionic, when bound to a metal, whereas Cor and Hcor are monoanions. These differences cause the cofactors to differ in the choice of metals, the number and types of preferred axial ligands, and ultimately in their function. The tetrapyrrole rings can be thought of as rigid equatorial frameworks, upon which additional functionality can be built via perturbations along the axis perpendicular to the ring plane.

How did the tetrapyrroles get their structures and why were these structures chosen? In the case of corrins, this matter has been discussed, but never quantified: It has been suggested that the ring cavity size is designed to fit both the Co^{III} ion and the Co^{II} ion [81] and that the Co^{II} radical in corrins may be the best way to design a directed (d_{z^2}) radical, which can be used reversibly during catalysis [82]. Another suggestion is that corrin has been designed as a flexible entity, which flips upwards upon response to steric strain from the lower axial ligand, transferring the strain via the corrin ring to the Co–C bond, which is thereby weakened (the so-called mechanochemical trigger mechanism) [83]. Flexibility, cavity size, and orbital alignment are properties of the particular tetrapyrrole designs that can be directly probed by theoretical calculations.

2.3.2
Spin States

One of the most important properties of the ground state of the cofactors is the spin. In general, a transition metal can exhibit several different spin states, e.g., high spin (HS), intermediate spin (IS), and low spin (LS), depending on the number of unpaired electrons in the 3d orbitals, and the relative energy of these states depends on the metal and the ring system. For example, it is experimentally known [84] that iron porphyrin systems can exhibit all these spin states. On the other hand, cobalt is exclusively LS in cobalamins, in contrast to what is observed with amino acid ligands [85, 86].

We have quantified these observations by calculating the energies of the optimized structures of the three spin states for various corrins, porphyrins, and hydrocorphins with Fe, Co, and Ni in the +I, +II, and +III oxidation states [63, 66]. It was found that the relative stability of the spin states was affected by both the metal and the ring system and that the two effects were approximately additive and not much affected by the axial ligand. For example, the splitting between the LS and IS states is much larger for $Co^{III}Cor$ than for $Fe^{III}Por$, and the metal contributes to this difference by 40 kJ mol^{-1} and the ring system by 50 kJ mol^{-1} [63]. In general, we find that Co is always LS in tetrapyrroles. Four-coordinate Fe^{II} is IS, whereas five-coordinate complexes of Fe^{II} and Fe^{III} typically are HS and six-coordinate complexes are LS, but we will see below that all three spin states are close in energy for several of these complexes. Nickel, in contrast, is ambiguous: Four-coordinate Ni^{II} is LS, but for the five- and six-coordinate complexes of both

Ni^{II} and Ni^{III}, the LS and HS states are close in energy and their relative stabilities depend both on the nature of the axial ligand and on the density functional method used (BP86 favors the LS state, whereas B3LYP favors HS). In addition, LS Ni^{II} is Jahn–Teller active and therefore normally dissociates possible axial ligands. Thus, a protein could select the HS state by providing an axial ligand, as in MCR [66].

The geometries of metal complexes, in particular the bond lengths to the metal, are sensitive to the spin state [84]: The HS state gives the longest equatorial bonds, whereas those of IS and LS are similar, because only the HS state has an occupied d$_{x^2-y^2}$ orbital [87]. However, our calculations have shown that the IS states have the *longest* axial bonds, i.e., longer than those in the HS state. This is because occupation of the d$_{x^2-y^2}$ orbital forces the iron ion out of the Por plane, which reduces the axial bond length via stereo-electronic effects (the equatorial Fe–N distances increases, leading to a shorter axial Fe–N bond). Thus, while the perturbations caused by the lower axial ligands affect spin, they also directly affect the geometry of the first coordination sphere [87].

2.3.3
Tetrapyrroles Prefer Their Native Ions

The most direct probe into the choice of metals and tetrapyrrole rings is the energy of substituting metals among rings: Which are the most stable combinations of metal ion and ring? Are there inherent electronic or steric effects that cause some ion not to fit in some ring? Surprisingly, it was found that the native combinations of metal ions and ring systems, CoCor and FePor, are stabilized compared with the combinations FeCor and CoPor [63]. This was true for all oxidation states, I, II, and III, and for all axial ligands except one case, the six-coordinate M^{II}MeIm complexes, which are not very biologically relevant.

Even more surprisingly, the same conclusion was reached when including combinations of Ni and Hcor [66]. When comparing Cor and Hcor, which both have a single negative charge and could be expected to be more similar choices, it was found that the native NiHcor and CoCor forms are preferred in all complexes except one, no matter what the axial ligands are. Altogether, these results indicate that there are strong inherent preferences of each ring for its particular metal ion, and that this is a general feature of tetrapyrrole chemistry [66]. These preferences may have been essential for the choice of the native complexes, which could then be modified to obtain a desired function, e.g., by axial perturbations.

2.3.4
Cavity Size and Flexibility of the Tetrapyrroles

It has been suggested that "a fundamental difference between porphyrins, hydroporphyrins, corrins, oxoporphyrins, and other tetrapyrrole macrocycles is their optimal hole size and the range of hole sizes that are readily accessible in their complexes." [88]. This is almost impossible to quantify experimentally (except indirectly by ion radii and cavity sizes), but the proposal can be addressed by computational chemistry.

Figure 2.3 shows the optimized potential energy surfaces for distorting the cavities of six tetrapyrroles: Por, Cor, Hcor, Chl, isobacteriochlorin (Ibc), and bacteriochlorin (Bchl) [66]. The results for the first time quantized not only the cavity size but also the *flexibility* of the ring, in terms of the energy needed to distort it. Both effects are important when discussing the design strategy of tetrapyrroles.

Clearly, from Figure 2.3, the cavity sizes of the various tetrapyrroles differ significantly, the order being Cor < Bchl ~ Chl < Por < Ibc < Hcor (note that Hcor has two conformational minima). The flexibility of the rings follows the trend Hcor > Cor > Por ~ Ibc > Chl > Bchl. These trends can be compared with the ionic radii of the various ions: LS Co^{III} ~ LS Fe^{III} < LS Ni^{III} < HS Ni^{III} < HS Co^{III} ~ LS Fe^{II} < HS Fe^{III} ~ LS Co^{II} < HS Ni^{II} < Mg^{2+} < HS Co^{II} < HS Fe^{II} [89]. However, these radii depend on the axial ligands and the type and charge of the equatorial ligands. Therefore, we have directly probed the ideal size of the various ring systems by using ring-broken models of the tetrapyrroles (Figure 2.4) [63, 66]. These models retain the charge, the number of bonds in the chelate rings, and the conjugation of the real tetrapyrroles, but they cannot enforce suboptimal M–N distances (M is the metal) by covalent strain within the ring system.

Calculations with these models showed that the cavity in Cor is ideal for LS Co^I, Co^{II}, and Co^{III}, because the Co–N bond lengths are the same in the normal and ring-broken models within 0.03 Å [63, 66]. In contrast, the central cavities in Por and Hcor are too large for all metals in their LS states. Thus, Hcor is ideal for

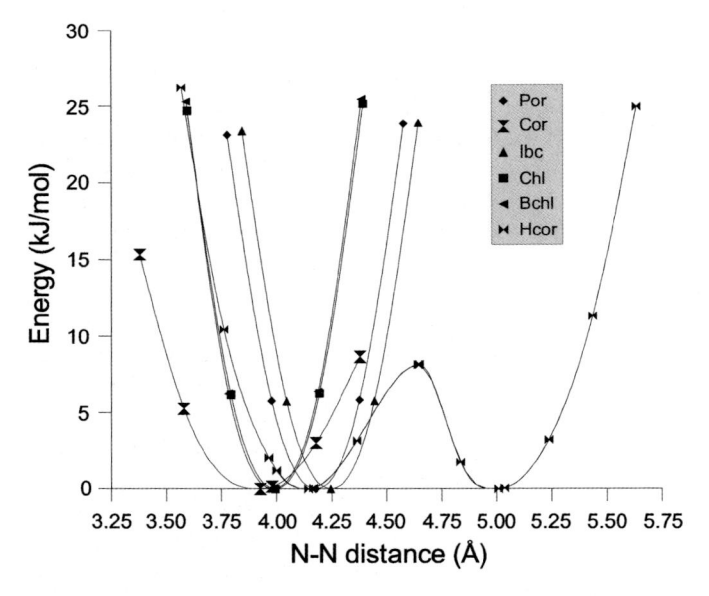

Figure 2.3 Potential energy curves for distortion of the cavities in Por, Cor, Ibc, Chl, Bchl, and Hcor [66].

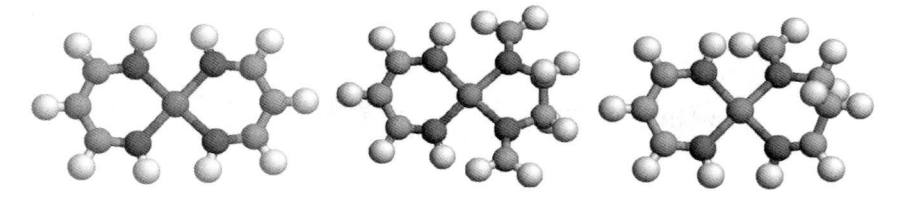

Figure 2.4 The three models used to estimate ring strain in Por, Cor, and Hcor, respectively [63, 66].

incorporating the large HS Ni^{II} ion, whereas the Por ring renders also the higher spin states of Fe accessible, an effect that can be further enhanced by using the right lower axial ligand, as discussed below.

2.3.5
Cytochrome-like Electron Transfer

In addition to the structural preferences outlined above, various functional aspects of tetrapyrroles have been compared. One important group of heme proteins is the cytochromes, whose function is to transfer electrons. According to the semi-classical Marcus equation:

$$k_{ET} = \frac{2\pi}{\hbar} \frac{H_{DA}^2}{\sqrt{4\pi \lambda RT}} \exp\left(-\frac{(E_0 + \lambda)^2}{4\lambda RT}\right) \quad (1)$$

the rate of electron transfer (k_{ET}) depends on three parameters, the electronic coupling element (H^2_{DA}), redox potential (E_0), and the reorganization energy (λ) [90]. Most of these depend on the detailed structure of the proteins involved in the electron transfer. However, the inner-sphere reorganization energy, i.e., the energy difference of the electron-transfer site in the geometry of its reduced and oxidized states, is almost entirely determined by the intrinsic properties of the electron-transfer site alone, and can therefore be studied directly by quantum mechanical methods [40].

We have calculated inner-sphere reorganization energies for several combinations of Fe/Co/Ni with Por/Cor/Hcor, using the most common set of axial ligands found in nature, viz. two imidazole (Im) ligands (as models of histidine) [63, 66]. It was found that these energies depend primarily on the type of metal. For example, the Co$^{II/III}$ pair gave reorganization energies of 179 and 197 kJ mol^{-1} in CoPorIm$_2$ and CoCorIm$_2$, respectively, whereas for the corresponding Fe complexes the values were 8 and 9 kJ mol^{-1}, a huge difference [63]. The reason for this effect is that the d$_{z^2}$ orbital is occupied for Co^{II} (which has seven d electrons), but not for the other ions (which have five or six d electrons). This orbital is directed towards the axial ligands and, therefore, causes a large change of the axial bond length upon reduction of Co^{III}. Thus, octahedral cobalt complexes cannot be functional electron carriers. The Ni complexes, for which the d_{z^2} orbital is occupied in both oxidation states, fall in between Co and Fe, and could in fact be decent electron carriers in tetrapyrrole complexes, in particular in Por (23 kJ mol^{-1}) [66].

The Por and Cor ring systems gave similar inner-sphere reorganization energies, and we found that they keep the reorganization energy low by restricting the change in the equatorial bonds of the metal [40]. Therefore, the more flexible Hcor ring always exhibited the largest reorganization energies with any metal ion (e.g., 23 kJ mol^{-1} for FeHcorIm$_2$). Thus, theory can explain why Fe is used as an electron carrier, but it does not explain why Por is used instead of Cor.

The reduction potential depends strongly on the surroundings, i.e., on the detailed structure of the surrounding protein. However, we can estimate the *intrinsic* potential of a certain combination of metal and tetrapyrrole ring (and axial ligands) using a continuum solvent. The resulting potentials cannot be directly compared with potentials in proteins, but they illustrate the effect of substitutions of the metals or ring systems. We have used this approach to study the reduction potential of the same complexes [63, 66]. These calculations showed the trend Co < Fe < Ni for the M$^{II/III}$ couple, with differences of $\sim$0.7 and 0.2 V. The potentials of the complexes of Por (with its double negative charge) were always lower than those of Cor and Hcor. However, we obtained the opposite trend for the I/II potentials in four-coordinated complexes: Ni < Fe < Co [66]. This confirms the stability of the Co^I state, which is found in corrin biochemistry, e.g., in the mechanism of methionine synthase [55]. In contrast, we saw no stabilization of the Ni^I state, which is the supposed active state of F430 in MCR.

2.3.6
Stability of a Metal–Carbon Bond

Cobalamin is taken up in the body as vitamin B$_{12}$, but its cyano ligand is replaced with other groups to form either 5′-deoxyadenosylcobalamin (AdoCbl) or methylcobalamin (MeCbl), which are the two biologically active cofactors. Both contain a unique organometallic Co–C bond, which is cleaved during catalysis. In the case of MeCbl, the cleavage is heterolytic, with both electrons of the organometallic Co–C bond staying on cobalt to form a Co^I intermediate [91]. In the case of AdoCbl, the Co–C bond is cleaved homolytically, with one electron ending up in Cob(II)alamin and one in the 5′-deoxyadenosyl (Ado) radical. The Ado radical subsequently initiates a radical mechanism by which chemical groups are subject to a 1,2-shift or elimination reactions. AdoCbl is a cofactor in many enzymes, including glutamate mutase, methylmalonyl-coenzyme A mutase (MCAM), diol dehydratase, ethanolamine ammonia lyase, and class II ribonucleotide reductase. MeCbl is coenzyme in methyl transferases, such as methionine synthase (MES), corrionoid Fe/S proteins, and coenzyme M methyl transferases [92].

Consequently, the properties of the organometallic Co–C bond have been extensively studied [27, 48–61]. Interestingly, we found that the homolytic M^{III}–C BDE is 10 kJ mol^{-1} larger for CoCorImMe than for FePorImMe [63], so this cannot be the reason why Co is used in these organometallic cofactors. Another idea is that the unwanted side reaction of hydrolysis is better prevented with Co–C bonds [81]. This was supported by our calculations: The Co–C bond was 33–48 kJ mol^{-1} more resistant to hydrolysis than the Fe–C bond [63].

Coenzyme F430 is also used for methyl-transfer reactions, and it has been suggested that methyl binds directly to Ni, although this has not yet been observed. Therefore, we have also studied a range of organometallic analogues of methylcobalamin, resembling possible methylated intermediates of MCR [66]. It turned out that Ni gave the weakest M–C bonds among the three studied metals, both in the +II and +III oxidation states. In fact, the bond is so weak that suggested homolytic mechanisms of MCR [93] can be ruled out [66, 68].

We have also studied the corresponding heterolytic reactions, i.e., the transfer of a methyl group from the metal to various acceptors (e.g., a deprotonated homocysteine in MES) [63, 66]. Our results showed that the M–C bond strength actually follows the trend Ni < Co < Fe, with differences of 10 and 80 kJ mol^{-1}, respectively. However, the reaction energies depend strongly on the methyl donor/acceptor, and it is therefore still possible that the methyl group binds directly to Ni in MCR, provided that the donor is properly activated [66].

2.3.7
Metallation Reaction

The use of Mg^{2+} in chlorophyll is quite unexpected, because Mg (in variance to Fe, Co, and Ni) strongly prefers O-ligands, rather than N-donors [85]. Therefore,

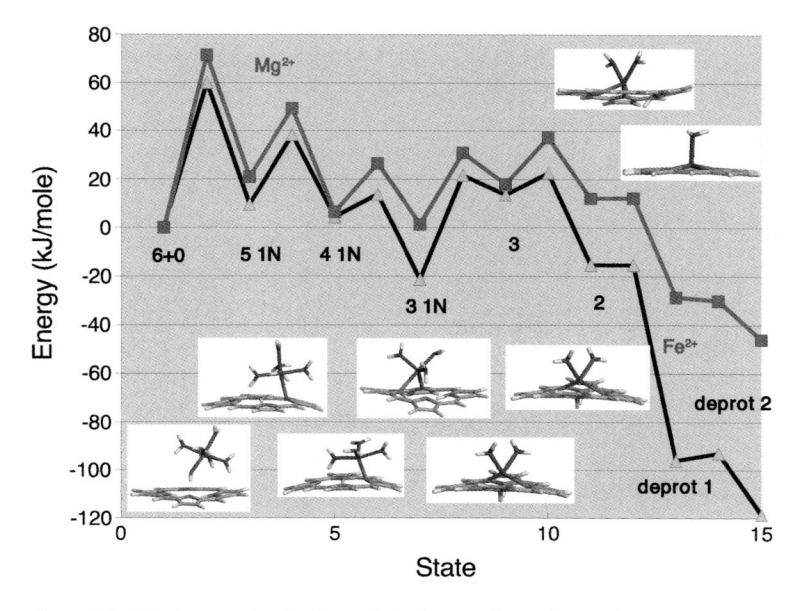

Figure 2.5 Relative energies for the various intermediates (inserted figures) and transition states for the metallation reaction of Por with Fe^{2+} (▲) and Mg^{2+} (■) [95].

the binding of Mg to Chl can be expected to be less favorable than the binding of Fe to Por. In fact, it is experimentally found that the incorporation of Mg into its tetrapyrrole precursor is ATP dependent – contrary to the corresponding reaction of Fe [94]. We have compared the various reaction steps of the incorporation of Fe^{2+} and Mg^{2+} into Por with density functional methods [95]. As can be seen in Figure 2.5, the reaction energies are mostly quite similar and the two curves run roughly parallel. However, in the final steps of the reaction, the deprotonation of the porphyrin, there is an appreciably larger gain in energy for Fe^{2+} than for Mg^{2+}, almost 80 kJ mol^{-1}, reflecting the larger affinity of Fe^{2+} for the Por ring (before these steps, the metal ion resides above the plane of the doubly protonated ring). Moreover, the first and rate-limiting step in the reaction mechanism, the formation of the first bond between the metal and the Por ring, has a 10 kJ mol^{-1} higher activation energy with Mg^{2+} than with Fe^{2+}.

2.4
Tuning of Tetrapyrrole Structure and Function by Axial Ligands

2.4.1
Introduction

Having discussed the inherent properties of the tetrapyrrole ring systems and the metals, we now study how these properties are tuned by axial ligands. The two axial coordination sites of tetrapyrroles are distinguished as the upper side and lower side. The upper (distal or β) side is where the substrate binds. The upper site is either open for binding of solvent, substrate, or other molecules, or is it occupied by a protein ligand. The lower (proximal or α) side is occupied by an amino acid residue or a group from the cofactor itself. In heme proteins, the lower axial ligand is typically histidine (His), methionine (Met), cysteine (Cys), or tyrosine (Tyr). In cobalamins, it is His or the 5,6-dimethylbenzimidazole group at the end of one of its side chains, whereas in F430 it is the oxygen of a glutamine (Gln) residue. In chlorophylls, many different axial ligands can be used. Thus, a great deal of variety is possible and the exact choice of lower ligand and its effect on the chemistry of the cofactor has intrigued people for decades [96–98].

2.4.2
Importance of the Lower Axial Ligand for B_{12} Chemistry

For AdoCbl in aqueous solution, the Co–C bond is cleaved with rates of 10^{-7}–10^{-9} s^{-1} at 37 °C ($\Delta G^{\ddagger} = 124$ kJ mol^{-1}), corresponding to a half-life of $\sim$22 years [99–101]. In contrast, several coenzyme B_{12} enzymes attain catalytic rates (k_{cat}) of 2–300 s^{-1} [102, 103]. Thus, the enzymes give rise to a 10^{9-13}-fold acceleration of Co–C bond cleavage [103, 104], corresponding to a reduction of $\Delta G^{\ddagger}$ by $\sim$70 kJ mol^{-1}. Moreover, the enzymes shift the equilibrium constant towards the homolysis products by a factor of 3×10^{12} (74 kJ mol^{-1}), giving an equilibrium constant close to unity [101, 105]. This cleavage initiates the subsequent radical reactions; therefore, understanding the reasons for this Co–C bond activation is arguably the most critical problem in the chemistry of B$_{12}$ [106].

It has long been hypothesized that the axial Co–N$_{ax}$ bond in cobalamins is used to labilize the Co–C bond by either steric or electronic effects [83]. Until the advent of DFT methods, these questions could not be addressed directly, but the so-called mechanochemical trigger hypothesis was believed by many authors to be the most reasonable mechanism of Co–C bond activation. This hypothesis asserted that upwards butterfly folding of the corrin ring could cause strain in the Co–C bond, thus lowering the bond strength enough to provide cleavage [83]. The effect can be attenuated by the fact that the axial His ligand typically forms a hydrogen bond to the carboxylate group of a Asp residue in most structures of coenzyme B$_{12}$ enzymes [107, 108].

The effect of the lower axial ligand has been studied by optimizing structures of B$_{12}$-models with a fixed Co–N$_{ax}$ bond length [51, 53]. It was found that the Co–N$_{ax}$ bond is extremely flexible: It can be varied over a range of 0.5 Å at an energy cost of less than 3 kJ mol^{-1}, quantifying the suggested "substantial energy cost" [109] of variations in this bond. This result explained the large variations in this bond length observed in crystal structures and also why theoretical calculations reproduced this bond length quite poorly [51]. Moreover, compression of the Co–N$_{ax}$ bond cannot provide any significant upward folding of the corrin ring, and variations in the Co–N$_{ax}$ bond lengths cannot change the Co–C BDE by more than a few kJ mol^{-1} (and typically the BDE is increased) [51].

In many coenzyme B$_{12}$ enzymes, the axial His ligand forms a hydrogen bond with the carboxylate side chain of an Asp residue. We have studied the effect of the His–Asp motif, because mutation of this Asp leads to a 15–1000-fold decrease in k_{cat} of glutamate mutase [110]. We studied the extreme case in which the His ligand is deprotonated to imidazolate [51]. This led to a strengthening of the Co–N$_{ax}$ bond and an elongation of the Co–C bond by 0.03 Å, but the BDE actually increased by 16 kJ mol^{-1}. The Co–C bond energy could be decreased by compressing the trans Co–N$_{ax}$ bond, but not by more than 15 kJ mol^{-1}. This energy corresponds to a $\sim$400-fold kinetic effect, resembling the experimental decrease in k_{cat} upon mutation of Asp [110]. The maximum of 15 kJ mol^{-1} of catalytic energy stored in the lower axial ligand is minor compared with the total 10^{9-13} rate enhancement [104] and is not even near an explanation of the catalytic power of B$_{12}$-dependent enzymes.

Work by other theoretical groups [54, 61] also confirmed that the effect of the axial ligand is modest. In particular, calculations of Co–C BDEs could directly probe the overstated potency of this ligand. The final conclusion from theory was that the Co–N$_{ax}$ bond is a low-energy mode of limited catalytic importance in mutases [51, 54, 61]. This observation was later repeated also in a protein system, by performing QM/MM calculations of bond length distortions of AdoCbl in MCAM [111]. Instead, it has been suggested that the axial His is relevant for the discrimination between homolysis and heterolysis of the Co–C bond [56, 111].

Then how is the Co–C bond labilized? Recent QM/MM calculations [58] have helped in understanding the much debated effects of the proteins on the Co–C bond dissociation reaction. Our calculations reproduced the main experimental observations [112, 113], including an equilibrium constant between the Co^{II} and Co^{III} states close to unity, and the fact that the protein removes the barrier of bond dissociation almost completely. The amazing ability of glutamate mutase to activate the Co–C bond to the point where this step is not rate-determining was understood from the calculations as well, pointing towards a strong influence of electrostatics, both directly, via the electrostatic field, and by changing the geometry of the cofactor [58].

It was shown by calculating the Co–C bond dissociation both with and without point charges of the protein that the direct electrostatic effect amounts to ~34 kJ mol^{-1} [58]. On the other hand, the electrostatically induced distortions of the coenzyme at short Co–C bond lengths reduced the BDE by 56 kJ mol^{-1}. However, the geometry was not changed for the lower axial ligand (Figure 2.6), but primarily for the polar ribose "handle" of the cofactor.

In particular, the cobalt ion moved out of the corrin plane and the C–Co–N$_{eq}$ angle was bent, in good agreement with the conformational changes that have been observed in crystallographic studies of both GluMut [114] and MCAM [115]. An extra 20 kJ mol^{-1} catalytic effect was caused by the fact that full dissociation is never accomplished in the protein, as compared with the isolated cofac-

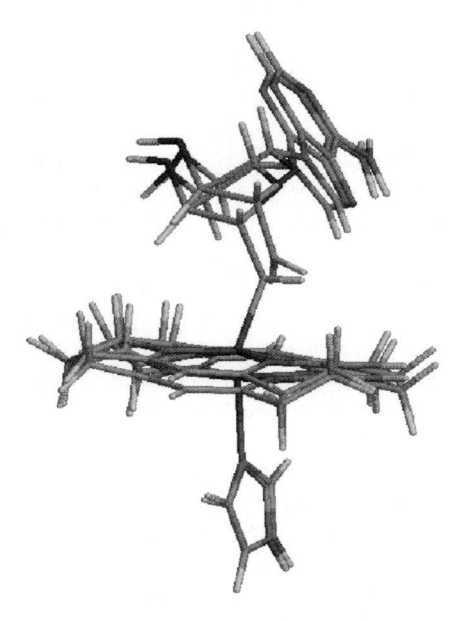

Figure 2.6 QM/MM optimized structures of the Co^{III} and Co^{II} states in glutamate mutase, showing no movement of the lower axial ligand and no trans steric effect as the Co–C bond is broken [58].

tor in solution [58, 116]. Thus, the entire mechanism of Co–C bond labilization was explained *without* any effect of the axial His ligand.

Recently, MCAM has been studied using a similar QM/MM procedure [117]. An authentic transition state was obtained for the cleavage of the Co–C bond at a distance of 2.67 Å with an energy of 42 kJ mol^{-1}, and an intermediate was located with a modified conformation of the Ado moiety at a Co–C distance of 2.19 Å and an energy of 29 kJ mol^{-1}. It was suggested that the change in conformation of the protein and the binding of the substrate are important for the catalysis. Unfortunately, the work did not analyze how the enzyme destabilizes the Co^{III} state relative to the Co^{II} state (by ~150 kJ mol^{-1}), and so the two QM/MM studies are hard to compare.

From these studies it became obvious why MeCbl cannot be used in catalysis involving homolysis of the Co–C bond: The methyl group is too small and non-polar to respond to the distortion forces [58]. The QM/MM calculated BDE for MeCbl in GluMut is only 27 kJ mol^{-1} smaller than in the isolated cofactor, whereas for AdoCbl the BDE is lowered by 135 kJ mol^{-1} by the protein, clearly showing this difference between the two cofactors [58]. Instead, it was found [55] that MES works by deprotonation of the methyl acceptor (homocysteine, which is bound to a zinc ion in MES [118]) and by securing a non-polar environment, which is in line with the hydrophobic environment in the crystal structure [119]. Furthermore, the Ado group cannot be subject to any nucleophilic attack, because the structure of the cofactor renders the C5′ atom inaccessible. Notably, the axial His ligand in MES detaches during formation of the transition state [55].

2.4.3
Lower Axial Ligand in Cofactor F430

The crystal structure of MCR shows that a Gln residue binds as an axial ligand to Ni in coenzyme F430. The role of this residue may be to stabilize the HS state of Ni^{II}, which is necessary during catalysis [120]. In our calculations, we found that four-coordinate Ni^{II} Hcor complexes have a LS ground state (by 100–157 kJ mol^{-1}), whereas five-coordinate complexes with strong methyl or OH ligands preferred a HS state by 13–86 kJ mol^{-1} [66]. With weaker ligands, such as an acetamide model of Gln, the LS state is still most stable, but the Gln ligand dissociates in this state, whereas it remains bound to the metal ion in the HS states. This indicates that the enzyme deliberately selects a HS state of Ni^{II} by providing the Gln ligand [66].

Therefore, it was concluded that the Gln residue serves as an ideal ligand for the HS Ni^{II} state: It binds weakly to the Ni^{I} state and can thus provide some guidance for changing the d-electron configuration upon reaction, in a manner only possible with the asymmetric structure of the Hcor ring [66]. The flexibility of the Hcor ring further adds to the cofactor's ability to deal with several large ionic states of Ni. It may be speculated that this flexibility in the ligand field of F430 is used directly during catalysis to minimize the otherwise severe effect of putting a ninth d-electron into an octahedral ligand field.

2.4.4
Importance of Axial Ligands for the Globins

One of the classical problems of biochemistry is the binding of dioxygen to hemoglobin or myoglobin. There are several issues involved [121]. First, oxygen has to bind reversibly, i.e., with a small binding energy, and fast, i.e., with a small activation energy. The experimental binding rate is ~ 15 M s^{-1} [122], but includes a multitude of sequential intermediates during the diffusion of O$_2$ towards the heme group. Second, O$_2$ is in a triplet spin state and deoxyheme is in the HS quintet state, whereas the oxyheme complex is a LS singlet. Therefore, spin crossover has to proceed [123]. We have studied how globins overcome this fundamental obstacle [41].

The ground state of oxyheme turned out to be a singlet, in accordance with experiments, with an uneven distribution of spin: It can be described as 75–80% Fe^{III}–O$_2^-$ and 20–25% Fe^{II}–O$_2$. A later multiconfigurational *ab initio* study found the two configurations to be of similar weight [124]. The electronic structure is in good agreement with Mössbauer experiments [125], where a mixture of $\frac{2}{3}$ ferric and $\frac{1}{3}$ ferrous forms explained the system well. Even the energy gap to the lowest triplet state agrees quite well with experiment [126].

Next, we studied the binding of O$_2$ to deoxyheme for the lowest states of each spin multiplicity [41]. Interestingly, it turned out that all of them had similar energy when the Fe–O distance was more than ~ 2.5 Å. Such a near-degeneracy can be important for at least three reasons [41]: First, both triplet and septet states can easily lead to products (the two unpaired spins of the incoming O$_2$ molecule can be either parallel or antiparallel to the four unpaired electrons of deoxyheme, giving rise to these two reactant states). Second, the slope in the crossing region of the four states is small, meaning that the crossing probability is large [127]. This may accelerate reversible binding by two orders of magnitude compared with non-heme FeO$^+$. Third, the shape of the binding curves indicates that activation energies are smaller than 15 kJ mol^{-1}, giving a rapid O$_2$ binding. Altogether, oxygen binding was found to be strongly accelerated and reversible, owing to the nature of the axial ligand, which together with the intrinsic effect of the porphine ring brings spin states close in energy in ferrous heme [41, 63].

2.4.5
Role of Axial Ligands for the Cytochromes

As mentioned above, the cytochromes are heme proteins whose function is to transport electrons. Cytochromes always have two axial ligands, to avoid binding of any unwanted ligands to iron. However, the axial ligands vary quite a lot, the most common combinations being two His ligands (found in most *b*-type cytochromes) or one His and one Met (found in most *c*-type cytochromes) [85].

We have calculated inner-sphere reorganization energies of cytochrome models with seven different combinations of axial ligands [40]. The results (Table 2.1) showed that neutral axial ligands (His, Met, and a neutral amino terminal) give

Table 2.1 Inner-sphere reorganization energies [40] and reduction potentials for seven cytochrome models. Amt is a neutral amino terminal.

Axial ligands		Reorganization energy (kJ mol^{-1})	Reduction potential (V)		
1	2		$\varepsilon = 1$	$\varepsilon = 4$	$\varepsilon = 80$
Met	Met	4.8	1.41	0.53	0.12
His	Met	8.3	0.57	−0.25	−0.63
His	His	8.2	0.39	−0.38	−0.72
His	Amt	8.6	0.45	−0.42	−0.82
His	Cys	20.0	−3.51	−2.23	−1.56
His	Tyr	47.0	−3.24	−2.16	−1.62
His	Glu	26.4	−3.11	−1.97	−1.40

a low inner-sphere reorganization energy, 5–9 kJ mol^{-1}. However, if one of the axial ligands is negatively charged (Cys, Tyr, or Glu), the reorganization energy is appreciably larger, 20–47 kJ mol^{-1}. Interestingly, such charged ligands are typically used only if the heme site also has other functions than electron transfer (e.g., catalysis). This illustrates the importance of the inner-sphere reorganization energy in electron transfer and it also explains the choice of axial ligands for the cytochromes.

Still, notably, even the reorganization energies with the charged ligands are quite low. In fact, the inner-sphere reorganization energies of the cytochromes are appreciably lower than those of the other two important groups of electron carriers found in nature, viz. the blue copper proteins and the iron-sulfur clusters (43–90 and 40–75 kJ mol^{-1}, respectively) [40, 128–131]. This is because the preferred geometry of Fe^{II} and Fe^{III} in the octahedral field of the cytochromes is very similar, and because the small difference in the preferred Fe–ligand distances is reduced by covalent strain in the rigid porphyrin unit by ∼0.05 Å, leading to a decrease in the reorganization energy of ∼8 kJ mol^{-1}. It is also important that both oxidation states of Fe are in the LS state – in the HS state, the reorganization energy would be 20–30 kJ mol^{-1} higher, owing to the longer Fe–N bond lengths. Finally, the compensation of two charges in the porphyrin ring and the delocalization of the charge over the rather soft N and S ligands are also important for the low reorganization energy in the cytochromes [40].

We have also studied the reduction potential, calculated in continuum solvents with different values of the dielectric constant (ε), ranging from 1 (vacuum) to 80 (water). The results of such calculations [40] are also shown in Table 2.1. The reduction potentials are much more sensitive to the choice of the axial ligands than the reorganization energies. Thus, the cytochromes can fine-tune their reduction potentials by the choice of axial ligands, without changing the reorganization energies significantly. The redox potentials strongly depend on the dielectric constant. However, the relative potentials of different sets of ligands are fairly constant (at least for the neutral ligands). For example, the His–Met set gives a 0.10–0.17 V higher potential than the His–His set, which is in fair accordance with the experimental difference of 0.17–0.30 V for these two sets of ligands in actual proteins [85, 132]. The Met–Met set causes an even higher potential, whereas the His–amino-terminal (Amt) set provides a potential similar to that of the His–His set. Thus, our results indicate that the axial ligand is used by the cytochromes to tune the reduction potential, rather than the reorganization energy.

2.4.6
Role of the Axial Ligand in Heme Enzymes

As mentioned above, the heme group is used in many different heme proteins. We have already discussed two important groups, viz. the globins and the cytochromes. A third important and diverse group is the oxidizing heme enzymes, including heme peroxidases, catalases, and cytochromes P450. All heme enzymes contain only one axial ligand from the protein, leaving the sixth coordination site free for the binding of a substrate. However, this axial ligand differs between the various types of heme enzymes: The cytochromes P450, as well as chloroperoxidase and NO synthase, have a Cys ligand, whereas all heme peroxidases use a His ligand, and all heme catalases utilize a Tyr ligand. Both the Cys and the Tyr ligands are expected to bind in their deprotonated, negatively charged forms.

Moreover, the properties of the axial ligand are tuned by its interactions with second-sphere ligands. Thus, the His ligand in the heme peroxidases invariably forms a hydrogen bond to the carboxylate group of an Asp residue, similar to

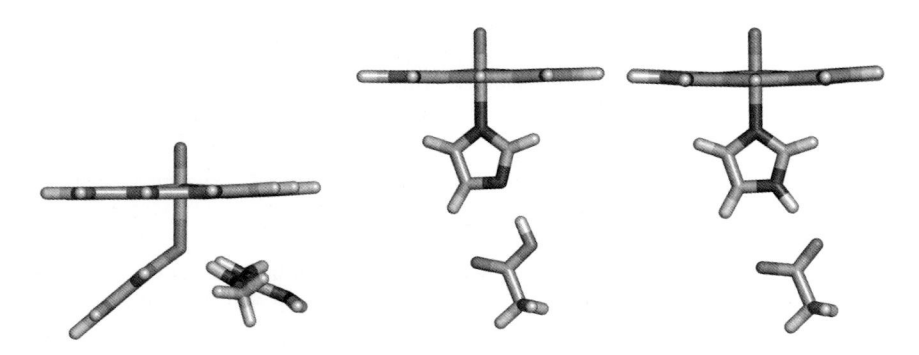

Figure 2.7 Three models with second-sphere hydrogen bonds (all showing compound I; see text): Tyr + Arg (left) and His + Asp with the proton either on Asp (middle, His + HAsp) or on His (right, HisH + Asp).

the His ligand in the coenzyme B_{12} enzymes (Figure 2.7). This is in variance with the globins, which do not possess such a hydrogen bond. Likewise, the Tyr ligand in the catalases invariably forms a hydrogen bond to the positively charged side chain of an Arg residue (Figure 2.7), and the Cys ligand in NO synthase forms a hydrogen bond with a tryptophan residue [133]. As it is likely that these hydrogen-bond interactions may tune the properties of these groups, they were also included in the theoretical studies.

Heme peroxidase, catalase, and cytochrome P450 are involved in similar catalytic cycles (Figure 2.8). The resting state of cytochrome P450 is the Fe^{III} state with a water molecule bound to Fe. When the enzyme binds the substrate, the water molecule is displaced and the spin state changes to HS. After reduction to Fe^{II}, O_2 binds to heme (note that these are the only two states used by the globins).

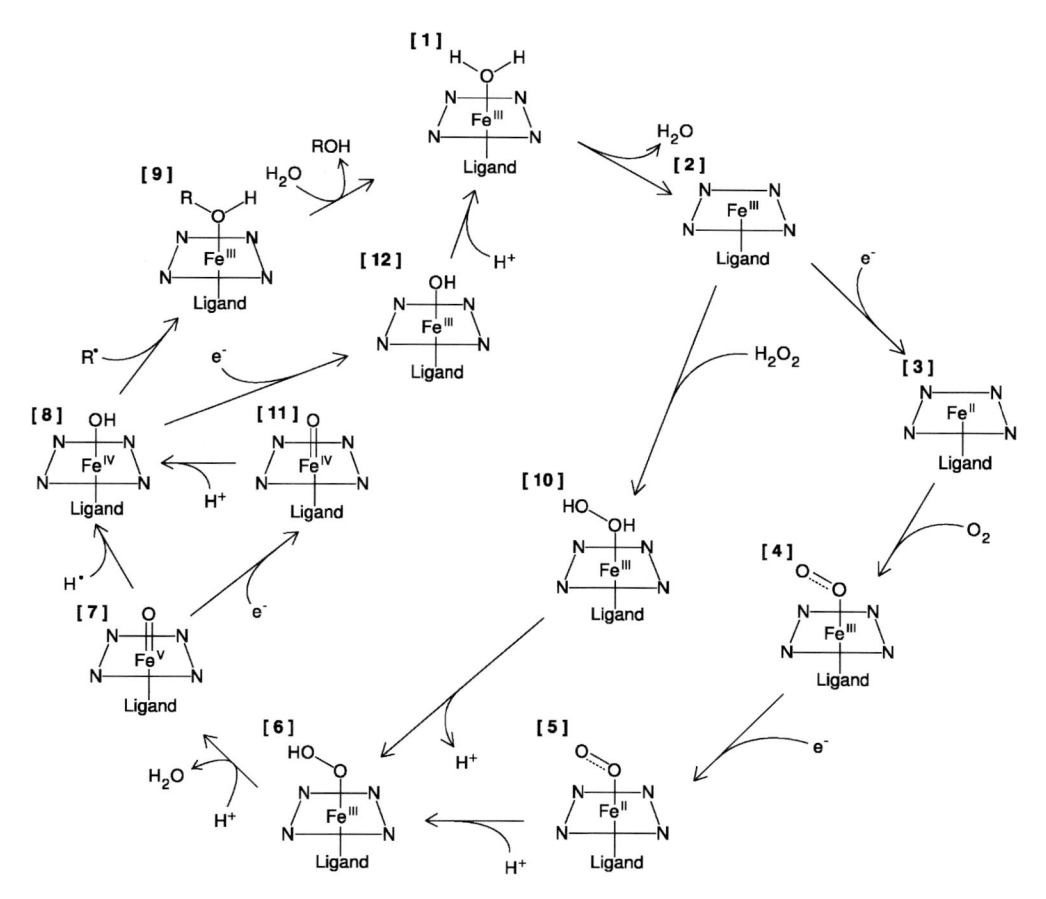

Figure 2.8 Reactions involved in the cycles of cytochrome P-450 (states 1–9), peroxidase (states 1, 2, 10, 6, 7, and 8 or 11), and catalase (states 1, 2, 10, 6, and 7).

After another reduction, a proton is bound, forming a Fe^{III}-hydroperoxide complex. This complex takes up another proton, which triggers cleavage of the O–O bond and the formation of a highly reactive state, called compound I, which formally is a Fe^V–O^{2-} complex. It can oxidize almost any compound, by hydrogen abstraction (and rebound of the formed radical to a hydroxide compound), epoxidation, or N, S, or SO oxidation. The intermediate after hydrogen abstraction is a Fe^{IV}–OH^- complex, whereas the product after the full reaction is the resting Fe^{III} state, possibly with the product, rather than a water molecule bound.

In contrast, the heme peroxidases bind hydrogen peroxide, which after deprotonation and reprotonation on the terminal oxygen forms compound I. However, in this class of enzymes, the active site is not large enough to bind a substrate larger than hydrogen peroxide. Instead, the substrates bind on the surface of the protein and are oxidized by one-electron transfer from the heme site, which first forms a $Fe^{IV}O$ state, called compound II, and then the Fe^{III} resting state after a second electron transfer. The reaction mechanism for catalases is identical up to the formation of compound I. However, this intermediate then reacts with another hydrogen peroxide molecule to form O_2. Thus, the catalases catalyze the disproportionation of two H_2O_2 molecules to water and O_2.

We have studied the influence of axial ligands for all of the 12 reaction intermediates shown in Figure 2.8, using five different sets of axial ligands: His, His + Asp, Cys, Tyr, and Tyr + Arg [47]. Naturally, the five types of axial ligands display extensive variation in their distances to the Fe ion. In general, the bond lengths follow the trend Tyr < Tyr + Arg ≈ His + HAsp < HisH + Asp < His < Cys. On the other hand, the Fe–O distances on the distal side follow the trend: His < Tyr + Arg < HisH + Asp < His + HAsp ≈ Tyr < Cys.

More interesting than these structural differences are their effect on the reaction energies. We have studied how the axial ligand influences all the reactions shown in Figure 2.8 [47]. For most reactions, the energies follow the trend Cys < Tyr < Tyr + Arg < His + Asp < His, reflecting the donor capacity of the various ligands. The most pronounced effects are seen for the reduction potentials. In particular, the $Fe^{II/III}$ reduction potential is negative and increases in solution for the negatively charged ligands (Tyr, Cys, and His + Asp), whereas it is more positive and decreases in solution for the two neutral ligands (His and Tyr + Arg). This is in agreement with the experimental potentials of catalase (<−0.5 V), cytochrome P450 (−0.30 V), horseradish peroxidase (−0.22 V), and myoglobin (+0.05 V) [47]. Thus, the choice of a neutral His ligand in the globins is important to avoid formation of the inactive Fe^{III} form.

Even more important are the next two steps of the reaction, the reduction and protonation of the O_2 complex, which appear to be concerted according to the calculations: The reduction potential is 0.3–0.4 V more negative for His than for the other four ligands, indicating that it is appreciably harder to reduce the Fe^{II}–O_2 complex in the globins (an unwanted side-reaction) than in the other enzymes (an essential step for cytochrome P450).

The reduction potential of the compound I/II couple exhibits a similar trend: Tyr < Cys < Tyr + Arg < His + Asp < His. This is also functionally appropriate:

The potential in cytochrome P450 and catalase should be as negative as possible to avoid the formation of compound II, whereas for the peroxidases the reduction potentials of compound I and II should be as similar as possible, because the two states should oxidize the same substrate. Experimentally, the reduction potentials are 0.90 and 0.87 V for horseradish peroxidase [134]. In the calculations, we obtain a 0.1–0.3 V lower potential for compound II with any type of ligand.

Moreover, there is a pronounced effect of the negative axial ligands (but not of Tyr + Arg) on the formation of compound I in a hydrophobic environment: the reaction from the hydroperoxide complex is 80–140 kJ mol^{-1} more exothermic for Cys, Tyr, and His + Asp than for His in a continuum solvent with $\varepsilon = 4$. This effect of the negatively charged axial ligand has often been termed a push effect [135]. However, the effect strongly depends on solvation and disappears in a water-like continuum solvent.

Finally, we observed that the Tyr and Tyr + Arg ligands cause a 8–20 kJ mol^{-1} more exothermic reaction energy for the reaction of H_2O_2 with compound I than the other three ligands do. This is relevant for the second step of the catalase reaction cycle, indicating that the choice of the Tyr ligand is also appropriate in this case.

In contrast, calculations did not reveal any effect of the axial ligand on the O_2 affinity or the hydrogen-atom abstraction. Likewise, there was no indication that negatively charged ligands should favor a heterolytic, rather than a homolytic, splitting of O–O, and we even found an opposite effect for the P450 oxidation step, i.e., that the His model gives a more favorable energy than the Cys model [47]. More recent studies have compared the activation energies for hydrogen abstraction using models with Cys, His, and Tyr ligands, but found small differences, sometimes indicating a lower barrier for His than for Cys [136, 137]. In particular, His models seem to prefer epoxidation, whereas Cys models favor hydrogen abstraction. For sulfoxidation, there is no difference between Cys and His, in terms of activation barriers [138].

2.4.7
Tuning the His Ligand by Hydrogen Bonds in Heme Proteins

We have seen that the axial His ligand in both heme peroxidases and coenzyme B_{12} enzymes is hydrogen bonded (by the non-coordinating N atom of the imidazole ring) to a carboxylate group (Figure 2.9 [139]) and that this may influence its properties. We have, therefore, studied this interaction in more detail and compared it with the limiting cases of no carboxylate group or a fully deprotonated imidazolate group, as well as to a case with a weaker hydrogen bond to a carbonyl group, encountered in the cytochromes and globins [87].

The results showed that a hydrogen bond to a carbonyl group had a small influence on the geometry and properties of the heme site. In contrast, hydrogen bonds to a carboxylate group could have a large influence, especially if the proton shared in this hydrogen bond moved to the carboxylate group.

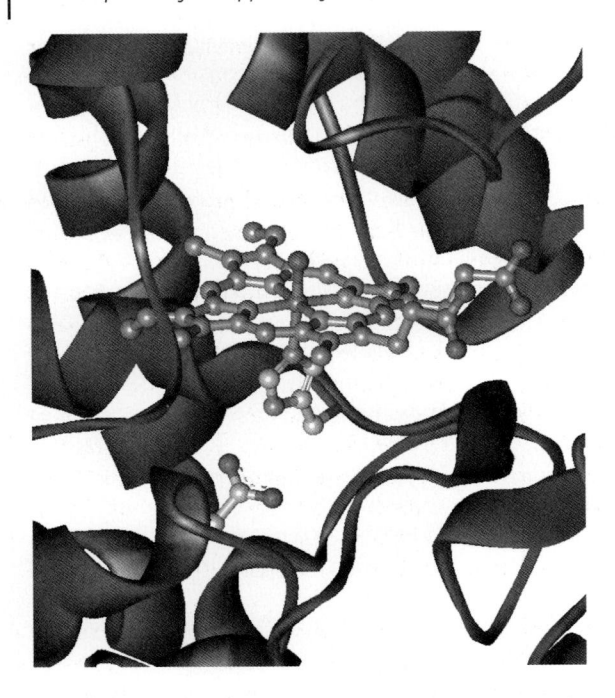

Figure 2.9 Arrangement of the amino acids His and Asp on the proximal side of heme in compound I of horseradish peroxidase [139].

It was also found that the spin splitting energy of the Fe^{II} heme states with neutral imidazole resembles those of Fe^{III} heme states with imidazolate, i.e., all spin states came close in energy [87]. We have seen that such degeneracy is important for the rapid binding of O_2 to the globins [41]. The interesting thing is that heme peroxidases bind the H_2O_2 substrate in the Fe^{III} state (cf. Figure 2.8), which requires a spin conversion similar to that in the globins. Thus, the choice of the His + Asp ligand may be a means to facilitate binding of small molecules to the heme group. These studies together put a new emphasis on the choice of axial ligand in heme proteins – a subject of very intense debate during the last 20 years [140, 141].

In contrast, the inner-sphere reorganization energies were not affected significantly by the perturbations, which indicates that proximal hydrogen bond networks can be used to tune redox potentials in heme groups without increasing reorganization energies [87].

Recently, a detailed QM and QM/MM study was presented of how various proximal and distal perturbations affect the O_2 affinity of Fe^{II}PorIm heme sites [142]. The authors studied the effect of rotations of the His ligand, the distance of the His ligand to the heme plane, as well as three different types of hydrogen bonds to the His ligand (His–carbonyl, His–Asp + Tyr, and His–H$_2$O–back-bone NH) and two actual proteins (myoglobin and leghemoglobin). Changes in the O$_2$ affinity were observed, of up to 15 and 30 kJ mol^{-1} for proximal and distal perturbations, respectively.

Considering the importance of the location of the shared proton in heme peroxidase, i.e., whether the proton resides on Asp (His + HAsp, middle of Figure 2.7, giving a deprotonated imidazolate group) or on His (HisH + Asp; right-hand side of Figure 2.7), we have used several different methods to determine its position in various states of the reaction mechanism [143]. First, we investigated the relative stability of the two protonation states in vacuum, using the small FePorImAc model (Ac = acetate) and various upper ligands. It turned out that, for most complexes, the HisH + Asp state is more stable, although the energy difference was small (1–6 kJ mol^{-1}). However, for the five-coordinate complex without any ligand, the oxidised complex with a water ligand, and for protonated compound II, the His + HAsp state was more stable (by 3–21 kJ mol^{-1}). Yet, these energies strongly depended on the solvation effect, and in a water-like continuum solvent the HisH + Asp state was stabilized, rendering it the ground state for most complexes.

Interestingly, these results also depend on the nature of the computational porphyrin model. For models with all the porphyrin side chains, the HisH + Asp state is more stable (by 1–19 kJ mol^{-1}) for all complexes, except the five coordinate complexes without any ligand. This is the only time that we have found a significant effect of the side chains.

Next, we studied the stability of the two states with QM/MM methods. They indicated that the HisH + Asp state is more stable for all six complexes studied (by 15–64 kJ mol^{-1}). However, QM/MM studies often do not treat solvation effects properly and may therefore overestimate electrostatic interactions. Therefore, we also studied the energy difference between the two protonation states using the QTCP [79, 80] method [143]. As expected, this reduced the energy difference but still the HisH + Asp state was more stable (by 12–44 kJ mol^{-1}) for all states.

Finally, we re-refined a crystal structure of cytochrome c peroxidase in the resting Fe^{III}H$_2$O state [144], using the quantum refinement procedure [145, 146], i.e., standard crystallographic refinement where the molecular mechanics potential used to supplement the experimental data is replaced by QM calculations for the heme site [143]. Unfortunately, the results were not conclusive: The HisH + Asp state gave slightly lower crystallographic R factors (0.18669 compared with 0.18673), but the His + HAsp state gave a 1–9 kJ mol^{-1} lower strain energy (i.e., the energy difference of the quantum system optimized in vacuum or in the crystal structure) and better electron density difference maps. Because of this, we concluded that, in the Fe^{III} resting form, the two protonation states are close in energy, whereas in all other catalytically significant states of the peroxidase reaction cycle the HisH + Asp state is more stable.

2.4.8
Axial Ligand in Chlorophylls

The Mg^{2+} ion in chlorophyll typically binds one additional axial ligand, but this ligand varies substantially. A survey of three crystal structures with a total of 308 chlorophylls showed that the most common ligands are His and water, but Asp/Glu, Tyr, methionine (Met), serine (Ser), asparagine/glutamine (Asn/Gln), and the back-bone amide groups are also observed, as well as phosphate and alcohol groups from non-protein molecules [44]. Fourteen of the chlorophylls do not have any fifth ligand, and none is six-coordinate.

In line with earlier work, we studied the influence of the axial ligand on the properties of the Chl and Bchl molecules, using eleven different models of axial ligands [44]. The Mg–ligand bond length was 1.90–1.99 Å for the negatively charged ligands (Ser^- < Asp < and phosphate), 2.10–2.21 Å for most neutral ligands (backbone < Asn < Ser < H_2O < Tyr < His), and even longer with two His ligands (2.35 Å) or with Met (2.74 Å).

The Mg–ligand bond length reflects the bond strength quite well: The BDE of the various axial bonds are 28–70 kJ mol^{-1} for the neutral and 170–260 kJ mol^{-1} for the negatively charged ligands, and they follow the same trends as the bond lengths, with the single exception that His gives the strongest bond among the neutral ligands [44]. Such bond energies are similar to what is found for other tetrapyrroles. However, the binding of a second axial ligand to Chl is unusually weak. In fact, there is no gain in energy of a second His ligand to Chl in a continuum solvent. In contrast, the corresponding energies for $Fe^{II}PorIm$ are 23–51 kJ mol^{-1}. This explains why chlorophyll almost invariably is five-coordinate in crystal structures (four-coordinate structures are almost only observed in low-resolution crystals, where no water positions are reported [44]). This also casts some doubt on most previous theoretical studies of Chl, which were performed on four-coordinate models. In fact, the calculated properties of Chl models are improved if a fifth ligand is included in the calculations [147].

The prime use of chlorophylls in nature is as pigments for the harvesting of light energy in photosynthesis. Therefore, we also studied the effect of axial ligands on the absorption spectra of chlorophylls [44]. The spectra were calculated with time-dependent DFT using the BP86 functional. In general, the absorption wavelength of both the Q and B bands (the major absorption bands around 660 and 430 nm for chlorophyll a) was increased by axial ligands. For the B band, the shift was 6–14 nm for the neutral ligands and 28–35 nm for the negatively charged ligands. For the Q band, the shift was half as much [44]. In the case of Bchl, the changes were somewhat smaller and negative for the Q band. Thus, the axial ligands can fine-tune the absorption properties of the chlorophylls, but the effect is quite restricted. In particular, axial ligands cannot alone explain the so-called red chlorophylls in photosystem I, which display spectral shifts of up to 50 nm [148]. This indicates that the protein surroundings and the possible formation of chlorophyll multimers are at least as important as the axial ligands for the absorption properties of the chlorophylls.

Furthermore, the effect of axial ligands on the reduction potentials of the chlorophylls was investigated [44], because chlorophylls are also involved in the electron transfer paths of the photosynthetic reaction centers. It was found that the axial ligands have quite strong effects on the potentials of chlorophyll, both for its reduction and its oxidation. Both potentials decreased, by 0.1–0.5 V for the neutral ligands and by up to 3 V for the charged ligands in vacuum, following approximately the same trends as the Mg–ligand bond lengths. However, the effect is strongly diminished if solvation effects are included, and in a water-like continuum solvent the reduction potentials of all models were the same within 0.2 V. However, the potentials are still 0.1–0.2 V lower than Chl without any axial ligand. Pheophytin was predicted to have a potential 0.2 V higher than the four-coordinate chlorophyll. Thus, it could be concluded, in line with the cytochrome work, that the axial ligands can be used to fine-tune the reduction potentials of the chlorophylls.

Interestingly, such tuning seems to occur in many proteins. In photosystem I, the electron-transfer path consists of the reaction center P700 (a special chlorophyll dimer), a chlorophyll molecule ligated by water, followed by a chlorophyll ligated by Met [149]. According to our calculations, the reduction potential of a neutral Chl model is more negative if the axial ligand is water than if it is Met (-1.87 and -1.79 V in a protein-like continuum solvent) [44]. This means that electron transport from the chlorophyll ligated by water to that ligated by Met is favorable. It has been proposed that this is a mechanism to prevent back-transfer of the electrons [150]. Interestingly, a similar arrangement is observed both in photosystem II and the bacterial reaction center, i.e., the second electron acceptor has a more negative potential than the third one (although the molecules and ligands are different) [44], indicating that this may be a general mechanism.

2.5
Concluding Remarks

In this chapter we have reviewed our computational efforts to understand why nature has selected certain combinations of metals, tetrapyrrole ring systems, and axial ligands in different proteins. Traditionally, biochemical investigations are performed by experimental methods, but we have seen that such hypothetical questions are appropriately studied with computational methods instead. With these methods, we obtain pure results about intrinsic differences, which are not biased by differences in the surrounding proteins or by the possibility of conformational changes in mutational studies, for example. Moreover, we can perform computational experiments that are hard and time consuming to perform in a laboratory.

These studies have pinpointed many important reasons for nature's choice of various metals, tetrapyrroles, and axial ligands:

- Corrin has the smallest central cavity, which fits ideally low-spin Co in all three relevant oxidation states (+I, + II, and +III).
- Porphyrin is selected to allow iron chemistry in both high-spin (substrate free) and low-spin state (with substrate bound). Its double negative charge has a strong influence on the reduction potentials, stabilizing the +IV and +V formal oxidation states, but making the +I state unavailable.
- The hydrocorphin ring has a larger central cavity and is more flexible than the other ring systems. Therefore, it fits the large Ni^{II} ion and the Ni^{I} ion with its singly occupied $d_{x^2-y^2}$ orbital.
- The $Fe^{II/III}$ couple is ideal for electron transfer in the cytochromes, giving a minimal reorganization energy, because the redox-active orbital is not directed towards any ligand.
- Co^{III} provides an ideal Co–C bond that is of intermediate strength: strong enough to resist hydrolysis but weak enough so that it can be readily broken in enzymes by both homolytic and heterolytic pathways.
- Mg^{2+} is an unexpected metal in tetrapyrrole chemistry. Its incorporation into a tetrapyrrole has a higher activation energy and a less favorable reaction energy than for Fe^{2+}.
- Neutral axial ligands have a small influence on the reorganization energies, but they can tune the redox potentials.
- Negatively charged axial ligands have a strong influence on the properties of heme sites, affecting both the redox potential and reaction energies, and therefore playing an important role in determining the reactivity of heme enzymes.
- The effect of axial ligands can be tuned by hydrogen bonds to nearby residues, especially if the latter are charged.
- The axial heme ligand affects the spin-splitting energies, a feature used by both globins and peroxidases to facilitate the spin-forbidden binding of substrates.
- The axial ligand of coenzyme B_{12} enzymes has a smaller influence on the reactions than previously supposed (but not smaller than the average effect of axial ligands in heme enzymes). Instead, the surrounding enzyme has a major influence on these reactions.
- The Gln ligand of coenzyme F430 is an ideal weak ligand, indifferent for the Ni^{I} state but enforcing a high-spin state for Ni^{II}.
- Chlorophyll can only bind one axial ligand, in variance to the other tetrapyrroles. This ligand can tune the absorption properties and reduction potentials of these sites in a functional way.

In conclusion, we have seen that in most cases there is a rationale behind the choices of metals, tetrapyrroles, and axial ligands for nature's design of cofactors, and that this rationale has been partly uncovered by theoretical calculations.

Acknowledgments

This investigation has been supported by funding from the research school in medicinal chemistry at Lund University (FLÄK) and the Swedish research council (VR). It has also been supported by the computer resources of Lunarc at Lund University.

References

1. D. Ghosh, V. L. Pecoraro, *Curr. Opin. Chem. Biol.* 2005, **9**, 97.
2. M. L. Ludwig, R. G. Matthews, *Annu. Rev. Biochem.* 1997, **66**, 269.
3. M. A. Halcrow, G. Christou, *Chem. Rev.* 1994, **94**, 2421.
4. U. Ermler, W. Grabarse, S. Shima, M. Goubeaud, R. K. Thauer, *Science* 1997, **278**, 1457.
5. A. D. Becke, *J. Chem. Phys.* 1993, **98**, 5648.
6. P. E. M. Siegbahn, *J. Biol. Inorg. Chem.* 2006, **11**, 695.
7. L. Noodleman, T. Lovell, W.-G. Han, J. Li, F. Himo, *Chem. Rev.* 2004, **104**, 459.
8. M.-H. Baik, M. Newcomb, R. A. Friesner, S. Lippard, *Chem. Rev.* 2003, **103**, 2385.
9. F. Himo, P. E. M. Siegbahn, *Chem. Rev.* 2003, **103**, 2421.
10. S. Shaik, D. Kumar, S. P. de Visser, A. Altun, W. Thiel, *Chem. Rev.* 2005, **105**, 2279.
11. T. A. Jackson, T. C. Brunold, *Acc. Chem. Res.* 2004, **37**, 461.
12. G. H. Loew, D. L. Harris, *Chem. Rev.* 2000, **100**, 407.
13. G. Frenking, N. Frohlich, *Chem. Rev.* 2000, **100**, 717.
14. P. E. M. Siegbahn, M. R. A. Blomberg, *Chem. Rev.* 2000, **100**, 421.
15. T. Ziegler, *Chem. Rev.* 1991, **91**, 651.
16. T. Ziegler, *Can. J. Chem.* 1995, **73**, 743.
17. P. E. M. Siegbahn, M. R. A. Blomberg, *Annu. Rev. Phys. Chem.* 1999, **50**, 221.
18. U. Ryde, *Dalton Trans.* 2007, 607–625.
19. F. Neese, *J. Biol. Inorg. Chem.* 2006, **11**, 702.
20. J. F. Harrison, *Chem. Rev.* 2000, **100**, 679.
21. J. Li, G. Schreckenbach, T. Ziegler, *J. Am. Chem. Soc.* 1995, **117**, 486.
22. M. Swart, A. R. Groenhof, A. W. Ehlers, K. Lammertsma, *J. Phys. Chem. A* 2004, **108**, 5479–5483.
23. Y. Zhao, N. E. Schultz, D. G. Truhlar, *J. Chem. Phys.* 2005, **123**, 161103.
24. N. E. Schultz, Y. Zhao, D. G. Truhlar, *J. Phys. Chem. A* 2005, **109**, 11127.
25. K. P. Jensen, B. O. Roos, U. Ryde, *J. Chem. Phys.* 2007, **126**, 014103.
26. A. Fougueau, S. Mer, M. E. Casida, L. M. L. Daku, A. Hauser, F. Neese, *J. Chem. Phys.* 2005, **122**, 044110.
27. K. P. Jensen, U. Ryde, *J. Phys. Chem. B* 2003, **107**, 7539.
28. J. Kuta, S. Patchkovskii, M. Z. Zgierski, P. M. Kozlowski, *J. Comput. Chem.* 2006, **27**, 1429.
29. A. Ghosh, *J. Am. Chem. Soc.* 1995, **117**, 4691.
30. A. Ghosh, *Acc. Chem. Res.* 1998, **31**, 189.
31. A. Ghosh, E. Gonzalez, T. Vangberg, *J. Phys. Chem. B* 1999, **103**, 1363.
32. T. G. Spiro, P. M. Kozlowski, M. Z. Zgierski, *J. Raman Spectrosc.* 1998, **29**, 869.
33. P. M. Kozlowski, T. G. Spiro, A. Bérces, M. Z. Zgierski, *J. Phys. Chem. B* 1998, **102**, 2603.
34. M. Wirstam, M. R. A. Blomberg, P. E. M. Siegbahn, *J. Am. Chem. Soc.* 1999, **121**, 10178.

35 M. R. A. Blomberg, P. E. M. Siegbahn, G. T. Babcock, M. Wikström, *J. Inorg. Biochem.* 2000, **80**, 261.
36 P. E. M. Siegbahn, *J. Comput. Chem.* 2001, **22**, 1634.
37 D. L. Harris, G. H. Loew, *J. Am. Chem. Soc.* 1998, **120**, 8941.
38 C. Rovira, K. Kunc, J. Hutter, P. Ballone, M. Parinello, *J. Phys. Chem. A* 1997, **101**, 8914.
39 F. Ogliaro, S. Cohen, S. P. de Visser, S. Shaik, *J. Am. Chem. Soc.* 2000, **122**, 12892.
40 E. Sigfridsson, M. H. M. Olsson, U. Ryde, *J. Phys. Chem. B* 2001, **105**, 5546.
41 K. P. Jensen, U. Ryde, *J. Biol. Chem.* 2004, **279**, 14561.
42 B. van Oort, E. Tangen, A. Ghosh, *Eur. J. Inorg. Chem.* 2004, 2442.
43 M. P. Johansson, M. R. A. Blomberg, D. Sundholm, M. Wikström, *Biochim. Biophys. Acta* 2002, **1553**, 183.
44 J. Heimdal, K. P. Jensen, A. Devarajan, U. Ryde, *J. Biol. Inorg. Chem.* 2007, **12**, 49.
45 H.-P. Hersleth, U. Ryde, P. Rydberg, C. H. Görbitz, K. K. Andersson, *J. Inorg. Biochem.* 2006, **100**, 460.
46 E. S. Ryabova, P. Rydberg, M. Kolberg, E. Harbitz, A.-L. Barra, U. Ryde, K. K. Andersson, E. Nordlander, *J. Inorg. Biochem.* 2005, **99**, 852.
47 P. Rydberg, E. Sigfridsson, U. Ryde, *J. Biol. Inorg. Chem.* 2004, **9**, 203.
48 T. Andruniow, M. Z. Zgierski, P. M. Kozlowski, *J. Phys. Chem. B* 2000, **104**, 10921.
49 T. Andruniow, M. Z. Zgierski, P. M. Kozlowski, *Chem. Phys. Lett.* 2000, **331**, 509.
50 K. P. Jensen, K. V. Mikkelsen, *Inorg. Chim. Acta* 2001, **323**, 5.
51 K. P. Jensen, U. Ryde, *J. Mol. Struct. Theochem.* 2002, **585**, 239.
52 K. P. Jensen, S. P. A. Sauer, T. Liljefors, P.-O. Norrby, *Organometallics* 2001, **20**, 550.
53 N. Dölker, F. Maseras, A. Lledos, *J. Phys. Chem. B* 2001, **105**, 7564.
54 T. Andruniow, M. Z. Zgierski, P. M. Kozlowski, *J. Am. Chem. Soc.* 2001, **123**, 2679.
55 K. P. Jensen, U. Ryde, *J. Am. Chem. Soc.* 2003, **125**, 13970.
56 P. W. Kozlowski, *J. Phys. Chem. B* 2004, **108**, 14163.
57 T. Andruniow, J. Kuta, M. Z. Zgiersky, P. M. Kozlowski, *Chem. Phys. Lett.* 2005, **410**, 410.
58 K. P. Jensen, U. Ryde, *J. Am. Chem. Soc.* 2005, **127**, 9117.
59 P. M. Kozlowski, T. Kamachi, T. Toraya, K. Yoshizawa, *Angew. Chem. Int. Ed.* 2006, **46**, 980.
60 A. J. Brooks, M. Vlasie, R. Banerjee, T. C. Brunold, *J. Am. Chem. Soc.* 2004, **126**, 8167.
61 N. Dölker, F. Maseras, A. Lledos, *J. Phys. Chem. B* 2003, **107**, 306.
62 D. Rutkowska-Zbik, M. Jaworska, M. Witko, *Struct. Chem.* 2004, **15**, 431.
63 K. P. Jensen, U. Ryde, *ChemBioChem* 2003, **4**, 413.
64 M. Jaworska, G. Kazibut, P. Lodowski, *J. Phys. Chem. A* 2003, **107**, 1339.
65 T. Wondimagegn, A. Ghosh, *J. Am. Chem. Soc.* 2000, **122**, 6375.
66 K. P. Jensen, U. Ryde, *J. Porph. Phthalocyan.* 2005, **9**, 581.
67 A. Ghosh, T. Wondimageng, H. Ryeng, *Curr. Opin. Chem. Biol.* 2001, **5**, 744.
68 V. Pelmenschikov, P. E. M. Siegbahn, *J. Biol. Inorg. Chem.* 2003, **8**, 653.
69 J. L. Craft, Y. C. Horng, S. W. Ragsdale, T. C. Brunold, *J. Biol. Inorg. Chem.* 2004, **9**, 77.
70 D. Sundholm, *Phys. Chem. Chem. Phys.* 2003, **5**, 4265.
71 A. Pandey, S. N. Datta, *J. Phys. Chem. B* 2005, **109**, 9066.
72 M. D. Elkin, O. D. Ziganshina, K. V. Berezin, V. V. Nechaev, *J. Struct. Chem.* 2004, **45**, 1086.
73 A. Wong, R. Ida, X. Mo, Z. H. Gan, J. Poh, G. Wu, *J. Phys. Chem. A* 2006, **110**, 10084.
74 E. Sigfridsson, U. Ryde, *J. Biol. Inorg. Chem.* 2003, **8**, 273.
75 B. Honig, *Science* 1995, **268**, 1144.
76 A. J. Mulholland, in *Theoretical and Computational Chemistry*, ed. L. A. Eriksson, Elsevier Science, Amsterdam, 2001, Vol. 9, pp. 597–653.
77 U. Ryde, *Curr. Opin. Chem. Biol.*, 2003, **7**, 136–142.
78 N. Källrot, K. Nilsson, T. Rasmussen, U. Ryde, *Int. J. Quantum Chem.* 2005, **102**, 520–541.

79 T. H. Rod, U. Ryde, *Phys. Rev. Lett.* 2005, **94**, 138302.
80 T. H. Rod, U. Ryde, *J. Chem. Theory Comput.* 2005, **1**, 1240.
81 J. M. Pratt, *Pure & Appl. Chem.* 1993, **65**, 1513.
82 R. J. P. Williams, *J. Mol. Catal.* 1985, **30**, 1.
83 J. Halpern, *Science* 1985, **227**, 869.
84 W. R. Scheidt, C. A. Reed, *Chem. Rev.* 1981, **81**, 543.
85 J. J. R. Frausto da Silva, R. J. P. Williams, *The Biological Chemistry of the Elements*, Clarendon Press, Oxford, 1994, p. 10.
86 R. J. P. Williams, *J. Mol. Catal.* 1985, **30**, 1.
87 K. P. Jensen, U. Ryde, *Mol. Phys.* 2003, **101**, 2003.
88 A. M. Stolzenberg, M. T. Stershic, *J. Am. Chem. Soc.* 1988, **110**, 6391.
89 R. H. Holm, P. Kennepohl, E. I. Solomon, *Chem. Rev.* 1996, **96**, 2239.
90 R. A. Marcus, N. Sutin, *Biochim. Biophys. Acta* 1985, **811**, 265.
91 B. Kräutler, in *Encyclopedia of Inorganic Chemistry*, ed. R. B. King, John Wiley & Sons, Chichester, England, 1994, Vol. 2, p. 697.
92 M. L. Ludwig, R. G. Matthews, *Annu. Rev. Biochem.* 1997, **66**, 269.
93 W. G. Grabarse, F. Mahlert, S. Shima, R. K. Thauer, U. Ermler, *J. Mol. Biol.* 2000, **303**, 329.
94 H. L. Schubert, E. Raux, A. A. Brindley, H. K. Leech, K. S. Wilson, C. P. Hill, M. J. Warren, *EMBO J.* 2002, **21**, 2068.
95 Y. Shen, U. Ryde, *Chem. Eur. J.* 2005, **11**, 1549.
96 H. A. O. Hill, J. M. Pratt, R. P. J. Williams, *Chem. Br.* 1969, **5**, 169.
97 J. H. Grate, G. N. Schrauzer, *J. Am. Chem. Soc.* 1979, **101**, 4601.
98 Z. Gross, *J. Biol. Inorg. Chem.* 1996, **1**, 368.
99 R. G. Finke, in *Vitamin B_{12} and the B_{12} Proteins*, eds. B. Kräutler, D. Arigoni, B. T. Golding, Wiley-VCH, Weinheim, 1998, Ch. 25.
100 B. P. Hay, R. G. Finke, *J. Am. Chem. Soc.* 1986, **108**, 4820.
101 K. L. Brown, X. Zou, *J. Inorg. Biochem.* 1999, **77**, 185.
102 R. G. Finke, B. P. Hay, *Inorg. Chem.* 1984, **23**, 3041.
103 R. Padmakumar, R. Padmakumar, R. Banerjee, *Biochemistry* 1997, **36**, 3713.
104 S. Chowdhury, R. Banerjee, *Biochemistry* 2000, **39**, 7998.
105 S. S. Licht, C. C. Lawrence, J. Stubbe, *Biochemistry* 1999, **38**, 1234.
106 R. Banerjee, *Chem. Rev.* 2003, **103**, 2083.
107 C. L. Drennan, R. G. Matthews, M. L. Ludwig, *Curr. Opin. Struct. Biol.* 1994, **4**, 919.
108 F. Mancia, P. Evans, F. Mancia, N. H. Keep, A. Nakagawa, P. F. Leadlay, S. McSweeney, B. Rasmussen, P. Bösecke, O. Diat, P. R. Evans, *Structure* 1996, **4**, 339.
109 K. L. Brown, S. Cheng, X. Zou, J. Li, G. Chen, E. J. Valente, J. D. Zubkowski, H. M. Marques, *Biochemistry* 1998, **37**, 9704.
110 H. P. Chen, E. N. G. Marsh, *Biochemistry* 1997, **36**, 7884.
111 M. Freindorf, P. M. Kozlowski, *J. Am. Chem. Soc.* 2004, **126**, 1928.
112 E. N. G. Marsh, *Bioorg. Chem.* 2000, **28**, 176.
113 K. Gruber, C. Kratky, *Curr. Opin. Struct. Biol.* 2002, **6**, 598.
114 K. Gruber, R. Reitzer, C. Kratky, *Angew. Chem. Int. Ed.* 2001, **40**, 3377.
115 F. Mancia, P. R. Evans, *Structure* 1998, **6**, 711.
116 N. Dölker, F. Maseras, P. E. M. Siegbahn, *Chem. Phys. Lett.* 2004, **386**, 174.
117 R. A. Kwiecien, I. V. Khavrutskii, D. G. Musaev, K. Morokuma, R. Banerjee, P. Paneth, *J. Am. Chem. Soc.* 2006, **128**, 1287.
118 K. Peariso, C. W. Goulding, S. Huang, R. G. Matthews, J. E. Penner-Hahn, *J. Am. Chem. Soc.* 1998, **120**, 8410.
119 C. L. Drennan, S. Huang, J. T. Drummond, R. G. Matthews, M. L. Ludwig, *Science* 1994, **266**, 1669.
120 M. W. Renner, J. Fajer, *J. Biol. Inorg. Chem.* 2001, **6**, 823.
121 L. Pauling, C. D. Coryell, *Proc. Natl. Acad. Sci. U.S.A.* 1936, **22**, 210.
122 G. N. Phillips, Jr, in *Handbook of Metalloproteins*, eds. A. Messerschmidt, R. Huber, K. Wieghardt, T. Poulos, John Wiley & Sons, Chichester, England, 2001.

123 M. Kotani, *Rev. Mod. Phys.* 1963, **35**, 717.
124 K. P. Jensen, B. O. Roos, U. Ryde, *J. Inorg. Biochem.* 2005, **99**, 45.
125 T. E. Tsai, J. L. Groves, C. S. Wu, *J. Chem. Phys.* 1981, **74**, 4306.
126 H. Eicher, D. Bade, F. Parak, *J. Chem. Phys.* 1976, **64**, 1446.
127 H. Eyring, J. Walter, G. E. Kimball, *Quantum Chemistry*, John Wiley & Sons, New York, 1944, p. 330.
128 M. H. M. Olsson, U. Ryde, B. O. Roos, *Protein Sci.* 1998, **7**, 2659.
129 U. Ryde, M. H. M. Olsson, *Int. J. Quantum. Chem.* 2001, **81**, 335.
130 M. H. M. Olsson, U. Ryde, *J. Am. Chem. Soc.* 2001, **123**, 7866.
131 E. Sigfridsson, M. H. M. Olsson, U. Ryde, *Inorg. Chem.* 2001, **40**, 2509.
132 F. A. Tezcan, J. R. Winkler, H. B. Gray, *J. Am. Chem. Soc.* 1998, **120**, 13383.
133 M. L. Fernández, M. A. Martí, A. Crespo, D. A. Estrin, *J. Biol. Inorg. Chem.* 2005, **10**, 595.
134 M. Gajhede, in *Handbook of Metalloproteins*, eds. A. Messerschmidt, R. Huber, T. Poulos, K. Wieghart, J. Wiley & Sons, Chichester, England, 2001, pp. 285–299.
135 T. L. Poulos, *Adv. Inorg. Biochem.* 1987, **7**, 1.
136 D. Kumar, S. P. de Visser, P. K. Sharma, E. Derat, S. Shaik, *J. Biol. Inorg. Chem.* 2005, **10**, 181.
137 Y.-K. Choe, S. Nagase, *J. Comput. Chem.* 2005, **26**, 1600.
138 D. Kumar, S. P. de Visser, P. K. Sharma, H. Hirao, S. Shaik, *Biochemistry* 2005, **44**, 8148.
139 A. Henriksen, A. T. Smith, M. Gajhede, *J. Biol. Chem.* 1999, **274**, 35005.
140 T. L. Poulos, *J. Biol. Inorg. Chem.* 1996, **1**, 356.
141 D. B. Goodin, *J. Biol. Inorg. Chem.* 1996, **1**, 360.
142 L. Capece, M. A. Marti, A. Crespo, F. Doctorovich, D. A. Estrin, *J. Am. Chem. Soc.* 2006, **128**, 12455.
143 J. Heimdal, P. Rydberg, U. Ryde, *J. Inorg. Biochem.*, submitted.
144 B. C. Finzel, T. L. Poulos, J. Kraut, *J. Biol. Chem.* 1984, **259**, 13027.
145 U. Ryde, L. Olsen, K. Nilsson, *J. Comput. Chem.* 2002, **23**, 1058.
146 U. Ryde, K. Nilsson, *J. Am. Chem. Soc.* 2003, **125**, 14232.
147 J. Linnanto, J. Korppi-Tommola, *PhysChemChemPhys* 2006, **8**, 663.
148 B. Bobets, I. H. M. van Stokkum, M. Rogner, J. Kruip, E. Schlodder, E. N. V. Karapetyan, J. P. Dekker, R. van Grondelle, *Biophys. J.* 2001, **81**, 407.
149 P. Jordan, P. Fromme, H. T. Witt, O. Klukas, W. Saenger, N. Krauss, *Nature* 2001, **411**, 909.
150 Y. M. Sun, H. Z. Wang, F. L. Zhao, J. Z. Sun, *Chem. Phys. Lett.* 2004, **387**, 12.

3
Modeling of Mechanisms for Metalloenzymes where Protons and Electrons Enter or Leave

Per E. M. Siegbahn and Margareta R. A. Blomberg

3.1
Introduction

Redox-active enzymes catalyze some of the most important and interesting processes in biology, such as photosystem II in the photosynthesis of plants and cytochrome c oxidase in the respiratory chain in mitochondria. For redox enzymes it is nearly always the case that electrons enter or leave the active site during catalysis. Experimental information on the energetics of these enzymes is generally available in terms of redox potentials and pK_a values. For about the last decade, quantum chemistry has taken part in the modeling of the reactivity of these enzymes by describing intermediates and transition states for bond-forming and bond-breaking processes. Practically all studies to date have been performed using density functional theory (DFT), most commonly using hybrid functionals. The description and elucidation of these processes have been very successful and have complemented the experimental information in an important and useful way. However, it is somewhat ironic that redox potentials and pK_a values are the most difficult properties to obtain from quantum chemical model calculations. These properties would otherwise be the most natural points of comparison with experiments.

The origin of the difficulty of obtaining redox potentials and pK_a values is the change of charge of the system in these processes. A change of charge gives rise to a long-range response from the surrounding medium, in this case the enzyme. The energetic response for a (spherical) complex with charge q and radius R in a surrounding with a dielectric constant ε is: $Eq = (\varepsilon - 1)/\varepsilon \times q^2/2R$.

With ε equal to 4, which is a common value for the interior of an enzyme, charge +1 and radius 5 Å, the energetic response is around 20 kcal mol^{-1}. A model of this size is a typical practical limitation for a quantum chemical calculation, which means that 20 kcal mol^{-1} is typically missing for a computed redox potential or pK_a value. To recover part of this energy loss, simple dielectric cavity models can be used, but they give results of only qualitative accuracy. Even more importantly, these models assume a homogeneous surrounding. It is well known

*Computational Modeling for Homogeneous and Enzymatic Catalysis.
A Knowledge-Base for Designing Efficient Catalysts.* K. Morokuma and D. G. Musaev (Eds.)
Copyright © 2008 WILEY-VCH Verlag GmbH & Co. KGaA, Weinheim
ISBN: 978-3-527-31843-8

that enzymes use inhomogeneously positioned charges to optimize redox potentials and pK_a values for their purposes [1]. Including dielectric effects, the errors can, therefore, still be 5–10 kcal mol^{-1}.

The common way of obtaining more accurate redox potentials and pK_a values is to somehow include a modeling of a large part of the enzyme, often using a molecular mechanics approach [1, 2]. Typically, the errors for redox enzymes are then reduced to 5 kcal mol^{-1} (0.2 V) or less. To combine this higher accuracy for redox and pK_a values with a modeling of bond formation or bond breaking requires a QM/MM model. For the enzymes discussed here, like photosystem II and cytochrome oxidase, this approach has so far not been applied successfully. The reason for this is that there are still general problems with the use of QM/MM, such as avoiding local minima and a proper description of the QM–MM interface region. Typical problems appearing with the use of QM/MM models are nicely illustrated in a recent application on C–H abstraction from camphor by the heme-oxo group of P450 [3].

The present chapter describes a different approach for obtaining the energetics of catalytic cycles, where bond formation or breaking occurs, and where redox potentials and pK_a values are also needed. The approach uses a minimal amount of experimental information and is, therefore, semi-empirical in its nature. The information needed is most often just the total driving force (or reaction energy) for the entire catalytic cycle. This driving force can usually be derived from known redox potentials, or it can even be guessed, as will be described below. The most important part of the approach is the realization that energies computed from a quantum chemical model are usually very accurate when the charge of the model is not changed. This means that thermodynamics and barriers for individual reaction steps are normally quite accurate, but even more importantly it means that the energy required to remove both an electron and a proton simultaneously is also quite accurate. Since the charge does not change in such a process, there is only a small response from the enzyme. The enzyme can therefore not use inhomogeneously distributed charges to modulate the energetics of these processes. The effects from the enzyme, outside a limited QM model, can be assumed to be rather small (<5 kcal mol^{-1}) and should be reasonably well described by a dielectric cavity model. If the energetics is required for steps where there is a change of charge, this can also be obtained, but then one additional empirical parameter has to be introduced. This parameter can, for example, be obtained from a minimization of the barriers in the catalytic cycle, as described in detail below. This procedure mimics nature, where evolution has optimized the enzyme with precisely this goal. Overall, an accuracy can be reached that is at least as high as from QM/MM models, without the pitfalls of those methods, and at a fraction of the cost.

3.2 Energy Diagrams

3.2.1 Photosystem II

Photosystem II is the enzyme that uses photons to make a charge separation, and forms the dioxygen molecule from water in the photosynthesis of plants. The overall reaction is:

$$h\nu + 2H_2O \rightarrow O_2 + 4e^- + 4H^+$$

The photons are absorbed by the antennas and the energy is transferred to the reaction center, where charge separation takes place in the chlorophyll dimer (or tetramer) P_{680}. P_{680} becomes ionized in the charge separation step, where the electron goes over a pheophytine to two quinones. P_{680}^+ is the strongest oxidant in biology with a redox potential of around 1.3 V, which is used to oxidize water molecules bound in the oxygen-evolving complex (OEC). The electrons from the OEC passes over a redox-active tyrosine, Tyr_Z, to avoid a short-cut in the process. The protons of the oxidized water molecules go directly to the lumen close to the OEC, on the outside of the membrane, building up a potential across the membrane to be used by another enzyme in the same membrane, ATP-synthase, for the production of ATP, the energy storage of living organisms. For each photon, one electron leaves the OEC, and after four photons O–O bond formation of dioxygen occurs. The five different states involved in these oxidizing transitions are labeled from S_0 to S_4. In the nomenclature used below, the states will also have an upper index for the charge of the complex, such as S_2^{-1}, which is the S_2 state for a model with a charge of -1.

A few X-ray structures of photosystem II (PSII) have been obtained only in recent years, although at a rather low resolution of 3.2–3.8 Å [4, 5]. In these structures the oxygen-evolving complex was identified as composed of a Mn_3Ca-cube with a dangling manganese outside. The metal atoms are connected by oxo or hydroxo bridges. Five to six amino acid ligands of the OEC were also assigned, but water ligands could not be identified at the low resolution.

In the quantum chemical calculations performed to study the dioxygen formation process, water-derived ligands first have to be added to complete the octahedral coordinations of the metals. To obtain a reasonable charge of the complex, which should be close to zero, some of these ligands have to be hydroxides. Possible protonations of the bridging oxo ligands also had to be investigated. Starting with an optimal S_0-state, protons and electrons were then removed to find optimal structures for each S-state. In each step, the proton with the lowest proton affinity (PA) was removed, or the electron with the lowest electron affinity (EA), until the S_4-state was reached, in which formation of the O–O bond was studied. Releasing dioxygen and binding two water molecules returns the cluster to its original S_0-state. Altogether, four optimal PAs and four optimal EAs were thus computed. Since it is reasonable to assume that every second transition is a proton release and every second one an electron release, an (H^+, e^-)-couple ("hydrogen atom") is then taken away from the OEC complex after every two transitions. The optimized structure for the S_2^{-1}-state, using the ligand assignment from the most recent X-ray structure [5], is shown in Figure 3.1.

Ideally, the computed PAs should be translated into pK_a values and the EAs into redox potentials, but a reasonable accuracy for these energies is very difficult to obtain because of the change of charge, as discussed above. Otherwise, from the redox potentials and the known redox potential of P_{680}^+, transition energies could have been obtained. Likewise, the calculated pK_as could have given the other transition energies, assuming that the pH on the outside of the membrane is 7. Instead, the present approach starts with the computed hydrogen atom (H^+, e^-) binding energies, obtained after every two transitions. With the computed optimal PAs and EAs, four hydrogen atom binding energies can be computed, using the energy of the free hydrogen atom with the same basis set of 315.1 kcal mol^{-1} (0.5022 a.u.). For complexes with structures like the one in Figure 3.1 these hydrogen atom binding energies become 83.9, 82.2, 89.1 and 105.8 kcal mol^{-1}, respectively. These energies include both zero-point and dielectric effects. The dielectric effects are $+2.0$, $+0.8$, -3.0, and $+1.4$ kcal mol^{-1}, respectively, showing

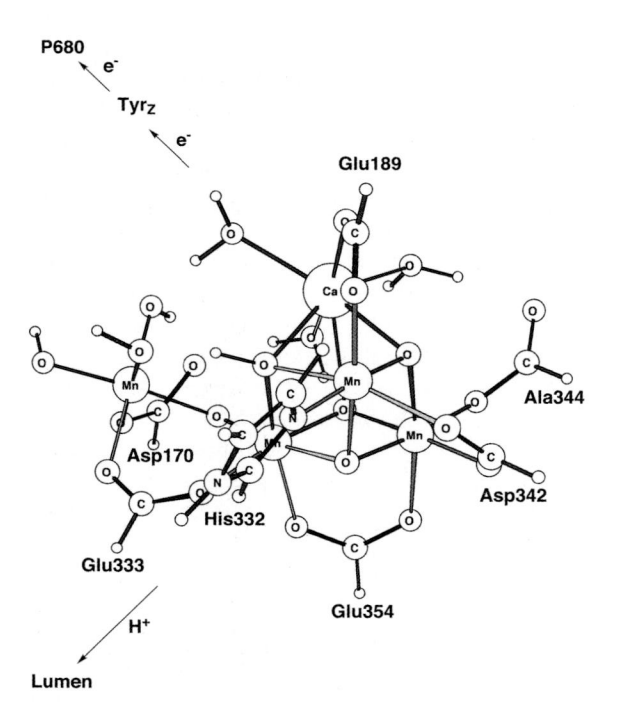

Figure 3.1 Optimized structure for the S_2^{-1} state of the oxygen-evolving complex in photosystem II.

that the surrounding enzyme has only a small effect on these binding energies. In contrast, the corresponding effects on the pK_a and redox potentials are on the order of 25 kcal mol^{-1}.

To make use of the computed hydrogen atom binding energies in a full energy diagram, the reaction energy in the step where dioxygen is made is also needed. This was computed to be -15.3 kcal mol^{-1}. The barrier for O–O bond formation is only 5.2 kcal mol^{-1} in this model. Another value needed is the energy for exchanging O_2 in S_4^{-1}-O_2 with two water molecules from bulk water, to come back to S_0^{-1}. This value was computed to be -7.8 kcal mol^{-1}, assuming a water molecule to be bound by 14 kcal mol^{-1} in bulk water [6].

With the above values, a full energy diagram can be obtained if the reaction energy of the full catalytic cycle is known. In this case it is known from the redox potential of P_{680}^+ of 1.3 V and the reduction potential of O_2 of 0.8 V. The total reaction energy then becomes $4(1.3 - 0.8)$ eV $= 2.0$ eV $= 46$ kcal mol^{-1}. The reference value for the combined transfer of one electron to P_{680}^+ and one proton to bulk water then becomes 96.1 kcal mol^{-1}, which is just fitted to the reaction energy of 46.0 kcal mol^{-1}. Using the computed binding energies, the energy released when a (H^+, e^-)-couple leaves the OEC in two subsequent transitions is then $(-96.1 + 83.9) = -12.2$ kcal mol^{-1} in the S_0^{-1} to S_1^{-1} transition; $(-96.1 + 82.2) = -13.9$ kcal mol^{-1} in the S_1^{-1} to S_2^{-1} transition; $(-96.1 + 89.6) = -6.5$ kcal mol^{-1} in the S_2^{-1} to S_3^{-1} transition; and finally $(-96.1 + 105.8) = +9.7$ kcal mol^{-1} in the S_3^{-1} to S_4^{-1} transition. These four energies, together with the reaction energy and barrier for forming O_2, and the exchange of O_2 with two waters, lead to the full energy diagram shown in Figure 3.2. It can be seen in the diagram that with the reference value of 96.1 kcal mol^{-1}, the reaction energy for the full cycle becomes 46.0 kcal mol^{-1}. It is of fundamental importance for the evolution of PSII that the energies in this diagram, which constitute the main part of dioxygen formation, can not be significantly affected by the enzyme surrounding the oxygen-evolving complex. The energies are defined almost solely by the properties of the local complex.

For a more complete picture of the energetics involved in forming the O–O bond of dioxygen, it is also of interest to obtain an energy diagram including every individual step of electron and proton transfer. A similar procedure to the one used above can be used also for this purpose, but requires the use of one additional parameter. As already mentioned, individual proton and electron affinities have been computed for each step of the cycle. This involves four proton affinities of 287.0, 291.4, 290.0 and 287.2 kcal mol^{-1}, and four electron affinities of 112.0, 105.9, 114.7 and 133.7 kcal mol^{-1}. Instead of just one reference value, 96.1 kcal mol^{-1} for the (H^+, e^-) binding, two reference values are now needed, one for the proton and one for the electron affinity. The sum of these reference values minus the energy of a free hydrogen atom (315.1 kcal mol^{-1}) should give back the reference value 96.1 kcal mol^{-1} for the (H^+, e^-) binding, which will then guarantee that the reaction energy for the full cycle is -46.0 kcal mol^{-1}, as above. In reality there is therefore just one additional parameter needed, which can be obtained by minimizing the barriers in the cycle. The best choice found

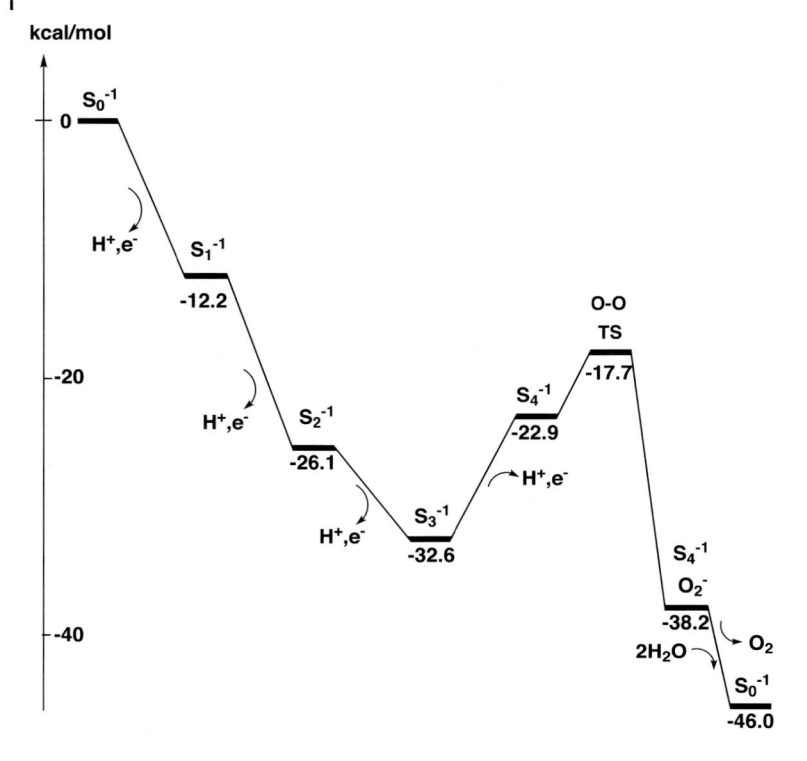

Figure 3.2 Energy diagram for dioxygen evolution in photosystem II, showing the transitions where both an electron and a proton have been released from the oxygen-evolving complex.

is 292.2 kcal mol^{-1} for the proton affinity and 119.0 kcal mol^{-1} for the electron affinity. Again, the sum of these values minus 315.1 kcal mol^{-1} is 96.1 kcal mol^{-1}, as it should be. With these values, the individual reaction step energies for proton transfer can be obtained as $(287.0 - 292.2) = -5.2$ kcal mol^{-1} for S_1^0 to S_1^{-1}, $(291.4 - 292.2) = -0.8$ kcal mol^{-1} for S_2^0 to S_2^{-1}, $(290.0 - 292.2) = -2.2$ kcal mol^{-1} for S_3^0 to S_3^{-1}, and $(287.2 - 292.2) = -5.0$ kcal mol^{-1} for S_4^0 to S_4^{-1}. For the electron transfer steps one obtains $(112.0 - 119.0) = -7.0$ kcal mol^{-1} for S_0^{-1} to S_1^0, $(105.9 - 119.0) = -13.1$ kcal mol^{-1} for S_1^{-1} to S_2^0, $(114.7 - 119.0) = -4.3$ kcal mol^{-1} for S_2^{-1} to S_3^0, and $(133.7 - 119.0) = +14.7$ kcal mol^{-1} for S_3^{-1} to S_4^0. With these energies the entire energy diagram for O–O bond formation in OEC is completed (Figure 3.3).

Since the redox potential of P_{680}^+ is 1.3 V, the individual redox potentials for the different S-states can be computed from the reaction step energies given above. They become $1.30 - 7.0/23.1 = 1.00$ V for S_0^{-1}; $1.30 - 13.1/23.1 = 0.73$ V for S_1^{-1}; $1.30 - 4.3/23.1 = 1.11$ V for S_2^{-1}; and $1.30 + 14.7/23.1 = 1.94$ V for S_3^{-1}. Similarly, the pK_a values for the different S-states can also be obtained by compar-

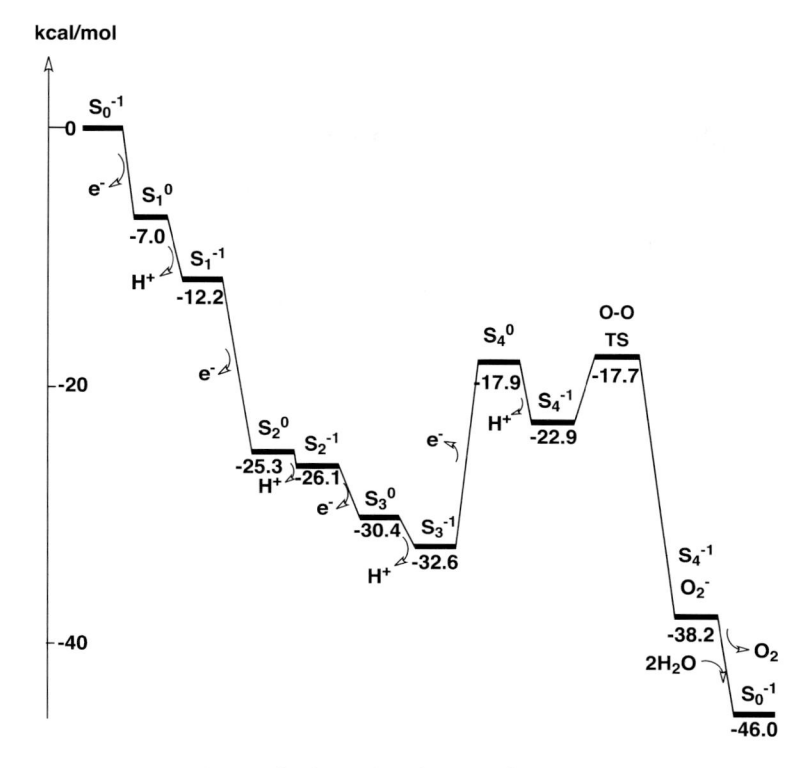

Figure 3.3 Energy diagram for dioxygen evolution in photosystem II, showing both the transitions where an electron has been released and those where a proton has been released from the oxygen-evolving complex.

ison with pH 7 for water on the outside of the membrane. For S_1^0 it becomes $7.0 - 5.2/1.4 = 3.3$, for S_2^0 $7.0 - 0.8/1.4 = 6.4$, for S_3^0 $7.0 - 2.2/1.4 = 5.4$, and for S_4^0 $7.0 - 5.0/1.4 = 3.4$. These values should be directly comparable with future experimental measurements.

Once an energy diagram like the one described above has been constructed, it is quite easy to investigate the effects of a membrane gradient. One goal with the chemistry in PSII is to create such a gradient, which should be used for energy storage by ATP synthesis. The membrane gradient is composed of both electrostatic and pH dependent parts. At full gradient the pH on the outside has decreased by about 3 units, which means that there will be an additional cost of about 4.1 kcal mol^{-1} every time a proton is released from the OEC. For the electrons there is no additional cost due to the membrane gradient since the electrons are sent to P_{680}^+ at the same side of the membrane. Since four protons are released in each cycle, the driving force for the full cycle with membrane gradient decreases by $4 \times 4.1 = 16.4$ kcal mol^{-1}, from 46.0 down to 29.6 kcal mol^{-1}, which is lost as heat.

Some conclusions about the catalytic cycle of OEC in PSII could be of general chemical interest. The dominant feature of the cycle is that most of the energy is lost in the first two cycles from S_0 to S_2. The origin of this waste of energy is that a redox-potential has to be used that is unnecessarily large for these simple transitions. The requirement of the very large potential of 1.3 V comes from the chemistry when the O–O bond is formed in the S_4 step. In fact, even though the redox potential is as large as it is, the electron transfer to P_{680}^+ is endergonic by 14.7 kcal mol^{-1}, which is close to the overall rate-limiting barrier for the full process. The rate is instead determined by the TS for O–O bond formation. In this case this point is 14.9 kcal mol^{-1} higher than the lowest point at S_3^{-1}. Experimentally, PSII turns around on the order of ms, which means that the computed barrier is slightly too high by a few kcal mol^{-1}. This discrepancy is typical for barriers computed by B3LYP. Interestingly, the rate-limiting barrier is the same in both diagrams, and therefore does not depend on the division of the energies for the (H^+, e^-)-couple into redox potentials and pK_a values. The important consequence of this is that the rate of dioxygen formation in PSII is almost unaffected by the enzyme surrounding the oxygen-evolving complex. The reason for this is that both an electron and a proton transfer are involved in the rate-limiting step from S_3^{-1} to the O_2^- product.

3.2.2
Cytochrome c Oxidase

Cytochrome c oxidase is the terminal enzyme in the respiratory chain, located in the mitochondrial or bacterial membrane. This is where molecular oxygen is reduced to water using electrons delivered by cytochrome c on the P-side (outside) of the membrane. The protons needed to form water are taken up from the N-side (inside) of the membrane, using one of two characterized proton channels. Reduction of molecular oxygen to water with electrons from cytochrome c is an exergonic process, releasing about 46 kcal mol^{-1} per O_2 molecule, as computed from the experimental reduction potentials (0.3 V for cytochrome c and 0.8 V for O_2). This exergonic chemical reaction is coupled to the pumping of protons across the entire membrane, one proton per electron, yielding the overall reaction:

$$O_2 + 4e^- + 8H_{in}^+ \rightarrow 2H_2O + 4H_{out}^+$$

Both the chemistry (forming water from molecular oxygen) and the proton pumping correspond to charge translocation across the membrane and therefore contribute to the build-up of an electrochemical gradient, found to be about 0.2 V in the working enzyme. Thus, when the enzyme is working under a gradient there is an extra cost for moving the charges perpendicular to the membrane, and at full gradient the overall reaction is exergonic by only about 9 kcal mol^{-1}. The gradient is used for ATP production.

The active site of cytochrome c oxidase, where the O_2 reduction occurs, is referred to as the binuclear center (bnc) and consists of an iron-porphyrin (heme a_3) and a copper complex (Cu_B). One of the histidine ligands on copper is covalently linked to a tyrosine residue, which is thought to be redox active together with the metal centers. Starting from the oxidized state, **O**, with Fe(III) and Cu(II), the first two electrons are transferred from cytochrome c to bnc, forming the reduced state **R**, with Fe(II) and Cu(I). These steps constitute the reductive part of the cycle. Two water molecules leave and molecular oxygen is taken up and weakly coordinated to the reduced iron, forming the observed compound **A**. In the next step the O–O bond is cleaved and the P_M state is formed, having Fe(IV)=O, Cu(II)–OH and a neutral tyrosyl radical. Two more electrons are transferred to bnc to complete the reduction and reform the oxidized state **O**. These steps constitute the oxidative part of the cycle. Two more intermediates are observed, the one-electron reduced state **E** in the reductive part, and the three-electron reduced state **F** in the oxidative part. Each electron transfer from cytochrome c to bnc is accompanied by a proton transfer from the N-side to bnc, resulting in the formation of a new O–H bond in each step, and the pumping of one proton across the entire membrane. This catalytic cycle, including the O–O bond cleavage, has been studied by quantum chemical calculations using the bnc model shown in Figure 3.4. To obtain the energetics of the catalytic cycle, including electron and proton transfer, electron affinities and proton affinities are calculated for all the steps in the cycle. A parametrization is made in the same way as described above for PSII. A reference O–H bond strength is determined to be 77.7 kcal mol^{-1}, to reproduce the experimental exergonicity of 46 kcal mol^{-1} for the whole cycle, in the case of no gradient across the membrane. To obtain this value the **R** to P_M step, the O–O bond cleavage, is assumed to be exergonic by 5 kcal mol^{-1}, a value taken from the calculations.

Using the calculated electron and proton affinities and the reference O–H bond strength the energy diagram shown in Figure 3.5 is constructed. The O–O bond cleavage barrier, as calculated using a similar model [7], is also introduced. The calculated barrier is 16.0 kcal mol^{-1} relative to compound **A**, which is assumed to be 1 kcal mol^{-1} below the reduced state **R**. This value is slightly higher than the experimental value of 12.4 kcal mol^{-1} [8], deduced from the life-time of compound **A** using transition state theory. It should be stressed here that the shape of this diagram is independent of the parametrization, which can be considered just as a scaling factor to reproduce the experimental value for the overall energetics of the catalytic cycle. From Figure 3.5 it can be noted that, while all four reduction steps are exergonic, the two steps in the oxidative part (P_M to **F** and **F** to **O**) are considerably more exergonic than the two steps in the reductive part of the cycle (**O** to **E** and **E** to **R**). The energy gain in the reductive part is only 8.5 kcal mol^{-1}, while it is 32.3 kcal mol^{-1} in the oxidative part. Furthermore, the proton and electron transfer steps are obviously associated with barriers, which are not indicated in this energy profile. Recent time-resolved spectroscopic and electrometric measurements indicate that the proton transfer barriers are in the

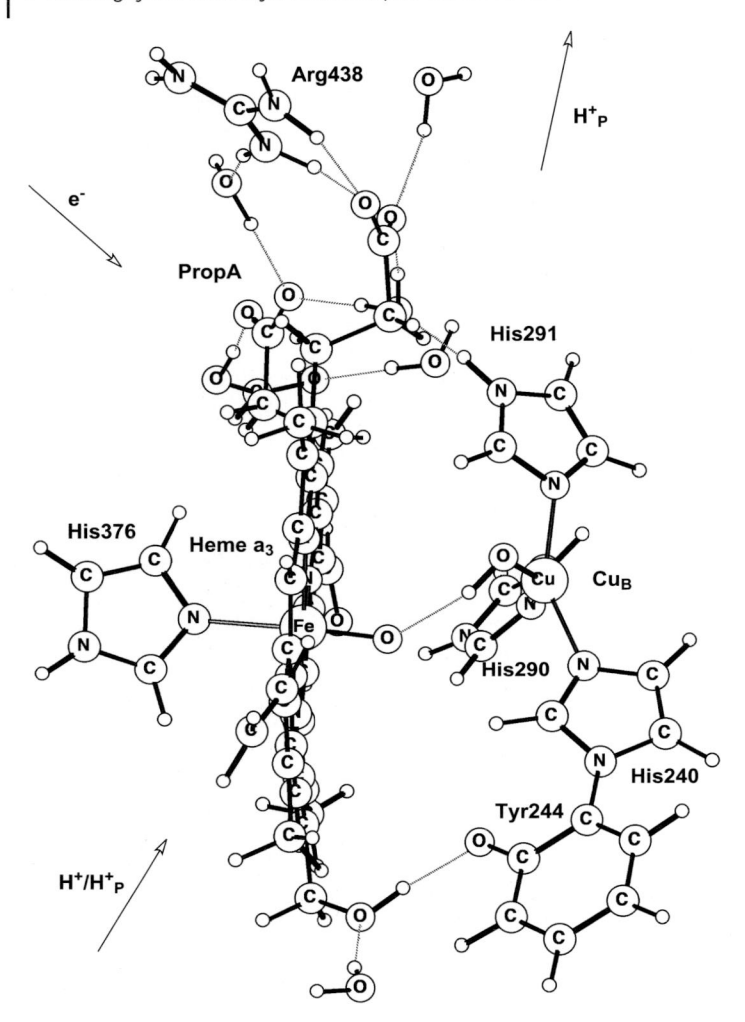

Figure 3.4 Model used in the calculations on cytochrome c oxidase.

12–14 kcal mol^{-1} region [9], while electron transfer is faster, and thus has lower barriers. Barriers of this size, as well as the O–O bond cleavage barrier, are compatible with the energy profile in Figure 3.5, and with the observation that the overall time scale for this reaction is in the millisecond regime. As mentioned above, each electron transfer is also connected with a proton pumping across the entire membrane, and this is indicated in the energy diagram by the H$^+$$_P$ notation. However, without a gradient across the membrane there is no net cost associated with this proton pumping.

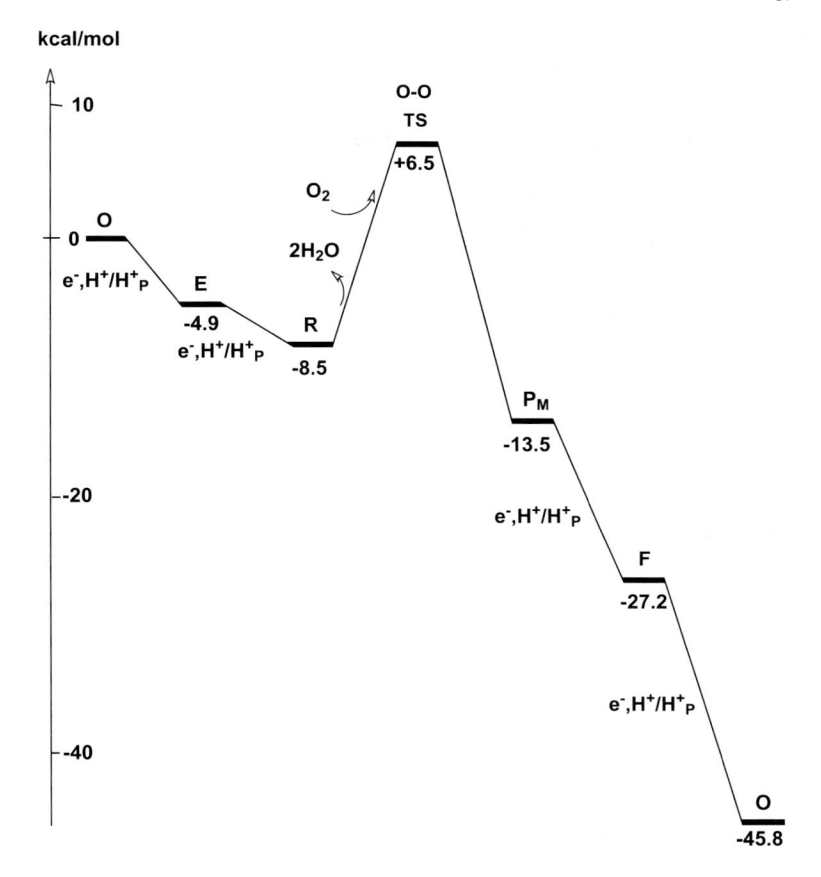

Figure 3.5 Energy diagram for dioxygen reduction in cytochrome c oxidase, showing the transitions where both an electron and a proton have been taken up.

As discussed above for PSII, the energetics of the individual electron and proton transfer steps can be related to redox and pK_a values if a parameter is introduced. This parameter can be determined in different ways and its value is slightly arbitrary, but in practice the degree of freedom is not very large. Figure 3.6 shows an energy profile for one parameter choice. Here the new energy levels obtained after uptake of one electron are labeled with the index R (reduced). As can be seen from the diagram the parametrization is made in such a way that each individual step (electron or proton transfer) is exergonic. In the case without membrane gradient this second type of diagram introduces little new insight as compared with the first one.

The most interesting energy profile in the case of cytochrome oxidase is obtained when the effect of the full electrochemical gradient across the membrane is applied. Each step corresponds to the transfer of two electrical charges across

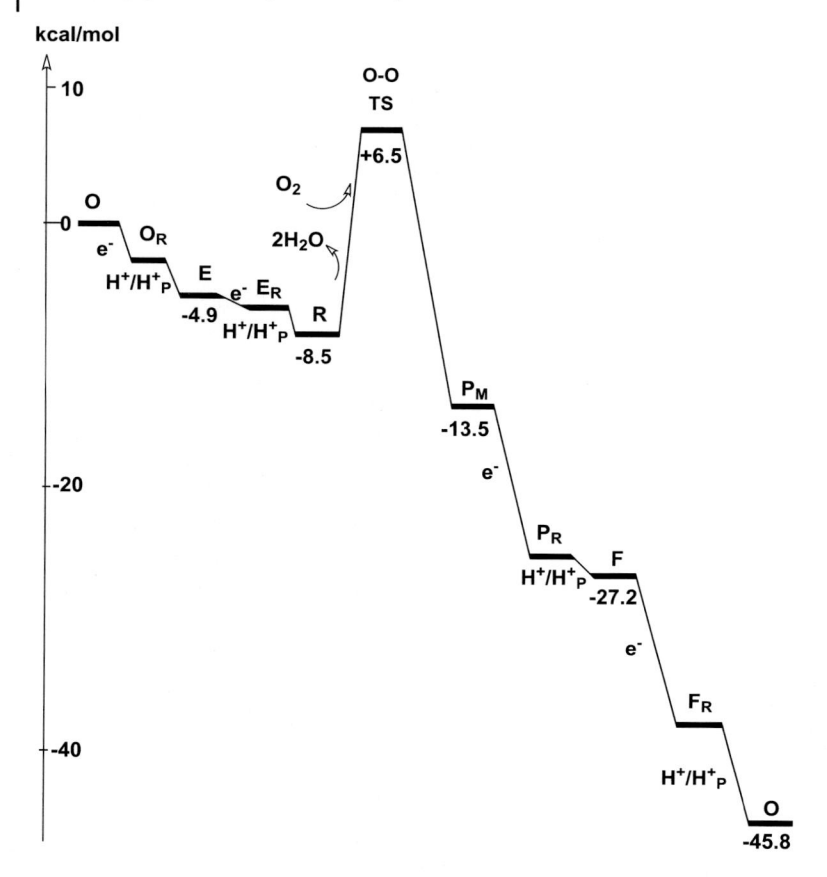

Figure 3.6 Energy diagram for dioxygen reduction in cytochrome c oxidase, showing the individual electron and proton uptakes.

the membrane, since besides the pumping of one proton across the membrane the electron and proton for the chemistry also contribute as they are taken from different sides of the membrane. Since the full gradient is measured as 0.2 V, this means that a total cost of 9.2 kcal mol^{-1} is introduced in each step due to the gradient. This leads to the energy profile shown in Figure 3.7. This energy diagram immediately elucidates a severe problem with the calculated energetics. Relative to the resting state (**O**) the barrier for the O–O bond cleavage is now 25 kcal mol^{-1}. Even if the experimental value for the barrier is used, the total barrier is 22 kcal mol^{-1}, which is much too high.

The source of this problem is the fact that the two steps in the reductive part are intrinsically much less exergonic than the two steps in the oxidative part, as discussed above, or in other words the uneven distribution of the free energy between the two halves of the cycle. It is very likely that the true energy profile

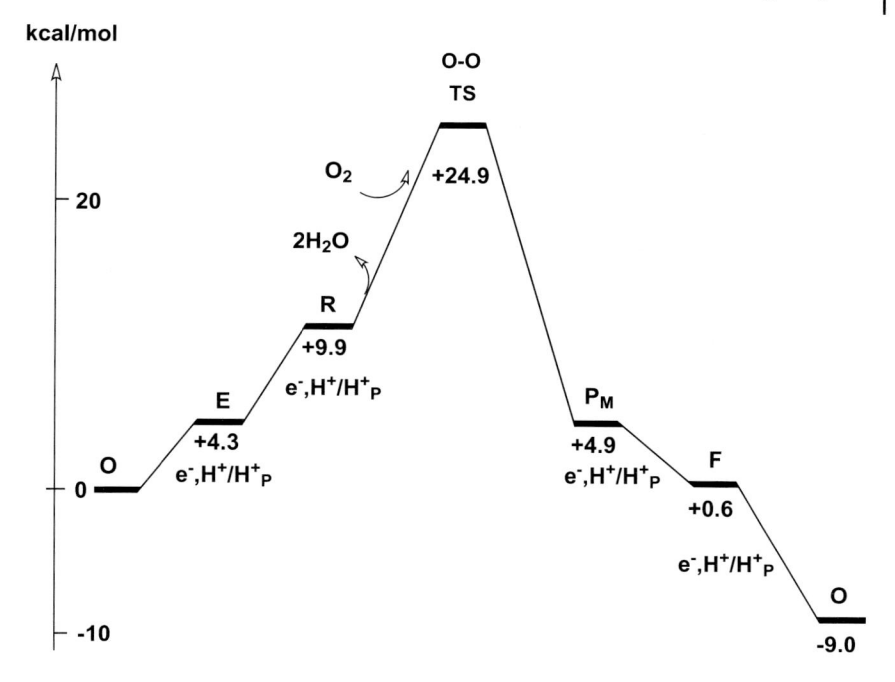

Figure 3.7 Energy diagram for dioxygen reduction in cytochrome c oxidase, showing the transitions where both an electron and a proton have been taken up, with full electrochemical gradient.

has an even distribution of the free energy among the four reduction steps in the catalytic cycle. One reason for this is the experimental observation that proton pumping is associated with each of the four steps. Another reason is that the average excess energy in each step is as low as 1 kcal mol^{-1} when the full gradient is present, and it is difficult to imagine that the pumping mechanism, which has to consist of a delicate balance between allowed and non-allowed paths, is not the same in all four steps when so little energy is available. All together this indicates that there is an error in the calculated relative O–H bond strengths, with two of them being 10–15 kcal mol^{-1} larger than the other two. A similar problem has also been noted on the basis of experimentally obtained reduction potentials for the redox components of the active site [10]. It is difficult to know to what extent this error is a result of an inaccuracy in the B3LYP method or if it is due to a problem with the model chosen. For a B3LYP error, it would be surprisingly large. Regarding the model, it has been successively increased since the first set of calculations [11] with only minor effects on the relative energies. In addition, notably, simple changes in the model with regard to its total charge will affect the individual charge transfer steps, but will only have minor effects on the O–H bond strengths.

3.2.3
Nitric Oxide Reduction

One step in the bacterial denitrification process is the reduction of nitric oxide (NO) to nitrous oxide (N_2O), which occurs in the enzyme nitric oxide reductase (NOR). There is no structure available for the bacterial NOR, but it is known that the active site is similar to the binuclear center in cytochrome oxidase, but with the Cu_B center replaced by a non-heme iron. It is furthermore known that NO reduction can occur also in some species of cytochrome oxidase. Therefore, a theoretical study was performed [12] for the NO reduction to N_2O in a ba_3 type of cytochrome oxidase, for which there is a structure available, and for which NO reduction has been studied experimentally, yielding a rate-limiting barrier of 19.5 kcal mol^{-1} [13]. The reaction studied is thus:

$$2NO + 2e^- + 2H^+ \rightarrow N_2O + H_2O$$

Figure 3.8 shows the model used. In this case it is not known from experiments at what points of the reaction the protons or electrons enter, and this is therefore

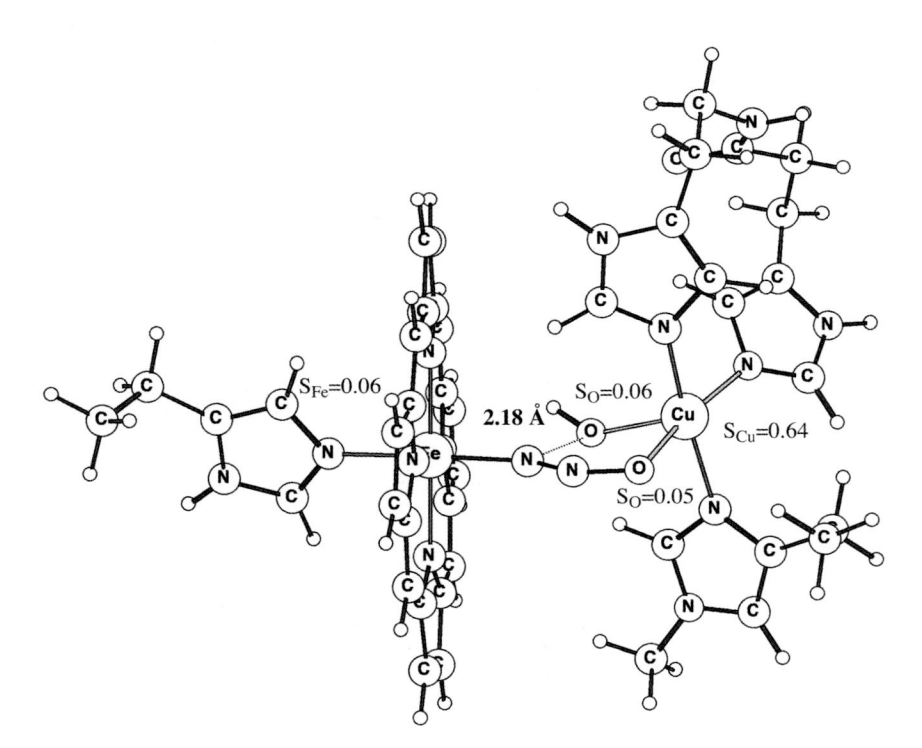

Figure 3.8 Optimized transition state for cleavage of N–O bond in the suggested mechanism for NO reduction.

3.2 Energy Diagrams

one of the questions of the theoretical study. As in the other studies discussed above, the cost of electron and proton uptake is parametrized to give the overall experimental energy of the reaction. For this reaction the total exergonicity is 45.6 kcal mol^{-1} when heme b is the electron donor. In this case partitioning of the electron and proton cost is fixed by using the experimentally determined difference in reduction potentials for the electron donor (heme b) and the electron acceptor in the second reduction step (heme a$_3$).

Several mechanisms for NO reduction were investigated. Figure 3.9 shows the calculated energy profile for the most likely mechanism. The first NO molecule coordinates to iron in the reduced active site. The bond is very weak, only −0.3 kcal mol^{-1}. The second NO molecule does not coordinate to the active site, but rather reacts directly with the first NO, forming the N–N bond in a very stable five-membered ring, at −12 kcal mol^{-1} (Figure 3.9). The activation energy in this step is 14 kcal mol^{-1}. After this one of the N–O bonds has to be cleaved. Since a copper-oxo compound is very unstable, it is not possible to cleave any of the N–O bonds without protonating one of the oxygens. Starting from the proto-

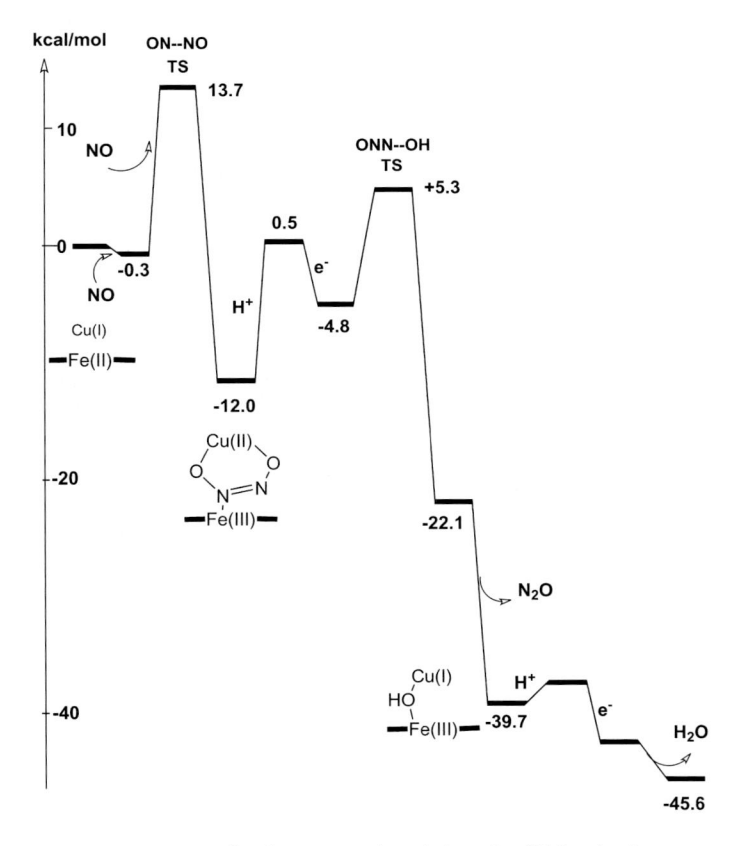

Figure 3.9 Energetics for the suggested catalytic cycle of NO reduction.

nated five-membered ring the N–O bond can be cleaved with an intrinsic barrier of 15.5 kcal mol^{-1}. However, the total barrier can not be calculated without knowing the cost of the protonation step. Furthermore, if an electron is transferred to the protonated five-membered ring, the intrinsic barrier for N–O bond cleavage is somewhat lower, 10.1 kcal mol^{-1}. However, in this case the cost of both protonation and reduction has to be known before the total barrier can be calculated. As it turns out, with the parametrization described above, proton transfer to the five-membered ring is endergonic by as much as 12.5 kcal mol^{-1}, which means that the total barrier for the first mechanism, N–O bond cleavage before reduction, becomes as high as 28 kcal mol^{-1}, which is much higher than the experimental value for the rate-limiting step of the reaction of 19.5 kcal mol^{-1}. In contrast, the electron transfer to the protonated five-membered ring, turns out to be exergonic by 5.3 kcal mol^{-1}, yielding a total barrier of only 17.3 kcal mol^{-1} for the mechanism where both protonation and reduction takes place before the N–O bond cleavage. This is compatible with a rate-limiting barrier of 19.5 kcal mol^{-1}, and it was therefore concluded that the N–O bond is cleaved only after the five-membered ring is both protonated and reduced. The rest of the reaction steps, releasing N_2O and water, and returning the enzyme to the reduced state, are essentially exergonic (Figure 3.9).

This study thus illustrates that, even if two different reaction mechanisms have local (intrinsic) barriers that are compatible with experimental rate measurements, it is only possible to make a comparison with experimental results after the costs of the proton and electron transfers are taken into account. For the suggested mechanism, Figure 3.9, it can also be noted that both an electron and a proton are taken up before the rate-limiting barrier, and therefore the height of this barrier is not very sensitive to the parametrization. Finally, the presently suggested mechanism should be possible to test experimentally, if the enzyme is prepared in such a form that there are no more electrons available than the one needed to produce the active site in its reduced form. Under these conditions, exposure to nitric oxide (NO) should lead to a stable, observable intermediate, i.e., the five-membered ring.

3.2.4
NiFe-hydrogenase

Hydrogenases are also enzymes where the transfer of protons and electrons play a fundamental role. There is at present substantial technical interest in these enzymes, since the use of photosynthesis to take protons and electrons from water, followed by hydrogen production with hydrogenases, is an attractive way to produce clean alternatives to fossil fuels. The most common group of hydrogenases has a NiFe-complex at its active site. Figure 3.10 shows an X-ray structure from *Desulfomicrobium bacalum* [14].

The most surprising feature of the Ni-Fe dimer in the hydrogenases is the ligand structure around iron with three diatomic ligands, where one is a carbonyl and the other two cyanides, which is unique for a biochemical system. There are

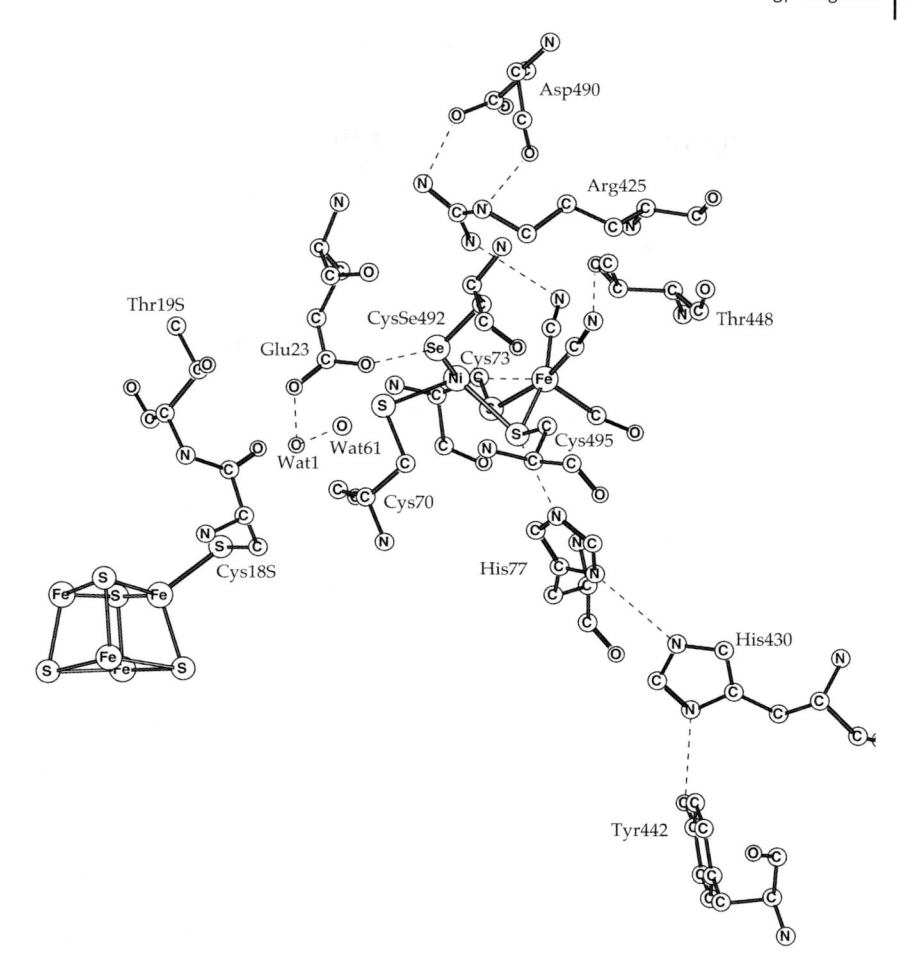

Figure 3.10 X-Ray structure for the reduced form of the [NiFe] hydrogenase from *Desulfomicrobium bacalum*.

also four cysteine ligands, where two are bridging between the metals and two are terminal nickel ligands. In the Ni-Fe-Se enzyme in the figure, selenium is substituted for sulfur in one of these terminal ligands, Cys492. The reaction catalyzed by hydrogenases is:

$$H_2 \leftrightarrow 2H^+ + 2e^-$$

The NiFe-hydrogenases are primarily utilized for hydrogen oxidation (the forward direction), while those with an all-Fe active site are primarily utilized for proton reduction (the reverse direction). Sine hydrogenases are enzymes where the purpose is to make a charge separation, there have to be pathways leading in different directions for the electrons and the protons produced. For the NiFe-hydrogenase shown in Figure 3.10 the electron transfer pathway is easy to identify, going over one or more Fe-S cluster. Only the closest Fe-S cluster is shown in the figure. At least two possible proton transfer pathways have been identified. One of these pathways starts at CysSe492 and goes over Glu23 and several conserved water molecules, of which two are shown in the figure, and leads to a Mg complex (not shown) near the surface [15]. Another pathway starts at the bridging Cys495 ligand and goes straight out to the surface via hydrogen bonds over His77, His430 and Tyr442. This pathway is similar to the one suggested for Ni-CO dehydrogenase [16, 17]. Other proton transfer pathways are also possible.

Several DFT studies have been made to investigate the mechanism of NiFe-hydrogenase, and these are all discussed in a recent review [18]. The reaction starts with the binding of dihydrogen in a local minimum on iron. The binding is quite strongly endergonic, by 11.2 kcal mol^{-1} when the entropy loss is included. The main chemical step in the hydrogenase catalytic cycle is the cleavage of the bond in H_2. Since there should be a charge separation, the cleavage is heterolytic, resulting in one hydride and one proton. The calculations indicate that the hydride ends up bridging between the metals and that the proton goes to a cysteine, most likely CysSe492. The calculated barrier is very small, only 1.4 kcal mol^{-1}, while the exergonicity is 13.8 kcal mol^{-1}.

Once the heterolytic cleavage has been made, the electron should leave for the Fe-S cluster. It is more likely that an electron leaves first than a proton since the NiFe-complex is negatively charged. The calculated electron affinity is 92.8 kcal mol^{-1}. This is followed by a transfer of the proton to the bulk from CysSe492, with a calculated proton affinity of 281.6 kcal mol^{-1}. Together this gives a binding of the (e^-, H^+)-couple of $(92.8 + 281.6 - 315.1) = 59.3$ kcal mol^{-1}. As above, the value of the free hydrogen atom is 315.1 kcal mol^{-1} (0.5022 a.u.).

The next step is a transfer of the hydrogen from the bridging hydride position to a position on a cysteine. Again, CysSe492 is the most likely proton acceptor. The computed barrier for this step is 3.5 kcal mol^{-1} and the reaction is endergonic by +1.6 kcal mol^{-1}. The final steps of the cycle are an electron transfer, with a computed electron affinity of 89.8 kcal mol^{-1}, and a proton transfer, with a computed proton affinity of 287.6 kcal mol^{-1}. This gives a binding energy of the (e^-, H^+)-couple of $(89.8 + 287.6 - 315.1) = 62.3$ kcal mol^{-1}.

To set up an energy diagram, like the ones above for PSII and CcO, a reference binding energy is needed for the (e^-, H^+)-couple that matches a predetermined reaction energy for the whole cycle. In this respect, hydrogenase differs from the enzymes discussed above in the sense that this reaction energy is not known directly from experiments, and a value therefore has to be assumed. Since the enzyme actually can perform the reverse reaction under slightly different circumstances, the reaction energy has to be small. For hydrogenase enzymes that work in metabolic processes, it is also clear that the driving force should be small under working conditions so as to minimize the loss of energy. A value of only −1.0 kcal mol^{-1} was, therefore, chosen for the driving force. This is a more uncertain value than those for PSII and CcO above, but should be qualitatively reasonable.

With a reference energy for the (e^-, H^+)-couple of 60.8 kcal mol^{-1}, the reaction energy becomes -1.0 kcal mol^{-1}, as desired. This means that the transfer of the first (e^-, H^+)-couple is exergonic by -1.5 kcal mol^{-1} and the second is endergonic by $+1.5$ kcal mol^{-1}. It appears reasonable that the reference energy should be close to the average of the binding energies for the two (e^-, H^+)-couples. With this a full energy diagram can be set up, and is shown in Figure 3.11.

The construction of a complete energy diagram, which includes all steps, is very important for many comparisons with experiments. The most important one is the comparison of the rate of the reaction. The rate of the reaction is determined by the barrier for the rate-limiting step, which starts at the lowest point before the highest barrier. For NiFe-hydrogenase, the lowest point is the one marked Ni$_a$-C* in Figure 3.11. This is the resting state and corresponds to the state of the enzyme observed in the X-ray structure. The highest point after that is the H–H bond cleavage in the next cycle, which is at $+11.6$ kcal mol^{-1} (since this cycle starts at -1 kcal mol^{-1}). The barrier is thus $(4.1 + 11.6) = 15.7$ kcal

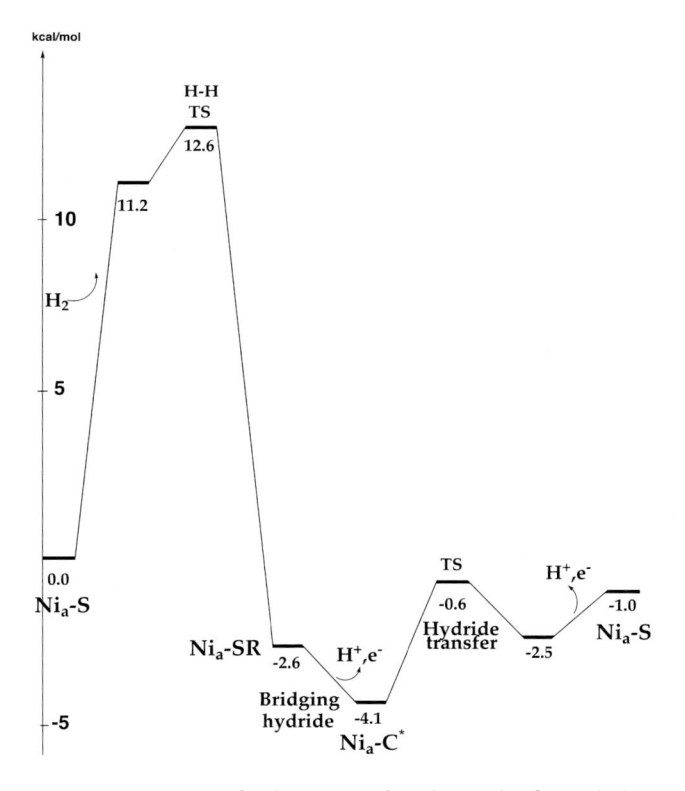

Figure 3.11 Energetics for the suggested catalytic cycle of Ni-Fe hydrogenase.

mol^{-1}. The experimental k_{cat} of 10^3 s^{-1} [15] can be translated to a free energy barrier of around 13 kcal mol^{-1} by using transition state theory.

The use of the B3LYP functional consequently overestimates the barrier by a few kcal mol^{-1}, which is common in bond-cleavage reactions [19]. The rate is thus not determined by the internal barrier for dihydrogen cleavage of 12.6 kcal mol^{-1}, as might have been expected. The identification of Ni$_a$-C* also depends on the access to the full energy diagram, since otherwise it would not be possible to determine which state is lowest in energy.

To realize the simplicity of the above scheme for obtaining an energy diagram, it should be compared with the alternative approach. In that approach the electron affinity of the FeS-cluster has to be obtained to get the electron transfer energy. As already mentioned, this is a property that is very sensitive to the enzyme surrounding the complex and therefore has to include a modeling of a major part of the enzyme. The electron affinity of the NiFe-complex also has to be computed in a similar way. These two calculations have to be perfectly balanced to compute the electron transfer energy. However, not even this may be enough, since the one with the lowest energy of the electron acceptors may not be the first FeS-cluster in the electron transfer chain. The other ones should, therefore, also be computed. The proton transfer energy is even more difficult to obtain in this way. However, it has to be realized that the accuracy of the present approach relies on the computed local chemistry, such as the exergonicity of the heterolytic cleavage of H$_2$. An error in that energy would transfer directly to the computed redox potentials and pK_a values. The accuracy of introducing just an estimated value for the reaction energy is a less severe approximation in this context.

3.2.5
Molybdenum CO Dehydrogenase

Molybdenum CO dehydrogenase (Mo-CODH) from the aerobic eubacterium *Oligotropha carboxidovarans* is a Cu-containing molybdo-iron-sulfur flavoprotein that catalyzes the oxidation of CO with H$_2$O, yielding CO$_2$, two electrons and two protons [20–22]:

$$CO + H_2O \rightarrow CO_2 + 2e^- + 2H^+$$

In its membrane-associated state, CO dehydrogenase transfers the electrons to the respiratory electron acceptor cytochrome b_{561} [22, 23]. In its soluble state, the enzyme transfers the electrons to O$_2$, yielding superoxide and hydrogen peroxide. The crystal structure of CO dehydrogenase has been solved at 1.1 Å resolution [24]. Figure 3.12 shows a computational model of the active site, which contains the unprecedented dinuclear heterometal (CuSMo(=O)OH) cluster. Molybdenum is, furthermore, coordinated to a dithiolene ligand having pterin and cytosin dinucleotide substituents (molybdopterin cytosine dinucleotide, MCD). The ligands around molybdenum form a distorted square pyramidal geometry. Residues in the second coordination shell around molybdenum include

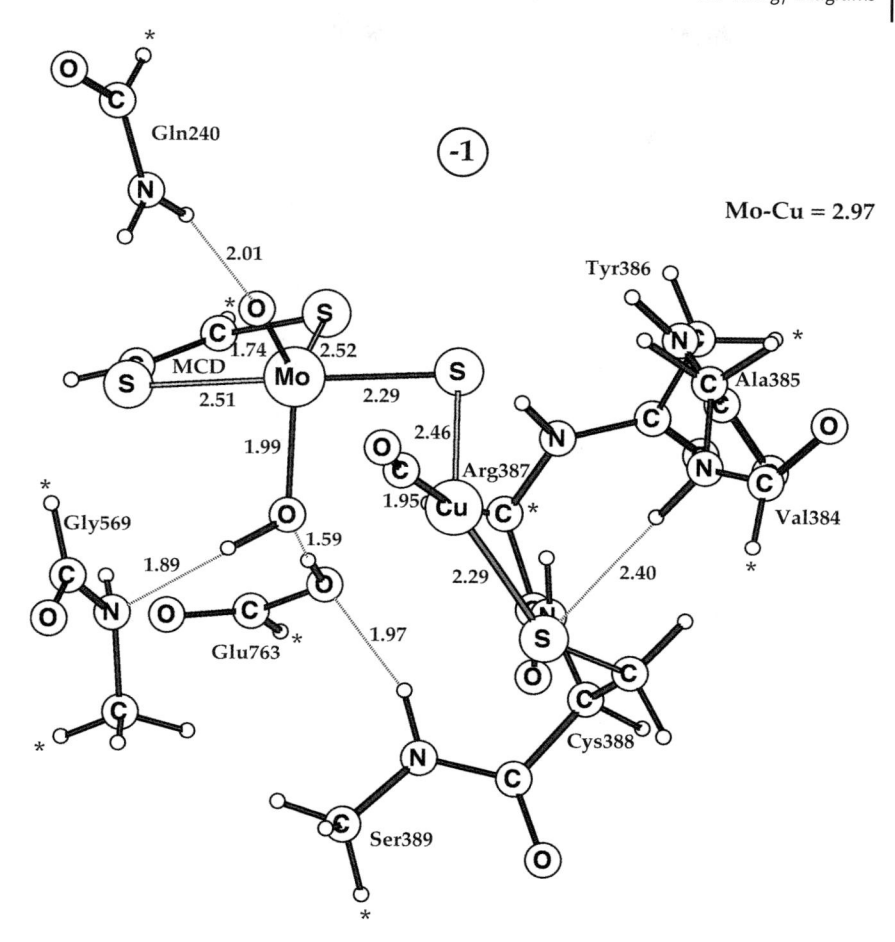

Figure 3.12 Optimized structure for the reactant of Mo-CODH, with bound CO (distances in Å). Atoms marked with * were frozen from the X-ray structure.

Glu763, Gln240 and Gly569. An unusual feature of the structure is that copper only has two ligands, apart from the sulfide also Cys388. Cys388 is part of a Val-Ala-Tyr-Arg-Cys-Ser chain of residues, which surrounds the copper complex.

A DFT study of the mechanism of Mo-CODH using the B3LYP functional has been performed in a similar manner as for the enzymes discussed above [25]. The goal was to obtain an energy diagram for all steps of the catalytic cycle, including the electron and proton transfers. The mechanism starts by binding CO to copper, with a very small binding energy of -0.7 kcal mol^{-1}. The optimized structure is shown in Figure 3.12. The bound CO should then attack an oxo ligand on molybdenum to form an anionic carboxylate bound to molybdenum. In the best mechanism found, a proton loss to the bulk from Glu763 precedes this attack. By transferring the proton on Glu763 to the outside (removing it from the model complex), the proton on the hydroxide ligand will automatically be transferred to the glutamate, leaving a doubly bonded oxo-ligand on molybdenum. After this, carboxylate formation occurs easily with a small local barrier of only 5.4 kcal mol^{-1}, and an endergonicity of +1.7 kcal mol^{-1}. In the next step CO_2 is released with a local barrier of 8.4 kcal mol^{-1} and an exergonicity of 15.0 kcal mol^{-1}. The remaining steps are electron and proton transfers, and binding of a new substrate water molecule, to return to the original reactant. The binding of the water molecule to molybdenum is exergonic by −5 kcal mol^{-1}. In the B3LYP study a more complicated mechanism was eventually found to be slightly advantageous, involving intermediate formation of an S–C bond with the bridging sulfide ligand. In order not to complicate the present description unnecessarily, this part of the mechanism is not be discussed here.

As already indicated by the description above, the electron and proton transfer steps are not sequential as they are for most of the other enzymes described so far. It is, thus, not meaningful to try to derive a diagram where the binding of (e^-, H^+)-couples appear. A complete diagram involving individual electron and proton transfer energies is therefore required. The proton affinity of the proton on Glu763 was calculated to be 285.4 kcal mol^{-1} at the stage before the C–O bond is formed. The computed electron affinity for the structure where CO_2 has just been released is 84.0 kcal mol^{-1}. After water has been bound on molybde-

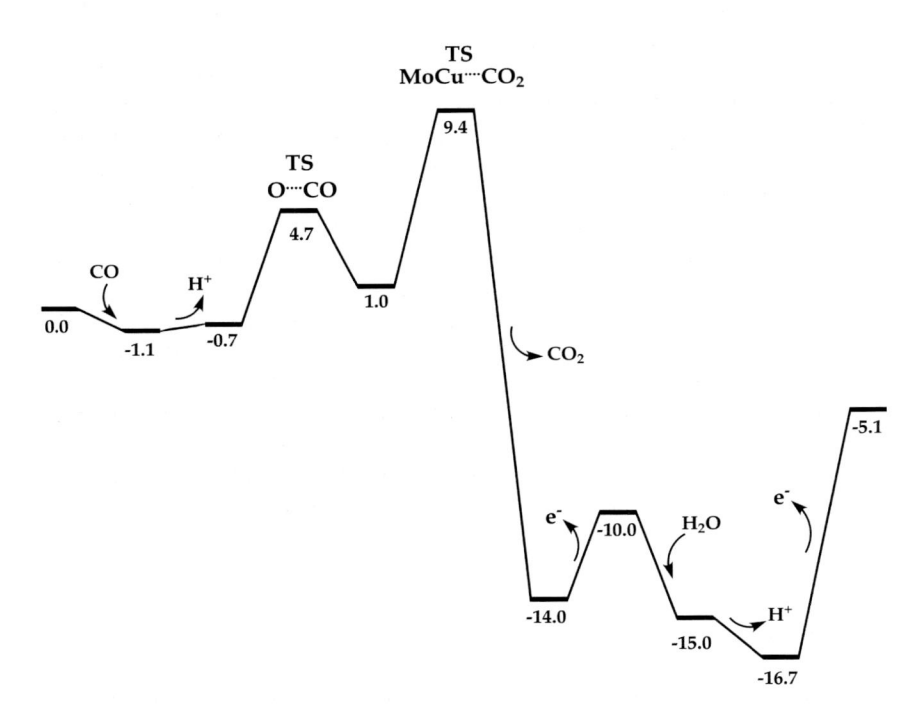

Figure 3.13 Energetics for the suggested catalytic cycle of molybdenum CO dehydrogenase.

num, the proton affinity of Glu763 is 283.3 kcal mol^{-1}. This value can be compared with the energy of 285.4 kcal mol^{-1} for removing the previous proton. These two energies are thus quite similar, which is reasonable since the protons should both end up in the surrounding medium. The electron affinity after the second proton has been released is 91.6 kcal mol^{-1}. These values, therefore, say that the second electron transfer has to be 7.6 kcal mol^{-1} more endergonic than the first transfer since the electrons are transferred to the same acceptor.

The study of Mo-CODH was one of the first ones where a full catalytic cycle was determined using the present procedure. The first reference value chosen in that study was the one for the proton affinity, which was set to the even value 285 kcal mol^{-1} based on similar models in previous studies. Similarly, a reference electron affinity of 80 kcal mol^{-1} was chosen. With these values the energy for the removal of an (e^-, H^+)-couple becomes around 50 kcal mol^{-1}, and the driving force for the full cycle is -5.1 kcal mol^{-1}. This is a reasonable, but perhaps somewhat too large, driving force. The proton transfer energies then become $+0.4$ kcal mol^{-1} for the first transfer, and -1.7 kcal mol^{-1} for the second. The electron transfer energies both become endergonic, $+4.0$ kcal mol^{-1} for the first one and $+11.6$ kcal mol^{-1} for the second. Figure 3.13 show the full cycle.

A few additional comments on how the catalytic cycle and mechanism was determined are of interest. First, the positions in the cycle where the protons and electrons are released were determined by comparing calculated absolute proton and electron affinities for all the different structures obtained. This comparison can be made with absolute values, since the relative magnitudes will remain the same as for the finally determined redox and pK_a values. This type of investigation led to removal of a proton after the step where CO became bound to copper, but before the C–O bond was formed with the molybdenum oxo ligand. Also, it was found to be optimal to bind the water substrate between the first electron transfer and the second proton transfer.

In the present case, the choice of the electron and proton affinity reference values contains a larger degree of arbitrariness than for the systems described above. For example, if the reference value for the proton affinity is increased by $+5$ kcal mol^{-1} to 290 kcal mol^{-1} and the one for the electron affinity decreased by -5 kcal mol^{-1} to 75 kcal mol^{-1} this would not affect the main features of the mechanism. The resting state would be the same and the rate-limiting barrier of 21.0 kcal mol^{-1} the same. The enzyme could use this upward flexibility of the proton reference value to adjust the environment of the MoCu-complex to match the redox potential to the acceptor to some extent, without affecting the rate. However, decreasing the reference proton affinity by even a small amount would increase the rate of the overall reaction, which is clearly undesirable.

3.3 Conclusions

In many important enzymatic processes, protons and electrons are released from, or accepted by, the catalytic site. While small models of 50–100 atoms are usually quite adequate for studying local processes where bonds are formed or broken, they have severe deficiencies for processes where there is a change of charge of the model. The effects of the enzyme surrounding the catalytic site are typically on the order of 20–30 kcal mol^{-1}. Qualitatively, some of the effects can be recovered by a simple dielectric cavity model, but the accuracy is usually not very high.

This chapter has outlined, and exemplified, a procedure where energy diagrams including proton and electron transfer steps can be obtained to high accuracy, using the small models commonly used in DFT studies. In the simplest version of this approach, only the energies for transfer of (e$^-$, H$^+$)-couples are included in the diagram. Since the transfer of these couples does not change the charge of the system, the effects of the enzyme surrounding is quite small (only on the order of a few kcal mol^{-1}) and the accuracy is therefore quite high. To obtain the energies required in a diagram of this type, the reaction energy for the full cycle is needed as a scaling factor. For many processes this energy is known from measured redox potentials, while for others it can be estimated with sufficient accuracy.

In the more complete form of the present approach, where the energies for each electron and proton transfer steps are included in the diagram, an additional parameter is needed. This parameter can, for example, be obtained from a minimization of the barriers in the diagram, but can also be obtained by comparison with a relevant redox potential that has been measured.

Several examples where the present approach has been successfully used have been discussed above. For both photosystem II and cytochrome c oxidase, it is shown that the main features of the energetics are already described by the simpler form of the diagram, without the determination of individual redox potentials and pK_a values. The interesting conclusion that can be drawn from this is that the most important parts of these processes are determined by the local properties of the catalytic sites. For example, in O–O bond formation in PSII, the rate-limiting barrier depends both on the preceding electron and proton transfer steps. If the charge distribution in the surrounding enzyme is optimized to decrease the energy required for one of them, the other one will be increased by essentially the same amount.

References

1 Warshel, A. *Computer Modeling of Chemical Reactions in Enzymes and Solutions*, Wiley, New York, 1991.
2 Ullmann, G.M., Noodleman, L., Case, D.A. *J. Biol. Inorg. Chem.* 2002, 7, 632–639.
3 Altun, A., Guallar, V., Friesner, R.A., Shaik, S., Thiel, W. *J. Am. Chem. Soc.* 2006, **128**, 3924–3925.
4 Ferreira, K. N., Iverson, T. M., Maghlaoui, K., Barber, J., Iwata, S. *Science* 2004, **303**, 1831–1838.
5 Loll, B., Kern, J., Saenger, W., Zouni, A., Biesiadka, J. *Nature* 2005, **438**, 1040–1044.
6 Siegbahn, P.E.M., Haeffner, F. *J. Am. Chem. Soc.* 2004, **126**, 8919–8932.
7 Blomberg, M.R.A., Siegbahn, P.E.M. *J. Comp. Chem.* 2006, **27**, 1373–1384.

8 Karpefors, M., Adelroth, P., Namslauer, A., Zhen, Y., Brzezinski, P. *Biochemistry* 2000, **39**, 14664.

9 Belevich, I., Bloch, D.A., Belevich, N., Wikstrom, M., Verkhovsky, M.I. *Proc. Natl. Acad. Sci. U.S.A.* 2007, **104**, 2685–2690.

10 Wikstrom, M. *Biochim. Biophys. Acta* 2004, **1655**, 241–247.

11 Siegbahn, P.E.M., Blomberg, M.R.A., Blomberg, M.L. *J. Phys. Chem. B* 2003, **107**, 10946–10955.

12 Blomberg, L.M., Blomberg, M.R.A., Siegbahn, P.E.M. *Biochim. Biophys. Acta* 2006, **1757**, 31–46.

13 Giuffre, A., Stubauer, G., Sarti, P., Brunori, M., Zumft, W.G., Buse, G., Soulimane, T. *Proc. Natl. Acad. Sci. U.S.A.* 1999, **96**, 14718–14723.

14 Garcin, E., Vernede, X., Hatchikian, E.C., Volbeda, A., Frey, M., Fontecilla-Camps, J.C. *Structure* 1999, **7**, 557–566.

15 Frey, M., Fontecilla-Camps, J.C. in *Handbook of Metalloproteins*, eds. Messerschmidt, A., Huber, R., Poulos, T., Wieghardt, K., John Wiley & Sons, Chichester, U.K., 2001, Vol. 2, pp. 880–896.

16 Dobbek, H., Svetlitchnyi, V., Gremer, L., Huber, R., Meyer, O. *Science* 2001, **293**, 1281–1285.

17 Drennan, C.L., Heo, J., Sintchak, M.D., Schreiter, E., Ludden, P.W. *Proc. Natl. Acad. Sci. U.S.A.* 2001, **98**, 11973–11978.

18 Siegbahn, P.E.M., Tye, J.W., Hall, M.B. *Chem. Rev.* 2007, **107**, 4414–4435.

19 Siegbahn, P.E.M. *J. Biol. Inorg. Chem.* 2006, **11**, 695–701.

20 Dobbek, H., Gremer, L., Meyer, O., Huber, R. in *Handbook on Metalloproteins*, eds. Bertini, I., Sigel, H.Marcel Dekker, New York, 2001, pp. 1136–1147.

21 Meyer, O. *J. Biol. Chem.* 1982, **257**, 1333–1341.

22 Meyer, O., Frunzke, K., Morsdorf, G. in *Microbial Growth on C1 compounds*, eds. Murrell, J.C., Kelly, D.P., Intercept Scientific Publication, Andover, U.K., 1993, pp. 433–459.

23 Cypionka, H., Meyer, O., *J. Bacteriol.* 1983, **156**, 1178–1187.

24 Dobbek, H., Gremer, L., Kiefersauer, R., Huber, R., Meyer, O. *Proc. Natl. Acad. Sci. U.S.A.* 2002, **99**, 15971–15976.

25 Siegbahn, P.E.M., Shestakov, A.F. *J. Comput. Chem.* 2005, **26**, 888–898.

4
Principles of Dinitrogen Hydrogenation: Computational Insights

Djamaladdin G. Musaev, Petia Bobadova-Parvanova, and Keiji Morokuma

4.1
Introduction

Reduced nitrogen is one of the most important nutrients and an essential component of nucleic acids and proteins [1–3]. It is extensively used in agriculture. Therefore, the reduction of atmospheric dinitrogen (dinitrogen makes up nearly 79% of the Earth's atmosphere [4]) has occupied the minds of many scientists over the centuries. However, the N_2 molecule is inert and its utilization has proven to be very difficult. The inertness of N_2 is associated with its strong and non-polar N≡N triple bond and with the large energy gap between its highest occupied (σ HOMO) and lowest unoccupied (π^*, LUMO) molecular orbitals [5–9]. Despite these harsh intrinsic properties of dinitrogen, both Nature and scientists have found several ways to reduce N_2.

In Nature there are several processes that utilize the N_2 molecule, among which we mention atmospheric and enzymatic N_2 reduction. In the Earth's atmosphere, a relatively small amount of N_2 oxidized under extreme conditions: the electrical discharge of a lightning breaks the strong N≡N triple bond and facilitates the reaction of N_2 with atmospheric oxygen [10]. However, the major dinitrogen fixation process in Nature is a biological process called nitrogenase that converts the N_2 molecule into ammonia under mild conditions via a sequence of electron- and proton-transfer reactions [11–14]:

$$N_2 + 8H^+ + 8e^- + 16\text{MgATP} \rightarrow 2NH_3 + H_2 + 16\text{MgADP} + 16\text{P(phosphate)} \quad (1)$$

Several industrial dinitrogen utilization processes, such as the Birkeland–Eyde [15, 16], Frank–Caro [15, 16], and Haber–Bosch [1, 2, 17, 18] processes, have also been reported. In the Birkeland–Eyde process, the N_2 and O_2 molecules form nitric oxide in the very high temperature of an electric arc. This process was very costly and energy-inefficient and, therefore, was soon abandoned for better pro-

Computational Modeling for Homogeneous and Enzymatic Catalysis.
A Knowledge-Base for Designing Efficient Catalysts. K. Morokuma and D. G. Musaev (Eds.)
Copyright © 2008 WILEY-VCH Verlag GmbH & Co. KGaA, Weinheim
ISBN: 978-3-527-31843-8

cesses. One such processes is the Frank–Caro process, where the reaction of nitrogen with calcium carbide at high temperatures forms calcium cyanamide, which later hydrolyzes to ammonia and urea. This process was utilized on a large scale before and during World War I, but was economically ineffective, too. Therefore, it replaced by Haber–Bosch process, at the beginning of 20th century.

Currently, almost all industrial dinitrogen fixation is due to the century-old Haber–Bosch process, which operates at high temperature and pressure, and uses a promoted metallic-Fe catalyst (in the first reaction chambers osmium and uranium catalysts were used) [19]. The reaction occurs by coordination of both N_2 and H_2 on the catalyst surface, followed by stepwise assembly of ammonia from these molecules via Reaction (2):

$$N_2(g) + 3H_2(g) \rightarrow 2NH_3(g) \tag{2}$$

Although Reaction (2) is thermodynamically favored under mild conditions (standard free-energy and enthalpy of this reaction at 298 K and 1 atm are -7.7 kcal mol^{-1} and -10.97 kcal mol^{-1}, respectively [7]) its realization is extremely difficult because of the kinetic stability of N≡N triple bond and requires an efficient catalyst.

Throughout the 20th century, scientists and engineers continuously tried to improve the famous Haber–Bosch process, model nitrogenase process under laboratory conditions, and develop a better (more efficient and less energy-intensive) catalytic process for the reduction of N_2. Extensive studies have clearly demonstrated that the coordination of N_2 to transition metal center(s) is the first condition to facile utilization of the N_2 molecule [20–44]. These studies received significant support after the preparation and analysis [45] of the first transition metal-N_2 complex, $[Ru(NH_3)_5N_2]^{2+}$. Detailed studies [46] on the structure of this complex have revealed that the N_2 molecule can coordinate to a transition metal center in two distinct fashions: (a) linear, with one of its N-ends – known as end-on and/or η^1-coordination mode, and (2) perpendicular, with both N atoms – known as side-on and/or η^2-coordination mode. Simple molecular orbital analysis shows that the M–N_2 bond has two components: donation (N_2 as a σ-donor) and back donation (N_2 as a π-acceptor) (Figure 4.1).

The donation component of M–N_2 favors a linear coordination of N_2 to a transition metal center, while its back-donation component is expected to favor a perpendicular coordination of N_2 to a transition metal center. Since during

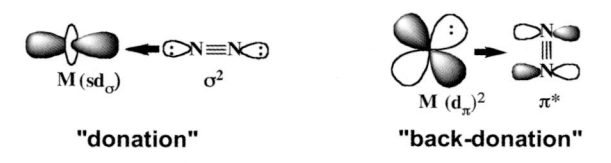

Figure 4.1 Schematic presentation of M–N_2 interactions.

the back-donation electron density transfers from d_π-orbitals of metal to the π^* orbitals of N_2, "back-donation" is expected to be more important for N≡N triple bond activation than the "donation" component of $M-N_2$ bond. Notably, during the "donation" an electron density of the lone-pair of one of N-atoms (that coordinated to transition metal) of N_2 transfers to the s and d_{zz}-orbitals of transition metal. However, the N_2 molecule is a poorer π-acceptor than σ-donor and, as a result, in *almost* all reported mononuclear transition metal–N_2 complexes the N_2 molecule is in its end-on, η^1-coordination mode.

As it could be expected, the N≡N triple bond of the η^1- and η^2-coordinated N_2 molecule can be utilized via different mechanisms. The η^1-coordinated N_2 molecule can be used via the Chatt mechanism [47], which proceeds via the coupled protonation (by electrophiles) and reduction (by the electrons flowing from the metal ion) of the coordinated N_2. Previously, we have demonstrated that direct hydrogenation of N_2 requires its side-on coordination to the transition metal center, which can be effectively achieved only on di- (or multi-) nuclear transition metal systems [40–42]. Thus, to hydrogenate of N_2 one should use a catalyst containing multi-transition metal centers (both under homogeneous and heterogeneous conditions) [41].

Furthermore, molecular orbital analysis indicates that the successful direct hydrogenation of the side-on coordinated N_2 molecule requires the availability of the double occupied π-bonding orbital of the $M-N_2-M$ fragment and the empty non-bonding σ (s and d)-orbitals of metal centers. Indeed, the HOMO, the π-bonding orbital of the $M-N_2-M$ fragment, is expected to donate electrons to the σ_u^* orbital of the incoming H_2 molecule, while the LUMO, σ (s and d)-orbitals of metal centers, is required to interact with the σ_g-bonding MO of the reacting H_2 molecule. This qualitative orbital picture is consistent with the "metathesis" transition state (involving one of the M and N atoms of the complex, and two H-atoms of H_2, see Figure 4.2) reported for H_2 addition to di-Zr-N_2 complexes [38, 39, 48].

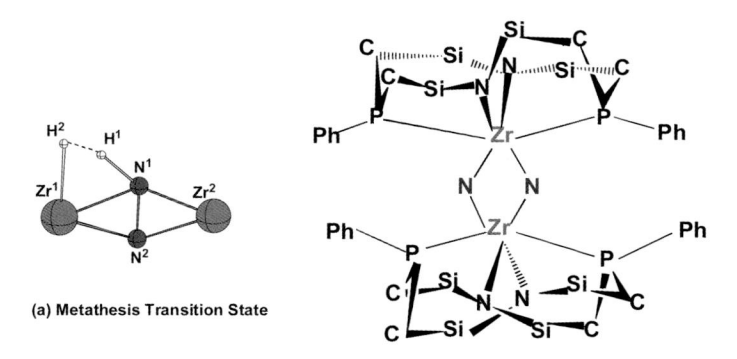

Figure 4.2 Schematic presentation of (a) the metathesis transition state for H_2 addition to di-nuclear-N_2 complex I and (b) the di-nuclear-N_2 complex I.

Studies of the reactivity of the η^1- and η^2-coordinated N_2 molecules were the subject of numerous excellent reviews [20, 25, 26, 40, 49]. Therefore, here we only briefly discuss the latest developments in terms of elucidating the fundamental principles and concepts of dinitrogen hydrogenation, and clarifying some specific issues puzzling researchers.

In the literature [25] numerous dinuclear transition metal–dinitrogen complexes, $[L_nM]_2(\mu_2,\eta^2,\eta^2\text{-}N_2)$, with a side-on, η^2, coordination mode of N_2 have been reported, among which those with group IV transition metals (Ti, Zr and Hf) look to be more promising.

In 1997, Fryzuk and coworkers [37] demonstrated that the dinuclear Zr complex $\{[P_2N_2]Zr\}_2(\mu_2,\eta^2,\eta^2\text{-}N_2)$, **I**, where $P_2N_2 = PhP(CH_2SiMe_2NSiMe_2CH_2)_2PPh$, adds H_2 at 22 °C and 1 atm of pressure and forms N–H bonds. Subsequent detailed computational studies [38–40, 48] have revealed that this reaction occurs via a "metathesis" transition state (Figure 4.2). More importantly, the computational studies have suggested the possibility of the addition of several (more than one) H_2 molecules to a di-Zr complex under appropriate conditions [38–40].

Four years after this theoretical prediction, in 2004, Chirik and coworkers [33] reported an addition of multiple H_2 molecule to a di-Zr-N_2 complex. The authors demonstrated that the side-on bound dinitrogen molecule of $[(\eta^5\text{-}C_5Me_4H)_2Zr]_2(\mu_2,\eta^2,\eta^2\text{-}N_2)$, **II** or **II_Zr**, adds two H_2 molecule at only 22 °C and 1 atm, affording $[(\eta^5\text{-}C_5Me_4H)_2ZrH]_2(\mu_2,\eta^2,\eta^2\text{-}N_2H_2)$, **III**. Subsequent warming of the complex to 85 °C even yielded a small amount of ammonia. Later, the same group reported a similar reactivity for the complexes $[(\eta^5\text{-}C_5H_2\text{-}1,2\text{-}Me_3\text{-}4\text{-}R)(\eta^5\text{-}C_5Me_5)Zr]_2(\mu_2,\eta^2,\eta^2\text{-}N_2)$, **IV**, where R = Me and Ph, [34] and for a Hf analog of **II_Zr**, $[(\eta^5\text{-}C_5Me_4H)_2Hf]_2(\mu_2,\eta^2,\eta^2\text{-}N_2)$, **II_Hf** [35]. It was reported that the Hf complex hydrogenates dinitrogen faster than the corresponding Zr complex.

These experimental studies of Chirik's group and Fryzuk's group have raised several intriguing questions, the elucidation of which could significantly enhance our understanding of the fundamental principles of dinitrogen hydrogenation.

The first intriguing question arises upon comparison of the reactivity of Chirik's complexes $[(\eta^5\text{-}C_5Me_4H)_2Zr]_2(\mu_2,\eta^2,\eta^2\text{-}N_2)$, **II_Zr**, and $[(\eta^5\text{-}C_5H_2\text{-}1,2,4\text{-}Me_3)(\eta^5\text{-}C_5Me_5)Zr]_2(\mu_2,\eta^2,\eta^2\text{-}N_2)$, **IV**, with the $[(\eta^5\text{-}C_5Me_5)_2Zr]_2(N_2)$, **V**, complex synthesized in 1974 by Manriquez and Bercaw [50]. Complex **II_Zr** with tetramethylated cyclopentadienyl ligands, $C_5Me_4H^-$, as well as complex **IV** with three Me substituents on two of its cyclopentadienyl (Cp) rings, and five Me substituents on the other two, react with H_2 molecules via the dinitrogen hydrogenation pathway. However, complex **V** of Manriquez and Bercaw, having $C_5Me_5^-$ ligands, loses its N_2 unit when exposed to dihydrogen. In other words, replacing only one Me group with hydrogen in each $C_5Me_5^-$ ligand dramatically alters the reactivity of the resulting complex! What causes this remarkable difference in reactivity?

The second question arises upon comparison of the reactivity of **II_Zr** and **II_Hf** complexes: $[(\eta^5\text{-}C_5Me_4H)_2Hf]_2(\mu_2,\eta^2,\eta^2\text{-}N_2)$, **II_Hf**, reacts faster with H_2 than does **II_Zr** [33, 35]. Why does complex **II_Zr** react more slowly than **II_Hf**? Will their Ti-analog show similar reactivity with H_2?

4.2 Reaction Mechanism of the Coordinated Dinitrogen Molecule in Di-zirconocene-N_2 Complexes

The third intriguing question arises upon comparison of reactivity of Fryzuk- and Chirik-types of di-Zr dinitrogen complexes. Fryzuk complex $\{[P_2N_2]Zr\}_2(\mu_2, \eta^2,\eta^2\text{-}N_2)$, **I**, adds [37] only one H_2 molecule to the coordinated N_2, while the Chirik complex **II_Zr** reacts with more than one H_2 molecule and even forms a small amount of ammonia [33]. What is the reason for this difference in the reactivity of these two classes of di-Zr-N_2 complexes?

Below, we intend to answer these questions using computational techniques (details are given in the Appendix) and extend of our understanding of the fundamental principles and concepts of direct dinitrogen hydrogenation.

4.2 Reaction Mechanism of the Coordinated Dinitrogen Molecule in Di-zirconocene-N_2 Complexes with a Hydrogen Molecule

To answer the first intriguing question related to the role of the number of Me substituents in cyclopentadienyl ligands of di-zirconocene-N_2 complex on the reactivity with dihydrogen, in this section we briefly report the computational findings on the mechanism of the reactions:

$$[(\eta^5\text{-}C_5H_5)_2Zr]_2(\mu_2,\eta^2,\eta^2\text{-}N_2) + mH_2, \quad m = 1\text{-}3 \qquad (3)$$

$$[(\eta^5\text{-}C_5Me_nH_{5-n})_2Zr]_2(\mu_2,\eta^2,\eta^2\text{-}N_2) + H_2, \quad n = 4, 5 \qquad (4)$$

and

$$[(\eta^5\text{-}C_5H_2\text{-}1,2,4\text{-}Me_3)(\eta^5\text{-}C_5Me_5)Zr]_2(\mu_2,\eta^2,\eta^2\text{-}N_2) + H_2 \qquad (5)$$

Complexes $[(\eta^5\text{-}C_5Me_4H)_2Zr]_2(\mu_2,\eta^2,\eta^2\text{-}N_2)$, **II**, and $[(\eta^5\text{-}C_5H_2\text{-}1,2,4\text{-}Me_3)(\eta^5\text{-}C_5Me_5)Zr]_2(\mu_2,\eta^2,\eta^2\text{-}N_2)$, **IV**, have been reported experimentally [33, 34], whereas the unmethylated $[(\eta^5\text{-}C_5H_5)_2Zr]_2(\mu_2,\eta^2,\eta^2\text{-}N_2)$, **VI**, and pentamethylated $[(\eta^5\text{-}C_5Me_5)_2Zr]_2(\mu_2,\eta^2,\eta^2\text{-}N_2)$, η^2-**V**, complexes have not.

4.2.1 Mechanism of the Reaction (3)

Figure 4.3 shows the calculated potential energy profile of Reaction (3). All energy values are in kcal mol^{-1} and include the zero-point-energy correction. This figure also includes the most important equilibrium structures and transition states of the reaction. For clarity only the core atoms are given. The transition states and intermediates are named in accordance with previous theoretical studies [38–40]: The letters **A**, **B**, and **C** correspond to the first, second, and third H_2 addition, respectively. Numbers given after the letters (e.g., as **A1**, **A2**, etc.) indicate the calculated intermediates. The name of each transition state contains the two equilibrium structures it connects.

The addition of the first H_2 molecule to **VI** (called here as **VI_A1**) results in diazenido complex, **VI_A3**. This occurs via a "metathesis-like" transition state, **VI_(A1-A3)**, where one of the $\pi(N-N)$ bonds and the H–H bond are breaking, and Zr–H and one N–H bonds are forming. A similar transition state was reported by Basch et al. [38] for the addition of H_2 to the model complex **I**. The calculated barrier of the first H_2 addition to **VI_A1** is 21.4 kcal mol^{-1} [42]. The overall reaction **VI_A1** + $H_2 \rightarrow$ **VI_A3** is exothermic by 6.9 kcal mol^{-1}. The resultant complex [$(\eta^5$-$C_5H_5)_2$Zr][$(\eta^5$-$C_5H_5)_2$ZrH][μ_2,η^2,η^2-(NNH)] has three different isomers. Isomer **VI_A3** with the two H-atoms *syn* to each other is found to be a 2.0 kcal mol^{-1} lower in energy than the isomer **VI_A5**, where the H atoms point in opposite direction. Another isomer, **VI_A7**, with the one of the H atoms in a bridging position between two Zr centers is 8.4 kcal mol^{-1} higher in energy than **VI_A3**. Our calculations show that the barrier separating **VI_A3** from **VI_A5** is only 2.5 kcal mol^{-1}. The estimated barrier for the **VI_A5** $\leftrightarrow$ **VI_A7** rearrangement is 6.6 kcal mol^{-1}.

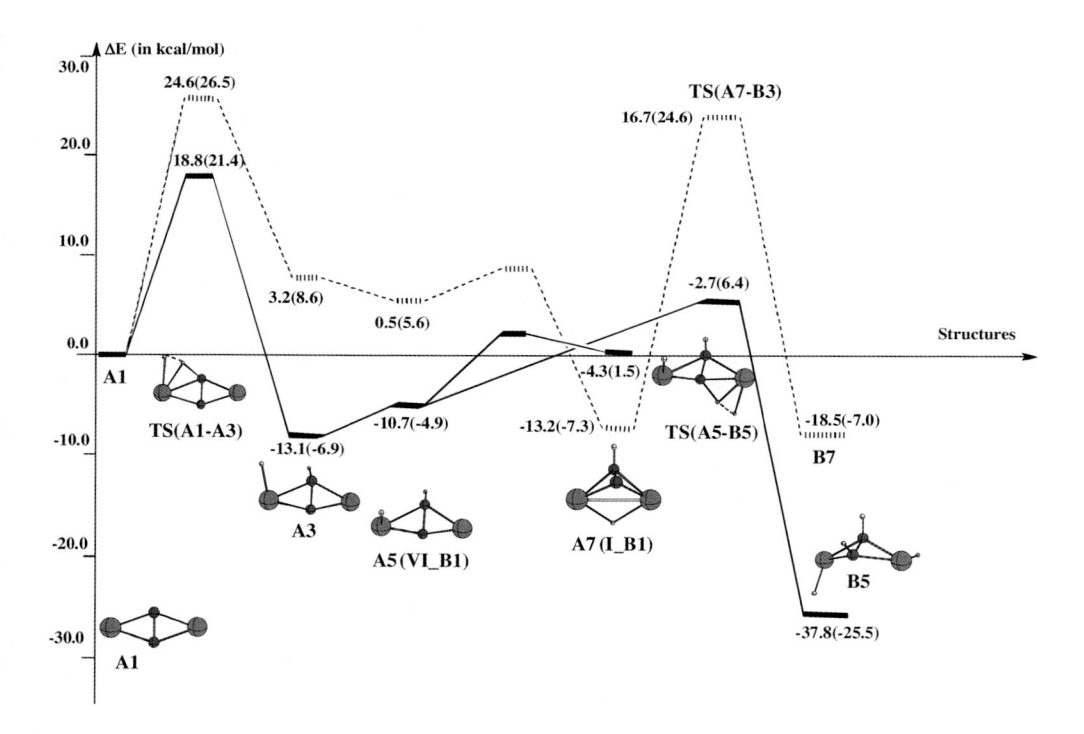

Figure 4.3 Comparison of the potential energy profiles of the first and second H_2 addition to the {[PhP(CH$_2$SiMe$_2$NSiMe$_2$CH$_2$)$_2$PPh]Zr}$_2$(μ_2,η^2,η^2-N_2) (**I**, dashed line) and unmethylated [$(\eta^5$-$C_5H_5)_2$Zr]$_2$(μ_2,η^2,η^2-N_2) (**IV**, solid line) complexes. Numbers given in parentheses include zero-point-energy correction, ZPC. See text for details.

Calculations show that the complexes $[(\eta^5\text{-}C_5H_5)_2Zr]_2[\mu_2,\eta^2,\eta^2\text{-}(NNH_2)]$ with N and NH_2 bridging ligands and $[(\eta^5\text{-}C_5H_5)_2Zr]_2[\mu_2,\eta^2,\eta^2\text{-}(NH)_2]$, with two bridging NH ligands lie energetically much lower on the potential energy surface of the reaction **VI_A1** + H_2. However, these thermodynamically most favorable structures are unlikely to be formed via intramolecular rearrangements of **VI_A3** and **VI_A5** because of the large, 42.8 and 36.4 kcal mol^{-1}, barriers, respectively.

Instead, the intermediate **VI_A3** (or its less stable **VI_A5** and **VI_A7** isomers) reacts with another, a second H_2 molecule, via a similar [to transition state **VI_(A1-A3)**] "metathesis-like" transition state to give the intermediate $[(\eta^5\text{-}C_5H_5)_2ZrH]_2[\mu_2,\eta^2,\eta^2\text{-}(NH)_2]$, which may also have three different isomers, **VI_B3**, **VI_B5** and **VI_B7**, among which structure **VI_B5** is energetically more stable and lies 25.5 kcal mol^{-1} lower than the reactants, **VI_A1** + 2H_2. Analysis of these isomers indicates that the energetically most stable isomer **VI_B5** could be formed either directly from **VI_A5** by addition of H_2 molecule, or by isomerization of **VI_B3**, the direct product of the reaction of **VI_A3** with a H_2 molecule. Since the **VI_A3** ↔ **VI_A5** isomerization (barrier of 2.5 kcal mol^{-1}) is an easier process than **VI_B5** ↔ **VI_B3** (barrier of 14 kcal mol^{-1}, reported by Miyachi and coworkers [51]), the pre-reaction complex for the second H_2 addition is expected to be isomer **VI_A5**. Our calculations show that the second H_2 addition to **VI_A5** occurs with a barrier of 11.3 kcal mol^{-1}, which is about two times lower than the first H_2 addition barrier (21.4 kcal mol^{-1}). The second H_2 addition is more exothermic than the first H_2 addition (20.6 versus 6.9 kcal mol^{-1}, respectively).

Miyachi et al. [51] have reported a similar potential energy surface for Reaction (3). However, they predicted the **VI_A3** + H_2 → **TS** (10.9 kcal mol^{-1}) → **VI_B3** → **TS** (13.8 kcal mol^{-1}) → **VI_B5** pathway for the second H_2 addition, leading to experimentally observed product **VI_B5**. Our calculations predict the **VI_A3** → **TS** (2.5 kcal mol^{-1}) → **VI_A5** + H_2 → **TS** (11.3 kcal mol^{-1}) → **VI_B5** pathway, which occurs with a barrier several kcal mol^{-1} lower.

We did not study in detail the third H_2 addition to **VI_A1**. However, our preliminary data indicate that the product of the second H_2 addition, **VI_B5**, adds the next (the third) H_2 molecule with a 26.2 kcal mol^{-1} barrier and leads to the **VI_C3** product. The reaction **VI_B5** + H_2 is calculated to be 11.7 kcal mol^{-1} endothermic.

4.2.2
Mechanisms of the Reactions (4) and (5)

As mentioned above, experimentally it was shown that the tetramethylated dizirconecene-N_2 complex $[(\eta^5\text{-}C_5Me_4H)_2Zr]_2(\mu_2,\eta^2,\eta^2\text{-}N_2)$, **II_Zr**, and the mixed ligand $[(\eta^5\text{-}C_5H_2\text{-}1,2,4\text{-}Me_3)(\eta^5\text{-}C_5Me_5)_2Zr]_2(\mu_2,\eta^2,\eta^2\text{-}N_2)$ complex, **IV**, hydrogenate the coordinated N_2-fragment at ambient conditions [33, 34].

As seen in Figure 4.4, the calculated important geometry parameters of complex **II_Zr** are in very good agreement with experimental values.

Previously, we also analyzed [44] different isomers of **II** and found the isomer of **II_Zr** given in Figure 4.4 to be the energetically most stable one, which is in

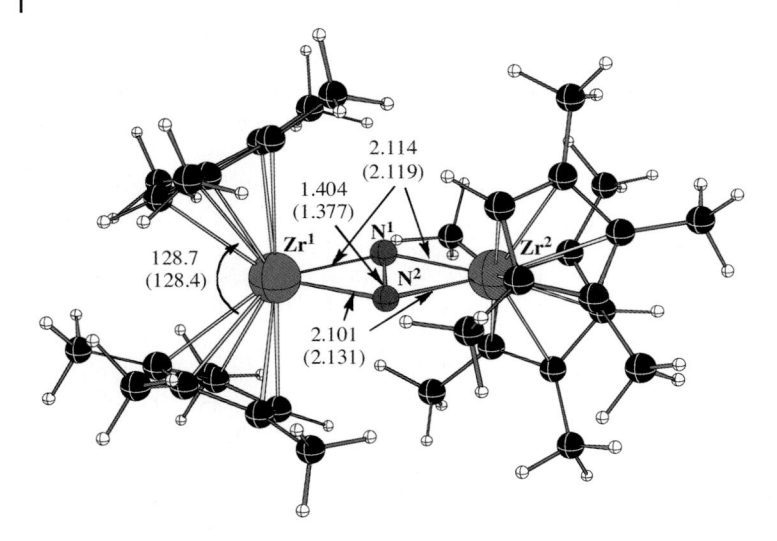

Figure 4.4 Calculated important geometry parameters (distances are in Å, angles are in degrees) of the energetically lowest isomer of tetramethylated [(η^5-C$_5$Me$_4$H)$_2$Zr]$_2$(μ_2,η^2,η^2-N$_2$), **II_Zr**, complex. Experimental values are given in parentheses.

good agreement with experimental observations. Therefore, we studied the mechanism of Reaction (4) only for this isomer of **II_Zr**.

As mentioned above, the pentamethylated [(η^5-C$_5$Me$_5$)$_2$Zr]$_2$(μ_2,η^2,η^2-N$_2$), **η^2-V**, complex with a side-on coordinated N$_2$ molecule is not experimentally reported. Despite that, here we study the potential energy surface of Reaction (4) for the hypothetical complex [(η^5-C$_5$Me$_5$)$_2$Zr]$_2$(μ_2,η^2,η^2-N$_2$) (**η^2-V**) to elucidate the role of Me substituents of Cp rings in the H$_2$ addition to the side-on coordinated N$_2$ molecule.

In our studies of the mechanisms of Reaction (4) for $n=4$ and 5, as well as of Reaction (5), we followed the pathway obtained for Reaction (3) (Section 4.2.1). Calculations show that the addition of H$_2$ molecule to complexes **II_Zr**, **η^2-V**, and **IV** proceeds with 19.6, 22.3 and 21.8 kcal mol^{-1} barriers, respectively. Comparison of these numbers with those for unmethylated complex **VI** (the reaction **VI + H$_2$** proceeds with a 21.4 kcal mol^{-1} barrier) reveals that complexes [(η^5-C$_5$Me$_n$H$_{5-n}$)$_2$Zr]$_2$(μ_2,η^2,η^2-N$_2$), ($n=0$, 4, 5) and [(η^5-C$_5$H$_2$-1,2,4-Me$_3$)(η^5-C$_5$Me$_5$)$_2$Zr]$_2$(μ_2,η^2,η^2-N$_2$) show almost the same reactivity toward H$_2$.

In the other words, the number of methyl groups does not significantly affect the barrier and reaction energy of the first H$_2$ addition to the side-on coordinated N$_2$ in all these complexes, including the hypothetical pentamethylated di-zirconocene-N$_2$ complex, [(η^5-C$_5$Me$_5$)$_2$Zr]$_2$(μ_2,η^2,η^2-N$_2$), **η^2-V**. Thus, *all* side-on N$_2$ complexes of di-zirconocene, in principal, should react with H$_2$ molecule via a dinitrogen hydrogenation pathway, and all of them should show similar activity.

However, the above conclusion is not consistent with the available experiments [50] showing that the pentamethylated di-zirconocene-N_2 complex [(η^5-C_5Me_5)$_2$Zr]$_2$(N_2) (V) does not hydrogenate the coordinated N_2 molecule. Conversely, for the di-Zr-N_2 complex [33] with tetramethylated cyclopentadienyl ligands, $C_5Me_4H^-$ reacts with several H_2 molecules. To answer the astonishing question of why the substitution of only one methyl group with hydrogen in each $C_5Me_5^-$ ligands dramatically alters the reactivity of the di-Zr-N_2 complex toward dihydrogen molecule, we have studied [43] the role of the Me substituents of the Cp rings in the structure and relative stability of the end-on η^1 and side-on η^2 coordination modes of N_2 molecule in [(η^5-$C_5Me_nH_{5-n}$)$_2$Zr]$_2$(N_2), ($n = 0$–5) complexes. Since the η^1 and η^2 isomers of transition metal-N_2 complexes can rearrange to each other, in the next section we also analyze the $\eta^1 \leftrightarrow \eta^2$ isomerization in various model and experimentally reported di-zirconocene-N_2 complexes with different numbers of methyl substituents on each Cp ring.

4.3
Factors Controlling the N_2 Coordination Modes in the Di-zirconocene-N_2 Complexes

We have divided our studies into two parts. First, we examined the role of the Me substituents of the Cp (cyclopentadienyl) rings in the coordination mode of N_2, in the mononuclear [($C_5H_{5-n}Me_n$)$_2$Zr](N_2) complexes for $n = 0$–5. Figure 4.5 shows the relative energies of the optimized η^1 and η^2 structures.

As could be seen from Figure 4.5, for all ns the η^2 isomer is thermodynamically more favorable than η^1 isomer. The calculated energy difference between these isomers slightly increases upon increasing the number of methyl groups, n, in Cp rings. Thus, in fact, the methyl substituent on the cyclopentadienyl rings favors the *side-on η^2 coordination* of N_2 molecule over its end-on η^1 coordination.

This trend can be understood by analyzing the NPAs (natural population atomic charges), N–N bond distances and N–N vibrational frequencies given in Table 4.1. The data show that the obtained trend in relative stability of side-on and end-on coordination of N_2 is due to electron donation from the Me groups.

For side-on coordinated complexes, the increasing number of Me groups increases the negative charge on the N atoms from -0.28e to -0.38e. For complexes with the η^1-N_2 ligand, this trend is also valid, but the charges are smaller in absolute value. The stronger back-donation in side-on complexes results in a higher degree population of the unoccupied π^* orbital of N_2. As a result, the negative charge of N atoms increases, and the N–N bond distance in the side-on complexes (1.20–1.21 Å) elongates. In the end-on complexes the calculated N–N bond distances are 1.16 Å. The calculated harmonic vibrational frequencies are also consistent with the above-presented trends in the calculated charges and bond distances.

In the second part of our investigations we examined the side-on and end-on coordination modes of N_2 in [($C_5H_{5-n}Me_n$)$_2$Zr]$_2$(N_2) dinuclear complexes for

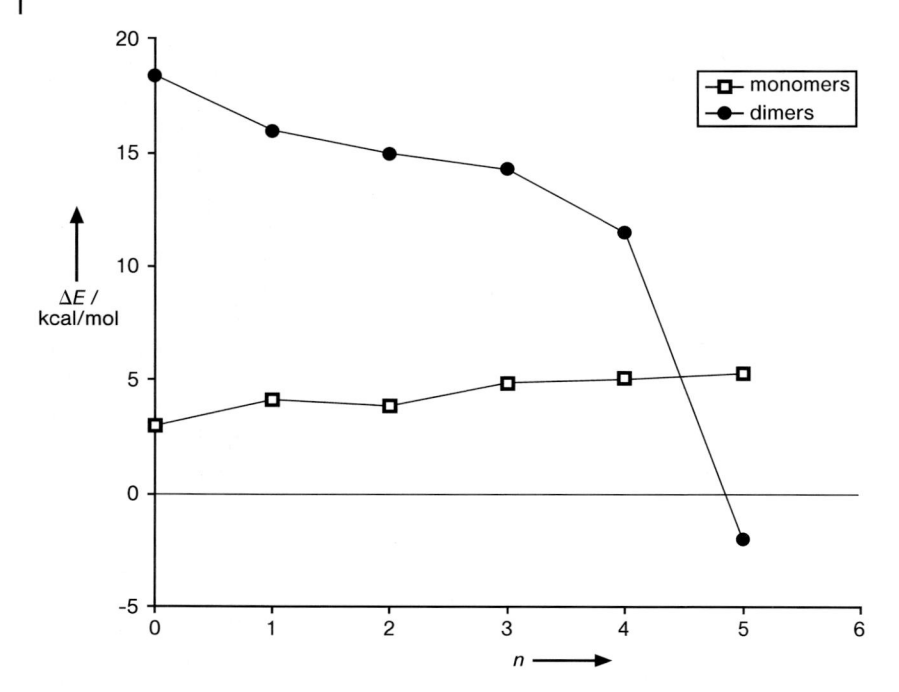

Figure 4.5 Energy difference (including ZPC), $\Delta E = E_{\text{end-on}} - E_{\text{side-on}}$, between the end-on and the side-on coordinated $(C_5H_{5-n}Me_n)_2Zr(N_2)$ (□) and $\{(C_5H_{5-n}Me_n)_2Zr\}_2(N_2)$ (•) complexes as a function of the number of methyl groups, n, in Cp rings.

Table 4.1 Nitrogen natural population atomic charges (NPAs) (Q_N, e), N–N bond lengths (r_{NN}, Å), and N–N stretching frequencies (ν_{NN}, cm^{-1}) in the side-on and end-on coordinated mononuclear $(C_5H_{5-n}Me_n)_2Zr(N_2)$ complexes, where $n = 0$–5.[a]

n	Side-on			End-on		
	Q_N	r_{NN}	ν_{NN}	$Q_N^{(\text{near})}/Q_N^{(\text{remote})}$	r_{NN}	ν_{NN}
0	−0.28	1.20	1805	−0.22/−0.03	1.16	2119
1	−0.29	1.20	1788	−0.22/−0.04	1.16	2108
2	−0.31	1.21	1781	−0.23/−0.05	1.16	2099
3	−0.33	1.21	1770	−0.25/−0.05	1.16	2084
4	−0.35	1.21	1763	−0.27/−0.07	1.16	2071
5	−0.38	1.21	1746	−0.29/−0.10	1.16	2054

a) Calculated Q_N (e), r_{NN} (Å), and ν_{NN} (cm^{-1}) are: 0.0, 1.10, and 2769 for N_2, −0.29, 1.23 and 1912 for N_2H_2, and −0.68, 1.46, and 1122, for N_2H_4, respectively.

$n = 0$–5. The relative energies of these two coordination modes are also shown in Figure 4.5.

As seen from these data, in contrast to the mononuclear complexes, in the dinuclear complexes the energy difference between the η^1 and η^2 coordinations of N_2 decreases with increasing of number of Me substituents, n, in Cp rings. For complexes with $n \leq 4$, the η^2 coordination of N_2, required for dinitrogen hydrogenation, is thermodynamically more favorable than its η^1 coordination. However, in the pentamethylated $\{(C_5Me_5)_2Zr\}_2(N_2)$ complex, the side-on η^2 coordination of N_2 is by 2.04 kcal mol^{-1} less stable than its end-on η^1 coordination.

Why does adding one more Me group to each Cp rings of $\{(C_5Me_4H)_2Zr\}_2(N_2)$ alter the stability of the η^2 and η^1 isomers in energy scale? In fact, data presented above for the mononuclear complexes $\{(C_5Me_4H)_2Zr\}(N_2)$ showed that increasing numbers of Me substituents in each Cp rings favors the side-on η^2 isomers. To answer this question we analyzed the roles of electronic and steric effects of Me substituents on the stability of η^2 and η^1 isomers of dinuclear $\{(C_5H_{5-n}Me_n)_2Zr\}_2(N_2)$ complexes.

As seen from Table 4.2, for such dinuclear complexes, increasing numbers of Me substituents in Cp rings increases (similar to the mononuclear complexes) the negative charges of N-atoms from $-0.72e$ to $-0.84e$ for side-on isomers and from $-0.53e$ to $-0.69e$ for end-on isomers. It only slightly affects the calculated N–N bond distances, which are longer in side-on (1.40–1.42 Å) than end-on (1.24–1.26 Å) isomers. The analysis showed that the reason for the opposite trend in the calculated energy difference between the η^2 and η^1 isomers, $\Delta E(\eta^2 - \eta^1)$, of mono- and di-nuclear complexes is a steric repulsion between the Me substituents of two monomers of dinuclear species. For $n=0$ the steric repulsion is minimal and $\Delta E(\eta^2 - \eta^1)$ has a largest value (Figure 4.5). The more methyl groups we add to the system, the stronger steric repulsion between the

Table 4.2 Nitrogen natural population atomic charges (NPAs) (Q_N), N–N bond lengths (r_{NN}), and N–N stretching frequencies (v_{NN}) of side-on and end-on coordinated dinuclear $\{(C_5H_{5-n}Me_n)_2Zr\}_2(N_2)$ complexes, where $n = 0$–5.

n	Side-on			End-on		
	Q_N (e)	r_{NN} (Å)	v_{NN} (cm^{-1})	Q_N (e)	r_{NN} (Å)	v_{NN} (cm^{-1})
0	−0.72	1.42	910	−0.53	1.25	1450
1	−0.74	1.41	926	−0.54	1.25	1447
2	−0.78	1.41	926	−0.52	1.24	1474
3	−0.82	1.42	905	−0.60	1.25	1410
4	−0.82	1.40	944	−0.66	1.26	1383
5	−0.84	1.41	948	−0.69	1.26	1370

Me substituents, which results in destabilization of η^2 isomer. The case of $n=4$ represents a boundary, and at $n=5$ the steric repulsion is so strong that it alters stability of the η^2 and η^1 isomers in energy scale.

In general, the η^1 and η^2 isomers of $\{(C_5H_{5-n}Me_n)_2Zr\}_2(N_2)$ can rearrange to each other, which may follow two distinct mechanisms: dissociation-and-addition and intermolecular rearrangement pathways. The former pathway for $\eta^2 \leftrightarrow \eta^1$ isomerization is expected to be much easier. Therefore, here we have investigated kinetic stability of the η^2 and η^1 isomers. Table 4.3 presents the calculated relative energies of η^2 and η^1 isomers, as well as barriers separating them, together with the barriers for H_2 addition to the η^2 isomers. As seen from this table, for all investigated complexes, except that with C_5Me_5 ligands, the $\eta^2 \leftrightarrow \eta^1$ isomerization barrier is higher than the H_2 addition barrier. In other words, under ambient experimental conditions, H_2 addition to the side-on coordinated N_2 of the complexes II, IV (with R = Me), and VI is a more favorable pathway than $\eta^2 \leftrightarrow \eta^1$ isomerization. Conversely, in the case of complex V with the C_5Me_5 ligands the barrier for $\eta^2 \leftrightarrow \eta^1$ isomerization, 12.4 kcal mol^{-1}, is smaller than that, 22.3 kcal mol^{-1}, for H_2 addition to the side-on coordinated N_2. Therefore, even if the side-on coordinated N_2 isomer of $[(\eta^5\text{-}C_5Me_5)_2Zr]_2(N_2)$, V, is available it will converge to its end-on isomer (which is un-suitable for hydrogenation of N_2) with a much smaller barrier than will adding H_2 to N_2.

The combination of these findings with the reported energetics (Sections 4.2.1 and 4.2.2) of Reactions (3–5) allows us to concluded that the tetramethylated $\{(C_5Me_4H)_2Zr\}_2(N_2)$, II, dinuclear complex reacts with H_2 via hydrogenation of N_2 while its pentamethylated $\{(C_5Me_5)_2Zr\}_2(N_2)$, V, analog does not because of the coordination mode of the N_2 molecule: In $\{(C_5Me_4H)_2Zr\}_2(N_2)$ the side-on η^2 coordination of N_2, required for hydrogenation of coordinated N_2, is energeti-

Table 4.3 Calculated energetics of (a) the reaction of di-zirconocene-N_2 complexes with the first H_2 molecule and (b) side-on ↔ end-on isomerization in these complexes for different numbers of methyl substituents in Cp rings. The presented energetics include zero-point energy correction.

Ligand, complex	H_2 addition (kcal mol^{-1})		(Side-on)	Side-on ↔ end-on (kcal mol^{-1})	
	A3	TS		TS (zig-zag)	End-on
C_5H_5, VI	−6.9	21.4	0	30.8	18.3
C_5HMe_4, II	−6.2	19.6	0	22.9	11.5
C_5H_2-1,2,4-Me_3, C_5Me_5, IV	−7.4	21.8	0	24.3	11.2
C_5Me_5, V	−12.3	22.3	0	12.4	−2.0

cally accessible and therefore, it reacts with H_2 via the dinitrogen hydrogenation pathway. In contrast, in the pentamethylated complex $[(\eta^5\text{-}C_5Me_5)_2Zr]_2(N_2)$, the side-on η^2 coordination of N_2 is *not* accessible both energetically and kinetically and, therefore, it does not react with a H_2 molecule via a hydrogenation mechanism.

This conclusion is supported by experimental [34] and computational (Section 4.2) studies of the structure and reactivity of the mixed-ligand complexes $[(\eta^5\text{-}C_5H_2\text{-}1,2,4\text{-}Me_3)(\eta^5\text{-}C_5Me_5)Zr]_2(\mu_2,\eta^2,\eta^2\text{-}N_2)$, **IV** (with R = Me), and $[(\eta^5\text{-}C_5Me_4H)(\eta^5\text{-}C_5Me_5)Zr]_2(\mu_2,\eta^1,\eta^1\text{-}N_2)$, **VII** [52]. Experimental studies indicate that in **IV** (R = Me) the N_2 molecule is in its side-on coordination mode, and reacts with H_2 via a hydrogenation mechanism. The calculations show that the side-on η^2 coordination of N_2 in **IV** (R = Me) is 11.2 kcal mol^{-1} more favorable than its end-on η^1 analog.

For complex **VII**, experiment shows the end-on η^1 bound N_2 structure, which loosens upon addition of H_2. This again is consistent with our conclusion.

Thus, the above presented data clearly show that the number of Me substituents in $[(\eta^5\text{-}C_5Me_nH_{5-n})_2Zr]_2(N_2)$ complexes does not substantially influence the energetics of their reaction with H_2, as long as the complex has a side-on η^2 coordinated N_2 molecule. However, the number of Me substituents is critical for accessibility of the side-on coordination mode of N_2 in these systems, required for hydrogenation of dinitrogen. The increase of steric repulsion between the methyl groups with increasing number of Me substituents, n, in Cp rings destabilizes the required side-on coordination of N_2 and makes it un-accessible in the case of $n = 5$. As a result, the reaction of $[(\eta^5\text{-}C_5Me_5)_2Zr]_2(N_2)$, **V**, with H_2 does not follow a dinitrogen hydrogenation pathway.

4.4
Why the $[(\eta^5\text{-}C_5Me_nH_{5-n})_2Ti]_2(\mu_2,\eta^2,\eta^2\text{-}N_2)$ Complex Cannot Add a H_2 Molecule to the Side-on Coordinated N_2, while its Zr- and Hf-analogs Can

In the previous sections we elucidated the mechanism of the reaction of $[(\eta^5\text{-}C_5Me_nH_{5-n})_2Zr]_2(\mu_2,\eta^2,\eta^2\text{-}N_2)$ with the dihydrogen molecule, and clarified the role of the number of Me substituents of Cp rings on the structure and reactivity of these complexes. In this section, we answer the second intriguing question that relates to the reactivity of the di-metalocene-N_2 complexes of Ti, Zr and Hf. Experimental studies [33, 35] have shown that both zirconocene, $[(\eta^5\text{-}C_5Me_4H)_2Zr]_2(\mu_2,\eta^2,\eta^2\text{-}N_2)$, **II_Zr**, and hafnocene, $[(\eta^5\text{-}C_5Me_4H)_2Hf]_2(\mu_2,\eta^2,\eta^2\text{-}N_2)$, **II_Hf**, complexes reacts with H_2 molecule via a hydrogenation mechanism. Furthermore, **II_Hf** complex reacts slightly (4.3×) faster than its Zr-analog. However, the reason for this difference in rates is still unclear. Also, it is not clear whether their Ti-analog will react with H_2 through similar pathways. To elucidate these issues we have studied in detail the addition of H_2 to the coordinated N_2 in $[(\eta^5\text{-}C_5Me_nH_{5-n})_2M]_2(\mu_2,\eta^2,\eta^2\text{-}N_2)$ complexes for M = Ti and Hf, and compared the obtained results with those for their Zr-analogs presented in Section 4.2. In our studies we followed the same pathway as that discussed for di-zirconocene-N_2 complexes (Section 4.2). Since the first-row transition metals tend to form complexes with high-spin states, here we report the reaction mechanisms of both singlet and triplet state reactants $[(\eta^5\text{-}C_5Me_nH_{5-n})_2M]_2(\mu_2,\eta^2,\eta^2\text{-}N_2)$ with H_2 molecule for $n = 0$ and 4.

Notably, the reaction of Ti complex $[(\eta^5\text{-}C_5Me_4H)_2Zr]_2(\mu_2,\eta^2,\eta^2\text{-}N_2)$ (or its derivatives) with H_2 molecule has not been reported. Several dinitrogen titanium sandwich complexes have been synthesized [53–55], but none of them were successful for the hydrogenation of N_2. Recently, a matrix isolation study showed that the "naked" Ti_2 dimer cleaves the N–N triple bond of N_2 and forms $(TiN)_2$ species without a significant energy barrier [56].

In our previous calculations, as well as in these studies, we used the DFT method with a double-ζ + polarization quality basis sets (see Appendix for details). However, hybrid density functionals such as B3LYP [57, 58] used in our studies overstabilize, while non-hybrid functionals (like BLYP [58, 59], PBE [60, 61], etc.) understabilize the high-spin states relative to the low-spin states [62–66]. Unfortunately, application of the best approaches for calculating the energy difference between the lowest electronic states, such as CCSD(T) and MRSD-CI with a very large basis sets, to transition metal complexes such as $[(\eta^5\text{-}C_5Me_nH_{5-n})_2M]_2(\mu_2,\eta^2,\eta^2\text{-}N_2)$ is not practical, yet. Therefore, to determine the energy gap between the lowest singlet and triplet states, $\Delta E(\text{S-T})$, of the reactants we utilized both hybrid (B3LYP) and non-hybrid (PBE) density functionals. We expected that comparison of the B3LYP- and PBE-calculated values of $\Delta E(\text{S-T})$ will enable us to qualitatively determine the ground electronic states of these species.

4.4.1
Relative Stability of the Lowest Singlet (S) and Triplet (T) Electronic States of the Complexes $[(\eta^5\text{-}C_5Me_nH_{5-n})_2M]_2(\mu_2,\eta^2,\eta^2\text{-}N_2)$, II_M (for M = Ti, Zr, and Hf, and $n = 0$ and 4)

In Table 4.4 we present the relative energies of the singlet and the triplet electronic states, $\Delta E(\text{S-T})$, of the complex **II_M** for M = Ti, Zr and Hf and $n = 0$ and 4. The PBE results are given in parentheses. Table 4.4 also presents the important geometry parameters of all structures, the calculated harmonic frequency of the N–N bond, and the nitrogen atomic charge computed using the NPA scheme [67].

As seen from Table 4.4, the ground electronic state of **II_Ti** complex is triplet state for both $n = 0$ and 4. At the B3LYP level, the singlet states are 23.5 and 20.8 kcal mol^{-1} higher in energy for $n = 0$ and 4, respectively. At the PBE level, the $\Delta E(\text{S-T})$ for the complex with $n = 0$ is found to be only 3.1 kcal mol^{-1}. By taking into account the fact that B3LYP overstabilizes and PBE understabilizes high-spin states, one can confidently conclude that the ground electronic states of **II_Ti** complex are triplets for both $n = 0$ and 4, with the singlet state lying several kcal mol^{-1} higher in energy.

4.4 Why the $[(\eta^5\text{-}C_5Me_nH_{5-n})_2Ti]_2(\mu_2,\eta^2,\eta^2\text{-}N_2)$ Complex Cannot Add a H$_2$ Molecule

Table 4.4 B3LYP/CEP-31G(+d$_N$) relative energies, geometrical and vibrational parameters, and nitrogen natural population atomic charges (NPAs) of the lowest singlet (S) and triplet (T) electronic states of $[(\eta^5\text{-}C_5Me_nH_{5-n})_2M]_2(\mu_2,\eta^2,\eta^2\text{-}N_2)$, II_M, complexes for M = Ti, Zr and Hf, and $n = 0$ and 4. The zero-point energy correction is taken into account. PBE results are given in parentheses.

Complex	M = Ti		M = Zr		M = Hf	
State	S	T	S	T	S	T
	$n = 0$					
ΔE (kcal mol^{-1})	0.0 (0.0)	−23.5 (−3.2)	0.0 (0.0)	0.8 (6.0)	0.0 (0.0)	10.5 (11.6)
r_{NN} (Å)	1.33	1.24	1.42	1.26	1.48	1.30
r_{MN} (Å)	1.98	2.14	2.09	2.25	2.04	2.17
ν_{NN} (cm^{-1})	1177	1621	910	1509	835	1291
Q_N (e)	−0.49	−0.34	−0.72	−0.43	−0.69	−0.43
$E_{\text{end-on}} - E_{\text{side-on}}$ (kcal mol^{-1})	−5.3	12.3	18.3	17.1	28.8	23.6
	$n = 4$					
ΔE (kcal mol^{-1})	0.0 (0.0)	−20.8 (−3.0)	0.0 (0.0)	−0.7 (4.2)	0.0 (0.0)	7.8 (9.7)
r_{NN} (Å)	1.33	1.24	1.40	1.27	1.46	1.30
r_{MN} (Å)	2.00	2.17	2.11	2.24	2.06	2.18
ν_{NN} (cm^{-1})	1180	1594	944	1422	818	1285
Q_N (e)	−0.52	−0.36	−0.82	−0.54	−0.85	−0.55

The picture is significantly different for analogous Zr- and Hf complexes. Unmethylated and tetramethylated Hf complexes have ground singlet states at both B3LYP and PBE levels of theory. In contrast, the singlet and triplet states of unmethylated and tetramethylated Zr complexes are almost degenerate.

Thus, titanocene-N$_2$ complexes **II_Ti** have triplet ground electronic states, while their Zr and Hf-analogs are diamagnetic species with singlet ground states. This is consistent with previous CASSCF studies of M$_2$N$_2$ complexes for different Ms [68, 69], concluding that the lower-spin state is the ground state for the second-row transition metals to the left in the periodic table (yttrium and zirconium), while the high-spin state is the ground state for all first-row transition metals.

98 | *4 Principles of Dinitrogen Hydrogenation: Computational Insights*

Also, data presented in Table 4.4 show that in the singlet states of these complexes the N_2 molecule is activated to the highest extent in **II_Hf** and to lowest extent in **II_Ti** complexes. The **II_Zr** complexes show an intermediate activation for bridging a N_2 molecule.

Figure 4.6 shows important frontier orbitals of the lower-lying singlet and triplet states of **II_M** complexes for M = Ti, Zr and Hf, and $n = 0$. As seen from this figure, at their singlet states the HOMO of these complexes is a bonding π-orbital of M–N_2–M fragment, which is suitable for interaction with the σ_u^* orbital of an incoming H_2 molecule [34, 41]. Their LUMO is a non-bonding σ-orbital of metals, which is appropriate for interaction with the σ_g-bonding orbital of the H_2 molecule [34, 41]. Thus, the frontier orbitals of all these complexes in their singlet electronic states are suitable for addition of a H_2 molecule via "metathesis" transition states.

As seen from Figure 4.6, the S → T excitation in these complexes corresponds to the promotion of an electron from the M–N_2–M bonding HOMO to the non-bonding LUMO. For the **II_Hf** complex, the resultant SOMO_2 and SOMO_1 of

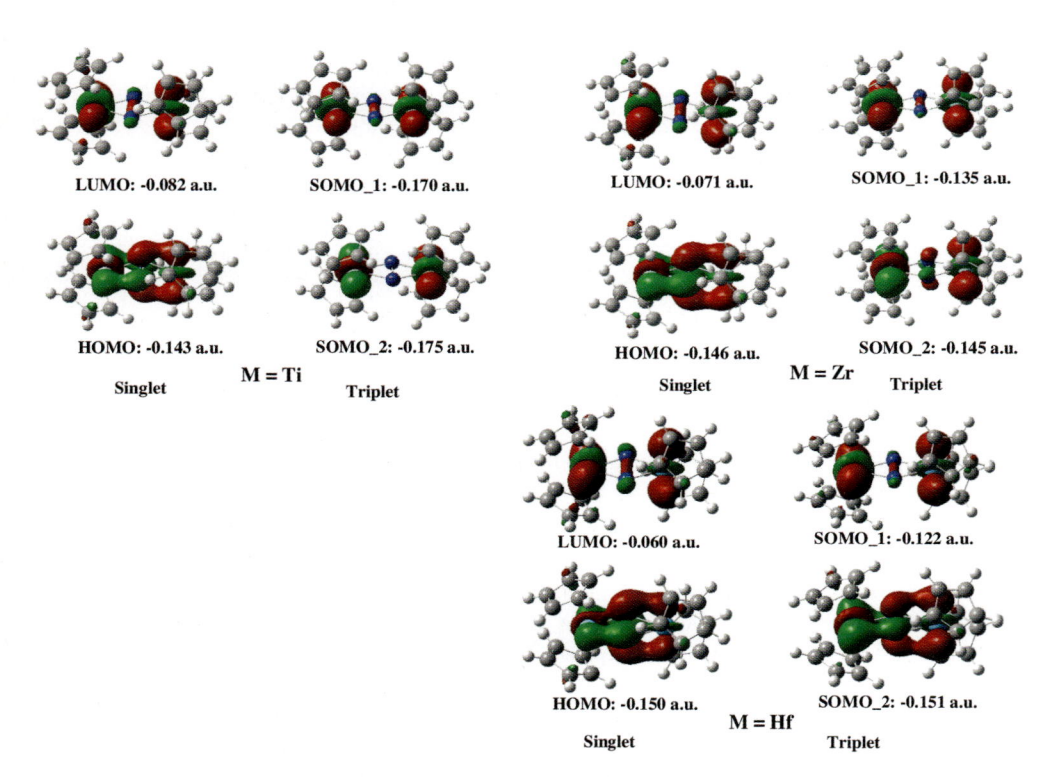

Figure 4.6 Schematic presentation of frontier orbitals of $[(\eta^5\text{-}C_5H_5)_2M]_2(\mu_2,\eta^2,\eta^2\text{-}N_2)$, **II_M**, for M = Ti, Zr and Hf in their lowest singlet (S) and triplet (T) electronic states.

the triplet state most closely resemble the HOMO and LUMO of the singlet state. However, for the triplet **II_Zr**, SOMO_2 has much less M–N_2–M bonding character than that for **II_Hf**. In the case of **II_Ti**, SOMO_2 has no M–N_2–M bonding character and is an antisymmetric combination of Ti d-orbitals. Thus, SOMO_2 orbitals of triplet state complexes posses the strongest M–N_2–M interaction in **II_Hf** and the lowest (or none) in **II_Ti**.

Furthermore, as seen in Table 4.4, the S→T excitation in **II_M** reduces the M–N_2 interaction, but increases the N–N bonding. The calculated M–N_2 and N–N bond distances are longer (by 0.12–0.17 Å) and shorter (by 0.09–0.18 Å) for triplet states than the corresponding singlet states, respectively.

4.4.2
Reactivity of the Lowest Singlet and Triplet States of the Complexes [(η^5-$C_5Me_nH_{5-n}$)$_2$M]$_2$(μ_2,η^2,η^2-N_2), II_M, (for M = Ti, Zr, and Hf, and n = 0 and 4) towards H_2 Molecules

First, we calculated the energy difference between the side-on η^2 and end-on η^1 isomers of these complexes at their lowest singlet and triplet electronic states. As discussed in Section 4.4.1, the side-on η^2 isomer is required for the successful hydrogenation of the coordinated N_2. As seen from Table 4.4, for M = Zr and Hf, the side-on coordination of N_2 is more favorable for the both singlet and triplet states of **II_M** complexes. At triplet state of the Ti complex, the side-on η^2 coordination of N_2 is 12.3 kcal mol^{-1} more favorable. However, at the singlet state of Ti complexes, the end-on η^1 coordination of N_2 becomes more favorable. In other words, the singlet state Ti complexes should not react with a H_2 molecule ecule. Despite this, below we investigate the reactivity of the side-on η^2 isomers of all **II_M** complexes at their both lowest singlet and triplet electronic states.

Figure 4.7 presents the calculated potential energy surfaces of the reaction of singlet and triplet electronic states of **II_M** complexes with side-on coordinated N_2 toward H_2, for M = Ti, Zr and Hf.

Since the singlet states all of these complexes have appropriate frontier orbitals (HOMO and LUMO), one may expect that these singlet state complexes add a H_2 molecule to the coordinated N_2 with a reasonable energy barrier. In fact, as seen in Figure 4.7, in their singlet electronic states, the reaction of **II_M** with H_2 proceeds with 26.7, 21.4 and 20.3 kcal mol^{-1} for n=0, and 18.7, 23.9 and 19.0 kcal mol^{-1} for n=4 barriers (relative to the singlet state reactants) for M = Ti, Zr, and Hf, respectively. Overall, the singlet state reaction **II_M** + H_2 is exothermic by 6.4, 6.9, and 7.8 kcal mol^{-1} for n=0, and 11.5, 6.2, and 9.3 kcal mol^{-1} for n=4, for M = Ti, Zr, and Hf, respectively. Thus, the reactivity of the singlet state complexes **II_M** with H_2 molecule increases in the order Ti < Zr ≤ Hf. This trend in reactivity of these singlet state complexes is consistent with that in the degree of activation of N–N bond in the reactant complexes, reported above. Furthermore, the above-mentioned trend in the reactivity of Zr- and Hf-complexes with H_2 correlates well with the available experimental results [33, 35].

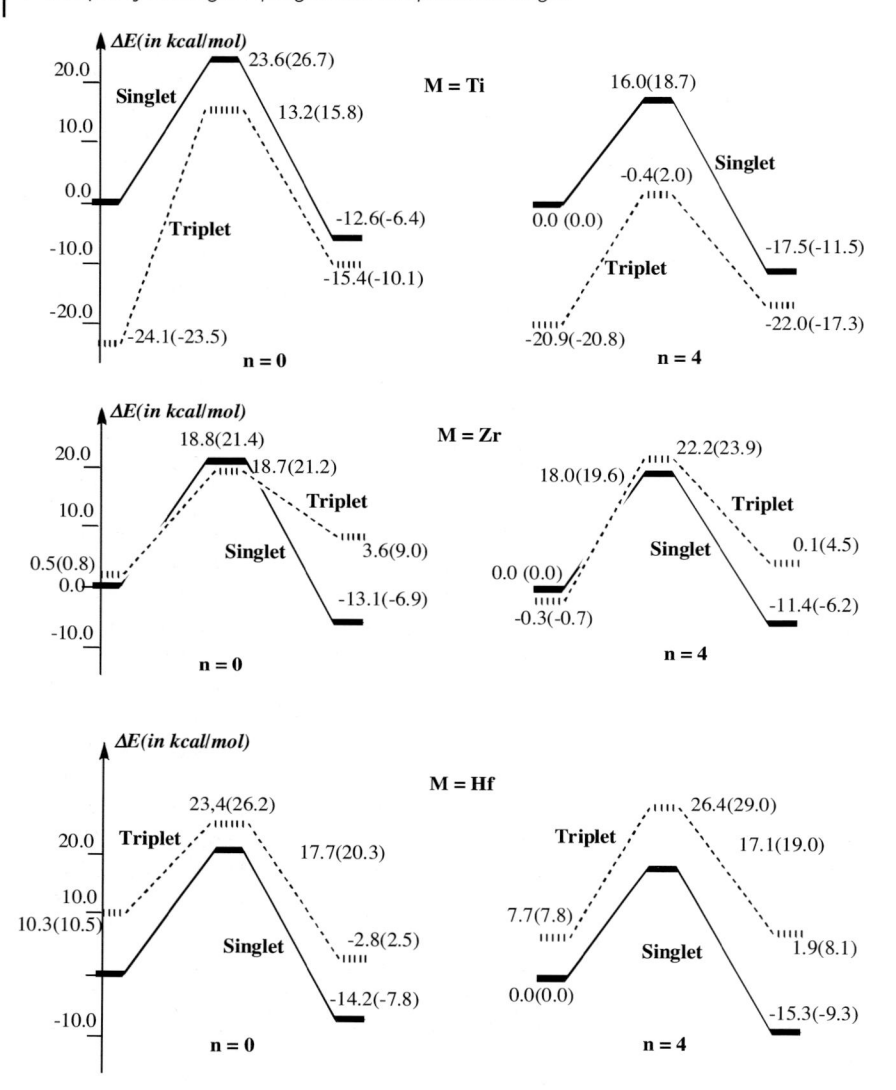

Figure 4.7 Schematic presentation (scaled to the zero-point corrected numbers) of the B3LYP-calculated relative energies of the reactants, transition states and products of the reaction of $[(\eta^5\text{-}C_5Me_nH_{5-n})_2M]_2(\mu_2, \eta^2,\eta^2\text{-}N_2)$ (for M = Ti, Zr and Hf, and $n = 0$ and 4) with a H_2 molecule at their lowest singlet (S) and triplet (T) electronic states.

However, as discussed above, for **II_Ti** complexes the singlet state *is not* the ground electronic state; it lies higher in energy than its ground triplet state. Therefore, we also studied the potential energy surface of the reaction of **II_M** complexes with H_2 molecule at their lowest triplet states. As seen in Figure 4.7, the calculated energy barriers for reaction **II_Ti** + H_2 (relative to the triplet state reactants) are very large, 39.3 kcal mol^{-1}, and 22.8 kcal mol^{-1}, for $n=0$ and 4, respectively. These values are larger than those for the analogous singlet-state complexes **II_M**, especially for the unmethylated cases. Furthermore, the first H$_2$ addition to the triplet **II_Ti** is endothermic by 13.4 and 3.5 kcal mol^{-1} for $n=0$ and 4, respectively. Based on these calculated barriers and reaction energies we conclude that at their ground triplet electronic state titanocene-N$_2$ complexes **II_Ti** will *not* react with H$_2$ via a dinitrogen hydrogenation mechanism under mild conditions. In other words, Ti complexes **II_Ti** are not expected to successfully hydrogenate dinitrogen, because they have triplet ground states and, consequently, a weak Ti–N$_2$ interaction and a strong N–N bond.

For the **II_Zr** complexes, the relative energies of the singlet and triplet state reactants, as the with transition states, are almost the same. However, the triplet state reaction is endothermic by 8.2 ($n=0$) and 5.2 ($n=4$) kcal mol^{-1}, while the singlet state reactions are exothermic for both $n=0$ and 4. Therefore, we expect the first H$_2$ addition to the **II_Zr** complexes to take place from their singlet electronic states.

The reaction of triplet state Hf complex **II_Hf** with H$_2$ molecule is exothermic by 8.0 and 0.3 kcal mol^{-1} for $n=0$ and 4, respectively. It has a 15.7 and 21.2 kcal mol^{-1} barrier (calculated from the triplet state reactants) for $n=0$ and 4, respectively. Since the ground states of the Hf complex are clearly the singlet states, the first H$_2$ addition to Hf complexes **II_Hf** also is expected to take place entirely in the singlet states.

Thus, data presented above allow us to conclude that both Zr and Hf complexes of **II** prefer the side-on coordination of N$_2$ in their both singlet and triplet states, and posses HOMO and LUMO orbitals with appropriate symmetry only at their ground singlet states. Therefore, their reaction with a H$_2$ molecule proceeds via a reasonably energy barrier and is exothermic only at their ground singlet states. Furthermore, the reaction of Hf complexes proceeds via a slightly lower barrier and is slightly more exothermic than the reaction of the corresponding Zr complexes. However, their Ti analogs, **II_Ti**, are unlikely react with H$_2$ via the dinitrogen hydrogenation pathway. Indeed, **II_Ti** complexes (for $n=0$ and 4) are found to have ground triplet electronic states, which posses the side-on coordinated N$_2$ molecule, but lack the appropriate frontier orbitals required for successful reaction with H$_2$ via the dinitrogen hydrogenation pathway. Their excited singlet electronic states lie higher in energy than the triplet ground states, and they lack the favorable side-on coordinated N$_2$ molecule required for dinitrogen hydrogenation.

4.5
Why Dizirconium-dinitrogen Complexes with bis(Amidophosphine) (P2N2) and Cyclopentadienyl (Cp) Ligands React differently with the Hydrogen Molecule: Role of Ligand Environment of the Zr Centers

Above, we have demonstrated that all side-on coordinated di-Zr-dinitrogen complexes can react with more than one H$_2$ molecule at their singlet electronic states. Furthermore, experiments have demonstrated [33] that the [(η^5-$C_5Me_4H)_2Zr]_2(\mu_2,\eta^2,\eta^2$-$N_2$) (**II_Zr**, below we will call it **II**) complex reacts with several H_2 molecules and produces ammonia. However, experimental and theoretical [37–39] studies revealed that the complex {[$P_2N_2]Zr\}_2(\mu_2,\eta^2,\eta^2$-$N_2$), (**I**), with bis(amidophosphine) ligand, P2N2 = [PhP($CH_2SiMe_2NSiMe_2CH_2$)$_2$PPh], reacts only with one H_2 molecule under mild experimental conditions (Figure 4.2). Elucidating the reason(s) for the difference in reactivity of these two classes of di-Zr-N_2 complexes is very intriguing and can help in designing more efficient di-Zr-(μ_2,η^2,η^2-N_2) complexes for utilization of the N≡N triple bond.

Intriguingly, comparison of the calculated potential energy surfaces of the reactions of model **I** (with P2N2 = [HP($CH_2SiH_2NSiH_2CH_2$)$_2$PH], below called **Im**) and **II** with H_2 molecules reveal that these two complexes show similar reactivity toward the first H_2 molecule: they proceed via a metathesis transition state (the calculated barriers are 26.5 and 21.4 kcal mol^{-1}, for **Im** and **II**, respectively) and produce the N–H bond. (Figure 4.3). Overall the reaction is found to be exothermic by 7.3 and 6.9 kcal mol^{-1}, for complexes **Im** and **II**, respectively. However, closer examination of the structure of the products and their reactivity with the next H_2 molecules reveals significant differences for these two reactions.

Indeed, as mentioned in Section 4.2.1, the product of the first H_2 addition, [L_2Zr][L_2ZrH][μ_2,η^2,η^2-NNH], may have three – *syn* (**A3**), *anti* (**A5**), and H-bridged (**A7**), different isomers, among which the isomers **A3** and **A5** with terminal Zr–H bonds are expected to be more favorable for the second H_2 addition. However, the isomers **A3** and **A5** are found to be more favorable in energy than the H-bridged **A7** isomer *only* for di-zirconocene complex, **II**. For the complex **Im**, the **A3** and **A5** structures are 12.9 and 15.9 kcal mol^{-1} higher in energy than the H-bridged **A7** isomer, respectively. Therefore, in the case of complex **I**, the pre-reaction compound for the second H_2 addition is expected to be the H-bridged isomer **A7**. Conversely, in the case of complex **II**, the pre-reaction compounds for the second H_2 addition are the non-H-bridged structures **A3** or/and **A5**. The calculated barrier for H_2 addition to **Im_A7** is 31.9 kcal mol^{-1}, which is significantly larger than the 13.3 kcal mol^{-1} found for the reaction of **II_A3** with H_2. The existence of such a large barrier explains why, under ambient conditions, the second H_2 addition to **I** has not been observed experimentally.

The above reported difference in the products of the first H_2 addition to complexes **I** and **II** is consistent with the available experiments: direct products of the reactions **II** + H_2 and **I** + H_2 were reported [33, 37] to be the non-H-bridged and H-bridged **A7** structures, respectively.

Thus, these calculations indicate that the observed difference in reactivity of the complexes **I** and **II** with H_2 molecules is due to the relative stability of their non-H-bridged **A3/A5** and H-bridged **A7** isomers of [L_2Zr][L_2ZrH][μ_2,η^2,η^2-NNH], which is the product of the first H_2 addition. What causes such a distinct difference in stability of different isomers of [L_2Zr][L_2ZrH][μ_2,η^2,η^2-NNH]?

To answer this question, we performed the Morokuma–Kitaura energy decomposition analysis (EDA) [70, 71] for the complexes **Im_A3**, **Im_A7**, **II_A3** and **II_A7**. Table 4.5 shows the results of this analysis, where *DEF* is a deformation

Table 4.5 Decomposition of the stabilization energy into deformation and interaction energy, $\Delta E = DEF + INT$, for H-bridged (**A7**) and non-H-bridged (**A3**) complexes of the di-zirconium complexes **I** and **II**. Given energies are calculated with respect to the reactants, **I** and **II**, respectively.

	I (kcal mol^{-1})			II (kcal mol^{-1})		
	ΔE	DEF	INT	ΔE	DEF	INT
A3	+3.2	23.9	−20.7	−13.1	19.6	−32.7
A7	−13.2	33.2	−46.4	−4.3	62.3	−66.6

energy required for the geometrical deformation of the two interacting fragments upon complexation, INT is an interaction energy between the deformed fragments, and $\Delta E = DE + INT$ is a stabilization energy. As can be seen, for both systems the interaction energy, INT, is approximately twice as high (in absolute value) for the H-bridged **A7** isomer than for the non-H-bridged **A3** isomer: −46.4 versus −20.7 kcal mol^{-1} for system **I**, and −66.6 versus −32.7 kcal mol^{-1} for system **II**. Thus, the interaction energy favors the isomer **A7** for both systems. However, the deformation energy changes the picture. For isomer **A3**, in both systems, DEF has approximately equal values, 23.9 versus 19.6 kcal mol^{-1}. For isomer **A7**, though, the required energy for geometry deformation of coordinated fragments is almost two times higher for system **I** than for **II** (62.3 kcal mol^{-1} versus 33.2 kcal mol^{-1}). In other words, complex **II** requires approximately two times more deformation energy for formation of the H-bridged isomer than complex **I**. Thus, one may conclude that complex **II** has a more rigid ligand environment than complex **I**, which makes the H-bridged **A7** isomer of **II** energetically unfavorable. In the case of complex **I**, the ligand environment of the Zr centers is more flexible, making the formation of H-bridging structures favorable.

Thus, the *rigid* ligand environment of Zr centers makes available the less stable non-H-bridged isomers of $[L_2Zr][L_2ZrH][\mu_2,\eta^2,\eta^2\text{-NNH}]$ (which relatively easily reacts with a H_2 molecule) for the addition of the second H_2 molecule, and, consequently, facilitates the hydrogenation of dinitrogen.

4.6
Several Necessary Conditions for Successful Hydrogenation of a Coordinated Dinitrogen Molecule

The above presented data, along with the available experiments and previous computational studies, clearly show that one of the important factors for successful hydrogenation of N_2 molecule is its side-on η^2 coordination to the transition metal center. The η^2 coordination of N_2 to a transition metal center is favored by the back-donation component of the M–N_2 interaction, during which the π^* orbitals of N_2 are partially populated, which facilitates activation of the N≡N triple bond. However, N_2 is a poor π-acceptor; therefore, in all calculated and reported mononuclear side-on coordinated-N_2 complexes it only weakly interacts with the metal center. Such a weak interaction only slightly activates the N≡N triple bond – insufficiently for its hydrogenation.

However, a stronger interaction of the η^2 coordinated N_2 molecule with transition metal centers and, consequently, a larger population of the $\pi^*(N_2)$ orbitals can be achieved in dinuclear (or multinuclear) transition metal complexes. In the literature many di-M-$(\mu_2,\eta^2,\eta^2$-$N_2)$ complexes have been reported, but not all them have a strong M–N_2–M interaction. For example, in the side-on coordinated $[rac$-$BpZr]_2(\mu_2,\eta^2,\eta^2$-$N_2)$ complex [72] the short N–N bond distance (1.241 Å) indicates a weak Zr–N_2–Zr interaction and less N≡N bond activation. As a result, this complex does not add H_2 to N_2 under mild experimental conditions [72]. These data indicate that there are other conditions for successful hydrogenation of side-on coordinated N_2 in dinuclear complexes.

One such condition is the availability of appropriate frontier orbitals of $[L_nM]_2(\mu_2,\eta^2,\eta^2$-$N_2)$: the HOMO, which is expected to donate electrons to the σ_u^* orbital of an incoming H_2 molecule, should be a π-bonding orbital of the M–N_2–M fragment, while the LUMO, which accepts electrons from the σ_g-bonding MO of the reacting H_2 molecule, should be a non-bonding combination of metal orbitals. This qualitative orbital picture is consistent with the "metathesis" transition state (involving one of the M and N atoms of the complex, and both H atoms of the H_2 molecule) reported for H_2 addition to all studied di-zirconium-N_2 complexes, $[L_nZr]_2(\mu_2,\eta^2,\eta^2$-$N_2)$.

The importance of suitable frontier orbitals for the successful hydrogenation of the coordinated N_2 molecule is clearly demonstrated upon comparing the reactivity of the complex $[(\eta^5$-$C_5Me_nH_{5-n})_2M]_2(\mu_2,\eta^2,\eta^2$-$N_2)$, **II_M**, for M = Ti, Zr and Hf. As discussed in Section 4.4.1, both **II_Zr** and **II_Hf** complexes prefer the side-on coordination of N_2 in both their singlet and triplet states, and posses HOMO and LUMO orbitals with appropriate symmetry *only* at their ground singlet states. Therefore, their reaction with a H_2 molecule proceeds via a reasonable energy barrier and is exothermic *only* at their ground singlet states. Conversely, their Ti analogs, **II_Ti**, are unlikely to react with H_2 via the dinitrogen hydrogenation pathway. Indeed, **II_Ti** complexes (for $n = 0$ and 4) have ground triplet electronic states, which posses the side-on coordinated N_2 molecule but lack the appropriate frontier orbitals required for successful reaction with H_2 via the dinitrogen hydrogenation pathway. Furthermore, their excited singlet electronic states lie higher in energy than the triplet ground states, and lack the favorable side-on coordinated N_2 molecule required for dinitrogen hydrogenation.

The above factors control the reactivity of transition metal-N_2 complexes toward addition of one H_2 molecule and the formation of N–H bonds. However, understanding the factors affecting the reactivity of these complexes towards second (and third) H_2 addition is also important. One such factor is demonstrated to be the rigidity of ligand environment of the metal centers and the aptitude for formation of H-bridged structures [42]. If the ligand environment of the Zr centers is rigid, reaction with the first H_2 molecule leads to the less stable non-H-bridged isomers of $[L_2Zr][L_2ZrH][\mu_2,\eta^2,\eta^2\text{-NNH}]$, which react relatively easily with the next H_2 molecule. Formation of the most stable H-bridged isomer of the first H_2 addition, $[L_2Zr][L_2Zr][\mu_2,\eta^2,\eta^2\text{-NNH}][\mu_2\text{-H}]$, makes the second H_2 addition thermodynamically and kinetically impossible under mild conditions.

Thus, future candidate complexes for effective N_2 hydrogenation should fulfill the following conditions: (1) The η^2-N_2 isomer should be thermodynamically and kinetically accessible. (2) They should have a strong M–N_2–M interaction. (3) The HOMO should be a π-bonding orbital of the M–N_2–M fragment, while the LUMO should be non-bonding orbitals of the metal centers. (4) The M centers should have a rigid ligand environment. Notably, other factors for successful N_2 hydrogenation may exist. Such factors are currently under investigation in our laboratory.

Appendix: Computational Details

All calculations were performed using the hybrid DFT B3LYP method [57–59] or the pure DFT PBE functional [60, 61] and the Stevens–Basch–Krauss (SBK) [73, 74] relativistic effective core potentials (for Ti, Zr, Hf, C, N, Si, and P) and the standard CEP-31G basis sets for H, C, N, Si, P, Ti, Zr and Hf atoms with additional d-type polarization functions with exponent of 0.8 for all N atoms. In the text this approach is denoted as B3LYP/CEP-31G(d_N) or PBE/CEP-31G(d_N). The reliability of this methodology has been tested in previous studies [38–40, 75]. We have shown that the use of larger basis sets such as the Stuttgart/Dresden effective core potential and associated SDD [76–79] basis set for the group IV atoms and the standard 6-31G(d) basis set for the remaining atoms has no significant effect on the calculated geometries and energetics of the reactants, transition states and products of the reactions. All structures were optimized without any symmetry constrains. The nature of all intermediates and transition states was confirmed by performing normal-mode analysis and IRC calculations. All data discussed in this text represent calculations in vacuum. All calculations were performed using the *Gaussian 03* program package [80].

Acknowledgments

This work was supported in part by a grant (CHE-0209660) from the U.S. National Science Foundation. Computer resources were provided in part by the Air Force Office of Scientific Research DURIP grant (FA9550-04-1-0321), by the William R. Wiley Environmental Molecular Sciences Laboratory, and by the Cherry Emerson Center for Scientific Computation.

References

1 J. R. Jennings, ed., *Catalytic Ammonia Synthesis*, Plenum, New York, 1991.
2 S. A. Topham, *Catalysis: Science and Technology*, eds. J. R. Anderson, M. Boudart, Springer-Verlag, Berlin, 1985.
3 B. K. Burgess, *Advances in Nitrogen Fixation*, eds. C. Veeger, W. E. Newton, Martinus Nijhoff, Boston, 1984.
4 See: (a) G. J. Leigh, *New J. Chem.* 1994, **18**, 157–161; (b) G. J. Leigh, *Science* 1998, **279**, 506–508, and references therein.
5 J. B. Howard, D. C. Rees, *Chem. Rev.* 1996, **96**, 2965.
6 M. P. Shaver, M. D. Fryzuk, *Adv. Synth. Catal.* 2003, **345**, 1061.
7 M. W. Chase, J. C. A. Davies, J. J. R. Downe, D. J. Flurip, R. A. McDonald, A. N. Syverue, *J. Phys. Chem. Ref. Data* 1985, **14** (Suppl. No 1).
8 H. M. Rosenstock, K. Draxl, B. W. Steiner, J. T. Herron, *J. Phys. Chem. Ref. Data* 1977, **6** (Suppl. No 1).
9 R. A. Alberty, *J. Biol. Chem.* 1994, **269**, 7099.
10 H. S. W. Massey, *Negative Ions*, Cambridge University Press, Cambridge, 1976.
11 P. L. Holland, Nitrogen Fixation, in *Comprehensive Coordination Chemistry II*, eds. J. McCleverty, T. J. Meyer, Elsevier, Oxford, 2004, Vol. 8, **4**, 69.
12 B. E. Smith, *Science* 2002, **297**, 1654.
13 J. Kim, D. C. Rees, *Science* 1992, **257**, 1677.
14 J. Kim, D. C. Rees, *Nature* 1992, **360**, 553.
15 A. J. Harding, *Ammonia Manufacture and Uses*, Oxford University Press, London, 1959.
16 N. N. Greenwood, A. Earnshaw, *Chemistry of the Elements*, Pergamon, New York, 1984.
17 R. Schlögl, *Angew. Chem. Int. Ed.* 2003 **42**, 2004.
18 D. Stolzenberg, *Fritz Haber: Chemist, Nobel Laureate, German, Jew*, Chemical Heritage Press, Philadelphia, 2004.
19 V. Smil, *Enriching the Earth; Fritz Haber, Carl Bosch, and the Transformation of World Food Production*, MIT Press, Cambridge, MA, 2001.
20 B. A. MacKay, M. D. Fryzuk, *Chem. Rev.* 2004, **104**, 385.
21 M. Hidai, Y. Mizobe, *Chem. Rev.* 1995, **95**, 1115.
22 M. D. Fryzuk, S. A. Johnson, *Coord. Chem. Rev.* 2000, **200–202**, 379.
23 G. J. Leigh, *Acc. Chem. Res.* 1992, **26**, 177.
24 R. R. Schrock, *Acc. Chem. Res.* 2005, **38**, 955.
25 E. A. MacLachlan, M. D. Fryzuk, *Organometallics* 2006, **25**, 1530, and references therein.
26 H. J. Himmel, M. Reither, *Angew. Chem. Int. Ed.* 2006, **45**, 6264.
27 D. V. Yandulov, R. R. Schrock, *J. Am. Chem. Soc.* 2002, **124**, 6252.
28 D. V. Yandulov, R. R. Schrock, *Science* 2003, **76**, 301.
29 D. V. Yandulov, R. R. Schrock, *Can. J. Chem.* 2005, **83**, 341.
30 D. V. Yandulov, R. R. Schrock, *Inorg. Chem.* 2005, **44**, 1103.
31 S. C. Lee, R. H. Holm, *Proc. Natl. Acad. Sci. U.S.A.* 2003, **100**, 3595.
32 Y. Nishibayashi, S. Iwai, M. Hidai, *Science* 1998, **279**, 540.
33 J. A. Pool, E. Lobkovsky, P. J. Chirik, *Nature* 2004, **427**, 527.
34 W. H. Bernskoetter, E. Lobkovsky, P. J. Chirik, *J. Am. Chem. Soc.* 2005, **127**, 14051.
35 W. H. Bernskoetter, A. V. Olmos, E. Lobkovsky, P. J. Chirik, *Organometallics* 2006, **25**, 1021.
36 (a) C. E. Laplaza, C. C. Cummins, *Science* 1995, **268**, 861; (b) C. E. Laplaza, M. J. A. Johnson, J. C. Peters, A. L. Odom, E. Kim, C. C. Cummins, G. N. George, I. J. Pickering, *J. Am. Chem. Soc.* 1996, **118**, 8623; (c) C. E. Laplaza, A. L. Odom, W. M. Davis, C. C. Cummins, *J. Am. Chem. Soc.* 1995, **117**, 4999.
37 M. D. Fryzuk, J. B. Love, S. J. Rettig, V. G. Young, *Science* 1997, **275**, 1445.
38 H. Basch, D. G. Musaev, K. Morokuma, *J. Am. Chem. Soc.*, 1999, **121**, 5754.
39 H. Basch, D. G. Musaev, K. Morokuma, *Organometallics* 2000, **19**, 3393.
40 D. G. Musaev, H. Basch, K. Morokuma, The N≡N Triple Bond Activation by Transition Metal Complexes, in *Computa-*

tional Modeling of Homogeneous Catalysis, eds. F. Maseras, A. A. Lledss, Kluwer, Academic Publishers, Dordrecht 2002, **13**, 325.

41 D. G. Musaev, *J. Phys. Chem. B* 2004, **108**, 10012.

42 P. Bobadova-Parvanova, D. Quinonero-Santiago, Q. Wang, K. Morokuma, D. G. Musaev, *J. Am. Chem. Soc.*, 2006, **128**, 11391.

43 P. Bobadova-Parvanova, Q. Wang, K. Morokuma, D. G. Musaev, *Angew. Chem. Int. Ed.*, 2005, **44**, 7101.

44 P. Bobadova-Parvanova, D. Quinonero-Santiago, K. Morokuma, D. G. Musaev, *J. Chem. Theory Comp.* 2006, **2**, 336.

45 A. D. Allen, C. V. Senoff, *J. Chem. Soc., Chem. Commun.* 1965, 621.

46 J. N. Armor, H. Taube, *J. Am. Chem. Soc.* 1970, **92**, 2560.

47 (a) J. Chatt, A. J. Pearman, R. L. Richards, *Nature* 1975, **253**, 39; (b) J. Chatt, A. J. Dilworth, R. L. Richards, *Chem. Rev.* 1978, **78**, 589; (c) J. Chatt, A. A. Diamantis, G. A. Heath, N. E. Hooper, G. J. Leigh, *J. Chem. Soc., Dalton Trans.* 1977, 688.

48 H. Basch, D. G. Musaev, K. Morokuma, M. D. Fryzuk, J. B. Love, W. W. Seidel, A. Albinati, T. F. Koetzle, W. T. Klooster, S. A. Mason, J. Eckert, *J. Am. Chem. Soc.* 1999, **121**, 523.

49 *Can. J. Chem.* 2005, **83** (special issue).

50 J. M. Manriquez, J. E. Bercaw, *J. Am. Chem. Soc.* 1974, **96**, 6229.

51 H. Miyachi, Y. Shigeta, K. Hirao, *J. Phys. Chem. A* 2005, **109**, 8800.

52 J. A. Pool, W. H. Bernskoetter, P. J. Chirik, *J. Am. Chem. Soc.* 2004, **126**, 14326.

53 T. E. Hanna, E. Lobkovsky, P. J. Chirik, *J. Am. Chem. Soc.* 2006, **128**, 6018.

54 T. E. Hanna, E. Lobkovsky, P. J. Chirik, *J. Am. Chem. Soc.* 2004, **126**, 14688.

55 J. E. Bercaw, L. G. Marvich, L. G. Bell, H. H. Brintzinger, *J. Am. Chem. Soc.* 1972, **94**, 1219.

56 H.-J. Himmel, O. Hubner, W. Klopper, L. Manceron, *Angew. Chem. Int. Ed.*, 2006, **45**, 2799.

57 A. D. Becke, *J. Chem. Phys.* 1993, **98**, 5648.

58 C. Lee, W. Yang, R. G. Parr, *Phys. Rev. B* 1988, **37**, 785.

59 A. D. Becke, *Phys. Rev. A* 1988, **38**, 3098.

60 J. P. Perdew, K. Burke, M. Ernzerhof, *Phys. Rev. Lett.* 1996, **77**, 3865.

61 J. P. Perdew, K. Burke, M. Ernzerhof, *Phys. Rev. Lett.* 1997, **78**, 1396.

62 I. V. Khavrutskii, D. G. Musaev, K. Morokuma, *Inorg. Chem.* 2003, **42**, 2606.

63 D. V. Koroshun, D. G. Musaev, T. Vreven, K. Morokuma, *Organometallics* 2001, **20**, 2007.

64 J. S. Sears, C. D. Sherrill, *J. Chem. Phys.* 2006, **124**, 144314.

65 S. Yanagisawa, T. Tsuneda, K. Hirao, *J. Chem. Phys.* 2000, **112**, 545.

66 S. Yanagisawa, T. Tsuneda, K. Hirao, *J. Comput. Chem.* 2001, **22**, 1995.

67 A. E. Reed, L. A. Curtiss, F. Weinhold, *Chem. Rev.* 1988, **88**, 899.

68 M. R. A. Blomberg, P. E. M. Siegbahn, *J. Am. Chem. Soc.* 1993, **115**, 6908.

69 P. E. M. Siegbahn, *J. Chem. Phys.* 1991, **95**, 364.

70 K. Kitaura, S. Sakaki, K. Morokuma, *Inorg. Chem.* 1981, **20**, 2292.

71 K. Morokuma, K. Kitaura, in *Chemical Applications of Atomic and Molecular Electrostatic Potentials*, eds. P. Politzer, D. G. Truhlar, Plenum Press, New York, 1981, **10**, 215.

72 P. J. Chirik, L. M. Henling, J. E. Bercaw, *Organometallics* 2001, **20**, 534.

73 W. J. Stevens, H. Basch, M. Krauss, *J. Chem. Phys.* 1984, **81**, 6026.

74 W. J. Stevens, M. Krauss, H. Basch, P. G. Jasien, *Can. J. Chem.* 1992, **70**, 612.

75 B. F. Yates, H. Basch, D. G. Musaev, K. Morokuma, *J. Chem. Theor. Comp.* 2006, **2**, 1298.

76 P. Fuentealba, H. Preuss, H. Stoll, L. V. Szentpaly, *Chem. Phys. Lett.* 1989, **89**, 418.

77 U. Wedig, M. Dolg, H. Stoll, H. Preuss, *Quantum Chemistry*: The Challenge of Transition Metals and Coordination Chemistry, ed. A. Veillard, Reidel, Dordrecht 1986, 79.

78 T. Leininger, A. Nicklass, H. Stoll, M. Dolg, P. Schwerdtfeger, *J. Chem. Phys.* 1996, **105**, 1052.

79 X. Y. Cao, M. Dolg, *J. Mol. Struct. (Theochem)* **2002**, *581*, 139.

80 *Gaussian 03, Revision C.01*, Frisch, M. J., Trucks, G. W., Schlegel, H. B., Scuseria, G. E., Robb, M. A., Cheeseman, J. R., Montgomery, Jr., J. A., Vreven, T., Kudin, K. N., Burant, J. C., Millam, J. M., Iyengar, S. S., Tomasi, J., Barone, V., Mennucci, B., Cossi, M., Scalmani, G., Rega, N., Petersson, G. A., Nakatsuji, H., Hada, M., Ehara, M., Toyota, K., Fukuda, R., Hasegawa, J., Ishida, M., Nakajima, T., Honda, Y., Kitao, O., Nakai, H., Klene, M., Li, X., Knox, J. E., Hratchian, H. P., Cross, J. B., Bakken, V., Adamo, C., Jaramillo, J., Gomperts, R., Stratmann, R. E., Yazyev, O., Austin, A. J., Cammi, R., Pomelli, C., Ochterski, J. W., Ayala, P. Y., Morokuma, K., Voth, G. A., Salvador, P., Dannenberg, J. J., Zakrzewski, V. G., Dapprich, S., Daniels, A. D., Strain, M. C., Farkas, O., Malick, D. K., Rabuck, A. D., Raghavachari, K., Foresman, J. B., Ortiz, J. V., Cui, Q., Baboul, A. G., Clifford, S., Cioslowski, J., Stefanov, B. B., Liu, G., Liashenko, A., Piskorz, P., Komaromi, I., Martin, R. L., Fox, D. J., Keith, T., Al-Laham, M. A., Peng, C. Y., Nanayakkara, A., Challacombe, M., Gill, P. M. W., Johnson, B., Chen, W., Wong, M. W., Gonzalez, C., Pople, J. A., Gaussian, Inc., Wallingford CT, 2004.

5
Mechanism of Palladium-catalyzed Cross-coupling Reactions

Ataualpa A. C. Braga, Gregori Ujaque, and Feliu Maseras

5.1
Introduction

Cross-coupling reactions are among the most widely used methods for the formation of carbon–carbon bonds, and are thus a fundamental tool in organic synthesis [1]. Most cross-coupling processes correspond to the general reaction:

$$R-X + R'-Y \xrightarrow{\text{catalyst}} R-R' + X-Y \qquad (1)$$

where R and R' are organic groups (often at least one of them is an aryl), X is a good leaving group (often halide or triflate), and the catalyst is a transition metal complex (often a palladium species with phosphine ligands). The Y group is usually attached to R' through an electropositive atom. The nature of Y decides the particular name of the cross-coupling reaction. In the Suzuki–Miyaura reaction, Y is an organoboronic acid [2]; in the Stille reaction it is an organotin compound [3]; in the Negishi reaction it is an organozinc compound [4], and there are other reactions where the electropositive atom is Al, Mg, Si, Cu, etc. Other related processes not corresponding exactly to the formula above are also classified as cross-coupling: the Heck reaction, where R'–Y is an alkene [5]; the Sonogashira reaction, where R'–Y is a terminal alkyne [6]; and the Buchwald–Hartwig reaction, where a carbon–nitrogen bond is formed [7, 8].

Cross-coupling reactions share this general label because of the use of similar catalysts and because they are considered to exhibit the same general mechanism (Figure 5.1). The R–X molecule is first oxidatively added to the metal. Then the transmetalation step follows, where the R' group is transferred from Y to the metal, and the X group leaves the metal coordination sphere. The final step is the reductive elimination, where R and R' are bound, and the metal catalyst recovers its initial state. The existence of this general scheme does not explain all the mechanistic details. Additional species not indicated in Figure 5.1 are often required, among them bases and salts. The catalyst is usually formed *in situ*, and its precise form is often unknown, in particular concerning the number and

---

Computational Modeling for Homogeneous and Enzymatic Catalysis.
A Knowledge-Base for Designing Efficient Catalysts. K. Morokuma and D. G. Musaev (Eds.)
Copyright © 2008 WILEY-VCH Verlag GmbH & Co. KGaA, Weinheim
ISBN: 978-3-527-31843-8

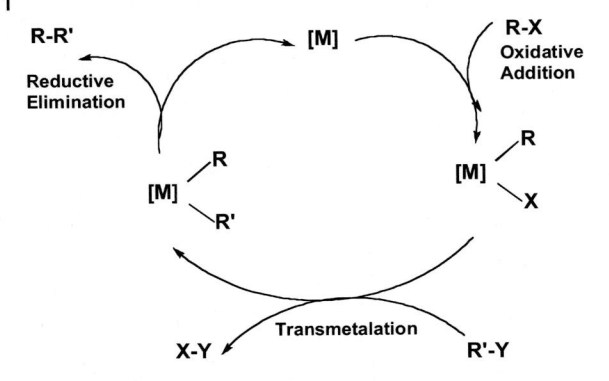

Figure 5.1 Generally accepted overall mechanism for cross-coupling reactions.

nature of ligands. The optimization of catalytic systems has often relied heavily on a trial-and-error approach, and a better mechanistic knowledge would be very helpful for a more rational design that would allow to better tackle several remaining challenges. There are still some substrates (aryl chlorides, alkyl halides) that are difficult to activate [9], and enantioselective cross-coupling remains largely unexplored. Several experimental studies on the mechanism of cross-coupling reactions have been carried out [10–12], although they are limited because of the multistep nature of the reactions and the elusiveness of the intermediates. It is in any case clear that different behaviors seem to be present for different reactions, the role of the Y group thus being critical.

The contribution of computational chemistry to this field is relatively recent. Some calculations of full catalytic cycles have appeared in the last few years for the Suzuki–Miyaura and Stille cross-coupling processes. The full catalytic cycle for the Suzuki–Miyaura reaction has been analyzed by Goossen, Thiel and coworkers for the coupling between acetic anhydride and phenylboronic acid [13, 14], and we have studied the coupling between vinyl bromide and vinylboronic acid [15]. The full catalytic cycle for the Stille reaction, assuming an associative cross-coupling, has been studied by Alvarez, de Lera and coworkers [16]. These works show complex mechanistic pictures, often with competing parallel pathways with similar energies. As an example, Figure 5.2 presents the simplified scheme obtained from our calculations for the cross-coupling between vinyl bromide and vinylboronic acid catalyzed by a palladium monophosphine species. Most of the steps shown in the scheme consist in fact of several microsteps, with different intermediates and transition states. In addition, this scheme also neglects the presence of (bis)phosphine species, which have a scheme of their own, interconnected with that presented in Figure 5.2.

Because of the complexity of the full mechanism, most theoretical works concentrate on particular steps of the catalytic cycle. This chapter is thus organized according to the different steps of the mechanism, rather than by reaction. Oxidative addition and reduction elimination have also been studied in the context

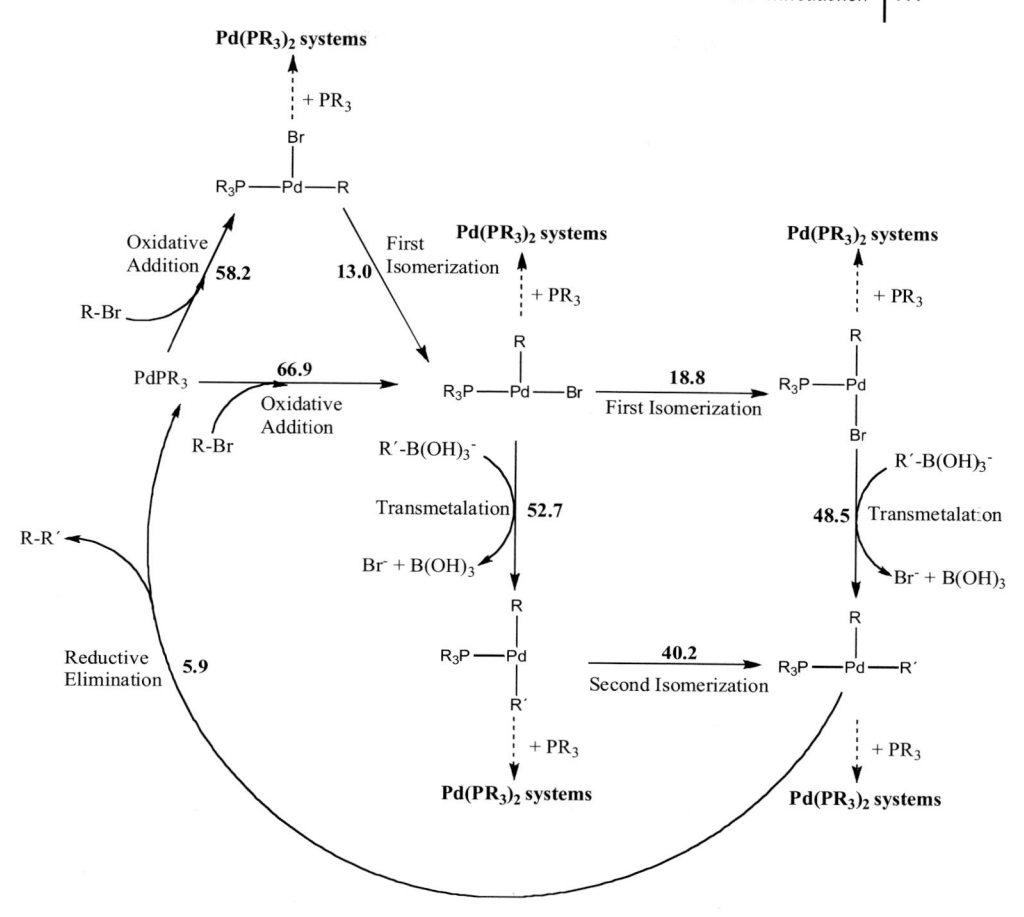

Figure 5.2 Computed mechanism for the cross-coupling between vinyl bromide and vinylboronic acid catalyzed by a monophosphine palladium species. The highest energy barrier involved in each reaction step is indicated in kJ mol^{-1}.

of other organometallic reactions. Transmetalation, though, is a specific cross-coupling processes and, because of this, it is discussed in more detail. This chapter reviews the present state of research in the area. As will be seen, a significant proportion of the articles mentioned have been published after 2003, indicating the current high activity in the topic. Consequently, it is impossible to provide a definitive mechanistic picture. This is, rather, a progress report on a very active research field.

Several of the articles discussed occupy over ten pages in a scientific journal. It is, therefore, out of question to present a detailed description of each of the intermediates and transition states that were computed. Such a description should be obtained directly from the source articles. The computational methods applied were not identical, but quite similar. The validity of the results do not seem to depend on the methodological details. The energy was, in almost all cases, computed using the density functional theory (DFT), with a generalized gradient approach (GGA). The functional applied more often was B3LYP, although some calculations also used BP86. Basis sets had at least a valence double-ζ plus polarization quality. Solvent effects were introduced, when mentioned, with the polarized continuum model (PCM).

5.2
Oxidative Addition

In the oxidative addition step, the bond between the organic group R and the leaving group X breaks, and two new bonds are formed with the metal, which in this way increases its oxidation state by two units. This is shown in Reaction (2):

$$[M] + R-X \rightarrow [M](R)(X) \qquad (2)$$

The oxidative addition of non-polar σ-bonds (H–H, C–C, C–H) to d^{10} ML_2 transition metal complexes was among the first organometallic reactions computationally studied [17–19], and was already established in the 1980s to take place in a concerted way with a mostly symmetric transition state. More recent studies on C–Ge, C–Si, Si–C, Si–H bonds confirm a similar picture [20, 21].

The specific case of polar σ-bonds, like that of aryl halides typically present in cross-coupling processes, has been the subject of several recent studies. The focus of computational attention is probably related to the fact that oxidative addition has been postulated to be the rate-limiting step in several cross-coupling reactions, although not in all of them.

Theoretical calculations by Senn and Ziegler [22] discuss the existence of two mechanisms (Figure 5.3). The concerted path is similar to that of non polar σ-bonds, and has simultaneous formation of the Pd–C and Pd–X bonds in the transition state. This is the only transition state that can be located in gas-phase calculations [23]. The second pathway is dissociative, and has two steps: first, the halide anion dissociates from the prereaction complex, leaving a cationic palladium species; subsequently, the two charged species collapse to the product. This second pathway has been also labeled as S_N2. Senn and Ziegler find the dissociative pathway to be the preferred one in their calculations in THF solution of the oxidative addition of chloro-, bromo- and iodobenzene with Pd(P-P) complexes, where P-P is a chelating diphosphine ligand of the type 1,2-bis(dimethylphosphino)ethane or (P)-2,2'-bis(dimethylphosphino)-1,1'-biphenyl. The barriers from the prereaction complex are in fact quite low: 31, 12 and 5 kJ mol^{-1} for the chloro, bromo and iodo species, respectively.

The preference for one of the two pathways seems to depend on the particular nature of the catalyst and the substrate. The role of anionic additive has been

5.2 Oxidative Addition

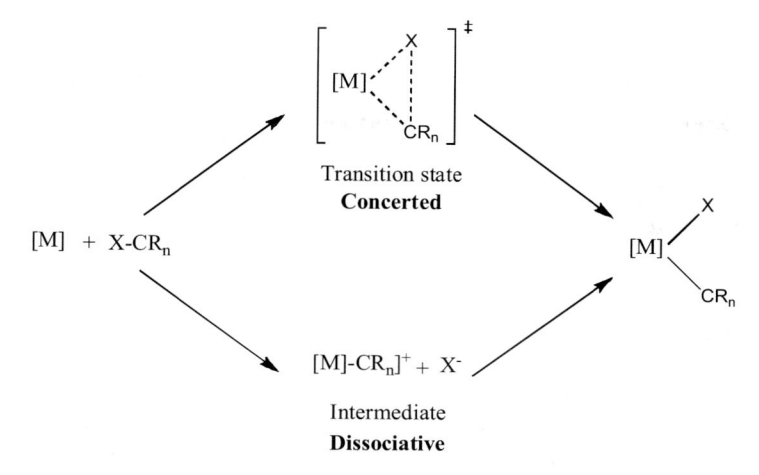

Figure 5.3 Concerted and dissociative pathways for oxidative addition.

analyzed by different authors. Jutand and coworkers have provided a significant amount of experimental evidence suggesting an anionic form for the catalyst, $[PdL_n(Cl)]^-$ [12, 24]. Calculations by Bickelhaupt and coworkers have shown that naked Pd favors a concerted path for the oxidative addition of $Cl-CH_3$, while $[PdCl]^-$ favors the dissociative path even in the gas phase [25]. Goossen, Thiel and coworkers have explored the reaction of $[Pd(PMe_3)_2(OAc)]^-$ with Ph-I [26]. The mechanism in this case is quite subtle. The acetate ligand moves away from palladium when iodobenzene coordinates, and returns afterwards to the metal to displace another ligand. Two different paths of similar energies are characterized for this process. The most favored of them ends up with departure of iodide, and resembles the dissociative pathway. The least favored pathway displaces a phosphine, and the transition state associated with C–I cleavage is of a concerted type.

Another factor affecting the oxidative addition is the number of phosphine ligands attached to the metal in the active form of the catalyst. With palladium systems, the catalyst is introduced as a precursor, often as $Pd_2(dba)_3$ (dba = dibenzylideneacetone) or $[Pd(PPh_3)_4]$, but this is clearly not the active form. The diphosphine form discussed above has been postulated in most experimental proposals and, consequently, used in calculations, because of the well-established stability of d^{10} ML_2 complexes, but the monophosphine form ML can be also envisaged. Its presence has in been fact proposed from experimental results on the activation of usually inert substrates by catalytic systems involving bulky phosphines [27, 28]. Computational studies have indeed shown that the barrier to oxidative addition of Cl–R bonds is lower for $Pd(PR_3)$ than for $Pd(PR_3)_2$ [29]. Norrby and coworkers have gone one step further and evaluated the oxidative addition for systems with a single ligand, either a phosphine or an anion [30, 31]. They compare the barriers for the oxidative addition of I-Ph to four different systems: $Pd(PPh_3)_2$, $Pd(PPh_3)$, $[PdCl]^-$ and $[Pd(OAc)]^-$. The results are conclusively in favor of the monocoordinated species, which present barriers of 2 kJ mol^{-1} at most, in contrast with a barrier of 13 kJ mol^{-1} for Pd(PPh$_3$)$_2$. Precise estimation of the energy required to access these more reactive catalytic forms is, however, not trivial.

A last issue is the nature of the substrate. As mentioned in the introduction, the prototypical substrates for oxidative addition are aryl halides, but fragments different from aryl are also relevant. In fact, alkyl groups are known to be less active, and there is considerable interest in the design of catalytic systems suitable for them. Ariafard and Lin have recently carried out calculations on the oxidative addition of different Br-R groups to Pd(PH$_3$)$_2$ and Pd(PPh$_3$)$_2$ [32]. Methyl, benzyl, phenyl, vinyl have been used for R group. The barriers relative to the separate fragments are lower for the systems involving sp^2 carbons than for those involving sp^3 carbons. The potential energy values are 39, 56, 77, 99 kJ mol^{-1} for vinyl, phenyl, benzyl and methyl, respectively. This result is explained by the existence of low-lying C-Br π^* orbitals in the unsaturated systems.

In summary, calculations carried out on the oxidative addition step show a large variety of mechanistic possibilities, all with reasonable energy barriers. The bond cleavage itself may take place through two different competitive pathways, either concerted or dissociative. The catalyst may be in different forms with regard to the number of ligands and the presence of additional anionic ligands. Finally, the nature of the organic group and the halide also affect the mechanism. The emerging global picture is that different mechanisms are likely to operate for different systems, and that there is no univocal choice of oxidative addition path for cross-coupling reactions.

5.3
Transmetalation

Transmetalation is the most characteristic process of the C–C cross-coupling reactions. This is the stage where the organic group R′ bound to an electropositive group Y is transferred to the palladium complex:

$$[M]-X + R'-Y \rightarrow [M]-R' + X-Y \tag{3}$$

Comprehension of this process has been computationally addressed in recent years by several research groups. The following two subsections summarize the available results on the Suzuki–Miyaura and Stille reactions, which are those studied in more detail. More limited studies have been also carried out on related processes, such as the Heck reaction [33] and direct arylation [34, 35].

5.3.1
Suzuki–Miyaura Reaction

The Suzuki–Miyaura reaction is one of the most used reactions in organic synthesis [1, 2]. The reaction takes place between an organic halide (or triflate),

R–X, and a boronic acid, R'-B(OH)$_2$:

$$R-X + R'-B(OH)_2 \rightarrow R-R' + X-B(OH)_2 \tag{4}$$

To understand the transmetalation process for this reaction several questions need to be answered: the role of the base that has to be added, the cis or trans arrangement of phosphines in four-coordinate systems, and the eventual role of three-coordinate monophosphine catalysts. These points are addressed in each of the following subsections.

5.3.1.1 Role of the Base

The transmetalation step of the Suzuki–Miyaura cross-coupling is known to require the presence of a base in solution. Several proposals have been put forward for the role of this base from experimental studies. Most of this information has been summarized with great clarity by Miyaura [36]. Two main pathways are proposed (Figure 5.4): either the base binds the boronic acid to form the organoboronate species (path A) or the base substitutes the halide ligand in the coordination sphere of the catalyst (path B). These two proposed pathways were theoretically evaluated by us with a model system, using trans-PdBr(CH$_2$=CH)(PH$_3$)$_2$, CH$_2$=CH-B(OH)$_2$ and OH$^-$ species as reactants [37].

We started our study with the direct mechanism in the absence of a base. The direct mechanism takes place in two steps with an uphill energy profile; the process is endothermic by more than 120 kJ mol^{-1}. The largest energy barrier is higher than 200 kJ mol^{-1}; the calculated energy barrier agrees with the experimental fact that the reaction does not proceed without addition of a base.

In path A (Figure 5.5), transmetalation takes place between the organoboronate species (obtained from the reaction between the boronic acid and OH$^-$) and the square planar palladium catalyst that comes from the oxidative addition of R–X to the PdL$_2$ catalyst. The reaction takes place in three steps: in the first step the organoboronate replaces the halide in the coordination sphere of the catalyst, sub-

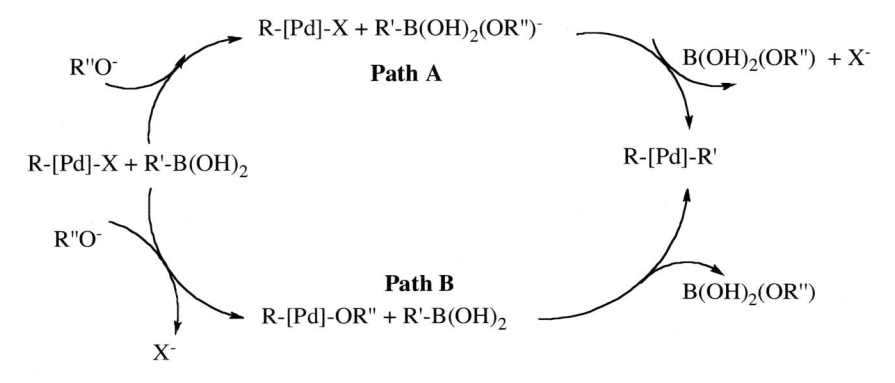

Figure 5.4 Proposed pathways for the role of the base in the transmetalation step of the Suzuki–Miyaura cross-coupling.

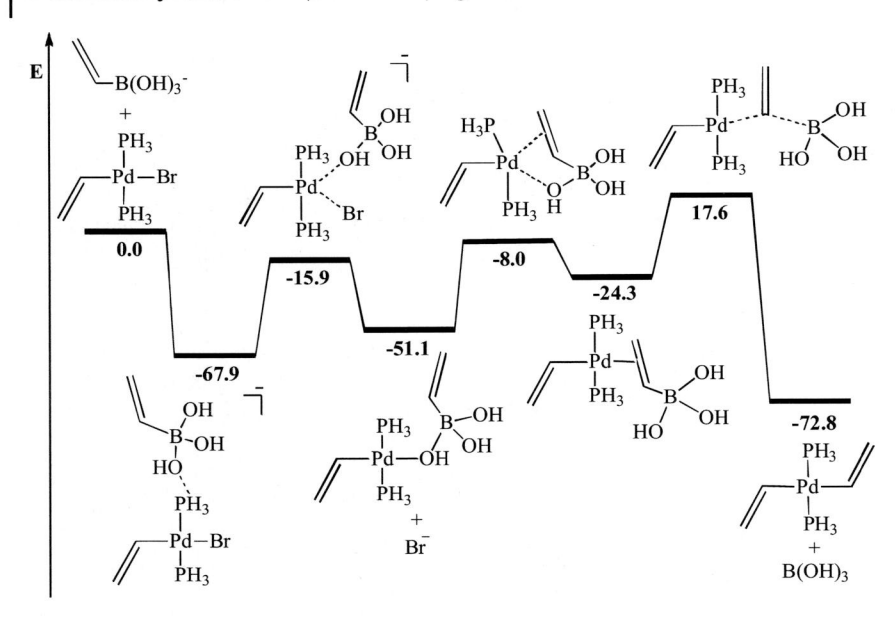

Figure 5.5 Energy profile of reaction path A for the transmetalation step in the Suzuki–Miyaura cross-coupling.

sequently there is an intramolecular substitution process where the vinyl group of the boronate replaces the OH group in the coordination sphere of the catalyst, and in the last step the proper transmetalation process takes place, generating the *trans*-Pd(CH$_2$=CH)$_2$(PH$_3$)$_2$ intermediate and the B(OH)$_3$ species. The energy difference between the lowest energy (-67.9 kJ mol^{-1} below reactants) and highest energy (17.6 kJ mol^{-1} above reactants) points is around 85 kJ mol^{-1}, therefore becoming a suitable pathway to explain the role of the base.

In the other proposed mechanism, labeled as path B, the base directly replaces the halide in the coordination sphere of the catalyst. Despite the computational effort devoted to looking for a transition state accounting for the direct replacement of the halide by the hydroxyl group, all the attempts were fruitless. This unsuccessful search was unexpected because the ligand substitution mechanism for a square-planar complex is a generally an accepted mechanism going through a trigonal bipyramid structure [38]. All the different optimizations drove to either high-energy chemically unreasonable structures or to two different structures: one with the OH$^-$ oxidizing one of the phosphine ligands, and the other where the phosphine was replaced by the hydroxyl group. Further analysis of the potential energy surface around this region allowed us to find an alternative pathway, labeled as path B'. In this new pathway the intermediate formed by reaction of the base with a coordinated phosphine forms the *trans*-[Pd(CH$_2$=CH)Br(PH$_3$OH)(PH$_3$)]$^-$ complex, and this intermediate can evolve to replace the halide ligand by the hydroxyl group by migration of the OH from

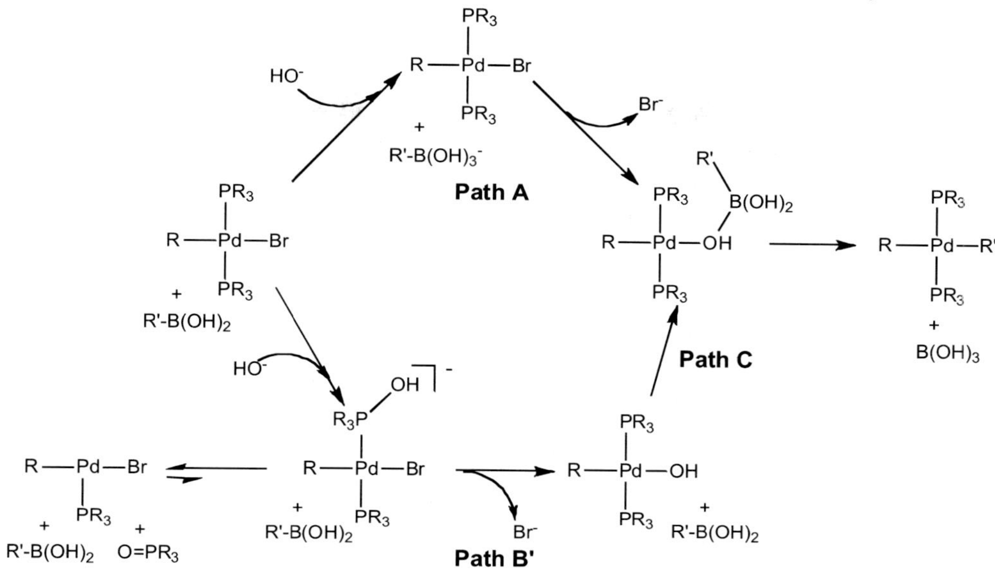

Figure 5.6 Schematic representation of the pathways accounting for the role of the base in the transmetalation step of the Suzuki–Miyaura reaction; path A is the most feasible one.

the phosphine to the Pd center; this complex could, in principle, give rise to the transmetalation process with the organoboronic acid (see below, path C). This pathway could be an alternative to path A; nevertheless, the presence of an oxidized phosphorus group makes the existence of this species in the reaction cycle quite unlikely. Indeed, phosphine oxidation is one of the known causes of catalyst destruction, and it seems to be connected to this compound (Figure 5.6).

Related experimental evidence is that transmetalation takes place in the absence of base if the halide is replaced by a hydroxo or alkoxo group in the starting palladium complex [39]. The reaction mechanism for this process (labeled as path C) was computationally studied, and the energy profile was found to be smooth. The highest energy barrier is quite low, around 80 kJ mol^{-1}. Path C is in fact identical to path A from the point where the organoboronate species is coordinated to the catalyst. Hence, computational results reproduce the experimental observation of a low-energy path C, and, furthermore, relate it to path A. Figure 5.6 presents the overall mechanistic picture resulting from these calculations.

These computational results were found to remain qualitatively valid when the PH$_3$ phosphine was replaced by PPh$_3$ and when vinyl groups were replaced by phenyl groups [40].

In total, these results indicate that, in the Suzuki–Miyaura reaction, one of the key problems of the transmetalation step is the displacement of the bromide ligand from the metal coordination sphere. The boronic acid itself is not able to undertake this replacement, as shown in the direct mechanism. A base is therefore necessary for transmetalation to take place. The direct replacement of the halide (Br⁻) by the base (OH⁻) is not feasible (path B). Nevertheless, the base may attack the boronic acid (path A) or the phosphine ligand (path B'). The latter must be discarded as the main mechanism because it easily leads to ligand oxidation and destruction of the catalyst. Therefore, the main catalytic cycle should proceed through path A, which has low-energy barriers and no obvious undesired products. Initial proposals about the transmetalation process suggested that it takes place though a four-center transition state [41], with two bridging group ligands (the halide and the base) between the two metal atoms. These results show that the transmetalation process does not take place in one concerted step but, rather, several steps are needed.

5.3.1.2 Cis versus Trans Species in bis(Phosphine) Systems

The calculations in the previous section assume that the reaction proceeds through an intermediate where the two phosphine ligands are in a trans disposition around the palladium metal center, as is often proposed in the textbook mechanism. Nevertheless, as shown Section 5.2 on the oxidative addition process, oxidative addition produces a cis intermediate. Thus, the transmetalation needs to be investigated for both intermediates having the cis and the trans arrangements of the ligands around the palladium center. This topic has been computationally

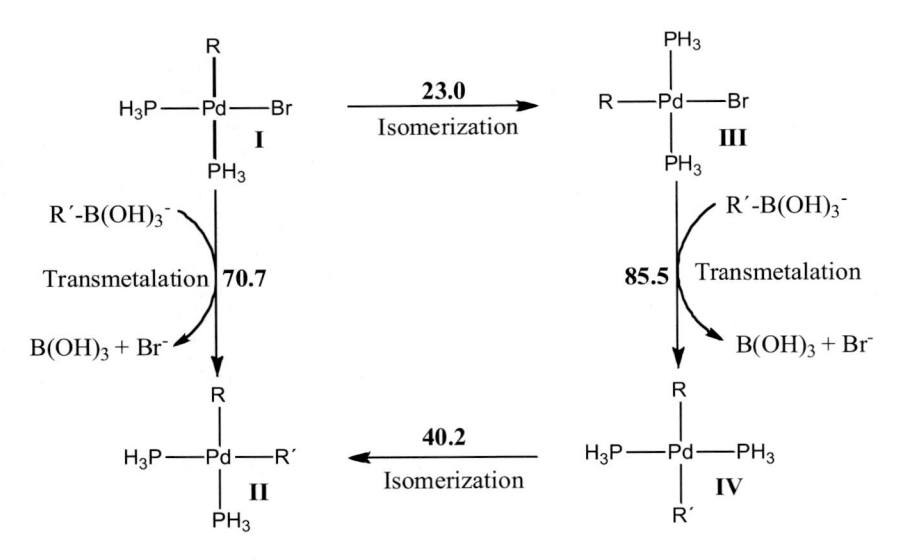

Figure 5.7 Proposed pathways for connecting complex I (arising from the initial oxidative addition) with complex II (ready to undertake reductive elimination), keeping two phosphine ligands in the Suzuki–Miyaura reaction.

analyzed by Goossen, Thiel and coworkers using cis-CH$_3$C(=O)Pd(OAc)(PMe$_3$)$_2$ [14], and by ourselves using cis-Pd(CH$_2$=CH)(Br)(PH$_3$)$_2$ as starting complex for the transmetalation [15].

We summarize our results in what follows. Transmetalation can take place starting from a cis or a trans intermediate (complexes I and III in Figure 5.7). The process starting from trans-Pd(CH$_2$=CH)(Br)(PH$_3$)$_2$ corresponds to that presented in the previous subsection, and summarized in Figure 5.5. The process starting from cis-Pd(CH$_2$=CH)(Br)(PH$_3$)$_2$ presents a similar energy profile (Figure 5.8). The process takes place in three steps: replacement of the halide by the organoboronate species, intramolecular ligand substitution between the vinyl and the OH groups, and a proper transmetalation step. Similarly to the trans alternative, the energy profile for this reaction is quite smooth. The overall process is exothermic, with an associated energy barrier near 85 kJ mol^{-1}. In this case, the intermediate generated after the transmetalation has the R, R' ligands in cis positions and, therefore, ready to undergo reductive elimination.

In addition to the mechanism discussed above for the cross-coupling reaction, which has been labeled as neutral mechanism (where the catalyst is supposed to be a PdL$_2$ species), Amatore and Jutand [12] also proposed another mechanism,

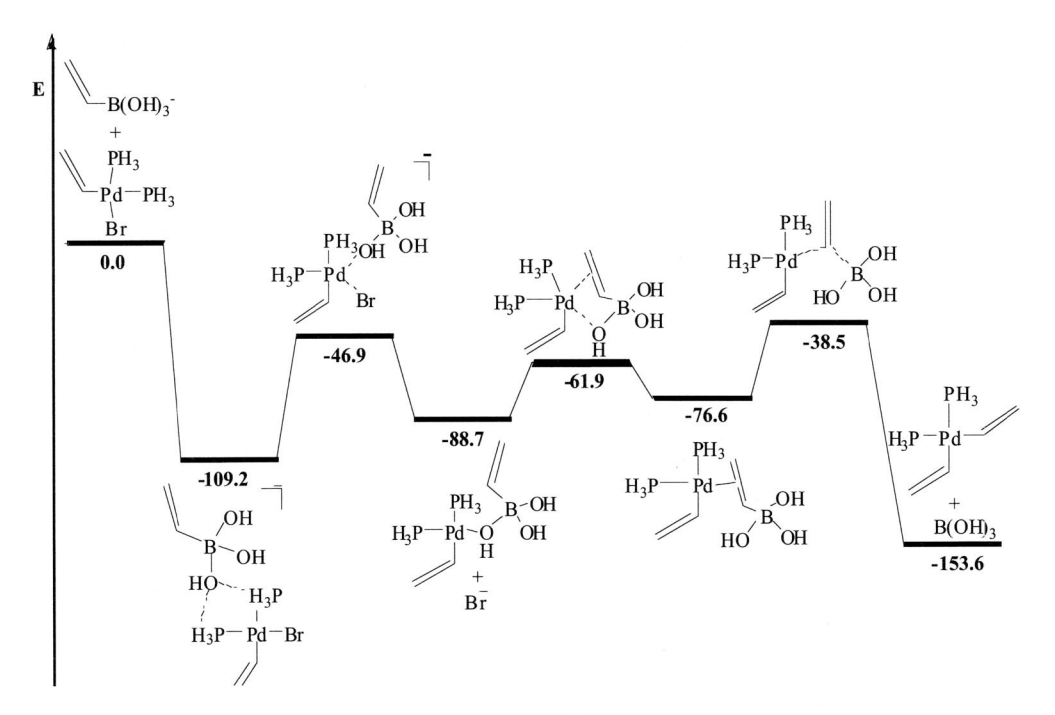

Figure 5.8 Energy profile of the cis alternative for the transmetalation step in the Suzuki–Miyaura cross-coupling catalyzed by (bis)phosphine systems.

labeled as anionic mechanism, where they suppose that the catalyst is an anionic [PdL$_2$X]$^-$ complex. The existence of this species was theoretically investigated by Shaik and coworkers [24, 42]. Goossen, Thiel and coworkers [13] studied computationally the cross-coupling reaction between phenylboronic acid and acetic anhydride using both the neutral Pd(PMe$_3$)$_2$ or the anionic [Pd(PMe$_3$)$_2$(OAc)]$^-$ complex as starting catalysts. According to their results the neutral and anionic mechanisms are both plausible for explaining the overall cross-coupling reaction, the preference depending on the experimental conditions. Nevertheless, the authors state that they did not find any evidence for the intermediacy of five-coordinate species during the catalytic cycle. Both pathways are dominated by square-planar palladium(II) diphosphine with a cis arrangement. Regarding the transmetalation process, it is qualitatively similar to that previously described, with an energy barrier around 85 kJ mol^{-1}. Therefore, theoretical calculations are unable to clearly distinguish between the cis or trans pathways, though the former seems to be slightly preferred. In all of their studies, however, these authors always contemplate mechanisms where the palladium complex loses a phosphine ligand, therefore becoming a monophosphine catalyst. The possibility of the reaction catalyzed by a monophosphine palladium catalyst is reviewed in the next subsection.

5.3.1.3 Monophosphine Systems

Whether the transmetalation occurs with two phosphines coordinated to the metal center or with only one has long been debated. The observation that when excess of phosphine is added the reaction is inhibited suggests a monophosphine catalyst [43]. Moreover, the use of bulky phosphines increases performance in some cases, and it has been proposed that in these cases a monophosphine species can act as a catalyst [44–47]. However, several chelating diphosphine complexes are active catalysts, suggesting the presence of (bis)phosphine complexes in the reaction mechanism. The problem is further complicated by the difficulty of computing reliable bond dissociation energies in solution, as has been discussed by ourselves in one of the works cited above [15]. In any case, even if the dissociation energy cannot be properly estimated, the intrinsic reactivity of monophosphine systems can be analyzed, and compared with that of bis(phosphine) systems. In this subsection, we present our computational study on monophosphine systems, which led to the overall mechanism for the catalytic cycle already presented in Figure 5.2.

As happened with the diphosphine systems, several isomers are possible (the R and X ligands in cis or trans arrangements) for starting the transmetalation, which suggests the existence of two different parallel pathways (Figure 5.9). We studied both paths [15] on the system described above involving vinyl bromide, vinylboronic acid and, in this case, a single PH$_3$ ligand. The two computed paths, cis and trans, were quite similar. Once the organoboronate species is initially coordinated through the double bond to the vacant site, one of the OH groups of the boronate substitutes the halide in the coordination sphere of the catalyst. In the next step, the transmetalation process itself takes place, with the vinyl group

5.3 Transmetalation | 121

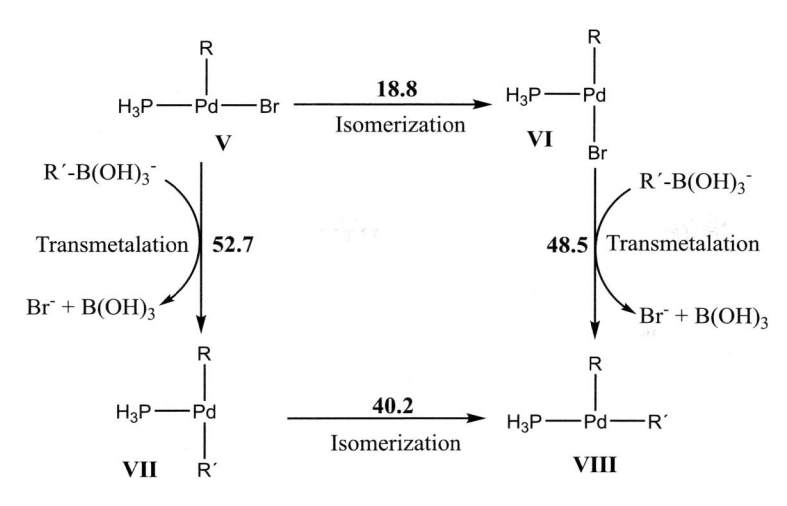

```
        R                                    R
        |                                    |
H₃P—Pd—Br      18.8        H₃P—Pd
        V      Isomerization       VI   |
                                        Br
```

V → (Transmetalation, 52.7; R'-B(OH)₃⁻; Br⁻ + B(OH)₃) → VII

VI → (Transmetalation, 48.5; R'-B(OH)₃⁻; Br⁻ + B(OH)₃) → VIII

```
        R                                    R
        |                                    |
H₃P—Pd         40.2           H₃P—Pd—R'
   |           Isomerization
   VII R'                              VIII
```

Figure 5.9 Proposed pathways for connecting complex V (arising from the initial oxidative addition) with complex VIII (ready to undertake reductive elimination), for a monophosphine system in the Suzuki–Miyaura reaction.

breaking the bond with the boron and becoming bonded to the palladium. The energy barriers in all these pathways are below 60 kJ mol⁻¹.

As commented in the previous subsection, Goossen, Thiel and coworkers have also analyzed the Suzuki–Miyaura cross-coupling reaction, starting from the anionic two-coordinated monophosphine [Pd(PMe₃)(OAc)]⁻ complex [14]. From their results, comparing this pathway with the neutral [Pd(PMe₃)₂] and the anionic three-coordinated [Pd(PMe₃)₂(OAc)]⁻ complex [13], they conclude that several catalytic pathways involving neutral and anionic palladium species may generally contribute to the catalytic turnover.

In theoretical work by Sakaki and coworkers [48] on the transmetalation process associated with the Pd-catalyzed borylation of iodobenzene reaction using diboron, the authors found that, during the reaction, one of the phosphine ligands was replaced by the diboron group. In their calculations, the authors assumed that the base (OH⁻) was initially coordinated to the catalyst. Then, the transmetalation was found to occur in two steps, an initial ligand substitution process (diboron by the phosphine), and the proper transmetalation. The authors evaluated associative and dissociative mechanisms for the ligand substitution process, obtaining the former as the most favorable. Afterwards, once the diboron is coordinated to the catalyst, transmetalation itself takes place and one of the boron groups (from the diboron species) is transferred to the palladium center.

In summary, a manifold of paths for the transmetalation process has been characterized with low energy barriers acceptable for a process taking place at room temperature. Then, the transmetalation can proceed smoothly with either one (dissociative mechanism) or two (associative mechanism) phosphine ligands at the palladium center. The reaction can also occur via the cis or trans arrangement of the ligands around the metal center. In addition, the anionic mechanism may also be operative in some cases. Therefore, these results strongly suggest that different a reaction mechanism for the transmetalation may occur for each particular system (depending on the phosphine, solvent, etc.), and even that several mechanism may contribute to the overall turnover of the catalytic cycle.

5.3.2
Stille Reaction

The Stille reaction is a general Pd-catalyzed cross-coupling reaction [3, 11], owing its popularity in part to the easy preparation of organotin compounds and their tolerance toward most functional groups (Reaction 5):

$$R-X + R'-SnR''_3 \rightarrow R-R' + X-SnR''_3 \qquad (5)$$

In this case, two different transmetalation mechanisms, labeled as *cyclic* and *open*, have been proposed to explain the experimental data (Figure 5.10). The cyclic mechanism was proposed by Espinet and coworkers [49] to account for the evidence of some Stille processes taking place with retention of configuration at the transmetalated carbon atom. This retention of configuration implies a transition state, or intermediate, where the X and R' ligands are bridging the two metal atoms (Figure 5.10, upper pathway). The open mechanism was proposed for those cases where the product presents inversion of configuration. This mechanism can produce either cis or trans geometries of the $L_2Pd(R)(R')$ species (Fig-

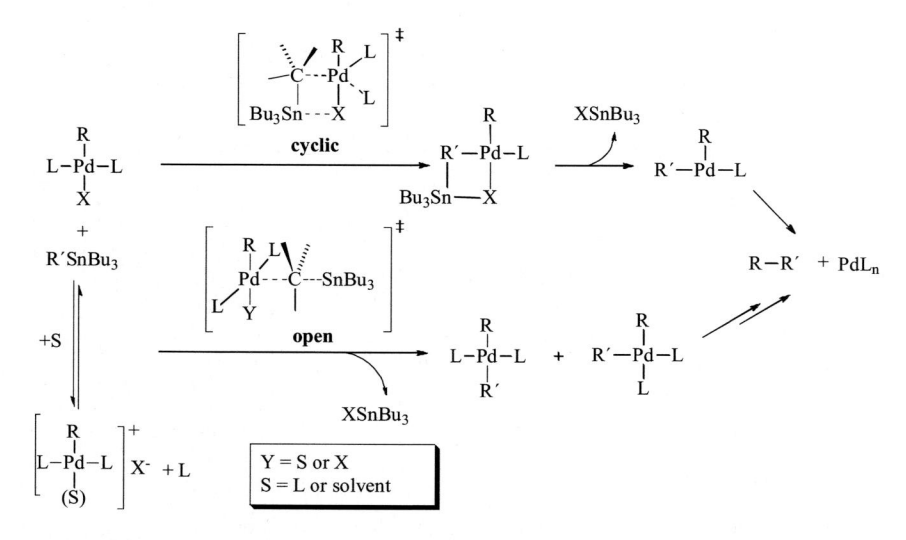

Figure 5.10 Schematic representation of the proposed cyclic and open pathways for the transmetalation in the Stille reaction.

ure 5.10, lower pathway). The open pathway seems to be more common and is expected to be favored for more electrophilic Pd centers, as in species with poor coordinating anions (like triflate), which are easily substituted by a neutral molecule (such as solvent or the ligand used in excess) [50–52].

Both mechanisms (cyclic and open) have been computationally analyzed by some of us, the cyclic mechanism having also been studied by other authors. Our computational study [53] used a model system where the coupling takes place between Br-Ph and CH_2=CH-$SnMe_3$, and is catalyzed by Pd(PH_3)$_2$ or Pd(AsH_3)$_2$. The effect of THF and chlorobenzene solvents were considered. In this work, we found that the cyclic mechanism did not take place in a concerted way. It was impossible to locate the proposed structure either as an intermediate or as a transition state. Instead, it was found that the transmetalation takes place in two steps (Figure 5.11). In the first one, with a barrier below 40 kJ mol^{-1}, there is a substitution of an L ligand by the incoming vinyl group of the stannane, whereas in the second the proper transmetalation takes place through a cyclic transition state. The cyclic four-membered ring transition state corresponding to the second step has the highest energy barrier, with a value between 62 and 102 kJ mol^{-1}, depending on L and the solvent, therefore becoming the rate-determining step. These results were in concordance with studies previously published by Napolitano and coworkers for alkynyl stannanes [54], and by Álvarez, de Lera and coworkers for alkenyl stannanes [16]. All these studies also agreed that this process takes place in an associative way – departure of the L ligand is coupled with the entry of the stannane in the palladium coordination sphere. Lin and coworkers [55] studied the effect of changing the halide ligand on the cyclic transmetalation step. Their results show that the activation barriers increase in

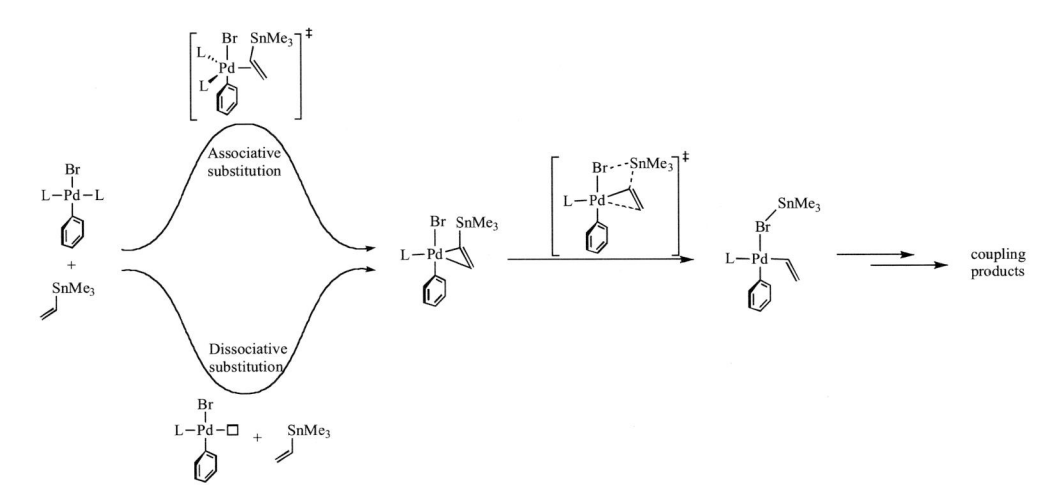

Figure 5.11 Schematic representation of the associative and dissociative pathways for the cyclic alternative of the transmetalation step in the Stille reaction.

the order Cl < Br < I. This trend is rationalized by observing that the Sn–X bond energy increases faster than the Pd–X bond energy in the opposite order I < Br < Cl.

For the open mechanism, the most characteristic features are the absence of a cyclic species, and the formation of a cationic palladium complex. Experimentally the *trans*-[PdL$_2$XR] complexes formed after the initial oxidative addition are found to be in equilibrium with other species such as *trans*-[PdL$_2$SR]$^+$ or [PdL$_3$R]$^+$, depending on the reaction conditions, especially on the nature of the X group (poor coordinating ligands are better leaving groups) and the solvent. In our theoretical study on the Stille reaction [53], we analyzed the formation of a cationic species by substitution of the X ligand (bromide or triflate) by a neutral species as a phosphine or arsine ligand or a solvent molecule. This ligand substitution is found to be, as expected, much more favorable (by about 60 kJ mol^{-1}) for the case of triflate than for the case of bromide.

The following steps on the transmetalation process were studied, starting from the cationic species, [PdL$_3$R]$^+$. Hence, the next step corresponds to the replacement of one of the L ligands by the incoming stannane, an R-for-L substitution step. This step may follow two different competitive pathways, depending on whether the stannane group goes cis or trans to the R substituent (Figure 5.12). Both possibilities were explored, and the energy barriers found for both pathways were quite similar, with values between 30 and 40 kJ mol^{-1} in solution. Once the stannane is coordinated to the palladium catalyst, a S_N2 substitution of the R group by a halide at the Sn center should follow. The energy barrier of this step

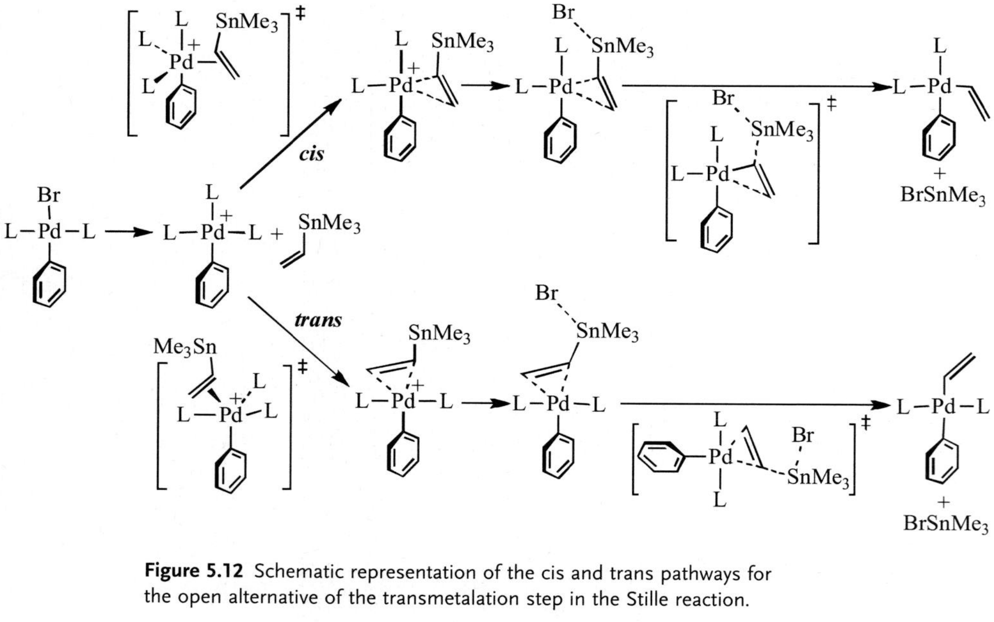

Figure 5.12 Schematic representation of the cis and trans pathways for the open alternative of the transmetalation step in the Stille reaction.

for both the cis and the trans pathways is quite low. This process produces an intermediate with both organic groups attached to the palladium center, thereby being ready to undergo the reductive elimination. Comparison of the two steps for the open mechanism shows that the barrier in the R-for-L substitution is higher than for the transmetalation itself, which represents an interesting difference with respect to the case of the cyclic mechanism.

The transmetalation step for the Stille reaction is thus quite different from that in the Suzuki cross-coupling. No base is required, and a Br–Sn bond is in fact formed after the transmetalation. However, the transmetalation step in the Stille reaction is also complex, with two major mechanisms being possible, one leading to retention of configuration and the other leading to inversion of configuration in the R' group. The specific mechanism taking place for a given catalytic system is mainly affected by the X group and the solvent, while the L ligand exerts a smaller (although kinetically significant) effect.

5.4 Reductive Elimination

In the reductive elimination step (Reaction 6), a bond between the two organic groups attached to the metal is made, and the two previously existing bonds with the metal are lost, which in this way decreases its oxidation state by two units.

$$[M](R)(R') \to [M] + R-R' \tag{6}$$

Reductive elimination is of course the microscopic reverse of oxidative addition, and the general studies mentioned above on the oxidative addition of non-polar σ-bonds (H–H, C–C, C–H) to d^{10} ML_2 transition metal complexes also apply. The transition state is in particular concerted, with simultaneous breaking of the two M–C bonds. Reductive elimination of C–C bonds is discussed in detail by Ananikov, Musaev and Morokuma Chapter 6.

Computational studies of the reductive elimination in systems related to cross-coupling are more scarce than those concerned with oxidative addition. This is in part explained by the dominance of the concerted mechanism and the lack of the alternative dissociative path shown in Figure 5.3. The dissociative mechanism is associated with the existence of a polar σ-bond. The bond in R–X is polar, but that in R–R' is not. Moreover, there are no experimental reports where reductive elimination is considered the rate-determining step, which also detracts computational attention from this process. Theoretical studies on the full catalytic cycle seem to confirm a lower barrier for reductive elimination than for oxidative addition. For instance, in our work on the Suzuki–Miyaura catalytic cycle, we found barriers in the range 55–85 kJ mol^{-1} for oxidative addition and barriers in the range 5–20 kJ mol^{-1} for reductive elimination. These results were obtained for the reaction between vinyl bromide and vinylboronic acid, and may not be general.

Computational studies by Ananikov, Musaev and Morokuma on C–C coupling in palladium and platinum bis(phosphine) complexes address the dependence of the barrier on the nature of the organic groups being reductively eliminated [56, 57]. In their calculations on reductive elimination from $Pd(PH_3)_2(R)_2$ complexes, they found the following order in potential energy barriers as a function of the R group: vinyl (28 kJ mol^{-1}) < phenyl (49 kJ mol^{-1}) < ethynyl (61 kJ mol^{-1}) < methyl (105 kJ mol^{-1}). The values are low for the unsaturated substrates, but the value for the sp^3 methyl system may be competitive with that of other reaction steps, and may be even related to the low reactivity of alkyl systems.

Reductive elimination may be thus a relevant step in some cross-coupling processes. It may be affected by some of the factors that have been discussed above for oxidative addition of R–X, like anionic additives and the number of ligands attached to the active form. However, analysis of these topics from a computational point of view is still limited. The main results can in fact be found in works not focusing on cross-coupling. Macgregor and coworkers have analyzed the reductive elimination process in a series of $M(PH_3)_2(CH_3)(R)$ and $M(PH_3)(CH_3)(R)$ complexes (M = Ni, Pd, Pt; R = CH$_3$, NH$_2$, OH) [58]. The work focused on the effect on the process of the nature of the metal and the R group, but it is worth mentioning that they found that the barrier for reductive elimination of CH$_3$-CH$_3$ is lower for the monophosphine system than for the (bis)phosphine system. A similar result has been found by us in the reductive elimination of two vinyl groups from $Pd(PH_3)_2(vinyl)_2$ [15]. Bo and coworkers have analyzed the effect of the P–Pd–P bite angle on the reductive elimination of acetonitrile in a series of diphosphine systems, $(R_2PXPR_2)Pd(CH_3)CN$ (R = H, Me, Ph; X = $(CH_2)_n$) [59]. Their work confirms that larger bite angles favor reductive elimination, and it is because of orbital, not steric, effects.

Reductive elimination seems thus a mechanistically simpler step than the oxidative addition and transmetallation processes discussed in the previous section. However, a better understanding of some details in reductive elimination would be welcome, and further computational work in this area seems necessary.

5.5
Isomerization

The intermediates obtained after the oxidative addition have been usually observed as *trans*-[PdL$_2$(R)(X)] complexes [2]. Consequently, it has been proposed that the catalytic cycle goes through this type of trans complex. This does not pose any complication to the transmetalation step, and in fact several calculations presented above use this trans arrangement. However, things are different for oxidative addition and reductive elimination. Both reactions require a cis arrangement of the ligands involved, as discussed in previous sections. Therefore, the isomerization steps shown in Reactions (7) and (8) must be postulated to close the catalytic cycle in some cases.

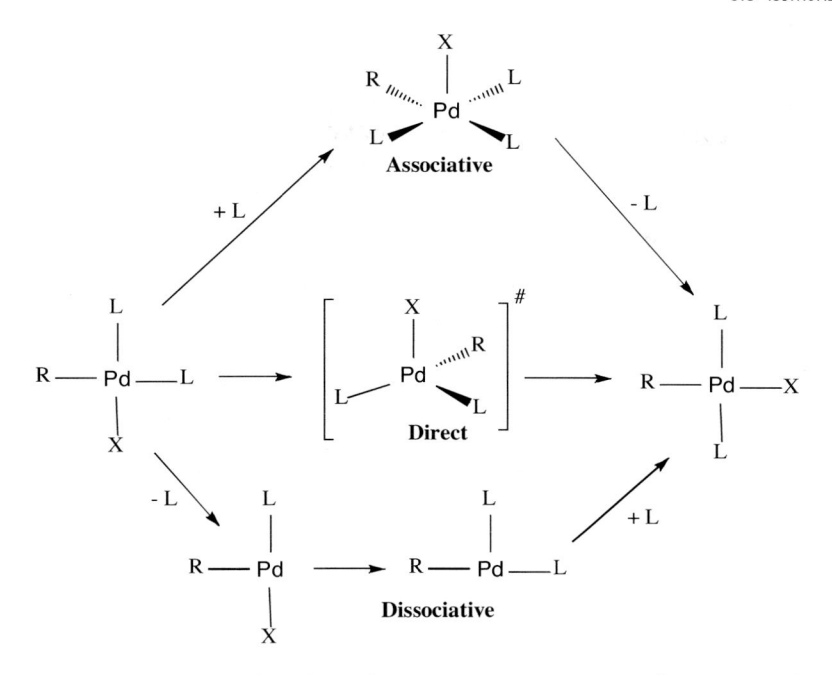

Figure 5.13 Three possible pathways for cis-trans isomerization in a PdL$_2$(R)(X) complex.

$$\text{cis-}[ML_2(R)(X)] \rightarrow \text{trans-}[ML_2(R)(X)] \quad (7)$$

$$\text{trans-}[ML_2(R)(R')] \rightarrow \text{cis-}[ML_2(R)(R')] \quad (8)$$

The cis-trans isomerization process has been studied experimentally by Casado and Espinet for the case of the complex formed by the oxidative addition of C$_6$Cl$_2$F$_3$I to [Pd(PPh$_3$)$_4$] [60]. They found up to four parallel pathways being operative, but concluded that the isomerization was always faster than transmetalation, and therefore not mechanistically relevant.

This step has been computationally analyzed in studies of the full catalytic cycles. In particular, in our study of the Suzuki–Miyaura cross-coupling [15] we considered the three possible pathways depicted in Figure 5.13. They differ in the coordination number of the intermediate/transition states involved.

The direct path is formally the simplest, with the coordination number of four being unchanged throughout the process. However, it has the highest barrier, 85 kJ mol^{-1} above the trans complex in the case of our particular set of PH$_3$ ligands and vinyl and bromide substituents. The barrier is even higher (142 kJ mol^{-1}) when bromide is replaced by vinyl for the isomerization step prior to reductive eliminations. These high barriers are consistent with the tetrahedral nature of the transition states, a geometry heavily disfavored for low spin d^8 ML$_4$ species due to the population of antibonding orbitals.

The two alternative paths, associative and dissociative, avoid the problem of the tetrahedral transition state through a change in the coordination number. In the associative path, a new ligand is attached, and the resulting five-coordinate complex rearranges through Berry pseudorotation. In the dissociative path, a phosphine ligand is lost, and the resulting T-shape three-coordinate species rearranges through a Y-shape transition state. For species with bromide and vinyl, our calculations showed both isomerization paths to proceed smoothly, with barriers of 23 kJ mol^{-1} at most. For the bis(vinyl)complexes, only the dissociative path was feasible, with a barrier of 50 kJ mol^{-1}.

Calculations are conclusive, indicating that the isomerization steps often proposed in some cross-coupling processes happen through low barrier mechanisms, and that these mechanisms can be diverse. This result is in agreement with experimental observation, and confirms the view that this step seems to have minor mechanistic relevance.

5.6
Concluding Remarks

Palladium-catalyzed cross-coupling reactions follow a common general scheme encompassing oxidative addition, transmetalation and reductive elimination steps. However, substantial differences exist with regard to significant mechanistic aspects for different catalytic systems. This experimental observation is confirmed and refined by the computational studies. These show, for example, that transmetalation follows substantially different paths, depending on whether organoboron (Suzuki–Miyaura) or organotin (Stille) reactants are involved. Theoretical studies also help to uncover a very rich mechanistic picture, with various parallel multistep pathways involving similar low energy barriers. It seems, thus, inappropriate to talk about *the* mechanism of cross-coupling reactions, or even of a given type of cross-coupling reactions, but rather of a manifold of mechanisms accessible to a certain type of system. For instance, our calculations on the full catalytic cycle of the Suzuki–Miyaura reaction [15] indicate the oxidative addition step to be rate-determining, in agreement with most experimental observations. However, they do not rule out a change in the nature of the rate-determining step for other particular combinations of catalyst and substrate.

Computational studies in this area have started recently and have been quite successful, but there is still plenty of work to do. The immediate future will likely bring the characterization of transmetalation processes for reactions that have not been yet considered. Completion of these general studies will be, however, only a part of the role of computational chemistry in the area. The main challenge ahead is the study of the reaction features for specific sets of catalyst, reactant and substrate. Only through such studies will computational chemistry be able to advance towards one of its final goals, the computational design of more efficient catalytic systems.

References

1. A. De Meijere, F. Diederich, (Eds.) *Metal-Catalyzed Cross-Coupling Reactions*, 2nd ed., Wiley-VCH, Weinheim, 2004.
2. N. Miyaura, A. Suzuki, *Chem. Rev.* 1995, **95**, 2457–2483.
3. J. K. Stille, *Angew. Chem. Int. Ed. Engl.* 1986, **25**, 508–524.
4. E. Negishi, L. Anastasia, *Chem. Rev.* 2003, **103**, 1979–2017.
5. I. P. Beletskaya, A. V. Cheprakov, *Chem. Rev.* 2000, **100**, 3009–3066.
6. K. Sonogashira, *J. Organomet. Chem.* 2002, **653**, 46–49.
7. J. F. Hartwig, M. Kawatsura, S. I. Hauck, K. H. Shaughnessy, L. M. Alcazar-Roman, *J. Org. Chem.* 1999, **64**, 5575–5580.
8. A. R. Muci, S. L. Buchwald, *Top. Curr. Chem.* 2002, **219**, 131–209.
9. A. F. Littke, G. C. Fu, *Angew. Chem. Int. Ed.* 2002, **22**, 4176–4221.
10. A. M. Echavarren, D. J. Cárdenas, *Metal-Catalyzed Cross-Coupling Reactions*, 2nd edn., eds. A. De Meijere, F. Diederich, Wiley-VCH, Weinheim, 2004, pp. 1–39.
11. P. Espinet, A. M. Echavarren, *Angew. Chem. Int. Ed.* 2004, **43**, 4704–4734.
12. C. Amatore, A. Jutand, *Acc. Chem. Res.* 2000, **33**, 314–321.
13. L. J. Goossen, D. Koley, H. L. Hermann, W. Thiel, *J. Am. Chem. Soc.* 2005, **125**, 11102–11114.
14. L. J. Goossen, D. Koley, H. L. Hermann, W. Thiel, *Organometallics* 2006, **25**, 54–67.
15. A. A. C. Braga, G. Ujaque, F. Maseras, *Organometallics* 2006, **25**, 3647–3658.
16. R. Alvarez, O. N. Faza, C. S. López, A. R. de Lera, *Org. Lett.* 2006, **8**, 35–38.
17. J. O. Noell, P. J. Hay, *J. Am. Chem. Soc.* 1982, **104**, 4578–4584.
18. J. J. Low, W. A. Goddard, *J. Am. Chem. Soc.*, 1984, **106**, 6928–6937.
19. S. Obara, K. Kitaura, K. Morokuma, *J. Am. Chem. Soc.*, 1984, **106**, 7482–7492.
20. S. Sakaki, N. Mizoe, Y. Musahi, B. Biswas, M. J. Sugimoto, *J. Phys. Chem. A*, 1998, **102**, 8027–8036.
21. T. Matsubara, K. Hirao, *Organometallics* 2002, **21**, 4482–4489.
22. H. M. Senn, T. Ziegler, *Organometallics* 2004, **23**, 2980–2988.
23. A. Sundermann, J. M. L. Martin, *Chem. Eur. J.* 2001, **7**, 1703–1711.
24. S. Kozuch, C. Amatore, A. Jutand, S. Shaik, *Organometallics* 2005, **24**, 2319–2330.
25. A. Diefenbach, G. T. de Jong, F. M. Bickelhaupt, *J. Chem. Theory Comput.* 2005, **1**, 286–298.
26. L. J. Goossen, D. Koley, H. L. Hermann, W. Thiel, *Organometallics* 2005, **24**, 2398–2410.
27. J. F. Hartwig, F. Paul, *J. Am. Chem. Soc.* 1995, **117**, 5373–5374.
28. A. G. Littke, C. Y. Dai, G. C. Fu, *J. Am. Chem. Soc.* 2000, **122**, 4020–4028.
29. T. R. Cundari, J. Deng, *J. Phys. Org. Chem.* 2005, **18**, 417–425.
30. M. Ahlquist, P. Fristrup, D. Tanner, P.-O. Norrby, *Organometallics* 2006, **25**, 2066–2073.
31. M. Ahlquist, P.-O. Norrby, *Organometallics* 2007, **26**, 550–553.
32. A. Ariafard, Z. Lin, *Organometallics* 2006, **25**, 403–4033.
33. D. Balcells, F. Maseras, B. A. Keay, T. Ziegler, *Organometallics* 2004, **23**, 2784–2796.
34. D. L. Davies, S. M. A. Donald, S. A. Macgregor, *J. Am. Chem. Soc.* 2005, **127**, 13754–13755.
35. D. García-Cuadrado, A. A. C. Braga, F. Maseras, A. M. Echavarren, *J. Am. Chem. Soc.* 2006, **128**, 1066–1067.
36. N. Miyaura, *J. Organomet. Chem.* 2002, **653**, 54–57.
37. A. A. C. Braga, N. H. Morgon, G. Ujaque, F. Maseras, *J. Am. Chem. Soc.* 2005, **127**, 9298–9307.
38. L. S. Hegedus, *Transition Metals in the Synthesis of Complex Organic Molecules*, University Science Books, Mill Valley, CA, 1994.
39. N. Miyaura, Y. Yamada, H. Suginome, A. Suzuki, *J. Am. Chem. Soc.* 1985, **107**, 972–980.
40. A. A. C. Braga, N. H. Morgon, G. Ujaque, A. Lledós, F. Maseras, *J. Organomet. Chem.* 2006, **691**, 4459–4466.

41 K. Osakada, *Fundamentals of Molecular Catalysis*, eds. H. Kurosawa, A. Yamamoto, Elsevier, Amsterdam, 2003, pp. 233–291.
42 S. Kozuch, S. Shaik, A. Jutand, C. Amatore, *Chem. Eur. J.* 2004, **10**, 3072–3080.
43 J. Louie, J. F. Hartwig, *J. Am. Chem. Soc.* 1995, **117**, 11598–11599.
44 F. Barrios-Landeros, J. F. Hartwig, *J. Am. Chem. Soc.* 2005, **127**, 6944–6945.
45 J. P. Stambuli, M. Bühl, J. F. Hartwig, *J. Am. Chem. Soc.* 2002, **124**, 9346–9347.
46 T. E. Barder, S. D. Walker, J. R. Martinelli, S. L. Buchwald, *J. Am. Chem. Soc.* 2005, **127**, 4685–4696.
47 U. Christmann, R. Vilar, *Angew. Chem. Int. Ed.* 2005, **44**, 366–374.
48 M. Sumimoto, N. Iwane, T. Takahama, S. Sakaki, *J. Am. Chem. Soc.* 2004, **126**, 10457–10471.
49 A. L. Casado, P. Espinet, *J. Am. Chem. Soc.* 1998, **120**, 8978–8985.
50 A. L. Casado, P. Espinet, A. M. Gallego, *J. Am. Chem. Soc.* 2000, **122**, 11771–11782.
51 V. Farina, B. Krishnan, D. R. Marshall, G. P. Roth, *J. Org. Chem.* 1993, **58**, 5434–5444.
52 A. L. Casado, P. Espinet, A. M. Gallego, J. M. Martínez-Ilarduya, *Chem. Comunn.* 2001, 339–340.
53 A. Nova, G. Ujaque, F. Maseras, A. Lledós, P. Espinet, *J. Am. Chem. Soc.* 2006, **128**, 14571–14578.
54 E. Napolitano, V. Farina, M. Persico, *Organometallics* 2003, **22**, 4030–4037.
55 A. Ariafard, Z. Lin, I. J. S. Fairlamb, *Organometallics* 2006, **25**, 5788–5794.
56 V. Ananikov, D. G. Musaev, K. Morokuma, *J. Am. Chem. Soc.* 2002, **124**, 2839–2852.
57 V. Ananikov, D. G. Musaev, K. Morokuma, *Organometallics* 2005, **24**, 715–723.
58 S. A. Macgregor, G. W. Neave, C. Smith, *Faraday Discuss.* 2003, **124**, 111–127.
59 E. Zuidema, P. W. N. M. van Leeuwen, C. Bo, *Organometallics* 2005, **24**, 3703–3710.
60 A. L. Casado, P. Espinet, *Organometallics* 1998, **17**, 954–959.

# 6
Transition Metal Catalyzed Carbon–Carbon Bond Formation: The Key of Homogeneous Catalysis

Valentine P. Ananikov, Djamaladdin G. Musaev, and Keiji Morokuma

6.1
Introduction

Transition metal complex promoted formation of carbon–carbon bonds is a field of much current interest since it can lead to the design of new highly selective and efficient synthetic procedures for organic chemistry [1–3]. In this chapter we focus on the theoretical studies of C–C bond formation reactions through the reductive elimination pathway. As far as possible, the conclusions made in the theoretical studies are verified by experimental data.

6.1.1
Catalytic C–C Bond Formation via the Reductive Elimination Pathway

Reductive elimination is the product formation stage of cross-coupling reactions [4–7]. Transition metal catalyzed cross-coupling methodology involves the oxidative addition of an organic electrophile, **I**, R–X (X = I, Br, Cl, OTf) to a low-valent transition metal complex (L_mM^{n-2}), leading to oxidative addition product **II** (Scheme 6.1). Transmetallation between **II** and organometallic reagent R′–Y (Y = in most cases a main group metal) affords intermediate **III** with two metal–carbon (M–C) σ-bonds. Reductive elimination from **III** yields the desired organic product R–R′ and regenerates low-valent metal complex (**IV**). Several fascinating practical tools for synthetic organic chemistry have been developed within this general framework (**I** → **II** → **III** → **IV**): the Suzuki–Miyaura reaction (Y = B) [8–11]; the Migita–Kosugi–Stille reaction (Y = Sn) [12–15]; the Sonogashira coupling reaction (Y = Cu) [7, 16, 17]; the Kumada–Tamao–Corriu reaction (Y = Mg) [18–20]; the Hiyama reaction (Y = Si) [21–24]; and the Negishi coupling (Y = Zn) [25–27]. Other organometallic and heteroatom species, Y = Al, Ge, In, Ag, Te, S, have been also tested as reagents for the transmetallation step [28–35].

Computational Modeling for Homogeneous and Enzymatic Catalysis.
A Knowledge-Base for Designing Efficient Catalysts. K. Morokuma and D. G. Musaev (Eds.)
Copyright © 2008 WILEY-VCH Verlag GmbH & Co. KGaA, Weinheim
ISBN: 978-3-527-31843-8

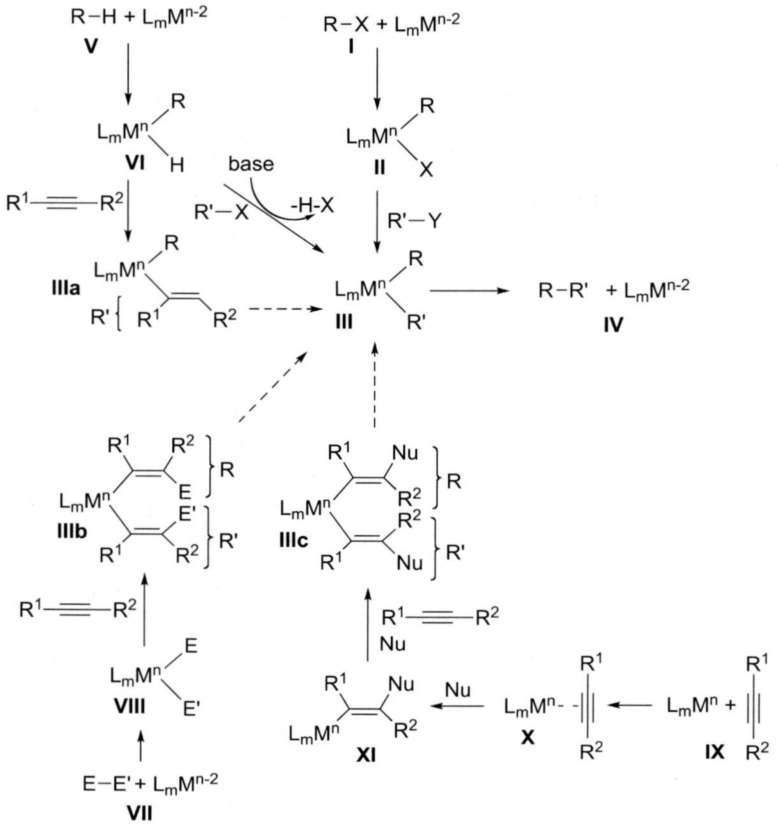

Scheme 6.1 Catalytic reaction involving C–C reductive elimination (complex **III** represents an abbreviated notation of the complexes **IIIa–IIIb**).

Palladium complexes are usually the catalysts of choice for performing cross-coupling reactions, while in some cases Ni and Fe compounds have also been used [4–7]. Usually, at least one of the organic ligands (R or R′) is an unsaturated group (vinyl, aryl or alkynyl), while cross-coupling of two alkyl groups is of rare practical importance [4–7]. Despite the simple general framework of the cross-coupling reactions (oxidative addition, transmetallation and reductive elimination), mechanistic studies have led to controversial conclusions about the rate-determining step. It was concluded that oxidative addition [7], transmetallation [4, 5] or C–C coupling [36] could be a rate-determining step, depending on the reaction type, conditions and substituents. Obviously, detailed mechanistic studies of all the elementary steps involved in the cross-coupling reaction are needed to gain more insight into the problem. In the present chapter we consider only the C–C bond formation stage of cross-coupling reactions. Other stages of the catalytic cross-coupling reactions are discussed in Chapter 5 by Braga, Ujaque, and Maseras.

Intermediate **III** or its derivatives can also be generated in several other catalytic transformations (Scheme 6.1). Indeed, oxidative addition of the R–H bond (**V**) to the low-valent transition metal complex (L_mM^{n-2}) followed by reaction of resultant **VI** with R′–X in the presence of base also leads to intermediate **III**, which finally produces **IV** (for M = Ir, Rh, Co) [37]. Reaction of **VI** with alkynes leads to alkenylation of the R–H bond (**V** → **VI** → **IIIa** → **IV**), which also proceeds through the sequence of R–H bond oxidative addition, alkyne insertion into the M–H bond of **VI** and reductive elimination (M = Ru, Rh, Ir) [38–43].

In addition, element–element bond (E–E′) addition to a low-valent transition metal followed by reaction with two alkyne molecules also gives a metal complex with two vinyl ligands (**IIIb**) suitable for C–C reductive elimination (for M = Pd, Rh, Pt, Ni; E = B, Si, Sn, S, Se, etc.) [44–46]. The final product obtained within this catalytic reaction (**VII** → **VIII** → **IIIb** → **IV**) retains cis geometry of the double bonds. The same approach is also applicable for reactions involving an E–H bond [45, 47–49] and dihydrogen molecule [50] (Scheme 6.1).

External nucleophile attack on the coordinated triple bond leads to bis-(σ-vinyl)-derivative **IIIc** with trans geometry of the double bonds with high selectivity (**IX** → **X** → **XI** → **IIIc**). Reductive elimination from the complex **IIIc** gives conjugated 1,3-dienes, preserving the trans geometry of the double bonds [51, 52].

As well as the synthetic methods discussed above, investigation of C–C coupling involving a vinyl group is an important part of mechanistic studies of some processes of industrial importance that are catalyzed by transition metal complexes [53, 54].

6.1.2
Reductive Elimination of Alkyl Groups (Alkyl–Alkyl Coupling)

Numerous mechanistic studies have been carried out for understanding reductive elimination from dialkyl or mixed metal complexes of platinum [55–62], palladium [58, 63–67], rhodium [55], ruthenium [60] and iridium [55].

Extensive theoretical investigations on C–C bond formation through reductive elimination of the methyl groups (as well as the reverse process: C–C bond activation of ethane) have been published for complexes of platinum [68–75], palladium [68–70, 72–76], rhodium [76–80] and iridium [79, 81].

Several reviews have summarized the results of theoretical studies of alkyl–alkyl coupling reactions [76, 82–85] and are not be repeated here. Our main interest is to analyze reductive elimination reactions of unsaturated ligands leading to conjugated products, which are of primary importance for modern transition metal catalyzed organic synthesis.

6.2
C–C Coupling of Unsaturated Ligands

In the text below we use the notation **n_a**, where **n = 1** [R = R′ = CH_3], **2** [R = R′ = ($CH=CH_2$)], **3** [R = R′ = C_6H_5], **4** [R = R′ = (C≡CH)], **5** [R = CH_3 and

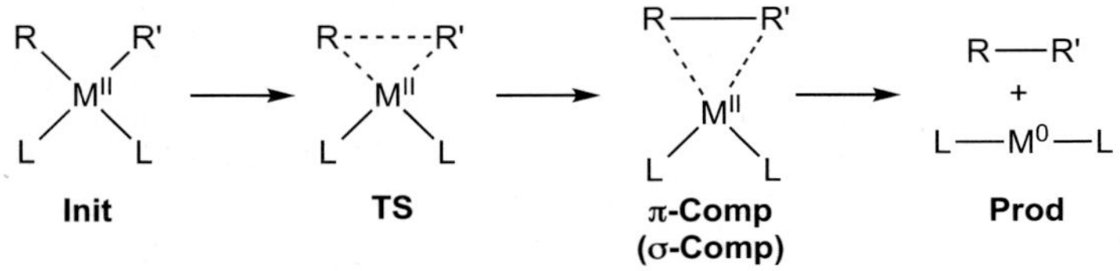

Scheme 6.2

R—M(L)(L)—R' → R⋯M(L)(L)⋯R' → R—R' over M(L)(L) → R—R' + L—M⁰—L

Init TS π-Comp (σ-Comp) Prod

M = Pd, Pt
R, R' = CH$_3$, CH=CH$_2$, C$_6$H$_5$, C≡CH

Scheme 6.2 General scheme of the C–C coupling reaction.

R' = (CH=CH$_2$)], **6** [R = CH$_3$ and R' = C$_6$H$_5$], **7** [R = CH$_3$ and R' = (C≡CH)], **8** [R = (CH=CH$_2$) and R' = C$_6$H$_5$)], **9** [R = (CH=CH$_2$) and R' = (C≡CH)], **10** [R = (C≡CH) and R' = C$_6$H$_5$], and stands for the complex, while **a** = **Init**, **TS**, **π-Comp** (or **σ-Comp**) and **Prod** marks the structures [initial reactant, transition state, π- (or σ-) complex and product, respectively] on the potential energy surface of the reaction (Scheme 6.2).

The following acronyms of organic ligands are used throughout: Me – methyl (CH$_3$), Vin – vinyl (CH=CH$_2$), Ph – phenyl (C$_6$H$_5$), and Eth – ethynyl (C≡CH).

6.2.1
Reductive Elimination from the Symmetrical R$_2$M(PH$_3$)$_2$ Complexes

Theoretical calculations at the B3LYP hybrid density functional level in conjunction with the standard 6-311G(d) basis set for C, P, H and the triple-ζ basis sets for main group elements and the Stuttgart/Dresden relativistic effective core potentials for the metals (B3LYP/BSI) have been carried out for a series of reactions involving unsaturated ligands and M = Pd, Pt [86]. Reductive elimination from these complexes proceeds through the three-centered transition states **1_TS**–**4_TS** (Figure 6.1); calculated energetics are listed in Table 6.1.

The reductive elimination of ethane (R = Me) requires overcoming a higher C–C coupling barrier than other ligands (Table 6.1). Namely, the calculated barrier for C–C coupling reaction decreases in the order: Me–Me > Eth–Eth > Ph–Ph > Vin–Vin.

Reductive elimination of two vinyl groups leading to buta-1,3-diene (**2_Prod**) requires a $\Delta E^{\neq} = 6.8$ and 19.3 kcal mol^{-1} coupling barrier for M = Pd and Pt, respectively, while reductive elimination of two methyl groups (**1_Prod**), leading to ethane, requires much higher barrier, $\Delta E^{\neq} = 25.2$ and 45.9 kcal mol^{-1} for M = Pd and Pt, respectively (Table 6.1).

Interestingly, reductive elimination of the vinyl ligands may proceed via the s-trans- or s-cis- pathways, depending on the conformation of the 1,3-diene skeleton [99, 100]. The former pathway was calculated to be energetically more favorable. To proceed with reductive elimination vinyl groups have to adopt the proper conformation via the rotation around M–C bond. If the rotation is inhibited by steric crowding, C–C reductive elimination does not take place, as was demonstrated by experimental and theoretical studies [43].

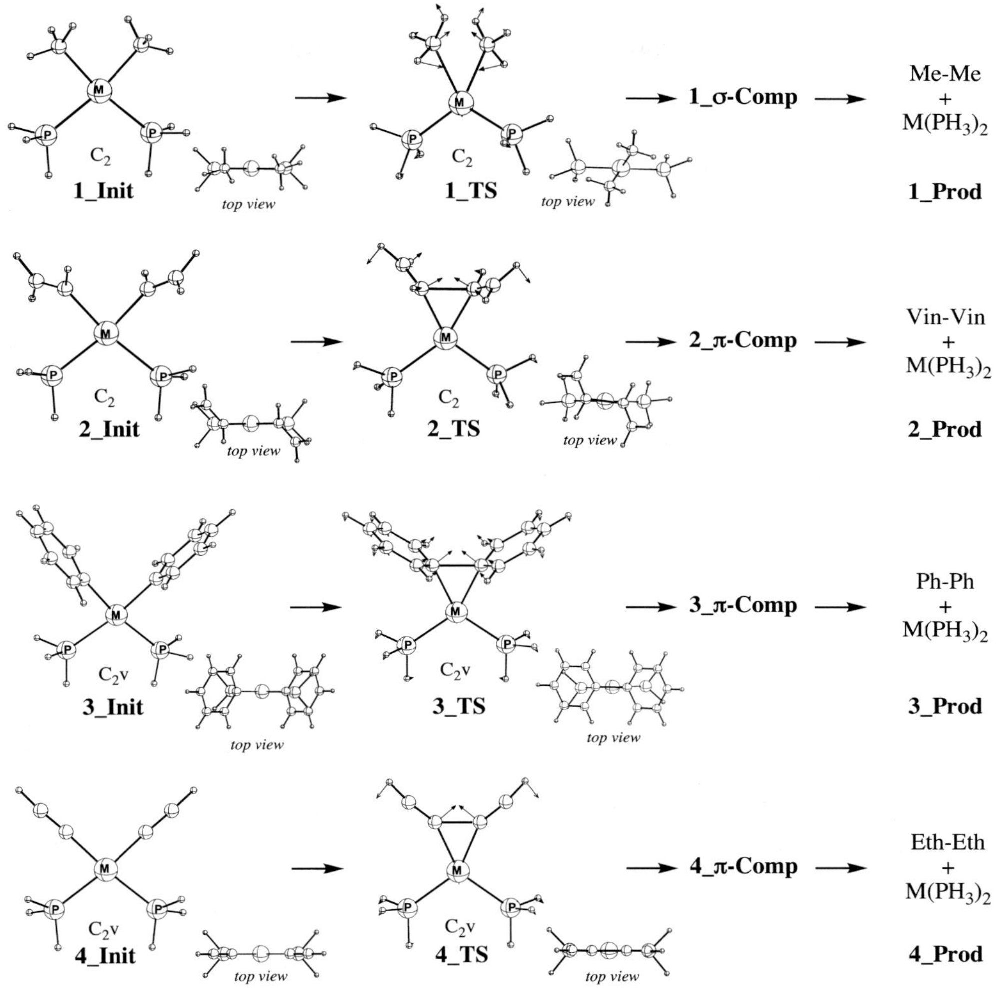

6.2 C–C Coupling of Unsaturated Ligands | 135

Figure 6.1 Schematic presentation of the symmetrical R–R coupling reactions from $R_2M(PH_3)_2$ complexes; molecular structures optimized at B3LYP/BSI level [86].

All reductive elimination reactions were calculated to be exothermic. For both metals (M = Pd, Pt) the exothermicity of the reaction decreases in the following order: Vin–Vin ≈ Ph–Ph ≈ Me–Me > Eth–Eth. Butadiyne formation is less exothermic by about 14 kcal mol^{-1} than the other symmetric C–C coupling reactions.

Overcoming the transition states leads to the weakly bound complexes. For the Me–Me coupling reaction, it is a weakly (<1 kcal mol^{-1}) bound σ-complex **1_σ-Comp** (Table 6.1). With unsaturated ligands, the corresponding π-complexes are

Table 6.1 Calculated activation energy ($\Delta E^{\neq}$, $\Delta H^{\neq}$, $\Delta G^{\neq}$)[a] and energy of reaction (ΔE, ΔH, ΔG)[b] for $RR'M(PH_3)_2 \rightarrow R-R' + M(PH_3)_2$ at B3LYP/BSI level (in kcal mol^{-1}, at 298.15 K and 1 atm) [86].[c]

	Complex	$\Delta E^{\neq}$ (kcal mol^{-1})	ΔE (kcal mol^{-1})	$\Delta H^{\neq}$ (kcal mol^{-1})	ΔH (kcal mol^{-1})	$\Delta G^{\neq}$ (kcal mol^{-1})	ΔG (kcal mol^{-1})
1	M(CH$_3$)$_2$(PH$_3$)$_2$	25.2 (45.9)	−31.4/−31.5 (−16.2/−16.3)	24.2 (44.8)	−30.4/−31.3 (−15.4/−15.5)	23.6 (45.0)	−41.7/−35.7 (−26.3/−19.9)
2	M(CH=CH$_2$)$_2$(PH$_3$)$_2$	6.8 (19.3)	−33.8/−42.5 (−17.8/−27.7)	5.9 (18.1)	−33.2/−41.2 (−17.5/−26.7)	6.0 (18.4)	−44.5/−42.2 (−29.0/−27.2)
3	M(Ph)$_2$(PH$_3$)$_2$	11.7 (28.1)	−31.4/−28.2 (−15.7/−7.5)	10.5 (26.5)	−31.1/−27.9 (−15.6/−7.1)	10.7 (26.2)	−41.9/−29.0 (−26.8/−8.2)
4	M(C≡CH)$_2$(PH$_3$)$_2$	14.6 (29.6)	−17.5/−28.8 (−0.3/−15.3)	13.0 (27.6)	−18.1/−28.8 (−0.8/−15.7)	11.3 (25.8)	−29.1/−30.5 (−11.6/−16.0)
5	M(CH$_3$)(CH=CH$_2$)(PH$_3$)$_2$	15.9 (33.4)	−30.6/−38.4 (−15.0/−24.6)	15.5 (32.2)	−29.4/−36.3 (−14.5/−23.3)	14.1 (33.5)	−42.0/−37.7 (−25.0/−22.2)
6	M^{tr}(CH$_3$)(Ph)(PH$_3$)$_2$	18.7 (37.8)	−30.4/−27.2 (−14.9/−6.7)	17.5 (36.4)	−30.0/−26.7 (−14.6/−6.2)	16.9 (37.5)	−41.9/−27.6 (−25.9/−6.1)
7	M^{tr}(CH$_3$)(C≡CH)(PH$_3$)$_2$	21.3 (39.8)	−21.3/−29.9 (−5.6/−18.3)	20.6 (38.3)	−20.7/−29.1 (−5.7/−17.5)	18.5 (38.1)	−33.0/−29.0 (−16.5/−16.5)
8	M^{tr}(CH=CH$_2$)(Ph)(PH$_3$)$_2$	9.8 (24.5)	−31.7/−40.7 (−15.9/−26.3)	8.6 (23.1)	−31.4/−39.6 (−15.7/−25.3)	8.7 (22.8)	−43.5/−40.3 (−27.6/−25.0)
9	M^{tr}(CH=CH$_2$)(C≡CH)(PH$_3$)$_2$	11.5 (25.9)	−22.6/−32.9 (−6.5/−19.5)	10.2 (24.3)	−22.8/−32.3 (−6.7/−18.8)	9.5 (23.4)	−34.7/−33.3 (−18.6/−19.1)
10	M^{tr}(Ph)(C≡CH)(PH$_3$)$_2$	12.2 (27.2)	−23.2/−32.3 (−7.2/−20.9)	10.8 (25.4)	−23.6/−32.9 (−7.6/−20.3)	10.4 (24.5)	−35.1/−29.7 (−18.9/−19.9)

a) $\Delta E^{\neq}/H^{\neq}/G^{\neq} = E/H/G(\mathbf{TS}) - E/H/G(\mathbf{Init})$.
b) $\Delta E/H/G = E/H/G(\mathbf{Prod}) - E/H/G(\mathbf{Init})$. Values after/correspond to $E/H/G(\mathbf{Comp}) - E/H/G(\mathbf{Init})$.
c) Values for M = Pd are given without parentheses, while those for M = Pt are in parentheses.

formed (**2_π-Comp** to **10_π-Comp**). For the Eth and Vin ligands the complexation energy is 8–9 (10–12) and 9–11 (13–15) kcal mol^{-1}, respectively (Table 6.1). The release of the final C–C coupling product from a π-complex may involve participation of an external ligand (ligand substitution).

In agreement with calculated energy trends, the difference in M–C bond distances between the transition state and the reactant [Δ(M–C)] shows that the reductive elimination in (Me)$_2$M(PH$_3$)$_2$ proceeds through a later transition state than the that for other R/R′ ligands. For the methyl ligand Δ(M–C) values are 0.098 and 0.162 Å (for M = Pd and Pt, respectively), while for Ph, Vin and Eth Δ(M–C) values are in the range of 0.006–0.044 and 0.041–0.083 Å (for M = Pd and Pt, respectively). As seen from the top view (Figure 6.1) the TS for the Me–Me coupling has a non-planar structure (the tilt angle > 50°), while for the other symmetric coupling processes it is almost planar (the tilt angle < 7°). The inclusion of an enthalpy correction only slightly changes the potential energy surface; inclusion of an entropy correction (Gibbs energy surface) reduces the activation barriers by 1–3 kcal mol^{-1} and increases the exothermicity of the reactions by 10–12 kcal mol^{-1} (Table 6.1).

Calculations at the B3LYP/6-31G(d)-Lanl2dz level performed for the s-trans pathway of Vin–Vin reductive elimination from the [Pd(Vin)$_2$(PH$_3$)$_2$] complex resulted in an activation barrier of 4.8 kcal mol^{-1} and reaction exothermicity of 36.2 kcal mol^{-1} [87, 88]. The results agree fairly well with the energy data obtained at B3LYP/BSI level (Table 6.1), indicating that an energy change of ~2 kcal mol^{-1} should be expected upon changing from a double-ζ to a triple-ζ basis set.

6.2.2
Reductive Elimination from the Asymmetrical RR′M (PH$_3$)$_2$ Complexes

As one would expect, the asymmetrical C–C coupling (R,R′ = Me, Vin, Eth, Ph; R ≠ R′) also proceeds through the three-centered transition states **5_TS–10_TS** (Figure 6.2), similar to the symmetrical one discussed above (here and later the notation "assymmetrical" refers to unsymmetrical ligand environment, rather than to chirality). The results presented in Table 6.1 clearly show that the calculated barriers and reaction energies for asymmetrical R–R′ coupling in RR′M(PH$_3$)$_2$ complex are close to an average between the corresponding barriers of the symmetrical R–R and R′–R′ coupling reactions in MR$_2$M(PH$_3$)$_2$ and R′$_2$M(PH$_3$)$_2$, respectively (M = Pd, Pt).

Using the simple Eqs. (1) and (2), the energetics of asymmetrical coupling reactions can be calculated from the corresponding data for the symmetrical coupling reactions with 1–2 kcal mol^{-1} accuracy.

$$\Delta H(\text{R–R}') \approx [\Delta H(\text{R–R}) + \Delta H(\text{R}'\text{–R}')]/2 \tag{1}$$

$$\Delta H^{\neq}(\text{R–R}') \approx [\Delta H^{\neq}(\text{R–R}) + \Delta H^{\neq}(\text{R}'\text{–R}')]/2 \tag{2}$$

For the asymmetrical coupling processes, the largest tilt angle, (9–18° for Pd, and 22–28° for Pt), corresponds to the Me–R coupling, R = Vin, Ph, Eth (see top

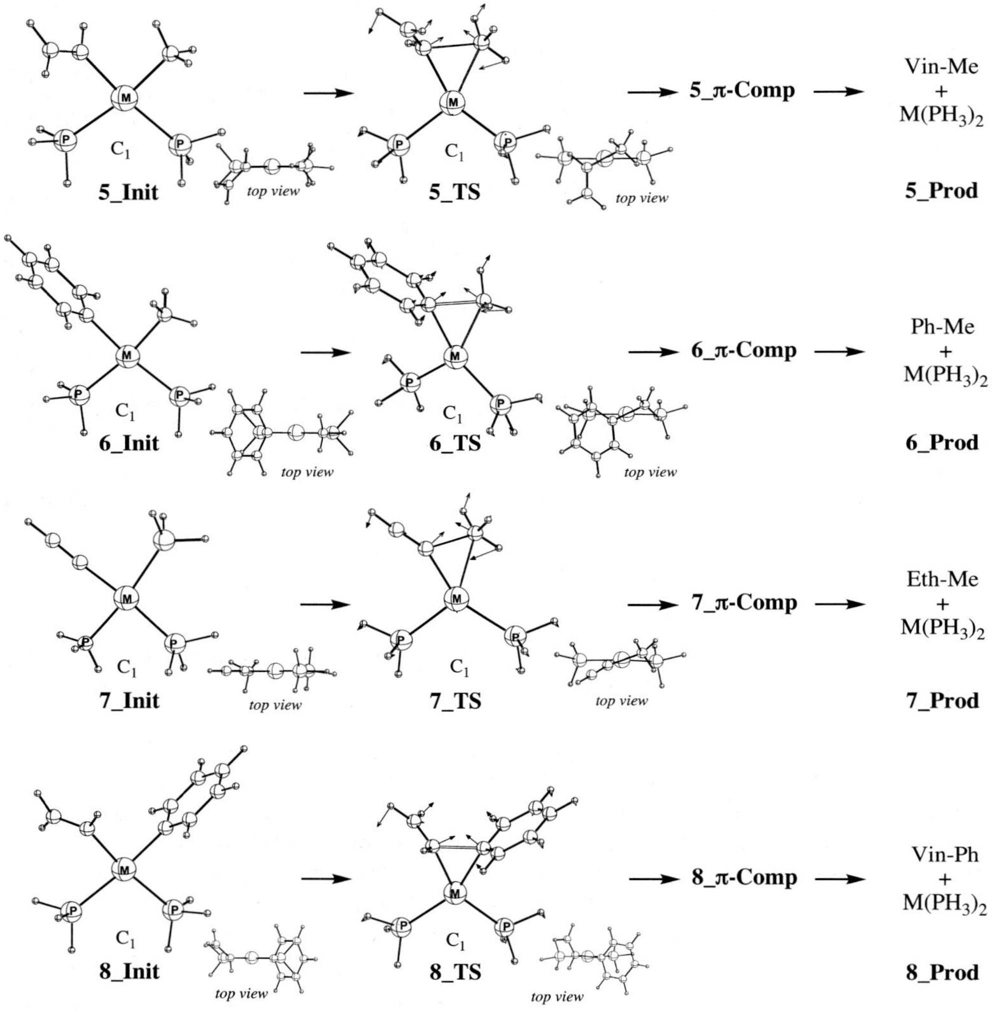

Figure 6.2 Schematic presentation of asymmetrical R–R′ coupling reactions from R′RM(PH$_3$)$_2$ complexes; molecular structures optimized at B3LYP/BSI level [86].

views; Figure 6.2). In other words, the process involving Me ligands always proceeds via a non-planar transition state, and the degree of the non-planarity is correlated with the number of Me ligands involved.

An interesting study of the Vin–Ph reductive elimination from dinuclear palladium complex has been carried out at the B3LYP/6-31G(d)-Lanl2dz level [89]. An activation energy of $\Delta E^{\neq} = 8.9$ kcal mol^{-1} was calculated for the phosphine com-

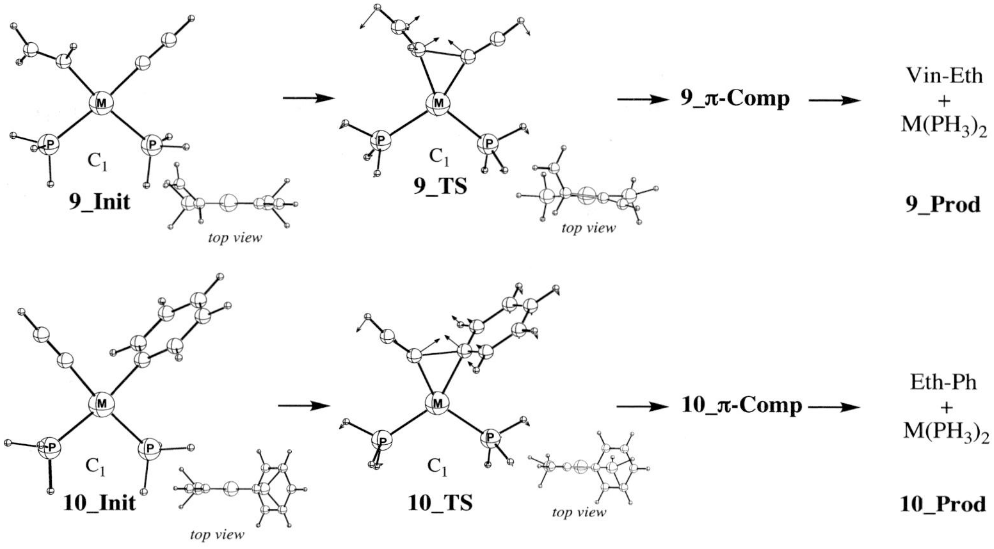

Figure 6.2 (cont.)

plex, which agrees well with the value predicted for the mononuclear palladium complex, $\Delta E^{\neq} = 9.8$ kcal mol^{-1} (Table 6.1).

Calculated energetic data for the symmetrical and asymmetrical coupling reactions are of particular use in analyzing the reactivity of mixed transition metal complexes containing various organic groups. For example, it was reported that mixed a (alkyl)(alkynyl)(alkenyl)Ir^{III} complex is stable towards the reductive elimination reaction (Scheme 6.3) [90, 91]. However, reaction with HCl converts it into the bis-(vinyl)-derivative, which undergoes reductive Vin–Vin coupling. The observed reactivity (Vin–Vin > Me–Vin, Vin–Eth) for the iridium complex is in line with calculations carried out for palladium and platinum analogs (Table 6.1).

Another study of the iridium complexes revealed an activity order of Vin–Vin > Vin–Me > Ph–Me > Me–Me [92] in the C–C coupling reaction, which is in excellent agreement with the calculated energetic trend for palladium and

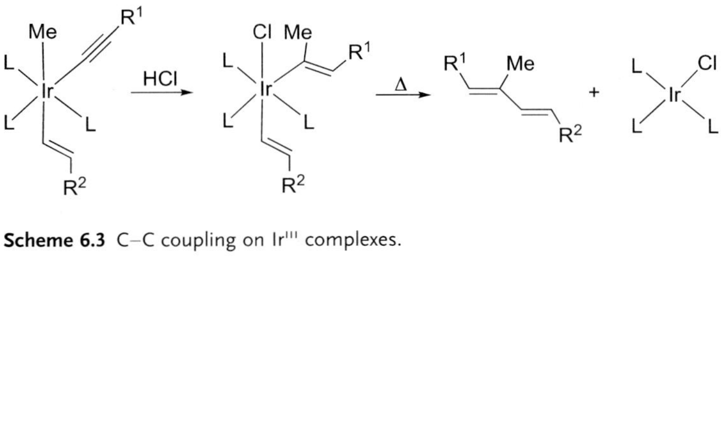

Scheme 6.3 C–C coupling on Ir^{III} complexes.

platinum complexes (Table 6.1). There are some other studies of C–C reductive elimination reactions performed for different metals, which are in line with calculated reactivity patterns [93–95].

6.2.3
Homocoupling versus Heterocoupling Pathways via C–C Reductive Elimination

Practical utilization of cross-coupling reactions sometimes results in a considerable amount of homocoupling impurities **XIII** and **XV** in addition to the desired heterocoupling product **IV** (Scheme 6.4) [4–11]. Theoretical studies discussed above (Sections 6.2.1 and 6.2.2) provide a flexible tool to predict relative reactivity in reductive elimination.

According to the calculated energetic parameters an intermediate reactivity for the heterocoupling reaction should always be expected, i.e., the reactivity is R–R > R–R′ > R′–R′. The activation and reaction energies given in Table 6.1 allow the selection of more reactive (R–R) and less reactive (R′–R′) C–C coupling pathways.

For example, if Ph–Vin heterocoupling reaction is performed the following relative reactivity on the reductive elimination step should be expected: Vin–Vin > Vin–Ph > Ph–Ph (Table 6.1). In this case the Vin–Vin by-product formation should be expected first (not Ph–Ph). Of course, a care should be taken to apply the calculated relative reactivity for the analysis of real practical systems, since the other steps (which influence the formation of complexes **III**, **XII** and **XIV**) may also change the overall reactivity in the reaction.

Experimental investigation of C–C reductive elimination reaction from (PNP)Rh(Ph)Me and (PNP)RhPh$_2$ complexes have revealed that the rate of elimination of Ph–Ph ($t_{1/2} = 7$ min) was greater than that of the elimination of Ph–Me ($t_{1/2} = 17$ min; both at 40 °C) [36]. The data corresponds to the case of R = Ph, R′ = Me (Scheme 6.4) and the observed relative reactivity is in total agreement with predicted energetic parameters (Table 6.1).

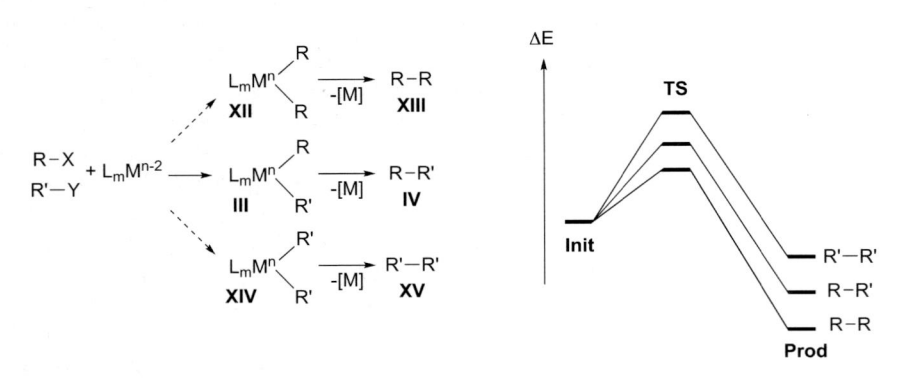

Scheme 6.4 Homocoupling and heterocoupling reactions (left) and schematic representation of their potential energy surfaces (right).

6.2.4
Metal Effect on C–C Reductive Elimination Reaction

The metal effect has been studied at the B3LYP/BSI level using a model reaction of Vin–Vin coupling from phosphine complexes of platinum group metals [100]. The activity of metal complexes in C–C bond formation reaction decreases in the order: Pd^{IV}, $Pd^{II} > Pt^{IV}$, Pt^{II}, $Rh^{III} > Ir^{III}$, Ru^{II}, Os^{II}.

Within each subgroup the reactivity of the metal complexes decreases via Pd > Pt, Rh > Ir and Ru > Os [100]. A noticeable effect of the metal oxidation state has been also predicted: $Pt^{IV} > Pt^{II}$ and $Pd^{IV} > Pd^{II}$ [100].

The former feature can be utilized to promote an oxidatively induced reductive elimination reaction [96, 97]. It was reported that reductive elimination from a rather inert cyclic Pt^{II} alkynyl complex can be facilitated by addition of I_2, which oxidizes it to the corresponding Pt^{IV} derivative (Scheme 6.5) [97]. This experimental observation proves the predicted reactivity of the platinum complexes $Pt^{IV} > Pt^{II}$ [100]. It was also proposed that oxidation by iodine of a bis-(σ-vinyl) complex of Pt^{II} promotes Vin–Vin reductive elimination [51].

Computational study of the Me–Me coupling from $[M(Me)_2(PH_3)_2]$ complex for M = Ni, Pd and Pt performed at the density functional level provided the information about the relative activity of nickel complexes in reductive elimination [98]. The activation barrier of the C–C bond formation reaction decreases in the order Pt > Pd > Ni, $\Delta E^{\neq} = 45.8$, 24.9 and 16.8 kcal mol^{-1}, respectively. However, the reaction energy does not follow the same trend: $\Delta E = -3.5$, -19.0 and -4.1 kcal mol^{-1} for M = Pt, Pd and Ni, respectively. According to this study, C–C bond formation on Ni and Pt complexes could be reversible due to the low barrier of the C–C bond oxidative addition (reverse transformation). C–C bond formation involving a Pd complex can be considered irreversible under standard conditions due to high exothermicity.

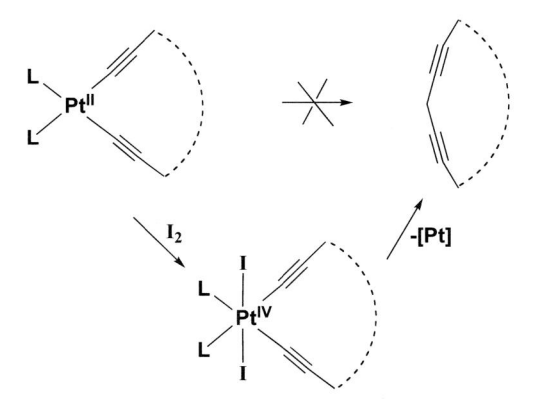

Scheme 6.5 C–C coupling of alkynyl groups on Pt^{IV} complex.

It was shown that the concept of lower lying electronic configurations provides reliable qualitative description of calculated reactivity of different metal complexes. This topic has been reviewed in detail recently [99].

6.2.5
Ligand Effect on C–C Reductive Elimination Reaction

The effect of different ligands (L = Cl, Br, I, NH_3, PH_3) was studied on the model reaction of Vin–Vin coupling for platinum and palladium complexes [100]. Calculations at B3LYP/BSI level have demonstrated that the reactivity of the $[M(Vin)_2L_n]$ complexes in C–C bond formation decreases in the order: L = PH_3 > I > Br, NH_3 > Cl. The conclusion is valid for both metals M = Pd and Pt in M^{II} and M^{IV} oxidations states. In all studied cases phosphine ligands decrease the activation barriers and increase the exothermicity of the reactions. The study revealed steric and electronic factors responsible for the calculated ligand effect [100]. An important part of the electronic effect of the ligands concerns stabilization of different electronic states of the metal center [99, 100].

Extended Hückel and density functional calculations were carried to probe possible influence of unsaturated groups on the metal center [101, 113]. It was indicated that not only the ligands (L) but also the organic groups (R, R′) attached to the metal effect the energetic cost of electronic reorganization.

6.2.6
Dissociative Mechanism of C–C Bond Formation

Ligand dissociation from an octahedral complex of Pt^{IV} substantially decreases the activation barrier of the C–C reductive elimination of vinyl groups (Scheme 6.6) [102]. Reductive elimination of vinyl groups from the six-coordinated complex **XVI** requires overcoming an activation barrier of $\Delta E^{\neq} = 28.0$ kcal mol^{-1}.

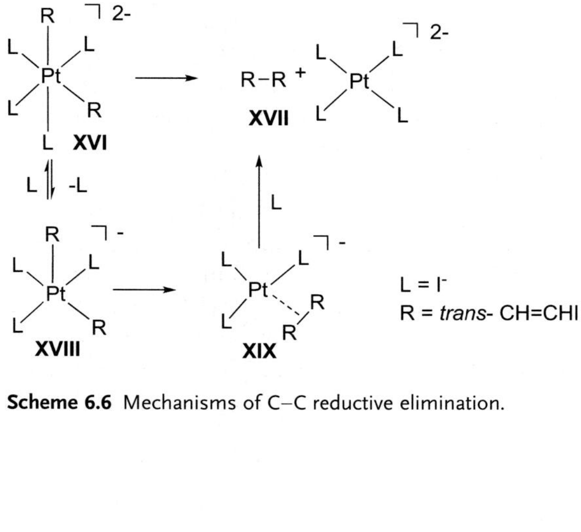

Scheme 6.6 Mechanisms of C–C reductive elimination.

After ligand dissociation, reductive elimination of the vinyl groups from the five-coordinate complex **XVIII** requires overcoming a much smaller activation barrier of $\Delta E^{\neq} = 7.8$ kcal mol^{-1}. Ligand replacement in π-complex **XIX** releases the final product R–R (**XVII**).

These calculations were carried out for the model reaction of 1,4-diiodobuta-1,3-diene formation employing B3LYP hybrid density functional method and the standard 6-311G(d,p) basis set for C and H and the triple-ζ basis set for main group elements and the Stuttgart/Dresden relativistic effective core potentials for the I and Pt (B3LYP/BSII) [102]. The results discussed above correspond to the s-trans-reductive elimination pathway; a similar decrease of activation energy upon changing from six-coordinate to five-coordinate metal complex was calculated for the s-cis-pathway as well [102].

A similar conclusion has been made in the experimental study of Ar–Ar coupling from octahedral complexes of Pt^{IV}: the presence of stronger binding ligands retards C–C reductive elimination, while the presence of labile ligands is essential to facilitate the reaction [103].

Rapid butadiene elimination was predicted for Vin–Vin coupling reactions from the three-coordinate complex [PdVin$_2$PMe$_3$]. An activation energy of 4.1 kcal was calculated at the B3LYP/6-31G(d)-SDD level [104]. This is 2.7 kcal mol^{-1} lower than for Vin–Vin reductive elimination from the four-coordinated **1_Init** complex calculated at the B3LYP/BSI level (Table 6.1) [86].

Calculations on Vin–Vin reductive elimination were also performed for the four-coordinate complex [PdVin$_2$(PH$_3$)$_2$] and the three-coordinate complex [PdVin$_2$PH$_3$] at the same theory level (B3LYP/6-31G(d)-Lanl2dz) [87], which allows direct comparison. The reaction involving [PdVin$_2$PH$_3$], $\Delta E^{\neq} = 1.4$ kcal mol^{-1} and $\Delta E = 25.5$ kcal mol^{-1}, was found to be kinetically favored compared with undissociated complex [PdVin$_2$(PH$_3$)$_2$], $\Delta E^{\neq} = 4.8$ kcal mol^{-1} and $\Delta E = 36.2$ kcal mol^{-1}.

An experimental study of a C–C coupling reaction involving a Ph group from a square-planar Pd^{II} complex has provided evidence for phosphine ligand dissociation prior to the reductive elimination step [105].

Therefore, according to theoretical and experimental studies, ligand dissociation facilitates C–C coupling of unsaturated organic ligands through the reductive elimination pathway for octahedral M^{IV} and square-planar M^{II} complexes.

A very interesting feature of C–C bond formation has been revealed upon comparing the relative reactivity of [M(Me)$_2$(PH$_3$)$_2$] for M = Ni, Pd, Pt [98]. According to calculated activation barriers, reductive elimination from a four-coordinate complex follows the trend: Ni > Pd > Pt ($\Delta E^{\neq} = 16.8$, 24.9 and 45.8 kcal mol^{-1}, respectively). However, reductive elimination from a three-coordinate complex [M(Me)$_2$PH$_3$] diminishes the difference between palladium and nickel: Ni, Pd > Pt ($\Delta E^{\neq} = 10.0$, 11.6 and 23.9 kcal mol^{-1}, respectively). Therefore, a smaller difference in reactivity of similar transition metal complexes would be expected for a dissociative mechanism, while the difference would be maximized in coordinatively saturated complexes.

6.2.7
Solvent Effect on C–C Reductive Elimination Reaction

Studies on the solvent effect on transition metal catalyzed reactions are of much current interest, especially taking into account the practical importance of catalytic reactions in aqueous media [106–110].

Comparing ΔG_{Solv} energy calculated at the PCM-B3LYP/BSI level, the following main trends were outlined for a model reaction of Vin–Vin coupling (M = Pd, Pt, Rh, Ir, Ru, Os) [100]: (a) solvation does not substantially effect reactions involving neutral transition metal complexes, the average energy change is 2–3 kcal mol^{-1}; (b) polar solvents (water, methanol, etc.) greatly facilitate reductive elimination from dianionic derivatives, reducing activation energies by ~8 kcal mol^{-1} and increasing reaction exothermicity by ~10 kcal mol^{-1}.

A detailed comparative study of the Vin–Vin coupling reaction at the PCM-B3LYP/BSII level has been carried out for a series of solvents (water, methanol, acetone, chloroform, benzene) and in the gas phase [102]. For reductive elimination pathways from both six- and five-coordinate complexes the reaction involving charged species is kinetically and thermodynamically favored in polar solvents. According to PCM calculations, electrostatic interactions make a dominant contribution to the solvation energy of charged species [100, 102].

An important issue of the solvent effect concerns stabilization of unusual dianionic M^0 complexes in polar solvents [100, 102]. It was suggested recently that anionic complexes of Pd^0 play an important role in cross-coupling and Heck reactions [111, 112].

6.2.8
Reductive Elimination of Unsaturated Organic Molecules Involving Cyano and Carbonyl Groups

A good comparative study has been published for the reductive elimination involving sp and sp^2 carbon atoms [CN, CH=CH$_2$ and C(O)CH$_3$ groups] on Pd^{II} phosphine complexes [113]. The activation barriers of C–C reductive elimination decrease in the order (the hybridization of carbon atoms is given in parenthesis): Me–CN (sp^3-sp) > Vin–CN (sp^2-sp); $\Delta E^{\neq} = 21.3$ and 12.4 kcal mol^{-1}, respectively.

The order predicted for the reductive elimination of the other unsaturated groups was: Me–Eth (sp^3-sp) > Vin–Eth (sp^2-sp); $\Delta E^{\neq} = 21.3$ and 11.5 kcal mol^{-1}, respectively (Table 6.1).

Replacing the Me group with C(O)CH$_3$ (sp^2 carbon atom) significantly increases the reactivity in C–C bond formation [113], again in excellent agreement with calculated trends for Me–Me and Me–Vin couplings (Table 6.1). The agreement between calculated values obtained for different organic groups suggests that it is a general property of reductive elimination involving unsaturated ligands.

6.3
Conclusions

The comparison made in this chapter suggests a similar nature of the C–C coupling reaction for different transition metal complexes and it would be worth employing the reactivity relationships obtained for well-studied Pd and Pt in the case of other metals.

For both symmetrical and asymmetrical coupling reactions it was shown [86] that the reactivity of the transition metal complexes in C–C coupling reactions depends on the following key factors: (a) directionality of the M–C bond; (b) steric interaction between the R and L ligands; and (c) exothermicity of the reaction.

Although Me–Me coupling is the simplest model for C–C reductive elimination it was clearly demonstrated that it does not account for key features of the reactions involving unsaturated ligands and can not be used as a model for studying such cross-coupling reactions.

The data summarized here indicates that joint analysis of theoretical calculations and experimental data provides a flexible and accurate tool for studying the mechanism of transition metal catalyzed reactions.

Finally, a care should be taken to compare the discussed results obtained for the reductive elimination process with the overall reactivity observed in multistep C–C coupling reactions. Obviously, in any multistep process, either catalytic or not, the overall reactivity may depend on the other factors as well. Outlined reactivity trends should be used to better understand reaction mechanisms and to reveal the rate-determining stage.

References

1. J. Tsuji, *Transition Metal Reagents and Catalysts: Innovations in Organic Synthesis*, John Wiley & Sons, Chichester, 2000.
2. M. Beller, C. Bolm, *Transition Metals for Organic Chemistry*, Wiley-VCH, Weinheim, 1998.
3. T. W. Wallace, *Org. Biomol. Chem.* 2006, **4**, 3197.
4. *Metal-catalyzed Cross-coupling Reactions*, eds. F. Diederich, P. J. Stang, Wiley-VCH, Weinheim, 1998.
5. *Handbook of Organopalladium Chemistry for Organic Synthesis*, ed. E. Negishi, John Wiley & Sons, New York, Chichester, 2002, Vol. 1, pp. 215–994.
6. I. P. Beletskaya, A. V. Cheprakov, Metal Complexes as Catalysts for C-C Cross-coupling Reactions, in *Comprehensive Coordination Chemistry II*, eds. J. A. McCleverty, T. J. Meyer, Vol. 9, Elsevier Ltd., Oxford, 2004, pp. 305–368.
7. S. P. Nolan, O. Navarro, C–C Bond Formation by Cross-Coupling, in *Comprehensive Organometallic Chemistry III*, ed. D.M.P. Mingos, R.H. Crabtree, Vol. 11, Elsevier Ltd., Oxford, 2007, pp. 1–37.
8. N. Miyaura, A. Suzuki, *Chem. Rev.* 1995, **95**, 2457.
9. A. Suzuki, *J. Organomet. Chem.* 1999, **576**, 147.
10. F. Bellina, A. Carpita, R. Rossi, *Synthesis* 2004, **15**, 2419.
11. N. Miyaura, *J. Organomet. Chem.* 2002, **653**, 54.
12. J. K. Stille, *Angew. Chem. Int. Ed. Engl.* 1986, **25**, 508.

13 J. K. Stille, *Pure Appl. Chem.* 1985, **57**, 1771.
14 V. Farina, V. Krishnamurthy, W. J. Scott, The Stille Reaction, in *Organic Reactions*, Vol. 50, John Wiley & Sons Inc., New York, 1997.
15 K. Kugami, M. Kosugi, *Top. Curr. Chem.* 2002, **219**, 87.
16 K. Sonogashira, *Comp. Org. Synth.* 1991, 3, 521.
17 K. Sonogashira, *Comp. Org. Synth.* 1991, 3, 551.
18 K. Tamao, K. Sumitami, M. Kumada, *J. Am. Chem. Soc.* 1972, **94**, 4374.
19 R. J. P. Corriu, J. P. Masse, *J. Chem. Soc., Chem. Commun.* 1972, 144.
20 K. Tamao, *J. Organomet. Chem.* 2002, **653**, 23.
21 T. Hiyama, Y. Hatanaka, *Pure Appl. Chem.* 1994, **66**, 1471.
22 T. Hiyama, *J. Organomet. Chem.* 2002, **653**, 58.
23 S. E. Denmark, R. F. Sweis, *Acc. Chem. Res.* 2002, **35**, 835.
24 T. Hiyama, E. Shirakawa, *Top. Curr. Chem.* 2002, **219**, 61.
25 E. Negishi, *Acc. Chem. Res.* 1982, **15**, 340.
26 E. Negishi, T. Takahashi, K. Akiyoshi, *J. Organomet. Chem.* 1987, **334**, 181.
27 E. Negishi, S.-Y. Liou, C. Xu, S. Huo, *Org. Lett.* 2002, **4**, 261.
28 I. Perez, J. P. Sestelo, L. Sarandeses, *J. Am. Chem. Soc.* 2001, **123**, 4155.
29 I. Perez, J. P. Sestelo, L. Sarandeses, *Org. Lett.* 1999, **1**, 1267.
30 M. Kosugi, T. Tanji, Y. Tanaka, A. Youshida, K. Fugami, M. Kameyama, T. Migita, *J. Organomet. Chem.* 1996, **508**, 255.
31 J. W. Faller, R. G. Kultyshev, *Organometallics* 2002, **21**, 5911.
32 S. Dillinger, P. Bertus, P. Pale, *Org. Lett.* 2001, **3**, 1661.
33 D. Gelman, D. Tsvelikhovsky, G. A. Molander, J. Blum, *J. Org. Chem.* 2002, **67**, 6287.
34 R. Cella, R. L. O. R. Cunha, A. E. S. Reis, D. C. Pimenta, C. F. Klitzke, H. A. Stefani, *J. Org. Chem.* 2006, **71**, 244.
35 B. W. Fausett, L. S. Liebeskind, *J. Org. Chem.* 2005, **70**, 4851.
36 S. Gatard, R. Celenligil-Cetin, C. Guo, B. M. Foxman, O. V. Ozerov, *J. Am. Chem. Soc.* 2006, **128**, 2808.
37 C. S. Chin, G. Won, D. Chong, M. Kim, H. Lee, *Acc. Chem. Res.* 2002, **35**, 218.
38 J. Navarro, E. Sola, M. Martın, I. T. Dobrinovitch, F. J. Lahoz, L. A. Oro, *Organometallics* 2004, **23**, 1908.
39 F. Kakiuchi, S. Murai, *Acc. Chem. Res.* 2002, **35**, 826.
40 V. Ritleng, C. Sirlin, M. Pfeffer, *Chem. Rev.* 2002, **102**, 1731.
41 G. Bhalla, J. Oxgaard, W. A. Goddard, III, R. A. Periana, *Organometallics* 2005, **24**, 5499.
42 J. Oxgaard, G. Bhalla, R. A. Periana, W. A. Goddard, III, *Organometallics* 2006, **25**, 1618.
43 R. Ghosh, X. Zhang, P. Achord, T. J. Emge, K.-J. Karsten, A. S. Goldman, *J. Am. Chem. Soc.* 2007, **129**, 853.
44 I. Beletskaya, C. Moberg, *Chem. Rev.* 2006, **106**, 2320.
45 A. Togni, H. Grutzmacher (eds.), *Catalytic Heterofunctionalization*, Wiley-VCH, Weinheim, 2001.
46 I. Beletskaya, C. Moberg, *Chem. Rev.* 1999, **99**, 3435.
47 M. Beller, J. Seayad, A. Tillack, H. Jiao, *Angew Chem. Int. Ed.* 2004, **43**, 3368.
48 F. Alonso, I. P. Beletskaya, M. Yus, *Chem. Rev.* 2004, **104**, 3079.
49 V. P. Ananikov, S. S. Zalesskiy, N. V. Orlov, I. P. Beletskaya, *Russ. Chem. Bull. Int. Ed.* 2006, No 11, 2030.
50 J. Navarro, M. Sagi, E. Sola, F. J. Lahoz, I. T. Dobrinovitch, A. Katho, F. Joo, L. A. Oro, *Adv. Synth. Catal.* 2003, **345**, 280.
51 V. P. Ananikov, S. A. Mitchenko, I. P. Beletskaya, *Russ. J. Org. Chem. (Engl. Transl.)* 2002, **38**, 636.
52 V. P. Ananikov, S. A. Mitchenko, I. P. Beletskaya, S. E. Nefedov, I. L. Eremenko, *Inorg. Chem. Commun.* 1998, **1**, 411.
53 P. M. Maitlis, *J. Mol. Cat. A* 2003, **204–205**, 55.
54 Z. Q. Wang, M. L. Turner, A. R. Kunicki, P. M. Maitlis, *J. Organomet. Chem.* 1995, **488**, C11.
55 B. Rybtchinski, D. Milstein, *Angew. Chem. Int. Ed.* 1999, **38**, 870.
56 L. M. Rendina, R. J. Puddephatt, *Chem. Rev.* 1997, **97**, 1735.
57 M. Albrecht, R. A. Gossage, A. L. Spek, G. van Koten, *J. Am. Chem. Soc.* 1999, **121**, 11898.

58 A. Baylar, A. J. Canty, P. G. Edwards, B. W. Slelton, A. H. White, *J. Chem. Soc., Dalton Trans.* 2000, 3325.

59 B. S. Williams, K. I. Goldberg, *J. Am. Chem. Soc.* 2001, **123**, 2576.

60 M. E. Van der Boom, H.-B. Kraatz, L. Hassner, Y. Ben-David, D. Milstein, *Organometallics* 1999, **18**, 3873.

61 J. Procelewska, A. Zahl, G. Liehr, R. Van Eldik, N. A. Smythe, B. S. Williams, K. I. Goldberg, *Inorg. Chem.* 2005, **44**, 7732.

62 D. M. Crumpton-Bregel, K. I. Goldberg, *J. Am. Chem. Soc.* 2003, **125**, 9442.

63 S. M. Reid, J. T. Mague, M. J. Fink, *J. Am. Chem. Soc.* 2001, **123**, 4081.

64 A. Moravskiy, J. K. Stille, *J. Am. Chem. Soc.* 1981, **103**, 4182.

65 M. K. Loar, J. K. Stille, *J. Am. Chem. Soc.* 1981, **103**, 4174.

66 F. Ozawa, T. Ito, Y. Nakamura, A. Yamamoto, *Bull. Chem. Soc. Jpn.* 1981, **54**, 1868.

67 D. A. Culkin, J. F. Hartwig, *Organometallics* 2004, **23**, 3398.

68 K. Tatsumi, R. Hoffmann, A. Yamamoto, J. K. Stille, *Bull. Chem. Soc. Jpn.* 1981, **54**, 1857.

69 J. L. Low, W. A. Goddard, *J. Am. Chem. Soc.* 1986, **108**, 6115.

70 J. L. Low, W. A. Goddard, *Organometallics* 1986, **5**, 609.

71 G. S. Hill, R. J. Puddephatt, *Organometallics* 1998, **17**, 1478.

72 S. Sakaki, N. Mizoe, Y. Musashi, B. Biswas, M. Sugimoto, *J. Phys. Chem. A* 1998, **102**, 8027.

73 S. Sakaki, M. Ieki, *J. Am. Chem. Soc.* 1993, **115**, 2373.

74 T. Kegl, L. Kollar, *J. Organomet. Chem.* 2007, **692**, 1852.

75 C. Michel, A. Laio, F. Mohamed, M. Krack, M. Parrinello, A. Milet, *Organometallics* 2007, **26**, 1241.

76 P. E. M. Siegbahn, M. R. A. Blomberg, in *Theoretical Aspects of Homogeneous Catalysis, Applications of Ab Initio Molecular Orbital Theory*, eds. P. W. N. M. van Leeuwen, J. H. van Lenthe, K. Morokuma, Kluwer Academic Publishers, Dordrecht, The Netherlands, 1995.

77 A. Sundermann, O. Uzan, J. M. L. Martin, *Organometallics* 2001, **20**, 1783.

78 A. Sundermann, O. Uzan, D. Milstein, J. M. L. Martin, *J. Am. Chem. Soc.* 2000, **122**, 7095.

79 Z. Cao, M. B. Hall, *Organometallics* 2000, **19**, 3338.

80 N. Koga, K. Morokuma, *Organometallics* 1991, **10**, 946.

81 K. Krogh-Jespersen, A. S. Goldman, in Transition State Modeling for Catalysis, eds. D. G. Truhlar, K. Morokuma, *ACS Symposium Series*, The American Chemical Society, Washington DC, 1999, pp. 151–162.

82 A. Dedieu, *Chem. Rev.* 2000, **100**, 543.

83 D. G. Musaev, K. Morokuma, *Top. Catal.* 1999, **7**, 107.

84 N. Koga, K. Morokuma, *Chem. Rev.* 1991, **91**, 823.

85 D. J. Cardenas, A. M. Echavarren, *New J. Chem.* 2004, **28**, 338.

86 V. P. Ananikov, D. G. Musaev, K. Morokuma, *Organometallics* 2005, **24**, 715.

87 A. A. C. Braga, G. Ujaque, F. Maseras, *Organometallics* 2006, **25**, 3647.

88 A. A. C. Braga, N. H. Morgon, G. Ujaque, F. Maseras, *J. Am. Chem. Soc.* 2005, **127**, 9298.

89 D. J. Cardenas, B. Martın-Matute, A. M. Echavarren, *J. Am. Chem. Soc.* 2006, **128**, 5033.

90 C. S. Chin, G. Won, D. Chong, M. Kim, H. Lee, *Acc. Chem. Res.* 2002, **35**, 218.

91 C. S. Chin, H. Cho, G. Won, M. Oh, K. M. Ok, *Organometallics* 1999, **18**, 4810.

92 P. M. Maitlis, H. C. Long, R. Quyoum, M. L. Turner, Z.-Q. Wang, *Chem. Commun.* 1996, 1.

93 A. Yahav-Levi, I. Goldberg, A. Vigalok, *J. Am. Chem. Soc.* 2006, **128**, 8710.

94 X. Li, L. N. Appelhans, J. W. Faller, R. H. Crabtree, *Organometallics* 2004, **23**, 3378.

95 K. Fagnou, M. Lautens, *Chem. Rev.* 2003, **103**, 169.

96 M. Sato, E. Mogi, *Organometallics* 1995, **14**, 3157.

97 G. Fuhrmann, T. Debaerdemaeker, P. Bauerle, *Chem. Commun.* 2003, 948.

98 S. A. Macgregor, G. W. Neave, C. Smith, *Faraday Discuss.* 2003, **124**, 111.

99 D. G. Musaev, K. Morokuma, Transition Metal Catalyzed σ-Bond Activation and Formation Reactions, in *Topics in Organometallic Chemistry*, Vol. 12, Springer Berlin, Heidelberg, 2005.
100 V. P. Ananikov, D. G. Musaev, K. Morokuma, *J. Am. Chem. Soc.* 2002, **124**, 2839.
101 M. I. Bruce, K. Costuas, J.-F. Halet, B. C. Hall, P. J. Low, B. N. Nicholson, B. W. Skelton, A. H. White, *J. Chem. Soc., Dalton Trans.* 2002, 383.
102 V. P. Ananikov, D. G. Musaev, K. Morokuma, *Organometallics* 2001, **20**, 1652.
103 A. Yahav, I. Goldberg, A. Vigalok, *Inorg. Chem.* 2005, **44**, 1547.
104 R. Alvarez, O. N. Faza, C. S. Lopez, A. R. de Lera, *Org. Lett.* 2006, **8**, 35.
105 W. J. Marshall, V. V. Grushin, *Organometallics* 2003, **22**, 1591.
106 *Aqueous-Phase Organometallic Catalysis: Concepts and Applications*, eds. W. A. Herrmann, B. Cornils, Wiley-VCH, Weinheim, 1998.
107 C.-J. Li, *Acc. Chem. Res.* 2002, **35**, 533.
108 B. Cornils, E. Wiebus, *Chemtech* 1995, **25**, 33.
109 C. J. Li, *Chem. Rev.* 1993, **93**, 2023.
110 A. Rossin, G. Kovacs, G. Ujaque, A. Lledos, F. Joo, *Organometallics* 2006, **25**, 5010.
111 S. Kozuch, C. Amatore, A. Jutand, S. Shaik, *Organometallics* 2005, **24**, 2319.
112 C. Amatore, A. Jutand, *Acc. Chem. Res.* 2000, **33**, 314.
113 E. Zuidema, P. W. N. M. van Leeuwen, C. Bo, *Organometallics* 2005, **24**, 3703.

7
Olefin Polymerization Using Homogeneous Group IV Metallocenes

Robert D. J. Froese

7.1
Introduction

Olefin polymerization using single-site metallocenes has been a field that has grown extensively since its inception [2]. The well-defined structure of transition metal catalysts and the ability to make minor adjustments, affecting both sterics and electronics, have major effects on the catalysts' performance and, thus, on the resulting polymer properties. Further understanding of polymerization using non-coordinating borate counterions has also been garnered [3a–d]. Much other work has aided the understanding of the mechanisms related to polymerization [3e–g]. In addition, extensive work in the areas of regio- and stereoselectivity of catalysis has contributed to our knowledge of the microstructure associated with the polymerization process [4]. As well as the extensive work done for early metal systems, some studies on late transition metal palladium and nickel systems have also been carried out [5].

To complement the experimental studies, much theoretical work has been performed on the mechanism of polymerization [6, 7]. Reasonable agreement between theory and experiment has been demonstrated for late transition metal nickel and palladium systems [8, 9], but for early metal zirconium, titanium, and hafnium catalysts it has been more difficult due to the lack of specific experimental data on activation barriers and parameters. Activation parameters ($\Delta H^\ddagger$ and $\Delta S^\ddagger$) for Group IV metallocenes were first estimated by Landis [10] and activation free energies ($\Delta G^\ddagger$) have been estimated by Erker [11].

The Cossee–Arlman mechanism [12] with adaptations by Brookhart and Green [13] is generally accepted as the mechanism for olefin polymerization (Figure 7.1).

The mechanism begins with a β-agostic species or the metal-alkyl with a coordinated counterion such as a borate. The olefin first binds, which displaces the agostic interaction or the counterion, then it inserts through a four-center transition state (TS) that possesses an α-agostic interaction. This stabilizing interaction

Computational Modeling for Homogeneous and Enzymatic Catalysis.
A Knowledge-Base for Designing Efficient Catalysts. K. Morokuma and D. G. Musaev (Eds.)
Copyright © 2008 WILEY-VCH Verlag GmbH & Co. KGaA, Weinheim
ISBN: 978-3-527-31843-8

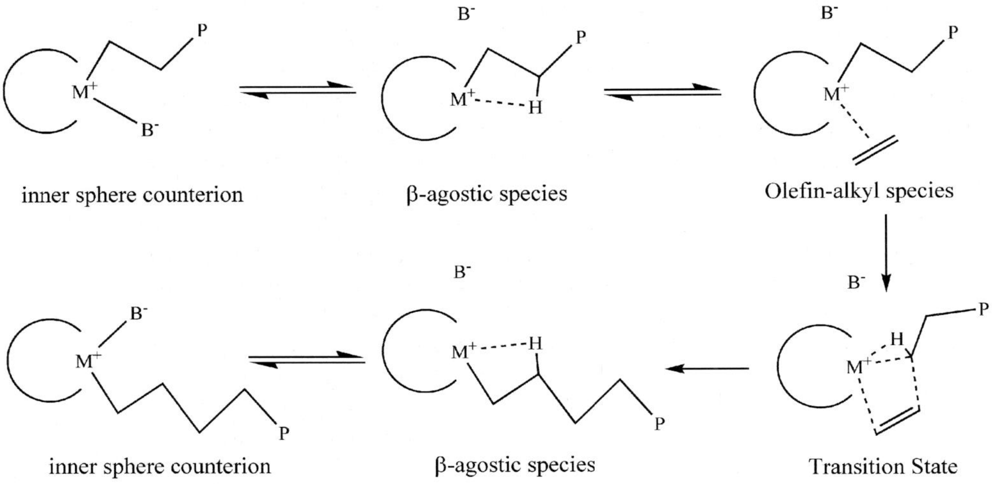

Figure 7.1 Mechanism for olefin propagation.

in the transition state has been suggested through experimental isotope studies [13, 14] and numerous computational studies [6–9]. Following the transition state, the direct product formed is, likely, the γ-agostic species, which then becomes either the more stable β-agostic species or the coordinated counterion complex (whatever the resting state is deemed to be). It is anticipated that the polymeryl chain swings back-and-forth after each insertion step and syndiotactic polypropylene formed from C_s symmetric catalysts supports this claim.

One important aspect of polymerization is the reversibility of olefin binding. Numerous experimental studies in the literature claim that olefin binding is reversible [10, 11, 15, 16]. Many of these reports study this reversibility under special circumstances. Jordan [16], for example, uses a metal alkoxide to prevent insertion, while Casey [15] tethers the olefin to the alkyl group. For olefin binding to be reversible, the reactant species (β-agostic or borate-coordinated), the olefin complex, and the transition state linking these two must all be lower in energy than the migratory insertion TS. This would imply that it is easier to coordinate and dissociate olefin than to insert it and, thus, can occur many times before an insertion event. Brookhart has suggested for certain late metal systems that olefin binding is reversible [17].

For reversible coordination, olefin binding enthalpies must be small; several previous computational papers, though, give contradictory results. For example, an early work on the ethyl cation of the constrained geometry catalyst suggested an ethylene binding enthalpy of 5 kcal mol^{-1} [18]. Another study on a series of Group IV catalysts (ethyl cation) predicted larger front-side enthalpic bindings in the range of ca. 16–26 kcal mol^{-1}, but many of the ligands used in that study were very small, such as CH_3, NH_2, or OH [6a]. In these cases, the binding energies would be expected to be larger. Three other computational papers have provided olefin binding energies for Ti/Zr methyl cations and propyl cations. In one of them, the olefin binding enthalpies to the methyl cations were in the 20–25 kcal mol^{-1} range (slightly higher than the present work), but this discrepancy is likely due to a stabilizing interaction in the reactant species [7a]. In the other works, olefin bindings to both methyl and propyl cations had a range of values, but most were similar to those here [19, 20]. It appears that the olefin binding equilibrium does depend on ligands, but for many of the systems this equilibrium appears to favor separated species, especially when one considers free energies at higher temperatures. In addition, notably, Brookhart suggests that the olefin-alkyl species is the resting state, but palladium binds olefins stronger than zirconium and his studies are at low temperature, which would favor the complex [5b]. In contrast, industrial processes using Group IV metallocenes polymerize at higher temperatures (100–150 °C), which favors the separated species. In addition, late transition metal olefin-alkyl complexes appear to be isolable, in sharp contrast to Group IV polymerization catalysts.

Here, interpretations of the polymerization process for early transition metal catalysts are provided. The inclusion of free energies and borate counterions (in selected cases) suggests that the equilibrium between the separated species and the olefin-alkyl complex favors the dissociated pair. Thus, activation parameters will be determined for the process beginning from the separated borate bound metal-alkyl + olefin (resting state) to the migratory insertion transition state. Further discussions on the implications of this pre-equilibrium will be made with respect to other transition states such as α-olefin incorporation, β-hydride elimination, chain-transfer-to-monomer, and hydrogenolysis. Computational results will be presented for the ethylene and propylene polymerization using the rac-C_2H_4(1-indenyl)$_2$ZrC$_4$H$_9^+$ catalyst (EBIZ, see Figure 7.2) with the tetrakis(perfluorophenyl)borate anion and comparisons with experiment are provided.

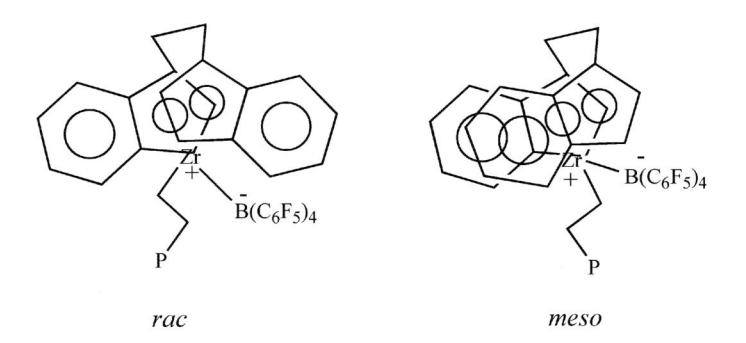

Figure 7.2 Structure of the expected resting states of the rac and meso forms of the EBIZ catalyst. The polymeryl chain (P) and the borate counterion are included.

7.2
Computational Details

Calculations were carried out using the *Gaussian03* program [21]. All computations utilized the B3LYP method [22]. For optimization of geometries and frequencies, the LANL2DZ basis set [23] was used for zirconium while the 6-31G* (5d) basis set [24] was used for all other atoms. Single point energies were recalculated using the LANL2DZ/6-311+G** combination [25]. The larger basis set single point energy predictions were used and the entropy corrections were added from the smaller basis set frequency calculations. Relative energies ($\Delta E°/\Delta E^{\ddagger}$, $\Delta G°/\Delta G^{\ddagger}$) are in kcal mol^{-1} and $\Delta S°/\Delta S^{\ddagger}$ are in cal mol^{-1} K^{-1}. As computed gas-phase entropies have been shown to be too large compared with solution, a scale factor of 0.6 was used [26]. This value is most critical for bimolecular reactions as the translational and rotational entropies in the gas phase are fully realized, whereas in solution they are damped. Hence the computed entropy of activation for a bimolecular reaction in the gas phase is significantly larger than in solution. In the text, the values quoted are the high-level single point energy predictions (B3LYP/6-311+G**), ΔE, which are similar to the relative enthalpic differences (ΔH) and are taken as the same. In addition, the scaled entropy differences (ΔS) are utilized from the lower level of theory (B3LYP/6-31G*). Overall, calculations from both levels of theory (B3LYP/6-31G* and B3LYP/6-311+G**) are included in the tables for comparison.

7.3
Results and Discussion

7.3.1
Chain Propagation

The polymer chain propagation step is fundamental to polymerization and involves the binding of olefin to the metal center followed by insertion into the metal–alkyl bond. Which of these two steps is rate-limiting was of some debate but it appears that, for Group IV transition metals, the overwhelming evidence supports the olefin binding step as being reversible and migratory insertion into the metal–alkyl bond as rate-limiting and selectivity-determining. We assume this is true and provide some evidence for this mechanism.

Erker's view (Figure 7.3) [11] depicts the important segment of the potential energy surface and one can see the rate-limiting step being given by $\Delta G^{\ddagger}_{\text{prop}}$. The enthalpy of olefin binding may be endothermic or exothermic, but when one includes free energies the bimolecular free energy of binding is most likely endothermic. Now, one must consider the role of the counterion. Presumably, it is in the resting state of the catalytic cycle and Landis' work [27] has suggested that the borate returns to the metal at each insertion step. His anion,

(Me)B$^-$(C$_6$F$_5$)$_3$, is not the most "non-coordinating" and it is expected that the B$^-$(C$_6$F$_5$)$_4$ counterion is even more weakly bound to the metal after each insertion step. If the counterion indeed returns to the metal at each step, the equilibrium it forms with the β-agostic species must favor the coordinated counterion complex – thus it is lower in energy. If the anion is completely non-coordinating, then the β-agostic species would be lower in energy. The polymerization barrier ($\Delta G^\ddagger_{prop}$) shown in Figure 7.3 depicts why more coordinating anions will impart slower rates since the coordinating borate anion will be lower in energy and, as the resting state, it will increase $\Delta G^\ddagger_{prop}$.

Presumably, upon olefin coordination, the borate is forced to the outer sphere and the difference between the olefin-alkyl complex and TS is not affected significantly by the counterion. This statement is not completely true as there are examples of counterion effects on tacticity, but one assumes the effects are minimal and with many catalysts that have been studied no effect is seen [27]. In the olefin-alkyl complex and the transition state, the borate exists in the "outer sphere" and plays little role in the polymerization. However, in the reactant species, a truly non-coordinating anion destabilizes the starting point relative to a more coordinating counterion, thus decreasing $\Delta G^\ddagger_{prop}$. Marks' computational work [28] also showed that olefin complexes could reside higher in energy than the alkyl species due to solvation and counterion effects, and the activation barrier must reflect this fact. Other works examining oligomerization [29] and a computational paper on polymerization [30] have also suggested similar features of the potential energy surface.

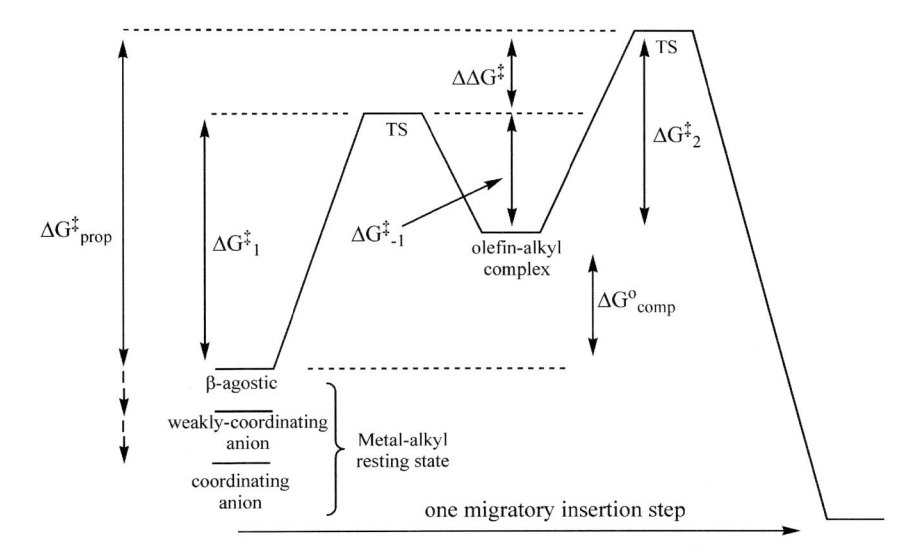

Figure 7.3 Depiction of olefin binding and migratory insertion (revised from Ref. [11]) for early metal systems.

The general propagation reaction is expressed in Eq. (1):

$$\text{catalyst} + \text{olefin} \underset{k_{-1}}{\overset{k_1}{\rightleftharpoons}} \text{olefin-alkyl complex} \xrightarrow{k_2} \text{propagation} \qquad (1)$$

A more accurate way of looking at the kinetics of propagation based on Eq. (1) and Figure 7.3 is shown in Eq. (2) [31]:

$$\frac{\partial[\text{product}]}{\partial t} = \frac{k_1 k_2}{k_{-1} + k_2}[\text{olefin}][\text{catalyst}] \qquad (2)$$

One can assume that the observed rate constant, k_{obs}, is the combination of the intrinsic rate constants given in Eq. (2). Thus, $\Delta G^\ddagger_{prop}$ or $\Delta G^\ddagger_{obs}$ can be determined based on forward and reverse free energy barriers of the individual steps and reflects the observed rate constant, k_{obs} or k_{prop}. This represents the experimentally observed barrier and is given in Eq. (3), with free energy barriers depicted in Figure 7.3 [31].

$$\Delta G^\ddagger_{prop} = \Delta G^\ddagger_1 + \Delta G^\ddagger_2 + RT \ln\{e^{-\Delta G^\ddagger_{-1}/RT} + e^{-\Delta G^\ddagger_2/RT}\} \qquad (3)$$

These derivations are true as long as the second step is irreversible and the resting state of the catalytic cycle is separated catalyst + olefin. One can now look at the extremes of this equation. If the first TS is higher than the second, $k_2 \gg k_{-1}$, Eq. (4) results:

$$e^{-\Delta G^\ddagger_{-1}/RT} \ll e^{-\Delta G^\ddagger_2/RT} \qquad (4)$$
$$\Delta G^\ddagger_{prop} = \Delta G^\ddagger_1$$

This simple result is easily interpreted as the turnover-limiting step being olefin binding. On the other hand, if $k_{-1} \gg k_2$ (first TS is lower than the second), this leads to Eq. (5):

$$e^{-\Delta G^\ddagger_2/RT} \ll e^{-\Delta G^\ddagger_{-1}/RT}$$
$$\Delta G^\ddagger_{prop} = \Delta G^\ddagger_1 + \Delta G^\ddagger_2 - \Delta G^\ddagger_{-1} \qquad (5)$$
$$= \Delta G^\circ_{comp} + \Delta G^\ddagger_2$$

In olefin polymerization, this is the expected result and leads to the view in Figure 7.3, where the rate-limiting step leads from the resting state + olefin to the migratory insertion TS. The inherent assumption in Eq. (5), $k_{-1} \gg k_2$, leads to olefin binding being reversible. Equation (5) can also be interpreted as an intrinsic barrier ($\Delta G^\ddagger_2$ based on k_2) plus a factor (ΔG°_{comp} based on K_{eq}) describing the fraction of catalysts in the productive state, the olefin-alkyl complex. One other case resulting from Eq. (3) is intriguing: if the two TSs lie at the same height, then $k_{-1} = k_2$, leading to Eq. (6):

$$e^{-\Delta G^{\ddagger}_{-1}/RT} = e^{-\Delta G^{\ddagger}_{2}/RT}$$

$$\begin{aligned}\Delta G^{\ddagger}_{prop} &= \Delta G^{\ddagger}_{1} + \Delta G^{\ddagger}_{2} + RT \ln\{2e^{-\Delta G^{\ddagger}_{2}/RT}\} \\ &= \Delta G^{\ddagger}_{1} + \Delta G^{\ddagger}_{2} + RT \ln 2 - \Delta G^{\ddagger}_{2} \\ &= \Delta G^{\ddagger}_{1} + RT \ln 2\end{aligned} \qquad (6)$$

Here the barrier becomes the difference between the resting state and either TS (since they are at the same height) plus a small term, $RT \ln 2$. This final term amounts to a degeneracy contribution and numerically equates to approximately 0.4 kcal mol^{-1} at 25 °C or 0.6 kcal mol^{-1} at 150 °C. Thus, the experimentally observed activation barrier leads from the resting state to whichever TS is higher on the potential energy surface plus a correction term not to exceed 0.4 kcal mol^{-1} (25 °C).

Overall, these data suggest that estimating the observed activation barrier is approximated well by determining which of the two TSs is higher and utilizing that TS relative to the resting state. Notably, when I refer to the heights of the two TSs, I am not referring to the intrinsic barriers ($\Delta G^{\ddagger}_{1}$ versus $\Delta G^{\ddagger}_{2}$) but to their actual heights relative to one another on the potential energy surface, as shown in Figure 7.3. Erker has shown that, for different catalysts and substrates, the second TS lies higher than the first one in all his examples [11]. Ultimately, if the olefin complex is not the resting state of the catalytic cycle, that species is not important in polymerization from the perspective of experimentally observed rates or selectivities. However, if one wants to utilize the kinetic expression of Eq. (2), the relative position of the olefin complex is needed, but for the critical activation free energy, $\Delta G^{\ddagger}_{prop}$, the olefin complex is not required.

Is it reasonable to assume the separated species are always the resting state and that migratory insertion is always the rate-limiting TS? To answer the question about the resting state, one can examine the experimental data that show the olefin pressure dependence on the rate of polymerization. For late transition metal palladium catalysts with much stronger olefin binding energies, propagation rates are often independent of olefin pressure as long as sufficient olefin is available to saturate all the catalyst sites [32]. This latter assertion must be correct, otherwise one could not make polyethylene catalytically. Group IV catalysts inevitably have an olefin pressure dependence, suggesting that the olefin complexes are not the resting state of the catalytic cycle. Notably, the more relevant industrial conditions of higher temperature (130–200 °C) entropically favors separated species to an even greater extent. In terms of the rate-limiting TSs, many studies (e.g., Erker [11] and Landis [27]) have shown that migratory insertion is rate-limiting, and the kinetic isotope effects of α-agostic interactions [14] have also predicted this TS to be most important. However, with large steric bulk, it is possible that one could push the first TS higher due to a difficult approach of the olefin to the metal. Few catalysts, though, display this trend although some computations have predicted that certain catalysts could [33].

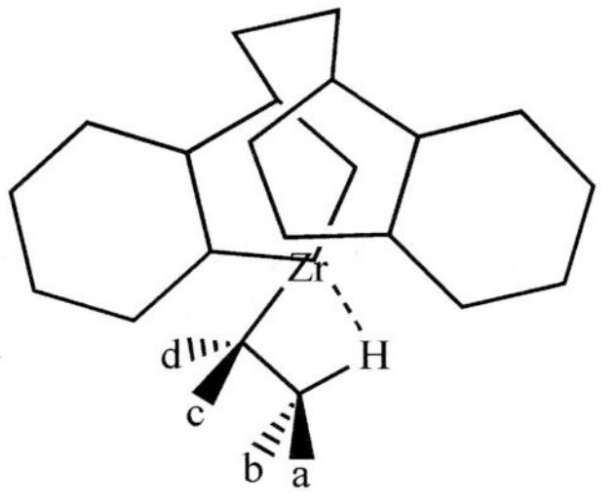

Figure 7.4 Structures of the β-agostic species.

Initially, we examined the transition states for propagation relative to the β-agostic species. While the β-agostic intermediates have never been shown to be the resting state for early transition metal polymerizations, they provide a convenient starting point and, later, we estimated the relative stability of the true resting state to these species. Also note that, because olefin binding is assumed to be reversible, the relative energies of the various TSs leading to different processes will be used to define the selectivities. Figure 7.4 depicts the labeling scheme of the precursor compounds studied.

To examine all possible scenarios, one needs to examine β-agostic species with polymer chains representing prior ethylene, 1,2-, and 2,1-insertions. These starting structures (Figure 7.4) can be viewed as following an inserted olefin: after ethylene, a = Et/b = H/c = H/d = H (**A0**) or a = H/b = Et/c = H/d = H (**A1**); after a 1,2-propylene insertion, a = Me/b = Me/c = H/d = H (**B0**); after a 2,1-propylene insertion, a = H/b = Me/c = Me/d = H (**C0**), a = Me/b = H/c = Me/d = H (**C1**), a = Me/b = H/c = H/d = Me (**D0**), or a = H/b = Me/c = H/d = Me (**D1**). After an α-olefin, the polymeryl chain only posses a methyl group at the β-carbon, whereas after an ethylene insertion there is an ethyl substituent at this position. This difference allows one to compare the relative stabilities of different polymer chain isomers, C_4H_9, n-butyl, i-butyl, and sec-butyl.

Seven precursor structures are possible, but only four become crucial because of the reversibility of olefin binding. Effectively, this reversibility equilibrates the a and b positions. Hence, compounds **A0** and **A1** are in equilibrium as well as the pairs **C0**, **C1** and **D0**, **D1**. Note that the position of the 2,1-inserted methyl (c or d) cannot be reversed once it has inserted and the chirality is fixed. The stereochemistry at this site can be inverted by a complex mechanism [34]. Normally, this chirality is also true after a 1,2-insertion, but, for our precursor, we are using an isobutyl polymeryl chain and no chirality is established. Thus, we are defining four unique, non-interconvertible potential energy surfaces: after an ethylene insertion (**A**), after a 1,2-propylene insertion (**B**), and after two different 2,1-propylene insertions having (R) (**C**) and (S) (**D**) stereochemistry about the secondary carbon bond.

Table 7.1 gives the energetic data for the β-agostic precursors and the transition states for β-hydride elimination. The two β-agostic species following an ethylene insertion show only a 0.3 kcal mol^{-1} difference in energy favoring **A0**. Because of

Table 7.1 Relative energies (see footnote) of β-agostic reactants and the TSs for β-hydride elimination. Labeling is based on the structures shown in Figure 7.4.

Compounds		6-31G*		6-311 + G**		Structure (see Figure 7.4)			
		$\Delta E°$	$\Delta S°$	$\Delta E°$	$\Delta S°$	a	b	c	d
β-agostic intermediates									
After ethylene (**A**):	A0	0.0	0.0	0.0	0.0	Et			
	A1	0.2	−0.4	0.3	−0.3		Et		
After 1,2-propylene (**B**):	B0	0.0	0.0	0.0	0.0	Me	Me		
After 2,1-propylene (**C**):	C0	0.0	0.0	0.0	0.0		Me	Me	
	C1	1.6	−0.8	1.5	−0.5	Me		Me	
After 2,1-propylene (**D**):	D0	0.0	0.0	0.0	0.0	Me			Me
	D1	1.8	0.3	1.4	0.2		Me		Me
		$\Delta E^\ddagger$	$\Delta S^\ddagger$	$\Delta E^\ddagger$	$\Delta S^\ddagger$	a	b	c	d
β-Hydride elimination TSs									
After ethylene (**A**):	A10	13.5	−3.5	13.5	−2.1	Et			
	A11	13.8	−2.5	13.9	−1.5		Et		
After 1,2-propylene (**B**):	B10	10.8	−3.4	10.8	−2.0	Me	Me		
After 2,1-propylene (**C**):	C10	12.0	−5.2	11.5	−3.1		Me	Me	
	C11	12.3	−1.2	12.0	−0.7	Me		Me	
After 2,1-propylene (**D**):	D10	9.1	−5.3	8.1	−3.2	Me			Me
	D11	11.8	0.3	11.4	0.2		Me		Me

Relative energies given across the table are $\Delta E°/\Delta E^\ddagger$ (B3LYP/6-31G*) in kcal mol^{-1}, $\Delta S°/\Delta S^\ddagger$ (B3LYP/6-31G*) in cal mol^{-1} K^{-1}, $\Delta E°/\Delta E^\ddagger$ (single point energy using B3LYP/6-311 + G**) in kcal mol^{-1}, the 0.6 scaled $\Delta S°/\Delta S^\ddagger$ (B3LYP/6-31G*) in cal mol^{-1} K^{-1}.

the four-center nature of the intermediate, the ethyl substituent at either a or b makes little difference. There is only one compound following a 1,2-propylene insertion, **B0**, with methyl groups at both the a and b positions. A comparison of the 2,1-precursors shows that **C0** is more stable than **C1** by 1.5 kcal mol^{-1} and **D0** is more stable than **D1** by 1.4 kcal mol^{-1}. The data for these 2,1-surfaces would seem to indicate that the driving force for stability is the preference for the methyl group at the β-position to be trans to that at the α-position.

While the potential energy surfaces are completely separate, one can compare the relative stabilities of each of the four β-agostic precursors to discern the relative stabilities of different hydrocarbyl chains on zirconium. The linear chain, **A0**, is the most stable, with **B0** lying 1.8 kcal mol^{-1} higher. Both the **A** and **B** surfaces lead from primary alkyl groups but the disubstituted β-carbon, **B0**, is slightly less stable due to steric interactions. **C0** resides 3.4 kcal mol^{-1} higher than **A0**, but note that, with c = Me/d = H, this isomer has the 2,1-insertion pointing advantageously away from the bulk of the ligand. This relative stability is typical of Group IV metallocenes where primary alkyls are more stable than secondary alkyl groups. **D0** is very unstable and is 6.7 kcal mol^{-1} higher than **A0** and 3.3 kcal mol^{-1} higher than **C0**. This significant difference between the **C** and **D** surfaces is important since the methyl group of the secondary carbon prefers to point away from the indenyl moiety. In comparing the **C** and **D** surfaces, the critical structures to examine are not the relative stabilities of the β-agostic species, but the propagation TS to form these unique stereoisomers. This topic is discussed later, but the trend in the TSs is similar to the stability in the β-agostic species, and thus the **C** surface with (R)-stereochemistry is the most important one and will be focused on.

For olefin binding, optimizations of several olefin-alkyl complexes were attempted and only the lowest energy ones for each class of polymeryl chains are considered. For each of the four unique potential energy surfaces, we need to determine ethylene and propylene binding energies and these are collected in Table 7.2.

Binding enthalpies are exothermic to **A0** by 4.8 and 7.0 kcal mol^{-1}, for ethylene (**A20**) and propylene (**A21**), respectively. The binding energies to **B0** are 3.9 (**B20**) and 6.5 kcal mol^{-1} (**B21**), to **C0** are 4.3 (**C20**) and 4.9 kcal mol^{-1} (**C21**), and to **D0** are 7.2 (**D20**) and 8.4 kcal mol^{-1} (**D21**). The computed (scaled) reaction entropies are all approximately -22 to -25 cal mol^{-1} K^{-1}, thus in the temperature range of 25–100 °C a 7–9 kcal mol^{-1} contribution to $T\Delta S°$ would be seen. These overall interactions are very weak and, in fact, are unlikely to overcome the entropy of the bimolecular reaction, especially at higher temperatures. This low reaction energy is one reason why olefin binding is reversible.

The transition state for polymer chain propagation is depicted in Figure 7.5 and energetic data are collected in Table 7.3.

The α-agostic interaction is a crucial structural feature of the TS. The four different surfaces have unique a and b substituents, while the nature of the approaching olefin defines c, d, e, and f. For the surface following ethylene and leading from β-agostic species **A0**, either a = ^{n}Pr or b = ^{n}Pr. Figure 7.6 depicts the equilibration of the β-agostic species by a reversible binding of the olefin followed by the two unique transition states with different α-agostic interactions. Clearly, a = ^{n}Pr will be more stable as the bulky polymeryl group will point away from the indenyl ligand and the TS for ethylene insertion (**A31**) is 2.1 kcal mol^{-1} lower in energy than the TS with polymeryl directed towards the ligand (**A30**).

The lowest 1,2-propylene TS (**A35**) possesses d = Me and resides 2.8 kcal mol^{-1} higher in energy than the lowest ethylene TS (**A31**). These energetics rep-

Table 7.2 Energies of olefin complexes on the four potential energy surfaces relative to the β-agostic starting structures shown in Table 7.1.

Compounds		6-31G*		6-311 + G**		Structure (see Figure 7.5)		
		ΔE° (kcal mol^{-1})	ΔS° (cal mol^{-1} K^{-1})	ΔE° (kcal mol^{-1})	ΔS° (cal mol^{-1} K^{-1})	a	b	c
After ethylene (A):								
Ethylene	A20	−9.0	−37.3	−4.8	−22.4	^{n}Pr		
Propylene	A21	−10.6	−38.6	−7.0	−23.2	^{n}Pr		Me
After 1,2-propylene (B):								
Ethylene	B20	−7.8	−38.8	−3.9	−23.3	^{i}Pr		
Propylene	B21	−10.3	−40.1	−6.5	−24.0	^{i}Pr		Me
After 2,1-propylene (C):								
Ethylene	C20	−8.0	−41.1	−4.3	−24.6	Et	Me	
Propylene	C21	−8.3	−41.7	−4.9	−25.0	Et	Me	Me
After 2,1-propylene (D):								
Ethylene	D20	−10.8	−40.3	−7.2	−24.2	Me	Et	
Propylene	D21	−11.7	−39.4	−8.4	−23.6	Me	Et	Me

resent the likelihood of a comonomer incorporation in a typical ethylene/α-olefin copolymerization. Propylene might not be a perfect representation of the reactivity of typical comonomers, 1-hexene or 1-octene, but could be considered the upper limit of reactivity as higher α-olefins typically react more slowly (rates of polymerization, propylene > 1-butene > 1-hexene) [11b]. The energy of the lowest 1,2-propylene TS with opposite stereochemistry (**A34**, c = Me) is 3.4 kcal mol^{-1} higher in energy. The lowest 2,1-propylene TS (**A39**, f = Me) also resides 3.4 kcal mol^{-1} higher than the lowest 1,2-propylene TS, defining the regioselectivity. This energetic difference between 1,2- versus 2,1-propagation TSs is similar to the rel-

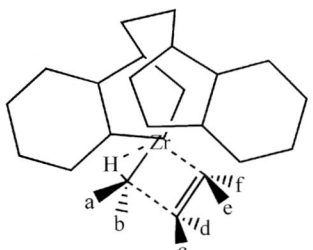

Figure 7.5 Structure of the transition state for olefin propagation.

Table 7.3 Energies of propagation TSs on the four potential energy surfaces relative to the β-agostic structures shown in Table 7.1.

Compounds		6-31G*		6-311 + G**		Structure (see Figure 7.5)					
		$\Delta E^\ddagger$ (kcal mol^{-1})	$\Delta S^\ddagger$ (cal mol^{-1} K^{-1})	$\Delta E^\ddagger$ (kcal mol^{-1})	$\Delta S^\ddagger$ (cal mol^{-1} K^{-1})	a	b	c	d	e	f
After ethylene (A):											
Ethylene	A30	0.5	−45.5	4.0	−27.3	^{n}Pr					
Ethylene	A31	−1.6	−45.0	1.9	−27.0	^{n}Pr					
1,2-Propylene	A32	5.9	−48.2	9.3	−28.9	^{n}Pr	Me				
1,2-Propylene	A33	5.9	−48.8	9.0	−29.3	^{n}Pr			Me		
1,2-Propylene	A34	4.8	−49.3	8.1	−29.6	^{n}Pr		Me			
1,2-Propylene	A35	1.5	−48.6	4.7	−29.1	^{n}Pr				Me	
2,1-Propylene	A36	13.0	−44.9	16.1	−26.9		^{n}Pr			Me	
2,1-Propylene	A37	7.4	−49.2	10.6	−29.5		^{n}Pr				Me
2,1-Propylene	A38	10.0	−48.8	13.3	−29.3	^{n}Pr				Me	
2,1-Propylene	A39	4.8	−48.0	8.1	−28.8	^{n}Pr					Me
After 1,2-propylene (B):											
Ethylene	B30	−1.0	−45.5	2.6	−27.3	^{i}Pr					
Ethylene	B31	−3.6	−44.5	0.0	−26.7	^{i}Pr					
1,2-Propylene	B32	4.3	−50.2	7.9	−30.1	^{i}Pr	Me				
1,2-Propylene	B33	4.3	−48.3	7.6	−29.0	^{i}Pr			Me		
1,2-Propylene	B34	3.1	−49.9	6.6	−29.9	^{i}Pr		Me			
1,2-Propylene	B35	−0.3	−50.8	3.1	−30.5	^{i}Pr				Me	
2,1-Propylene	B36	11.5	−47.6	14.8	−28.6		^{i}Pr			Me	
2,1-Propylene	B37	6.2	−46.7	9.6	−28.0		^{i}Pr				Me
2,1-Propylene	B38	8.1	−46.6	11.5	−27.9	^{i}Pr				Me	
2,1-Propylene	B39	2.9	−47.7	6.3	−28.6	^{i}Pr					Me
After 2,1-propylene (C):											
Ethylene	C30	1.3	−43.6	4.6	−26.2	Me	Et				
1,2-Propylene	C31	5.5	−51.4	8.9	−30.8	Me	Et	Me			
1,2-Propylene	C32	6.5	−50.3	9.7	−30.2	Me	Et		Me		
2,1-Propylene	C33	13.1	−47.7	16.1	−28.6	Me	Et			Me	
2,1-Propylene	C34	7.1	−51.1	10.3	−30.6	Me	Et				Me
After 2,1-propylene (D):											
Ethylene	D30	0.1	−47.1	3.2	−28.3	Et	Me				
1,2-Propylene	D31	2.3	−51.6	5.6	−31.0	Et	Me	Me			
1,2-Propylene	D32	2.6	−49.4	5.7	−29.6	Et	Me		Me		
2,1-Propylene	D34	3.4	−48.9	6.4	−29.3	Et	Me				Me
2,1-Propylene	D33	10.0	−47.7	12.8	−28.6	Et	Me			Me	

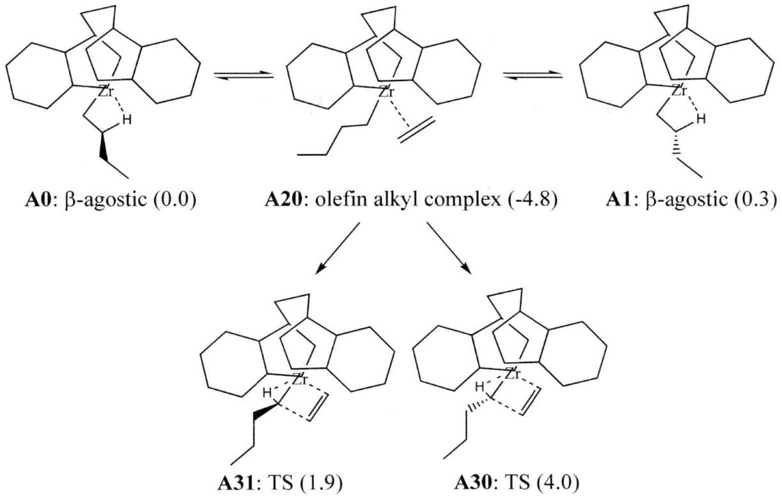

A0: β-agostic (0.0)   **A20**: olefin alkyl complex (-4.8)   **A1**: β-agostic (0.3)

A31: TS (1.9)   **A30**: TS (4.0)

Figure 7.6 Depiction of the equilibration of the β-agostic precursors and how the olefin complex leads to two unique transition states. Relative enthalpies are in parenthesis (kcal mol^{-1}).

ative stabilities of the products. The primary alkyl and secondary alkyl polymeryl groups based on structures **A0** and **C0**, respectively, have similar energy differences and these are manifested in the TSs leading to these β-agostic products.

The reactions following a 1,2-comonomer (**B0**) would be more representative of a homopolymerization of propylene. In this case, position a or b of the TS shown in Figure 7.5 is occupied with an ^{i}Pr group and, as with the prior case of the ^{n}Pr group (surface **A**), there is a preference for this group being at position a. In examining the effect of the α-agostic interaction, one sees that the two corresponding ethylene insertion TSs (**B30**, **B31**) differ by 2.6 kcal mol^{-1}, 0.5 kcal mol^{-1} greater than that on surface **A** due to the additional bulk of the isopropyl group. The likelihood of a regio-misinsertion of a propylene unit (the difference between **B35** and **B39** is 3.2 kcal mol^{-1}) appears as likely as a stereo-misinsertion (the difference between **B34** and **B35** is 3.5 kcal mol^{-1}). As stereo- and regio-control in polypropylene are very important, they are discussed further here. Eight different TSs define the possible insertion modes with a = ^{i}Pr or b = ^{i}Pr and for these two possibilities c = Me, d = Me, e = Me, or f = Me.

Table 7.4 provides the structural definition of these eight TSs as well as their relative energies and the interactions associated with them.

Entry 2 (**B35**) represents the lowest energy TS and the additional interactions of the other TSs are defined as:

α: defines whether a cis interaction exists between the polymeryl chain and the 1,2-inserting propylene [α = cis, zero = trans].

7 Olefin Polymerization using Homogeneous Group IV Metallocenes

Table 7.4 Relative energies (B3LYP/6-311 + G**) of the eight TSs associated with propylene inserting into an isobutyl polymeryl chain. See the text for the definition of the interactions.

	Structure (see Figure 7.5)						ΔE (kcal mol^{-1})	Interactions
	a	b	c	d	e	f		
B34	^{i}Pr		Me				3.5	$\alpha + \varepsilon$
B35	^{i}Pr			Me			0.0	
B38	^{i}Pr				Me		8.4	$\beta + \delta + \varphi$
B39	^{i}Pr					Me	3.2	φ
B32		^{i}Pr	Me				4.8	$\gamma + \varepsilon$
B33		^{i}Pr		Me			4.5	$\alpha + \gamma$
B36		^{i}Pr			Me		11.7	$\gamma + \delta + \varphi$
B37		^{i}Pr				Me	6.5	$\beta + \gamma + \varphi$

β: defines whether a cis interaction exists between the polymeryl chain and the 2,1-inserting propylene (β = cis, zero = trans).

γ: defines whether an interaction exists between the polymeryl group and the ligand (γ = polymeryl pointing towards ligand, zero = polymeryl pointing away from ligand).

Δ: defines whether the 2,1-inserting olefin points up towards the ligand (e = Me) or down (Δ = methyl points towards ligand, zero = methyl points away from ligand).

ε: defines whether the 1,2-inserting olefin points up towards the ligand (c = Me) or down (ε = methyl points towards ligand, zero = methyl points away from ligand).

φ: defines whether the olefin inserts with 1,2- or 2,1-regiochemistry (φ = insertion is 2,1-, zero = insertion is 1,2-).

Utilizing these six interactions and the relative energies from Table 7.4, we get seven equations and six unknowns and can fit the parameters in an effort to deconvolute which interactions are leading to the observed selectivities. Upon solving the equations, we find that $\alpha = 1.5$, $\beta = 0.0$, $\gamma = 3.1$, $\Delta = 5.2$, $\varepsilon = 1.8$, $\varphi = 3.3$ kcal mol^{-1}. The first thing to notice is that $\beta = 0.0$ kcal mol^{-1}, as might be expected since the cis/trans interaction for a 2,1-insertion should be negligible given the additional –CH$_2$– unit between the eclipsed groups. The desire for the polymeryl chain to point away from the indenyl group ($\gamma = 3.1$ kcal mol^{-1}) is significant, indicating a substantial preference for a particular hydrogen to be involved as the α-agostic interaction in the TS. With the polymer chain preferentially pointing in one direction, the 1,2-insertion prefers to reside trans to the polymeryl group ($\alpha = 1.5$ kcal mol^{-1}). This result leads to "chain-end control" for the influence of tacticity. This outcome coupled with the 1,2-olefin's desire to place its methyl group opposite the indenyl moiety ($\varepsilon = 1.8$ kcal mol^{-1}) leads to the observed stereoselectivity of approximately 3.5 kcal mol^{-1}. The fact that $\alpha = 1.5$ kcal mol^{-1} and $\varepsilon = 1.8$ kcal mol^{-1} suggests that both "chain-end control" and "site-control" are important. It is somewhat surprising that $\varepsilon = 1.8$ kcal mol^{-1} as this methyl group resides near the central location of the four-center TS and one would not think that the sterics of the ligand would affect this position to any significant degree. One has to be careful with the term "chain-end control". Here the implication is that how the polymeryl chain resides influences the stereochemistry of the next insertion. One has to realize that the position of the polymeryl chain is completely influenced by the ligand structure. Thus our definition of parameter α is more aptly termed "indirect site-control". The 2,1-insertions are inherently disfavored by 3.3 kcal mol^{-1} (φ). This difference is likely due to the greater stability of primary alkyl over secondary alkyl chains. The only parameter not discussed was $\Delta = 5.2$ kcal mol^{-1}, the largest value calculated. Not surprisingly, the large magnitude describes the directing influence of the ligand on a 2,1-propylene insertion. Once one has the eight TSs, determining these parameters is not of the utmost importance, but it provides a way to understand substituent effects in an effort to enhance the desired selectivity.

On examining the two 2,1-propagating surfaces, we focus on the surface leading from **C0** since this has been shown to be significantly lower in energy than that leading from **D0**. While the TSs for propagation might differ little, the formation of **C0** will be much more likely than **D0** and, thus, will be more prevalent in the polymerization cycle. In general, the ethylene TSs on **B** are lower in energy relative to the β-agostic species than on the **A** surface. In contrast, the 2,1-propylene insertion TSs appear somewhat higher on the **C** surface compared with the **A** or **B** potential energy surfaces. This effect can be understood on observing Figure 7.5. On the **C** surface, a and b are both alkyl groups (one is a Me and one is an Et), thus the inserting 1,2-propylene cannot adopt a trans conformation and the 1.5 kcal mol^{-1} energy difference determined for α in the homopolymerization of propylene is not realized. However, to truly understand these numbers, we need to determine resting states such that the rates of reactions on the different surfaces can be properly compared. This is carried out below, when the borates are included in the calculations to determine an estimate of the stability of the true resting state. Overall, on the 2,1-surface, there is little stereo- or regioselectivity observed, with the three crucial structures defining these selectivities all being within 1 kcal mol^{-1} of each other.

One final comparison of the propagation reaction is the relative rate of ethylene versus propylene polymerization after an ethylene and after a propylene insertion. On surface **A** (after ethylene), the lowest propylene TS (**A35**) resides 2.8 kcal mol^{-1} higher than the lowest ethylene TS (**A31**). This difference is 3.1 kcal mol^{-1} following a propylene insertion (**B31** versus **B35**), indicating that the relative rates of these reactions differ little. However, following a 2,1-insertion, the lowest ethylene TS (**C30**) is 4.3 kcal mol^{-1} more stable than the lowest 1,2-inserting propylene (**C31**).

7.3.2
β-Hydride Elimination

β-Hydride elimination has been studied previously in computations, but some complications arise with its application. First, the barrier appears very well-defined as the difference between the β-agostic species and TS for β-hydride elimination, but if the β-agostic species is not the true resting state, then some compound must be more stable and, hence, the calculated barrier is only a lower limit to the true barrier. The second thing to consider is that because this process is intramolecular it is inherently much faster than the other principal mechanism of molecular weight control, chain-transfer-to-monomer (discussed in the next section). One of the key factors for examining the relevance of the mechanism is understanding that the resting state is not important if one wants to understand selectivity. In this case, the selectivity is the branching between propagation and β-hydride elimination. As depicted in Figure 7.7, the probability of chain-transfer by β-hydride elimination depends only on the relative heights of the two TSs. Ultimately, these relative rates depend on the difference in free energies, thus one must be careful in interpreting data from computational sources.

We can estimate the difference in the TSs following an ethylene insertion (difference between **A10** and **A31**) as $\Delta\Delta H^{\ddagger} = 11.6$ kcal mol^{-1} and $\Delta\Delta S^{\ddagger} = 24.9$ cal mol^{-1} K^{-1}. At 25 °C, this translates to $\Delta\Delta G^{\ddagger} \approx 4$ kcal mol^{-1}. Experimental data

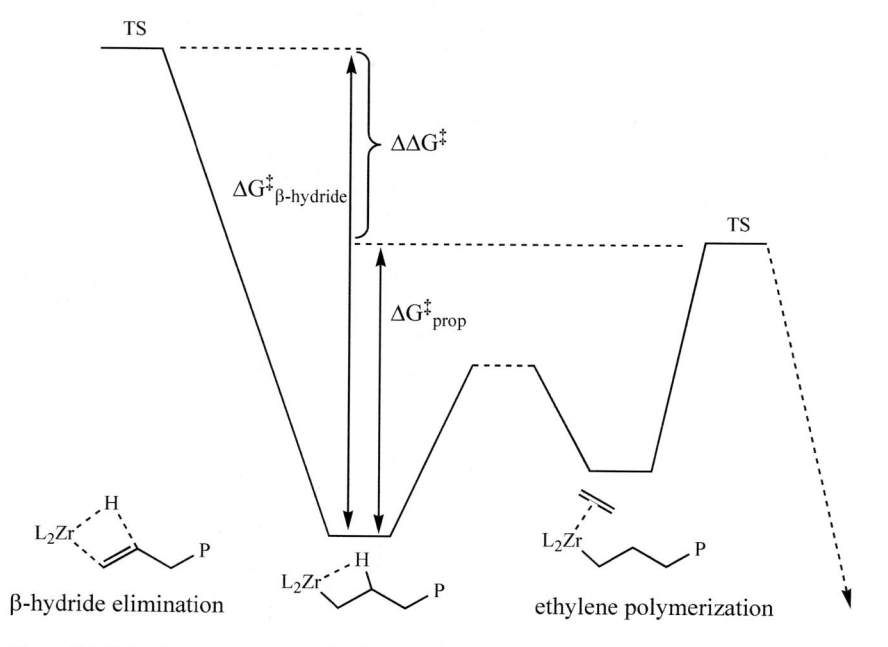

Figure 7.7 Potential energy surface for the propagation of ethylene and the β-hydride elimination step.

are not available for these numbers, but if we consider the difference between β-hydride elimination and propagation for the homopolymerization of propylene, experimental data are available. Here the computed values (**B10**, **B35**) are $\Delta\Delta H^{\ddagger} = 7.7$ kcal mol^{-1} and $\Delta\Delta S^{\ddagger} = 28.5$ cal mol^{-1} K^{-1}. The experimental values are $\Delta\Delta H^{\ddagger} = 9.5$ kcal mol^{-1} and $\Delta\Delta S^{\ddagger} = 19$ cal mol^{-1} K^{-1} [10b]. The overall agreement is satisfactory, especially for the enthalpic differences. Despite the reasonable agreement, though, the assumption that both the propagation and β-hydride elimination TSs have no borate present in the inner sphere is not likely a good postulate in this case. Landis [35] has shown that, for the β-hydride elimination process, the borate probably remains in the inner sphere and, thus, on moving from the resting state to the TS it is not displaced.

If one examines the differences in the enthalpies of activation following each of the different insertion modes and compares them with a propagating ethylene moiety, interesting trends are seen. As stated above, $\Delta\Delta H^{\ddagger} = 11.6$ kcal mol^{-1} following an ethylene insertion (**A10**, **A31**), while following a 1,2-propylene this value is reduced to 10.8 kcal mol^{-1} (**B10**, **B31**). This reduction is small but significant and appears to be generally true. Hence, if one can incorporate more comonomer in an ethylene/α-olefin copolymerization, the molecular weight is likely to be reduced if β-hydride elimination is the predominant chain-transfer mechanism. For the two 2,1-surfaces, β-hydride elimination appears to be significantly faster than propagation. For the predominant 2,1-surface, **C**, this difference is $\Delta\Delta H^{\ddagger} = 6.9$ kcal mol^{-1} and $\Delta\Delta S^{\ddagger} = 23.1$ cal mol^{-1} K^{-1} (**C10**, **C30**). Thus, at normal polymerization temperatures, intramolecular β-hydride elimination is faster than propagation. While experimental data are not available for ethylene propagating on the secondary-branched polymeryl, it is available for the propagation of a 1,2-propylene monomer. Computed values are $\Delta\Delta H^{\ddagger} = 2.6$ kcal mol^{-1} and $\Delta\Delta S^{\ddagger} = 27.7$ cal mol^{-1} K^{-1} for structures **C10** versus **C31**, implying that at any regular temperature, such as 25 °C, β-hydride elimination is faster than propagation. These data imply that, after each 2,1-propylene insertion, chain-transfer by β-hydride elimination should occur. Experimentally, this result is observed as the chain-transfer reaction appears first order in olefin, which is often indicative of chain-transfer-to-monomer as the dominant pathway. However, in this case, β-hydride elimination is so fast that the misinsertion effectively is the rate-limiting step to chain-transfer. Figure 7.8 shows the additional stability gained in the TS

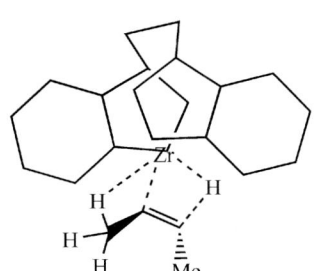

Figure 7.8 Additional stability gained in the β-hydride elimination TS of the trans 2,1-species.

by an agostic interaction with the 2,1-methyl group, which stabilizes the TS to a greater extent than similar structures on the surfaces following ethylene (**A**) and 1,2-α-olefin insertions (**B**). This stabilization only occurs with the trans TSs on surfaces **C** and **D**.

7.3.3
Chain-transfer-to-monomer

Chain-transfer-to-monomer is a second important chain-transfer mechanism. Computationally, it is very simple to understand molecular weight with this approach. Since this mechanism involves the binding of an olefin and the transfer of a β-hydrogen from the growing polymer chain to the olefin, it is first order in olefin. Thus, direct comparison of the TSs for propagation and chain-transfer-to-monomer gives us a view of the likelihood of this process. To grow a polymer chain of MW ≈ 250 000 g mol^{-1}, one needs the chain-transfer barrier to reside approximately 7 kcal mol^{-1} higher than the propagation barrier (100 °C). Since chain-transfer-to-monomer has basically the same entropic dependence, one can examine the differences in the $\Delta\Delta H^{\ddagger}$ to determine if this is a viable mechanism.

The general structure of the chain-transfer-to-monomer TS is shown in Figure 7.9 and the energetic data are collected in Table 7.5.

For all four surfaces, the order of olefin to transfer to is (faster to slower transfer): ethylene > 1,2-α-olefin > 2,1-α-olefin. The interesting comparison is between chain-transfer-to-monomer and propagation. Following an ethylene insertion and transfer to another ethylene, these differences (**A31**, **A40**) are $\Delta\Delta H^{\ddagger} = 7.2$ kcal mol^{-1} and $\Delta\Delta S^{\ddagger} = -2.3$ cal mol^{-1} K^{-1}. If one examines the experimentally known chain-transfer barriers following a 1,2-propylene insertion and contrasts this with the propagation of a 1,2-propylene, a good comparison can be made with the β-hydride elimination reaction described earlier. Here, one finds (**B35**, **B42**) that $\Delta\Delta H^{\ddagger} = 9.4$ kcal mol^{-1} and $\Delta\Delta S^{\ddagger} = 0.9$ cal mol^{-1} K^{-1}). This difference in barriers is slightly too high and supports the hypothesis that β-hydride elimination is dominant.

Chain-transfer following 2,1-insertions appears to be extremely fast with barriers only slightly higher than propagation ($\Delta\Delta H^{\ddagger} = 2.3$ kcal mol^{-1} for the difference between **C31** and **C43**). Here, the β-hydride elimination barrier was faster

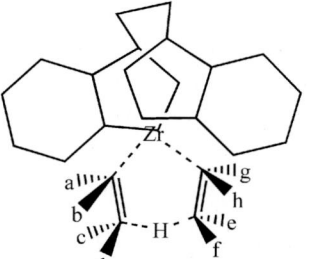

Figure 7.9 General structure of the different chain-transfer-to-monomer TSs examined in this work. a–d Represent substituents symbolizing the previously inserted monomer while e–h represent the approaching olefin.

Table 7.5 Energies of chain-transfer-to-monomer TSs on the four potential energy surfaces relative to the β-agostic structures shown in Table 7.1.

Compounds		6-31G*		6-311 + G**		Structure (see Figure 7.9)							
		ΔE‡ (kcal mol⁻¹)	ΔS‡ (cal mol⁻¹ K⁻¹)	ΔE‡ (kcal mol⁻¹)	ΔS‡ (cal mol⁻¹ K⁻¹)	a	b	c	d	e	f	g	h
After ethylene (A):													
Ethylene	A40	6.0	−48.9	9.1	−29.3				Et				
Ethylene	A41	6.3	−48.9	9.6	−29.3		Et						
1,2-Propylene	A42	11.2	−50.8	14.2	−30.5				Et		Me		
1,2-Propylene	A43	7.9	−51.4	10.8	−30.8				Et	Me			
1,2-Propylene	A44	−7.1	−49.5	14.3	−29.7			Et		Me			
1,2-Propylene	A45	9.3	−51.1	12.6	−30.7			Et			Me		
2,1-Propylene	A46	19.7	−48.4	22.4	−29.0				Et				Me
2,1-Propylene	A47	12.2	−50.6	15.0	−30.4				Et			Me	
2,1-Propylene	A48	12.9	−49.5	15.8	−29.7			Et				Me	
2,1-Propylene	A49	19.7	−48.2	22.5	−28.9			Et					Me
After 1,2-propylene (B):													
Ethylene	B40	4.6	−48.2	8.0	−28.9			Me	Me				
1,2-Propylene	B41	9.2	−50.5	12.7	−30.3			Me	Me		Me		
1,2-Propylene	B42	9.2	−49.3	12.5	−29.6			Me	Me	Me			
2,1-Propylene	B43	17.8	−48.8	20.8	−29.3			Me	Me				Me
2,1-Propylene	B44	10.7	−50.7	13.9	−30.4			Me	Me			Me	
After 2,1-propylene (C):													
Ethylene	C40	6.3	−48.6	9.3	−29.2	Me			Me				
Ethylene	C41	7.0	−48.6	10.4	−29.2	Me	Me						
1,2-Propylene	C42	11.6	−49.6	14.5	−29.7	Me			Me		Me		
1,2-Propylene	C43	8.4	−49.3	11.2	−29.6	Me			Me	Me			
1,2-Propylene	C44	12.7	−49.2	15.8	−29.5	Me	Me			Me			
1,2-Propylene	C45	10.1	−51.9	13.5	−31.2	Me	Me				Me		
2,1-Propylene	C46	20.4	−46.6	23.1	−27.9	Me			Me				Me
2,1-Propylene	C47	12.4	−50.9	15.2	−30.6	Me			Me			Me	
2,1-Propylene	C48	13.4	−51.3	16.5	−30.8	Me	Me					Me	
2,1-Propylene	C49	20.6	−50.0	23.6	−30.0	Me	Me						Me
After 2,1-propylene (D):													
Ethylene	D40	10.8	−45.3	14.0	−27.2	Me			Me				
Ethylene	D41	10.1	−47.3	13.2	−28.4	Me		Me					
1,2-Propylene	D42	16.9	−49.2	19.8	−29.5	Me			Me		Me		
1,2-Propylene	D43	13.0	−48.3	16.0	−29.0	Me			Me	Me			
1,2-Propylene	D44	14.3	−47.2	17.1	−28.3	Me		Me		Me			
1,2-Propylene	D45	12.5	−48.6	15.5	−29.1	Me		Me			Me		
2,1-Propylene	D46	22.5	−47.2	25.6	−28.3	Me			Me				Me
2,1-Propylene	D47	16.8	−50.6	19.7	−30.4	Me			Me			Me	
2,1-Propylene	D48	17.1	−46.8	19.7	−28.1	Me		Me				Me	
2,1-Propylene	D49	23.7	−47.0	26.3	−28.2	Me		Me					Me

than propagation and, thus, every 2,1-misinsertion is immediately terminated by β-hydride eliminating. While chain-transfer-to-monomer is very fast in this case, it does not compete with the β-hydride elimination reaction.

In general, one can compare chain-transfer-to-ethylene on each of the surfaces. After ethylene (**A31** and **A40**), this difference ($\Delta\Delta H^{\ddagger}$) is 7.2 kcal mol^{-1} and after a 1,2-propylene (**B31** and **B40**) it is 8.0 kcal mol^{-1}. Here, chain-transfer becomes slightly more difficult after an α-olefin incorporation whereas it is typically easier to β-hydride eliminate after a 1,2-insertion. This may not be the case with all catalysts but, computationally, this result has been observed before. Hence, one would expect similar or slightly higher molecular weights upon incorporating more comonomer if this was the primary chain-transfer mechanism. If one examines ethylene propagation and chain-transfer following a 2,1-insertion (**C30** versus **C40**), one observes the expected further reduction in $\Delta\Delta H^{\ddagger}$ to 4.7 kcal mol^{-1}.

7.3.4
Chain-transfer-to-hydrogen

A widely used method of molecular weight control in polymerization is through the addition of hydrogen. Two different types of TSs have been found: (a) attack from the frontside and displacing the β-agostic interaction (left-hand side of Figure 7.10) and (b) attack from the backside while maintaining the β-agostic interaction (right-hand side of Figure 7.10). When one compares the TSs on each of the four surfaces, the latter (backside attack) are always lower in energy than the former (frontside) and these will be the ones discussed further. Table 7.6 collects the energetic data.

In looking at the potential energy surface following ethylene insertion (**A**), we can compare directly the ethylene propagation TS with the hydrogen TS. While binding energies for H$_2$ were not calculated they will certainly be very weak and both olefin and hydrogen will be reversibly bound. The differences between the

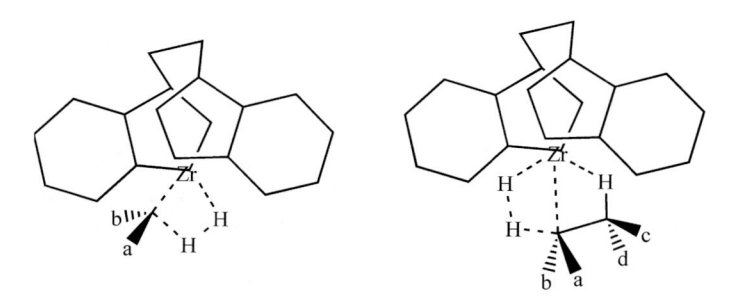

Figure 7.10 Two possible modes by which chain-transfer to hydrogen can occur; no β-agostic interaction on the left (frontside attack) and with a β-agostic interaction on the right (backside attack).

Table 7.6 Energies of chain-transfer-to-hydrogen TSs on the four potential energy surfaces relative to the β-agostic structures shown in Table 7.1.

Chain-transfer-to-hydrogen		6-31G*		6-311+G**		No β-agostic				β-agostic	
		$\Delta E^\ddagger$ (kcal mol^{-1})	$\Delta S^\ddagger$ (cal mol^{-1} K^{-1})	$\Delta E^\ddagger$ (kcal mol^{-1})	$\Delta S^\ddagger$ (cal mol^{-1} K^{-1})	a	b	c	d	a	b
After ethylene (A):											
Without β-agostic	A50	9.1	−28.2	6.4	−16.9					^{n}Pr	
Without β-agostic	A51	10.4	−27.9	7.7	−16.7						^{n}Pr
With β-agostic	A52	6.4	−32.1	4.0	−19.3			Et			
With β-agostic	A53	5.6	−32.5	3.0	−19.5				Et		
After 1,2-propylene (B):											
Without β-agostic	B50	7.3	−26.7	4.6	−16.0					^{i}Pr	
Without β-agostic	B51	8.5	−26.9	5.8	−16.1						^{i}Pr
With β-agostic	B52	6.3	−32.3	3.8	−19.4			Me	Me		
After 2,1-propylene (C):											
Without β-agostic	C50	10.9	−30.4	7.9	−18.2					Me	Et
With β-agostic	C51	6.1	−32.3	3.4	−19.4	Me		Me			
With β-agostic	C52	4.1	−30.1	1.4	−18.0	Me			Me		
After 2,1-propylene (D):											
Without β-agostic	D50	8.0	−29.6	4.8	−17.8					Et	Me
With β-agostic	D51	2.3	−29.9	−0.4	−17.9	Me	Me				
With β-agostic	D52	1.8	−32.8	−0.9	−19.7		Me		Me		

TSs (**A31**, **A53**) are $\Delta\Delta H^\ddagger = 1.1$ kcal mol^{-1} and $\Delta\Delta S^\ddagger = 7.5$ cal mol^{-1} K^{-1}, suggesting that hydrogen reacts somewhat slower, but can help to control the MW given the appropriate pressure. One interesting aspect of these calculations is that, despite both reactions being bimolecular, there is a difference in the entropies of activation. If one assumes the β-agostic species as the resting state, the scaled $\Delta S^\ddagger$ for ethylene propagation (**A31**) of -27.0 cal mol^{-1} K^{-1} has a greater magnitude than the -19.5 cal mol^{-1} K^{-1} value for the hydrogen reaction (**A53**). Of the separated reactant species in the hydrogenolysis reaction, $H_2 + \text{cat} \rightarrow \text{TS}$, hydrogen is linear and, thus, possesses $3N - 5$ vibrational modes with five rotational and translational degrees of freedom. This value differs from the ethylene reactant, which possesses six rotational/translational modes. Thus, as one moves from the reactants to the TS, more entropy is lost in the bimolecular ethylene reaction from the critical rotations and translations.

It is interesting to compare the hydrogenolysis reactions relative to ethylene propagation for the three different types of potential energy surfaces. Following a 1,2-insertion (**B**), $\Delta\Delta H^\ddagger = 3.8$ kcal mol^{-1} (**B31**, **B52**), which is a less probable reaction than after an ethylene insertion (**A**). For the more stable of the 2,1-surfaces (**C**), $\Delta\Delta H^{\ddagger} = -3.2$ kcal mol^{-1} (**C30**, **C52**); in other words, it reacts *much* faster than ethylene. This result is in line with Landis' data suggesting that secondary alkyls react much faster with hydrogen than primary alkyls [35]. While we are not quoting absolute rate constants, we are claiming that, for primary alkyls, hydrogen reacts slightly slower but comparable with ethylene, while for secondary alkyls it reacts significantly faster.

7.3.5
Absolute Rates of Reactions

In previous sections, the selectivities between different TSs were examined as reversibility in olefin/hydrogen binding created a Curtin–Hammett regime where only the relative heights of the TSs is necessary to know which processes are most likely. However, if one wants to know the absolute rates of these reactions, especially comparing the different surfaces, then one must be able to ascertain the resting state of the catalytic cycle for each of the surfaces. Most researchers believe that the counteranion, a borate, e.g. tetrakis(perfluorophenyl)borate, takes the position of the olefin in the open coordination site, leading to the resting state. Landis has shown that the borate gets displaced into the outer sphere in the TS but returns after each olefin insertion [10c, 36]. Our assumption is that the borate does coordinate to the metal center (inner sphere), but is pushed into the outer sphere in any of the olefin/hydrogen complexes or TSs. This suggests that comparing the selectivities of different TSs in the absence of a borate is acceptable. However, determination of the absolute rate will require a borate. Figure 7.11 shows the way we propose to estimate this correction factor for the resting state.

Olefin binding from the β-agostic species is enthalpically exothermic (**A0** $\rightarrow$ **A20**). This olefin binding from the borate-bound species is likely to also be enthalpically exothermic (**A0'** $\rightarrow$ **A20'**). Since the intrinsic step from the olefin-alkyl complex to the TS (**A20'** $\rightarrow$ **A31'**) has the borate in the outer sphere, this barrier should be the same whether a counterion is included or not. In other words, the barrier **A20'** $\rightarrow$ **A31'** should equal **A20** $\rightarrow$ **A31**. Thus, we can estimate this value as the difference between **A20** and **A31** with no borate present (similarly, all other TSs such as α-olefin insertion, β-hydride elimination, chain-transfer-to-monomer, and chain-transfer-to-hydrogen can be compared on the same scale). If we can now estimate the difference between the olefin dissociated and olefin bound species, **A0'** $\rightarrow$ **A20'**, we would have an enthalpic estimate of the difference between **A0'** and **A31'**. The key is being able to determine the difference between compounds **A0'** and **A20'** where the borate is present.

The resting state, **A0'**, is not difficult to locate as one of the fluorine atoms interacts directly with the metal, thereby stabilizing the resting state. For **A20'**, 16 different species were optimized for each of the four potential energy surfaces, starting from different orientations/coordination modes of the borate, ethylene, and the polymeryl chain. For the surface following an ethylene insertion ("Bu

7.3 Results and Discussion

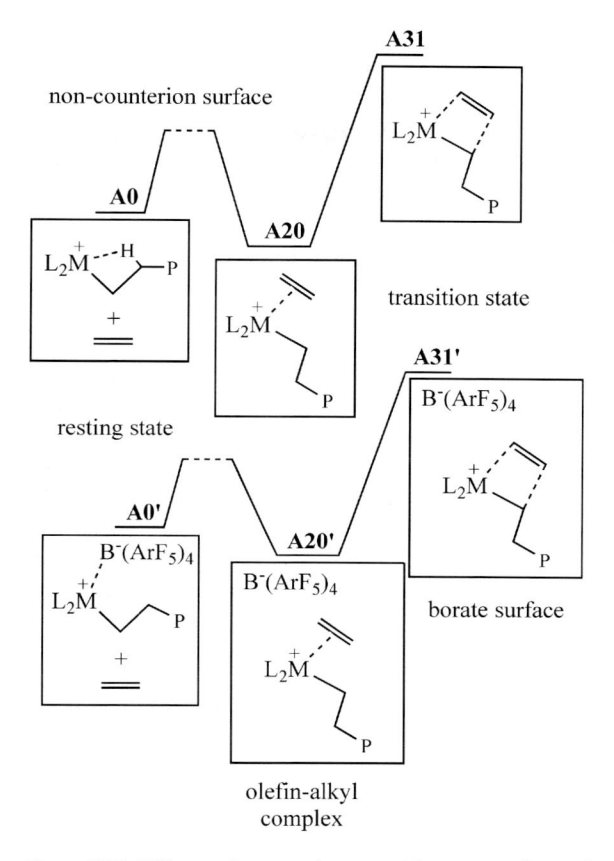

Figure 7.11 Difference between the potential energy surfaces where a borate is included and where it is excluded.

chain, **A**), the 16 different optimizations led to three different types of compounds: ten species where the borate was pushed to the outer sphere, three species where ethylene was ejected, and three species where both the ethylene and borate remained bound. Of these, the lowest energy of the species with ethylene ejected was within 2 kcal mol^{-1} of the calculated species in **A0'**, the separated ethylene + borate-bound catalyst. The lowest energy isomer of the group where both olefin and borate remained bound (both interactions are less than 3 Å) was higher in energy than the other two structural motifs. Thus, the only compounds of great interest were those with the borate in the outer sphere. The lowest energy isomer (**A20'**) resides 6.5 kcal mol^{-1} below the separated species, **A0'**. Thus the resting state resides 6.5 kcal mol^{-1} in enthalpy above the separated species, olefin + borate-bound catalyst. This statement may seem unusual since the resting state should be lower in energy, but if one were to include a reasonable entropic term of 10 kcal mol^{-1} ($T\Delta S°$ at 130 °C) then the separated species would be more stable on the free energy surface and, hence, be the true resting state.

These data now allow a correction of the olefin insertion barriers predicted earlier. If one examines the difference between the lowest energy ethylene complex (**A20**) and the propagation TS (**A31**), one sees an intrinsic enthalpic barrier of 6.7 kcal mol^{-1}. However, we have just shown that the resting state resides with an enthalpy 6.5 kcal mol^{-1} above the olefin complex. Hence, we can predict an enthalpy of activation ($\Delta H^\ddagger$) of 0.2 kcal mol^{-1}. A couple of things to note:

(1) The calculations are not accurate to the nearest kcal mol^{-1}, thus the barriers could be higher or lower than stipulated here.

(2) Landis calculates $\Delta H^\ddagger = 3$ kcal mol^{-1} for this catalyst with propylene and with a more coordinating counterion, MeB – (C$_6$F$_5$)$_3$. It is well known that ethylene polymerizes faster than propylene and the B – (C$_6$F$_5$)$_4$ anion imparts faster rates than the stickier MeB – (C$_6$F$_5$)$_3$ counterion. Both these changes could increase the rate by an order of magnitude or more. To increase this reaction rate by 100-fold, one must reduce $\Delta G^\ddagger$ by approximately 3 kcal mol^{-1}, and, since the reaction is still bimolecular, the entropy of activation is little changed by these effects. Thus, $\Delta H^\ddagger$ must take the brunt of the reduction in barrier and, thus, one could envision that this value is approaching zero. Consequently, our barrier appears to be slightly high relative to experiment.

One can now compare the 1,2-propagation TSs with the ethylene propagation TSs to determine an absolute $\Delta H^\ddagger$ for propylene polymerization. Here the lowest 1,2-propylene TS (**A35**) resides 2.8 kcal mol^{-1} above the ethylene TS (**A31**), leading to an activation enthalpy of 3.0 kcal mol^{-1} on the **A** surface. For a proper comparison with experiment, one needs to examine the **B** surface for the homopolymerization of propylene. Following a 1,2-propylene insertion (**B**), the resting state lies 6.0 kcal mol^{-1} higher than the ethylene complex. The barrier for ethylene insertion from the olefin complex (**B20**) to the TS (**B31**) with no borate present is 3.9 kcal mol^{-1}; thus, correcting this value leads to an absolute $\Delta H^\ddagger$ of −2.1 kcal mol^{-1}. The lowest 1,2-propylene insertion TS on the **B**-surface resides 3.1 kcal mol^{-1} higher than the ethylene TS, leading to an absolute $\Delta H^\ddagger$ of 1.0 kcal mol^{-1}. This value is in reasonable agreement with the experimental value of 3.4 kcal mol^{-1} [10b] since the counterion utilized here is tetrakis(perfluorophenyl)borate whereas the experimental anion was methyltris(perfluorophenyl)borate. These data also suggest that ethylene inserts faster into an isobutyl group than into a linear polymeryl group.

Furthermore, one can compare these results with the 2,1-surface (**C**) where the correction value for the counterion is somewhat lower at 3.1 kcal mol^{-1}. Here ethylene has an intrinsic barrier (**C20** to **C30**) of 8.9 kcal mol^{-1}, which leads to an absolute $\Delta H^\ddagger$ of 5.8 kcal mol^{-1}. This value is somewhat larger than that on the other surfaces and disagrees with experiment, suggesting that the reactivity is similar for primary and secondary alkyls [37]. Notably, also, despite the optimization from numerous starting structures with the inclusion of the counterion, it is uncertain whether the most stable species was located. Solvation of the separated ion pair has not been considered although one could argue that these effects are similar on the different surfaces.

The absolute β-hydride elimination and chain-transfer-to-monomer barriers can now be determined but their relative values to propagation will remain the same and the selectivities associated with molecular weight will not change. However, it is interesting to examine the absolute hydrogenolysis barriers and compare the activation enthalpies on the different surfaces. Including the counterion correction on each of the surfaces, one finds that the barriers ($\Delta H^{\ddagger}$) are 4.0 (**A**), 6.8 (**B**), and 2.6 kcal mol^{-1} (**C**). Hydrogenolysis on the 2,1-surface (**C**) is faster than on the primary alkyl surface (**A**), in agreement with the 100-fold difference predicted by experiments [37].

7.3.6
Kinetic Considerations

An interesting aspect of the extremely low activation enthalpies is the dependence of the rate constant on temperature. Based on the Eyring equation, one can easily derive Eq. (7):

$$k = cTe^{-\Delta G^{\ddagger}/RT}$$

$$k = cTe^{-\Delta H^{\ddagger}/RT}e^{\Delta S^{\ddagger}/R} \tag{7}$$

$$\frac{\partial k}{\partial T} = \frac{k}{T}\left(1 + \frac{\Delta H^{\ddagger}}{RT}\right)$$

One assumes that increasing temperature increases the rate of reaction, implying $\partial k/\partial T > 0$. Clearly k, R, and T are always positive, and one always assumes $\Delta H^{\ddagger}$ to be positive, but as can be seen from the activation parameter data presented earlier $\Delta H^{\ddagger}$ can get very near to zero. From the above equation, $\partial k/\partial T < 0$, when $\Delta H^{\ddagger} < -RT$. Thus, as long as $\Delta H^{\ddagger} \geq 0$, the rate constant will increase with increasing temperature. However, since the slope of the $\ln(k/T)$ versus $1/T$ plot relates to the rate constant, the more horizontal the slope the less the rate constant depends on temperature. Thus, clearly, for the more active polymerization catalysts the intrinsic propagation rate will increase little with temperature.

The apparent catalyst productivity depends on the relative rate of propagation versus the catalyst deactivation process. The value of $\Delta H^{\ddagger}$ for deactivation must be higher than propagation or no polymer would be formed. Olefin polymerization catalysts usually show an inverse relationship in that the productivity is reduced as the temperature is increased. Highly active catalysts have a temperature profile of propagation that is relatively flat, but how does the rate constant for catalyst deactivation differ with temperature? One can derive the relative rate constants for any two process, e.g., deactivation versus propagation, at two different temperatures as shown in Eq. (8):

$$\frac{\left(\dfrac{k_{deact1}}{k_{prop1}}\right)}{\left(\dfrac{k_{deact2}}{k_{prop2}}\right)} = \exp\left[\left(\frac{\Delta H^{\ddagger}_{deact} - \Delta H^{\ddagger}_{prop}}{R}\right)\left(\frac{1}{T_2} - \frac{1}{T_1}\right)\right] \tag{8}$$

The relative rate constant of deactivation divided by propagation (k_{deact1}/k_{prop1}) can be defined at a temperature (T_1) and similarly defined at a second temperature, T_2. If one assumes that ($\Delta H^{\ddagger}_{deact} - \Delta H^{\ddagger}_{prop}$) is temperature independent, which is reasonable for the temperatures examined here, then the terms involving $\Delta S^{\ddagger}$ drop out and the relative change in two rate constants at different temperatures depends only on the difference in $\Delta H^{\ddagger}$. In general, one assumes that deactivation should increase relative to propagation as the temperature increases. On looking at the right-hand part of Eq. (8), a few things can be realized. If we set $T_2 > T_1$, then $(1/T_2 - 1/T_1) < 0$. As we also know that $\Delta H^{\ddagger}_{prop} < \Delta H^{\ddagger}_{deact}$, the first bracket is positive, and with the second bracket being negative the product will always be less than zero. The exponential of a negative number returns a value between 0 and 1. Thus the relative rate constants shown on the left is less than 1 as long as $T_2 > T_1$. Thus, (k_{deact1}/k_{prop1}) < (k_{deact2}/k_{prop2}). This demonstrates that the amount of catalyst dying relative to propagation at T_1 is less than at T_2. The activation enthalpy term depends only on the difference in the values; thus, a bigger effect is seen as $\Delta\Delta H^{\ddagger} = (\Delta H^{\ddagger}_{deact} - \Delta H^{\ddagger}_{prop})$ becomes larger. However, for the temperature term, not only is the difference important but so is the magnitude of the values. The larger the temperature terms are the smaller the fraction will be. This term can also be expressed as shown in Eq. (9):

$$\left(\frac{1}{T_2} - \frac{1}{T_1}\right) = \frac{T_1 - T_2}{T_1 T_2} \tag{9}$$

The alternate expression in Eq. (9) clearly shows that the magnitude of this term becomes larger as the difference between T_1 and T_2 is increased, but also is greater if the overall temperatures utilized are lower. For example, the temperature range 25–50 °C has a value of -2.6×10^{-4} K^{-1}, whereas the same 25 °C range of 150 – 175 °C has a value with half the magnitude, -1.3×10^{-4} K^{-1}. This makes physical sense if one looks at extremes. Approaching absolute zero, this term freezes out the higher energy pathway, whereas at very high temperatures both pathways can be more easily accessed, even if the same temperature difference is used. Table 7.7 attempts to show how this ratio changes with temperature ranges and with $\Delta\Delta H^{\ddagger}$.

For a bimolecular deactivation reaction, catalyst death and propagation have similar rate expressions, and one could estimate $\Delta\Delta H^{\ddagger} \approx 10$ kcal mol^{-1}; thus, increasing the temperature from 25 to 100 °C results in only 3% of the efficiency at the higher temperature. If one views catalyst termination to be an intramolecular process, then $\Delta\Delta H^{\ddagger} \approx 20$ kcal mol^{-1} and the effect of increasing the temperature the same 75 °C as above affects the relative productivity significantly, by more than an order of magnitude at 0.1%.

Table 7.7 Ratio $(k_{deact1}/k_{prop1})/(k_{deact2}/k_{prop2})$ shown with different $\Delta\Delta H^{\ddagger}$ and temperature ranges.

$\Delta\Delta H^{\ddagger}$ (kcal mol^{-1})	Temperature, T_1/T_2 (°C)			
	$-250/-225$	0/25	25/100	100/200
3	2×10^{-15}	0.63	0.36	0.43
8	6×10^{-40}	0.29	0.07	0.10
10	1×10^{-49}	0.21	0.03	0.06
18	6×10^{-89}	0.06	0.002	0.01
20	9×10^{-99}	0.05	0.001	0.003

The expressions in Eqs. (8) and (9) are general for any competing reactions and one can consider comonomer incorporation versus ethylene reactivity. Here, one might estimate $\Delta\Delta H^{\ddagger} \approx 3$ kcal mol^{-1}, and thus a temperature increase from 25 to 100 °C increases the comonomer rate by a factor of almost 3. This obvious kinetic result appears to disagree with the conventional wisdom of α-olefin incorporation, which often decreases with increasing temperature.

As an example of why the mechanism of the reaction is crucial for determining the temperature effect, let us examine molecular weight control by β-hydride elimination or chain-transfer-to-monomer. For propagation, let us use $\Delta H^{\ddagger}_{prop} = 3$ kcal mol^{-1} and $\Delta S^{\ddagger}_{prop} = -30$ cal mol^{-1} K^{-1}, which leads to $\Delta G^{\ddagger}_{prop} = 16$ kcal mol^{-1} at 150 °C. For a given molecular weight of 400 000, the activation free energy of chain-transfer must be ≈ 8 kcal mol^{-1} higher than propagation, thus $\Delta G^{\ddagger}_{CT} = 24$ kcal mol^{-1}. For a bimolecular reaction such as chain-transfer-to-monomer, the same entropy component as indicated for propagation can be included; thus, $\Delta H^{\ddagger}_{CTM} = 11$ kcal mol^{-1}, $\Delta S^{\ddagger}_{CTM} = -30$ cal mol^{-1} K^{-1}, leading to $\Delta G^{\ddagger}_{CTM} = 24$ kcal mol^{-1} at 150 °C. However, for β-hydride elimination, $\Delta G^{\ddagger}_{\beta H}$ still must be 24 kcal mol^{-1} and with an estimate for $\Delta S^{\ddagger}_{\beta H}$ of -8 cal mol^{-1} K^{-1} this leads to $\Delta H^{\ddagger}_{\beta H} = 21$ kcal mol^{-1}. Thus, $\Delta\Delta H^{\ddagger}_{CTM} \approx 8$ kcal mol^{-1} for chain-transfer-to-monomer and $\Delta\Delta H^{\ddagger}_{\beta H} \approx 18$ kcal mol^{-1} for β-hydride elimination. Table 7.7 shows the temperature effect on going from 25 to 100 °C when $\Delta\Delta H^{\ddagger}$ equals these two values. Clearly, the molecular weight will be affected much more drastically (35-fold) on increasing temperature if the intramolecular β-hydride elimination is the predominant chain-transfer mechanism. Overall, this concept is not that revolutionary. If one examines competing pathways, one finds that at elevated temperatures it becomes easier to access the higher energy one.

Notably, one can estimate the relative rate constants for a process, e.g., molecular weight control, by plotting the temperature effect: $\ln(k_{CT}/k_{prop})$ versus $1/T$. Ultimately, this results in a plot leading to an estimate of the relative rate constants with a slope of $(\Delta H^{\ddagger}_{CT} - \Delta H^{\ddagger}_{prop})$ and an intercept of $(\Delta S^{\ddagger}_{CT} - \Delta S^{\ddagger}_{prop})$.

7.4
Conclusions

Homogeneous olefin polymerization using the C_2-symmetric metallocene, rac-C_2H_4(1-indenyl)$_2$ZrC$_4$H$_9^+$ (EBIZ), has been studied using density functional theory. Four pathways, including migratory insertion of the olefin, β-hydride elimination, chain-transfer-to-monomer, and chain-transfer-to-hydrogen, were examined. Since the counterion exists in the outer sphere in the various TSs, the selectivities of these processes can be directly compared without the use of a counterion. However, to understand the absolute rates of these reactions, the true resting state of the catalytic cycle must be found. Numerous species were probed on the potential energy surface and the lowest energy metal-alkyl and olefin-alkyl complexes with the tetrakis(perfluorophenyl)borate counterion were located. These species allowed for the correction of the true resting state of the catalytic cycle. Four different potential energy surfaces were utilized, representing prior ethylene, 1,2-propylene, and two 2,1-propylene surfaces, (R) and (S).

The propagation TSs indicate that α-olefins inherently insert more slowly than ethylene by a couple of kcal mol^{-1}. In addition, 2,1-insertions are disfavored relative to 1,2-insertions into a primary alkyl chain (after an ethylene or after a 1,2-propylene insertion). This difference may be because secondary alkyl groups are not as stable as primary alkyl chains for Group IV transition metals and this energetic difference is manifested in the TS for addition. If one compares the addition of 1,2- and 2,1-α-olefins to a secondary alkyl group, the selectivity is less pronounced as the substituent on the 1,2-inserting substrate cannot avoid an eclipsed interaction with the secondary polymeryl group. Owing to an α-agostic interaction in the transition state, eight unique TSs define the stereo- and regioselectivities in the homopolymerization of propylene. Various interaction parameters have been defined in an effort to deconvolute various steric and electronic effects in these TSs.

One can compare the relative heights of the TSs for propagation with those for β-hydride elimination to determine if this mechanism is feasible. One must take into account that the intramolecular reaction is much faster than the bimolecular propagation step. However, one finds that chain-transfer via β-hydride elimination relative to propagation becomes more prevalent after an α-olefin insertion and is even more dominant following a 2,1-insertion. These data suggest that, if β-hydride elimination is the dominant chain-transfer mechanism, increasing comonomer incorporation in ethylene/α-olefin copolymers should lead to a decrease in the molecular weight of the polymer. Experimental data suggest that, following 2,1-insertion, β-hydride elimination is so fast that chain-transfer appears first order in olefin.

Chain-transfer-to-monomer has also been examined as a possible chain-transfer mechanism. The TSs for chain-transfer and for propagation can be compared directly as both reactions are bimolecular and the entropic components are very similar. The difference in the activation enthalpies for the homopolymerization of propylene is 9.4 kcal mol^{-1}. An interesting comparison is how the chain-transfer versus propagation barriers change following ethylene, 1,2-, and 2,1-

insertions. After a 1,2-α-olefin insertion, chain-transfer is more difficult relative to propagation than after an ethylene insertion. This contrasts with β-hydride elimination, which is easier following a 1,2-insertion. Chain-transfer-to-monomer following a 2,1-insertion is faster relative to propagation (compared with the surface following an ethylene insertion), but still does not compete with the β-hydride elimination of a secondary alkyl. In general, chain-transfer to ethylene is favored over transfer to other α-olefins while chain-transfer after an ethylene is favored over chain-transfer after a 1,2-propylene insertion.

Two different types of TS were located for the hydrogenolysis reactions. The transition states maintaining a β-agostic interaction (backside attack) were found to be lower in energy than those without the interaction. Following an ethylene insertion (primary alkyl polymeryl chain), the hydrogen TSs were slightly higher in energy than the subsequent ethylene propagation TS, suggesting that hydrogen pressure could be used to control molecular weight. Interestingly, following a 1,2-propylene insertion, hydrogen reacts more slowly than ethylene propagation. However, following a 2,1-insertion, hydrogen reacts significantly faster than the subsequent propagation of an ethylene substrate.

The major part of this work examined the TSs of various processes relative to propagation, which allowed the determination of selectivities. However, in an effort to determine absolute rates, we utilized the tetrakis(perfluorophenyl)borate counterion to estimate the stability of the true resting state of the catalytic cycle. This process is difficult due to the many different conformations possible as the olefin displaces the borate into the outer sphere. Our estimate of the absolute enthalpy of activation of ethylene inserting into a polymeryl chain following an ethylene is 2.9 kcal mol^{-1}. Interestingly, $\Delta H^{\ddagger} = 1.0$ kcal mol^{-1} for propylene inserting after a 1,2-propylene, which seems to agree reasonably well with Landis' data, but the fact that propylene following propylene is faster than ethylene following ethylene is very surprising. Clearly, this is not due to the ease of inserting propylene over ethylene, but rather related to the relative stabilities of the linear versus ^{i}Bu-like polymeryl chains. Overall, the propagation reactions appear somewhat slower on the secondary alkyl surface. The absolute hydrogenolysis rates can also be compared and they proceed in the order: after 2,1-propylene > after ethylene > after 1,2-propylene.

Acknowledgments

I would like to acknowledge the helpful contributions and discussions of Dr. Peter Margl and Dr. Tim Wenzel of The Dow Chemical Company, and Professor Clark Landis of the University of Wisconsin.

References

1 (a) L. H. Shultz, M. Brookhart, Organometallics 2001, **20**, 3975. (b) D. J. Arriola, E. M. Carnahan, P. D. Hustad, R. L. Kuhlman, T. T. Wenzel, Science 2006, **312**, 714. (c) G. W. Coates, R. M. Waymouth, Science 1995, **267**, 217.

2 H. Sinn, W. Kaminsky, *Adv. Organomet. Chem.* 1980, **18**, 99.

3 (a) R. F. Jordan, R. E. LaPointe, C. S. Bajgur, S. F. Echols, R. Willett, *J. Am. Chem. Soc.* 1987, **109**, 4111. (b) M. Bochmann, L. M. Wilson, M. B. Hursthouse, R. L. Short, *Organometallics* 1987, **6**, 2556. (c) A. D. Horton, J. H. G. Frijns, *Angew. Chem., Int. Ed. Engl.* 1991, **30**, 1152. (d) J. A. Ewen, M. J. Elder, *Makromol. Chem., Macromol. Symp.* 1993, **66**, 179. (e) R. F. Jordan, *Adv. Organomet. Chem.* 1991, **32**, 325. (f) H. H. Brintzinger, D. Fischer, R. Mulhaupt, B. Rieger, R. Waymouth, *Angew. Chem., Int. Ed. Engl.* 1995, **34**, 1143. (g) E. Y. X. Chen, T. J. Marks, *Chem. Rev.* 2000, **100**, 1391.

4 (a) L. Resconi, L. Cavallo, A. Fait, F. Piemontesi, *Chem. Rev.* 2000, **100**, 1253. (b) G. W. Coates, *Chem. Rev.* 2000, **100**, 1253. (c) A. L. McKnight, R. M. Waymouth, *Chem. Rev.* 1998, **98**, 2587. (d) H. G. Alt, A. Koppl, *Chem. Rev.* 2000, **100**, 1205.

5 (a) L. K. Johnson, C. M. Killian, M. Brookhart, *J. Am. Chem. Soc.* 1995, **117**, 6414. (b) S. D. Ittel, L. K. Johnson, M. Brookhart, *Chem. Rev.* 2000, **100**, 1169. (c) G. J. P. Britovsek, V. C. Gibson, B. S. Kimberly, P. J. Maddox, S. J. McTavish, G. A. Solan, A. J. P. White, D. J. Williams, *Chem. Commun.* 1998, 849.

6 For example, (a) P. Margl, L. Deng, T. Ziegler, *Organometallics* 1998, **17**, 933. (b) P. Margl, L. Deng, T. Ziegler, *J. Am. Chem. Soc.* 1999, **121**, 154.

7 For example, (a) R. D. J. Froese, D. G. Musaev, T. Matsubara, K. Morokuma, *J. Am. Chem. Soc.* 1997, **119**, 7190. (b) R. D. J. Froese, D. G. Musaev, K. Morokuma, *Organometallics* 1999, **18**, 373.

8 (a) L. Deng, P. Margl, T. Ziegler, *J. Am. Chem. Soc.* 1997, **119**, 1094. (b) P. Margl, T. Ziegler, *Organometallics* 1996, **15**, 5519.

9 (a) D. G. Musaev, R. D. J. Froese, M. Svensson, K. Morokuma, *J. Am. Chem. Soc.* 1997, **119**, 367. (b) D. G. Musaev, M. Svensson, K. Morokuma, S. Stromberg, K. Zetterberg, P. E. M. Siegbahn, *Organometallics* 1997, **16**, 1933.

10 (a) Z. X. Liu, E. Somsook, C. B. White, K. A. Rosaaen, C. R. Landis, *J. Am. Chem. Soc.* 2001, **123**, 11193. (b) D. R. Sillars, C. R. Landis, *J. Am. Chem. Soc.* 2003, **125**, 9894. (c) C. R. Landis, K. A. Rosaaen, D. R. Sillars, *J. Am. Chem. Soc.* 2003, **125**, 1710.

11 (a) M. Dahlmann, G. Erker, K. Bergander, *J. Am. Chem. Soc.* 2000, **122**, 7986. (b) J. Karl, M. Dahlmann, G. Erker, K. Bergander, *J. Am. Chem. Soc.* 1998, **120**, 5643.

12 (a) P. Cossee, *J. Catal.* 1964, **3**, 80. (b) E. J. Arlman, P. Cossee, *J. Catal.* 1964, **3**, 99.

13 M. Brookhart, M. Green, L. L. Wong, *Prog. Inorg. Chem.* 1988, **36**, 1.

14 (a) M. Brookhart, A. F. Volpe, Jr., D. M. Lincoln, I. T. Horvath, J. M. Millar, *J. Am. Chem. Soc.* 1990, **112**, 5634. (b) M. K. Leclerc, H. H. Brintzinger, *J. Am. Chem. Soc.* 1995, **117**, 1651. (c) W. E. Piers, J. E. Bercaw, *J. Am. Chem. Soc.* 1990, **112**, 9406. (d) W. D. Cotter, J. E. Bercaw, *J. Organomet. Chem.* 1991, **417**, C1. (e) H. Krauledat, H. H. Brintzinger, *Angew. Chem., Int. Ed. Engl.* 1990, **29**, 1412. (f) R. H. Grubbs, G. W. Coates, *Acc. Chem. Res.* 1996, **29**, 85. (g) L. Clawson, J. Soto, S. L. Buchwald, M. L. Stelgerwald, R. H. Grubbs, *J. Am. Chem. Soc.* 1985, **107**, 3377.

15 (a) C. P. Casey, D. W. Carpenetti II, H. Sakurai, *J. Am. Chem. Soc.* 1999, **121**, 9483. (b) C. P. Casey, D. W. Carpenetti II, *Organometallics* 2000, **19**, 3970. (c) C. P. Casey, T.-Y. Lee, D. W. Carpenetti II, *J. Am. Chem. Soc.* 2001, **123**, 10762. (d) C. P. Casey, D. W. Carpenetti, II, H. Sakurai, *Organometallics* 2001, **20**, 4262. (e) C. P. Casey, J. F. Klein, M. A. Fagan, *J. Am. Chem. Soc.* 2000, **122**, 4320. (f) C. P. Casey, J. A. Tunge, M. A. Fagan, *J. Am. Chem. Soc.* 2003, **125**, 2641.

16 (a) Z. Wu, R. F. Jordan, J. L. Peterson, *J. Am. Chem. Soc.* 1995, **117**, 5867. (b) E. J. Stoebenau, III, R. F. Jordan, *J. Am. Chem. Soc.* 2003, **125**, 3222.

17 (a) D. J. Tempel, L. K. Johnson, R. L. Huff, P. S. White, M. Brookhart, *J. Am. Chem. Soc.* 2000, **122**, 6686. (b) M. D. Leatherman, S. A. Svejda, L. K. Johnson, M. Brookhart, *J. Am. Chem. Soc.* 2003, **125**, 3068.

18 L. Fan, D. Harrison, T. K. Woo, T. Ziegler, *Organometallics* 1995, **14**, 2018.

19 S. F. Vyboishchikov, D. G. Musaev, R. D. J. Froese, K. Morokuma, *Organometallics* 2001, **20**, 309.

20 R. D. J. Froese, D. G. Musaev, K. Morokuma, *J. Mol. Struct. (Theochem)* 1999, **461**, 121.

21 *Gaussian 03, Revision C.02*, M. J. Frisch, G. W. Trucks, H. B. Schlegel, G. E. Scuseria, M. A. Robb, J. R. Cheeseman, J. A. Montgomery, Jr., T. Vreven, K. N. Kudin, J. C. Burant, J. M. Millam, S. S. Iyengar, J. Tomasi, V. Barone, B. Mennucci, M. Cossi, M., G. Scalmani, N. Rega, G. A. Petersson, H. Nakatsuji, H. Hada, M. Ehara, M. K. Toyota, R. Fukuda, J. Hasegawa, M. Ishida, T. Nakajima, Y. Honda, O. Kitao, H. Nakai, M. Klene, X. Li, J. E. Knox, H. P. Hratchian, J. B. Cross, C. Adamo, J. Jaramillo, R. Gomperts, R. E. Stratmann, O. Yazyev, A. J. Austin, R. Cammi, C. Pomelli, J. W. Ochterski, P. Y. Ayala, K. Morokuma, G. A. Voth, P. Salvador, J. J. Dannenberg, V. G. Zakrzewski, S. Dapprich, A. D. Daniels, M. C. Strain, O. Farkas, D. K. Malick, A. D. Rabuck, K. Raghavachari, J. B. Foresman, J. V. Ortiz, Q. Cui, A. G. Baboul, S. Clifford, J. Cioslowski, B. B. Stefanov, G. Liu, A. Liashenko, P. Piskorz, I. Komaromi, R. L. Martin, D. J. Fox, T. Keith, M. A. Al-Laham, C. Y. Peng, A. Nanayakkara, M. Challacombe, P. M. W. Gill, B. Johnson, W. Chen, M. W. Wong, C. Gonzalez, J. A. Pople, Gaussian, Inc., Wallingford, CT, 2004.

22 (a) A. D. Becke, *J. Chem. Phys.* 1993, **98**, 5648. (b) C. Lee, W. Yang, R. G. Parr, *Phys. Rev B* 1988, **37**, 785. (c) B. Miehlich, A. Savin, H. Stoll, H. Preuss, *Chem. Phys. Lett.* 1989, **157**, 200.

23 (a) T. H. Dunning, Jr., P. J. Hay, *Modern Theoretical Chemistry*, ed. H. F. Schaefer, III, Plenum, New York, 1976, Vol. 3, p. 1. (b) P. J. Hay, W. R. Wadt, *J. Chem. Phys.* 1985, **82**, 270. (c) W. R. Wadt, P. J. Hay, *J. Chem. Phys.* 1985, **82**, 284. (d) P. J. Hay, W. R. Wadt, *J. Chem. Phys.* 1985, **82**, 299.

24 (a) R. Ditchfield, W. J. Hehre, J. A. Pople, *J. Chem. Phys.* 1971, **54**, 724. (b) W. J. Hehre, R. Ditchfield, J. A. Pople, *J. Chem. Phys.* 1972, **56**, 2257. (c) M. S. Gordon, *Chem. Phys. Lett.* 1980, **76**, 163.

25 (a) A. D. McLean, G. S. Chandler, *J. Chem. Phys.* 1980, **72**, 5639. (b) R. Krishnan, J. S. Binkley, R. Seeger, J. A. Pople, *J. Chem. Phys.* 1980, **72**, 650.

26 (a) B. O. Leung, D. L. Reid, D. A. Armstrong, A. Rauk, *J. Phys. Chem. A* 2004, **108**, 2720. (b) W. L. Jorgensen, D. Lim, J. F. Blake, *J. Am. Chem. Soc.* 1993, **115**, 2936.

27 C. R. Landis, K. A. Rosaaen, J. Uddin, *J. Am. Chem. Soc.* 2002, **124**, 12062.

28 G. Lanza, I. L. Fragalà, T. J. Marks, *Organometallics* 2002, **21**, 5594.

29 (a) Z. Yu, K. N. Houk, *Angew. Chem. Int. Ed.* 2003, **42**, 808. (b) T. J. M. de Bruin, L. Magna, P. Raybaud, H. Toulhoat, *Organometallics* 2003, **22**, 3404. (c) S.-Y. Yang, T. Ziegler, *Organometallics* 2006, **25**, 887.

30 S.-Y. Yang, T. Ziegler, *Organometallics* 2006, **25**, 887.

31 I. Silanes, J. M. Mercero, J. M. Ugalde, *Organometallics* 2006, **25**, 4483.

32 (a) D. J. Tempel, L. K. Johnson, R. L. Huff, P. S. White, M. Brookhart, *J. Am. Chem. Soc.* 2000, **122**, 6686. (b) S. D. Ittel, L. K. Johnson, M. Brookhart, *Chem. Rev.* 2000, **100**, 1169.

33 Z. Flisak, T. Ziegler, *Proc. Natl. Acad. Sci. U.S.A.* 2006, **103**, 15338.

34 (a) V. Busico, R. Cipullo, *J. Am. Chem. Soc.* 1994, **116**, 9329. (b) M. B. Harney, R. J. Keaton, J. C. Fettinger, L. R. Sita, *J. Am. Chem. Soc.* 2006, **128**, 3420.

35 C. R. Landis, Z. Liu, C. B. White, *Polym. Prepr. (Am. Chem. Soc., Div. Polym. Chem.)* 2002, **43**, 301.

36 (a) C. R. Landis, K. A. Rosaaen, *Polym. Mater., Sci. Eng.* 2002, **87**, 41. (b) Z. Liu, E. Somsook, C. R. Landis, *J. Am. Chem. Soc.* 2001, **123**, 2915.

37 (a) C. R. Landis, D. R. Sillars, J. M. Batterton, *J. Am. Chem. Soc.* 2004, **126**, 8890. (b) C. R. Landis, M. D. Christianson, *Proc. Natl. Acad. Sci. U.S.A.* 2006, **103**, 15349.

8
Group Transfer Polymerization of Acrylates with Mono Nuclear Early d- and f-block Metallocenes. A DFT Study

Simone Tomasi and Tom Ziegler

8.1
Introduction

The traditional way of obtaining poly(acrylates) and poly(methacrylates) involving radical initiators affords polymers having rather broad molecular weight distributions, and therefore is not as adequate as a living polymerization technique for the challenge of synthesizing high molecular weight and low polydispersity polymers [1–4].

The group transfer polymerization (GTP) of methacrylates by Group IV metallocenes and related compounds has attracted much attention in the scientific community since the seminal and independent works of Yasuda et al. [3–5], Collins et al. [1, 6] and Boffa et al. [2], dating back to the beginning on the 1990s. It offers a valid alternative to traditional, radical polymerizations because of the improved control it provides on important polymer properties like number-average molecular weight, molecular weight distribution and tacticity [7–12].

Numerous metallocene systems based on early d-block and f-block metals have shown interesting properties in catalyzing polymerizations of acrylates and methacrylates with the group transfer mechanism (GTP) [7, 9]. Although several metallocenes have been tested, the most successful systems are based on Zr and Sm.

Both experimental [13, 14] and theoretical [15–17] studies have shown that acrylate and methacrylate GTP using Group IV and related metallocenium ester enolates can proceed via a mononuclear or a binuclear mechanism. It can, further, be living [7, 13, 14, 18], thus also providing access to block copolymers when suitable catalyst systems are employed [19–24]. While it has been shown that a relatively straightforward mononuclear GTP takes place with Sm-based systems, a very effective binuclear mechanism has been demonstrated to be responsible for the polymerization of acrylates and methacrylates with the Zr-based bicomponent systems proposed by Collins et al. [6, 13]. Under these conditions the so-called bimetallic mechanism is preferred to the standard GTP. Under appropriate experimental conditions, monocomponent cationic zirconocenes can also catalyze the polymerization following the GTP mechanism postulated for the isoelectronic samarocene ester enolate first studied by Yasuda and coworkers [20, 25], albeit generally at a slower rate.

To the best of our knowledge, there are no reports on stereoregular polymerizations using bimetallic systems, whereas, depending on the reaction conditions and the ligands, syndio or iso poly(acrylates) and poly(methacrylates) have been obtained with neutral systems based on Sm or with cationic species based on Zr. Our study, therefore, considers only the mononuclear, because of the better chances it offers of controlling the stereochemistry.

The GTP of methacrylates has been studied much more than that of acrylates, mainly because the latter suffers from several unwanted side reactions caused by the presence of a mobile proton on the acrylate α-carbon. Collins has shown that when n-butyl acrylate is polymerized using the bicomponent system there are at least two side processes at higher temperatures, namely backbiting and proton transfer, and that backbiting occurs even at low temperatures. Backbiting causes termination, while proton transfer does not, although it contributes to widening the molecular weight distribution and, likely, slows down the overall rate of polymerization. Similar processes have been detected by Chen and coworkers in their study on the polymerization of n-butyl acrylate catalyzed by cationic, mononuclear racemic zirconocenium ester enolates [26].

Using neutral samarocene ester enolates only circumvents partially these shortcomings because, while Yasuda and coworkers have shown that it is possible to obtain poly(n-butyl acrylate) with high molecular weight and relatively narrow molecular weight distribution, also in this case it has been shown that the process is not stereospecific, nor living.

The first aim of this work is to explore the forces driving mononuclear group transfer polymerizations, as well as the factors that determine whether a stereoregular polymer is obtained, by means of density functional calculations.

Investigation of side reaction pathways in the GTP of acrylates is the subject of the second part of this chapter. To our knowledge, only an experimental study on this topic has been published [26], while no theoretical reports, other than ours [27], are yet available. Understanding the fine details of these processes is the first necessary step towards the design of modified catalysts, capable of polymerizing acrylic substrates more efficiently, without incurring low conversions and lack of control of the polymer properties. Possible beneficial modifications to the catalyst system will be considered in the final part of this account.

8.2
Computational Details

Density Functional Theory (DFT) calculations on the systems of interest were carried out with the program *ADF* [28, 29], version 2004.01 [30], using the Becke–Perdew exchange-correlation functional (BP86) [31–33]. Double-ζ STO basis sets with a polarization function were employed for H, C and O atoms, while for Zr and Sm atoms triple-ζ STO basis sets with one p-type polarization function were employed. The 1s electrons of C and O, as well as the 1s-3d electrons of Zr and the 1s-4d electrons of Sm, were treated as frozen core. First-order scalar relativistic corrections were applied to the systems studied [34–36]. To ensure SCF convergence, an ionic Sm^{3+} ion fragment was created with a restricted calculation, in which a fractional, even occupation of the 4f orbitals was imposed (five electrons in seven orbitals). All calculations on systems containing Sm were based on atomic fragments, plus a Sm^{3+} ion fragment (in all the systems studied Sm is in the +3 oxidation state). Note that the choice of fragments might help the SCF convergence, but it has no influence on the final outcome of the calculation.

Additionally, again to achieve SCF convergence, it was necessary to turn off the DIIS procedure [37] and to reduce the mixing parameter from the standard value to values ranging from 0.02 to 0.1.

All calculations were carried out in the gas phase without any symmetry constraint. Approximate transition states were located through reaction coordinate studies in which all degrees of freedom were minimized, while keeping a specific internal coordinate, or linear combination of internal coordinates, fixed. For the C–C coupling reaction, in all the mechanisms studied (Zr-based, both neutral and cationic, as well as Sm-based) the coordinate in question is the distance between the α-carbon of the enolate and the β-carbon of the acrylate. For the ring-opening reactions, the chosen coordinate was the difference between the distances of the ligands (carbonyl oxygen of the incoming acrylate and carbonyl oxygen of the leaving ester group, respectively) from the metal center. The leading coordinate for the backbiting reactions was the distance between the α-carbon of the enolate and the carbonyl carbon involved in the reaction. For reactions in which transfer of a proton was involved, the reaction coordinate has been chosen as the length of the bond being broken minus the length of the bond being formed. All stationary points were then fully optimized as minima or transition states, starting from the constrained geometries. The systems studied are very floppy, especially the transition states in the ring-opening reactions, and frequency calculations have yielded numerous low frequency modes. Under these conditions the harmonic oscillator approximation, under which vibrational frequencies are calculated, fails, thus affording unreliable values for the corrections to Gibbs free energies. For this reason reaction energies and not free energies are reported throughout the chapter.

Hybrid quantum-mechanical (QM) and molecular-mechanical (MM) models (QM/MM) have been applied to the modified catalysts described in the last part of this study, using the IMOMM scheme of Morokuma and Maseras [38], as implemented in ADF by Woo et al. [39]. The substituents added to the base Cp ligands were modeled as MM atoms and H atoms were used to cap the QM system. The MM atoms were described using the SYBYL/TRIPOS 5.2 force field constants [40].

No counteranions are present for the neutral samarocene systems. Compared with the metallocene-catalyzed polymerization of olefins, the anion plays a much less important role in the GTP of acrylates, since the two available metal coordination sites are occupied at all times by oxygen-based strong σ-donor ligands and the anion, if existing, is always relegated to an outer shell. In the light of such considerations, and to be able to compare more directly the results, the study with cationic zirconocenes was conducted without considering the counteranion [which experimentally is a non-coordinating anion such as $B(C_6F_5)_4^-$].

All energies are given in kcal mol^{-1} relative to the parent reactants ($Cp_2ZrMeTHF^+$, Cp_2ZrMe_2, MA, THF, $Cp_2SmMeTHF$).

8.3
Discussion

8.3.1
Polymerization Mechanism

8.3.1.1 Initiation: Formation of a Metallocene-enolate Complex

In many cases the precursors of the catalytically active species is a metallocene to which a molecule of tetrahydrofuran (THF) is complexed. When these systems are employed for the polymerization, usually an induction period is observed. Removal of the THF molecule frees a coordination site, which can accommodate an acrylate molecule and activate it for an internal 1,4 conjugated nucleophilic attack, yielding an enolate (Scheme 8.1). Removal of the THF molecule is, most likely, the rate-determining step for the initiation.

As expected, because of its high oxophilicity, the zirconium center interacts more strongly with THF than the samarium in the corresponding species (41.5 kcal mol^{-1} dissociation energy compared with 18.6 kcal mol^{-1}, see Table 8.1).

The formation of the metallocene-MA complex is in both cases only a few kcal mol^{-1} endothermic. It is possible that substitution of THF by MA occurs via a relatively low barrier associative mechanism, having a lower barrier than that of THF dissociation. Experimentally, the initiation step can be avoided using preformed enolates. With cationic zirconocenes, this is a necessary step for those systems, which otherwise are not reactive, or to avoid the conditions under which the competing binuclear mechanism prevails. Another advantage is that the molecular weight dispersion of the polymers so-obtained is lower because all the chains start at the same time. For this reason, the mechanism of the THF-MA exchange was not further investigated.

Transfer of the methyl group attached to the metal center to the β-carbon of the α,β-unsaturated MA is exothermic and yields a neutral enolate for the Sm-based system or a cationic enolate for the Zr-based complex (Scheme 8.1 and Table 8.1).

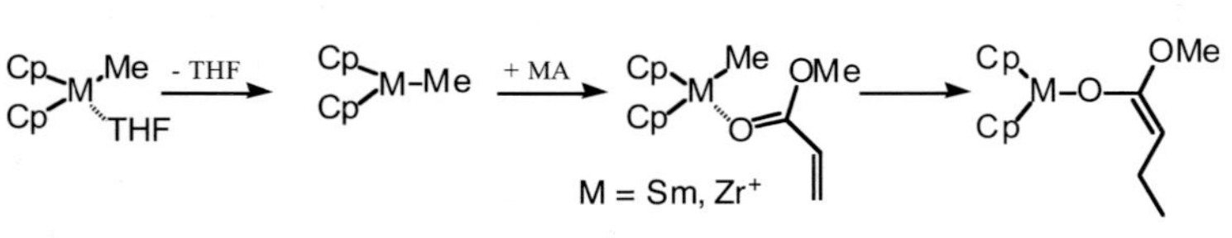

Scheme 8.1 The initiation step: formation of a metallocene-enolate complex.

Table 8.1 Energies of species involved in the formation of samarocene and zirconocene enolates (initiation reactions).

Species	E_{rel} (kcal mol^{-1})
MA	0
THF	0
Cp$_2$ZrMeTHF$^+$	0
Cp$_2$ZrMe$_2$	0
Cp$_2$ZrMe$^+$	41.5
Cp$_2$ZrMeMA$^+$	2.4
Cp$_2$ZrEno$^+$	−15.8
Cp$_2$ZrMeEno	−32.0
Cp$_2$SmMeTHF	0
Cp$_2$SmMe	18.6
Cp$_2$SmMeMA	5.5
Cp$_2$SmEno	−28.4

8.3.1.2 Conformations of the Metallocene-acrylate-enolate Complexes

Coordination of a MA molecule to the metallocene-enolate results in a tetracoordinated complex, which can then give a conjugated 1,4 intramolecular addition, by which the polymer chain is increased by one unit. Four stable conformations of the complexes, determined by the relative positions of the substituents of the α-carbons of the acrylate and enolate groups, are possible (Scheme 8.2). The stereochemical outcome depends on such conformations, because the side of the

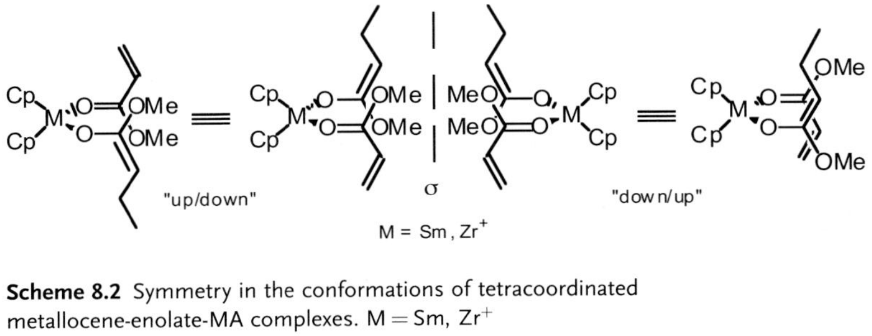

Scheme 8.2 Symmetry in the conformations of tetracoordinated metallocene-enolate-MA complexes. M = Sm, Zr$^+$

enolate facing the MA ligand determines the configuration of the stereocenter formed in the coupling reaction (*vide infra*).

In Scheme 8.2 and in the following discussion, "up" refers to a methoxy group above the plane defined by Zr and the coordinated oxygen atoms, when the molecule is seen in such a way that the methyl group is away from the viewer (the enolate is in the foreground), whereas "down" refers to the opposite situation.

The number of conformations that must actually be studied can be reduced to two on the basis of symmetry: it can be seen from Scheme 8.2 that the "up/up", "down/down" and "up/down", "down/up" complexes are two pairs of mirror conformations. Since enantiomers have the same energies, only one complex from each of the two enantiomeric pairs was considered.

Complexation energies range from −9.4 to −14.3 kcal mol^{-1} (Tables 8.1 and 8.2). As expected, the distance from the enolate oxygen to Zr is shorter than that from the MA carbonyl oxygen (Table 8.3), indicating that the former interaction is stronger (because the enolate is a stronger donor). Enolate–Zr distances range from 1.98 to 2.07 Å, while MA–Zr distances are on average 0.2 Å longer (2.15–2.26 Å).

The stereochemical outcome of the coupling reaction depends on the enantiotopic face of the enolate attacking MA. The "up/up" and "up/down" complexes will both form a (S) stereocenter because the enolate is facing MA with the same prochiral side, while for the same reason "down/down" and "down/up" geometries give rise to a (R) stereocenter. The orientation of MA is important for the formation of the stereocenter in the following step, since after the coupling it is transformed into an enolate. The implications that this has on stereoregularity are discussed in Section 8.3.1.4 on ring-opening reactions.

Neutral samarocene complexes are isoelectronic to cationic zirconocene complexes, and also require the study of only two complex conformations. Samarocene-enolate-MA complexes share with the analogous zirconocene complexes the pseudo-tetrahedral arrangement of the ligands, which are, however, more distant

Table 8.2 Relative energies for species involved in GTP[a] catalyzed by a monometallic cationic zirconocene.

Geometries	E_{rel} (complex)[b,c] (kcal mol^{-1})	E_{rel} (metallacycle)[b,c] (kcal mol^{-1})
"up/down"[d]	−25.2	−47.9
"down/down"	−30.1	−46.3

a) The process is displayed in Scheme 8.3.
b) No transition states have been found on the potential energy surface.
c) Reference state corresponds to the reactants (Cp$_2$ZrMe$_2$, Cp$_2$ZrMeTHF+, MA and THF).
d) See Scheme 8.2.

Table 8.3 Selected interatomic distances for species involved in GTP[a] catalyzed by a monometallic cationic zirconocene.

Species[b]	C=C(MA) (Å)	C–C (Å)	C=O (Å)	=O···Zr (Å)	–O–Zr (Å)	C–O (Å)	C=C(eno) (Å)	C···C (Å)
Complex								
"up/down"[c]	1.36	1.43	1.272	2.15	2.07	1.32	1.37	2.27
"down/down"[c]	1.34	1.47	1.256	2.26	1.98	1.34	1.34	7.44
Products								
"up/down"[c]	1.50	1.35	1.345	2.04	2.22	1.250	1.50	1.54
"down/down"[c]	1.50	1.36	1.341	2.04	2.21	1.255	1.50	1.56

a) Process is displayed in Scheme 8.3.
b) Bond labels have been retained for clarity although bond orders in the products are different (refer to text).
c) See Scheme 8.2.

from Sm because of its larger size and lower positive charge. The O–Sm distances are 2.37–2.45 Å for MA (Table 8.4), roughly 0.2 Å longer than the corresponding O–Zr ones. MA complexation is exothermic by 10.1–10.7 kcal mol^{-1}, which is comparable to the analogous cationic zirconocene complexes (Tables 8.1 and 8.5).

Table 8.4 Selected interatomic distances for species involved in GTP[a] catalyzed by a monometallic neutral samarocene.

Species[b]	C=C(MA) (Å)	C–C (Å)	C=O (Å)	=O·Sm (Å)	–O–Sm (Å)	C–O (Å)	C=C(eno) (Å)	C··C (Å)
Complexes								
"down/down"[c]	1.39	1.45	1.24	2.37	2.23	1.30	1.37	3.20
"down/up"[c]	1.34	1.46	1.24	2.45	2.20	1.30	1.36	5.36
Products								
"down/down"[c]	1.50	1.37	1.30	2.19	2.46	1.24	1.51	1.57
"down/up"[c]	1.49	1.37	1.30	2.22	2.43	1.24	1.50	1.61
Exp.[d]	1.52	1.39	1.31	2.188	2.39	1.23	1.53	1.61

a) Process is displayed in Scheme 8.3.
b) Bond labels have been retained for clarity although bond orders in the products are different (refer to text).
c) See Scheme 8.2.
d) Ref. [3].

Table 8.5 Relative energies for species involved in GTP[a] catalyzed by a monometallic neutral samarocene.

Geometries	E_{rel} (complex)[b,c] (kcal mol^{-1})	E_{rel} (metallacycle)[b,c] (kcal mol^{-1})
"down/up"[d]	−38.5	−55.5
"down/down"[d]	−39.1	−57.9

a) Process is displayed in Scheme 8.3.
b) No transition states have been found on the potential energy surface.
c) Reference state corresponds to the reactants (Cp$_2$SmMeTHF, MA and THF).
d) See Scheme 8.2.

8.3.1.3 Transition States of the Coupling Reaction

Scheme 8.3 shows the mechanism for GTP by monometallic cationic zirconocene and neutral samarocene. Simple rotation along selected torsion angles lead the π systems of the acrylate and of the enolate to a distance and a relative orientation suitable for the 1,4-conjugated nucleophilic addition. The first conjugated addition of the enolate α-carbon to the acrylate β-carbon in the cationic system has no barrier on the potential energy surface (Figure 8.1). This result has been confirmed scanning the reaction coordinate forward and backward. If a barrier actually exists, it is likely due to entropic factors not considered in our calculations and it could only be detected by studying the reaction on the free energy surface, which is presently computationally too expensive for the information it would

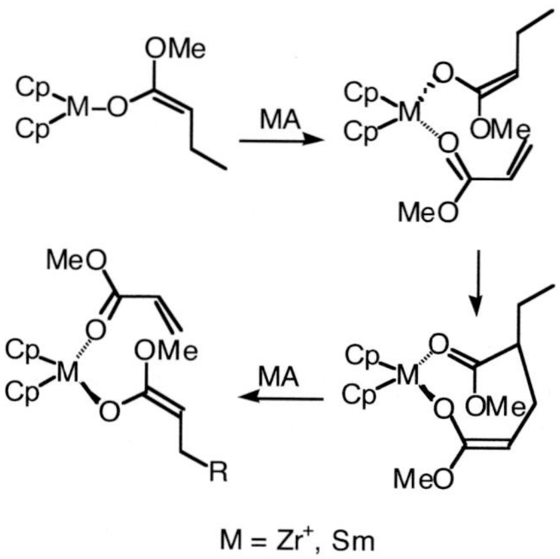

Scheme 8.3 Mechanism for GTP catalyzed by a monometallic cationic zirconocene or a monometallic neutral samarocene.

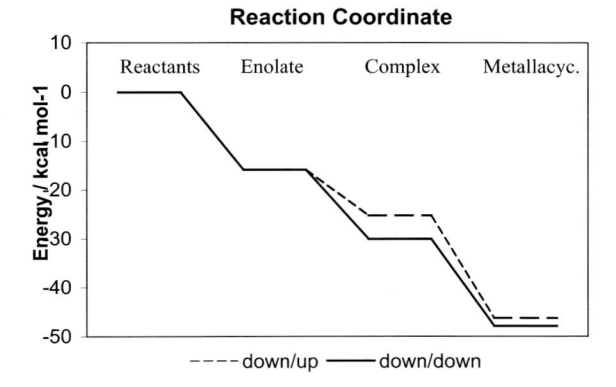

Figure 8.1 Energy profile for GTP catalyzed by a monometallic cationic zirconocene. The process is illustrated in Scheme 8.3. Geometrical parameters are shown in Table 8.3 and energies in Table 8.2.

afford. The product of the coupling reaction is a stable eight-membered metallacyclic species, which lies 16.2–22.7 kcal mol^{-1} below the corresponding complex. The polymerization, therefore, is strongly exothermic, and the resting state is the metallacycle.

As expected, the neutral samarocene behaves similarly to the cationic zirconocene, showing no activation barrier for the C–C coupling (Figure 8.2). The shape of the potential energy surface at the first addition step could be different from subsequent ones, in which the bulk of the growing polymer chain might contribute to a modest activation energy. As with the Zr-based system, the resting state is a metallacycle, which is 17.0–18.8 kcal mol^{-1} more stable than the parent react-

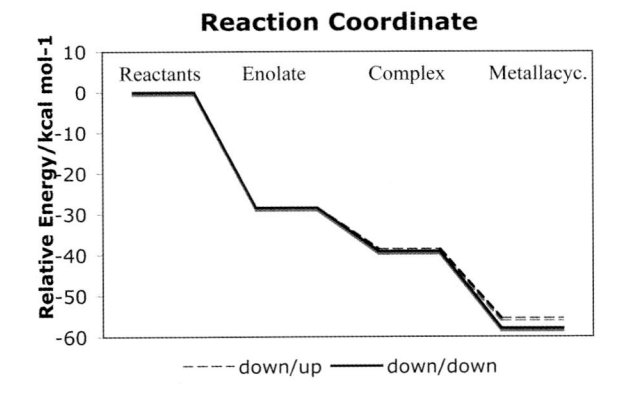

Figure 8.2 Energy profile for GTP catalyzed by a monometallic neutral samarocene. The process is illustrated in Scheme 8.3. Geometrical parameters are shown in Table 8.4 and energies in Table 8.5.

ing complex (Table 8.5). The computed metallacycle bond lengths of the model neutral samarocene compare very well with the crystallographic measurements performed by Yasuda et al. on an analogous system containing pentamethylated Cp rings (Table 8.4) [3].

8.3.1.4 Ring-opening Reactions

While isotactic polymers have been obtained with two Cp rings having different substituents [8, 10–12], most symmetric systems afford non-stereoregular polymers, or syndiotactic polymers if the polymerizations are carried out at very low temperatures.

Stereocontrol has been attributed to enantiomorphic site control (iso-specific) or chain-end control (syndio-specific). Since the systems studied in this work bear two identical Cp ligands, they are not expected to polymerize MA isospecifically.

Höcker et al. have shown that cationic zirconocenes similar to our model system, but containing a bridge between the Cp rings, are good catalysts for the production of syndio PMMA [poly(methyl methacrylate)] at low temperatures. Similar systems in which a Cp ring has been substituted with an indenyl ring are iso-specific at somewhat higher temperatures [10].

Syndio-enriched polymers are usually obtained only at very low temperatures, since, for the chain-end stereocontrol to be effective, the weak influence exerted by the chiral center must not be overcome by thermal fluctuations. It is reasonable to assume that the closer the last formed stereocenter is to the active site the greater will be its influence on the formation of syndio products. The metallacycle is the resting state in GTP catalysis by both cationic zirconocene and neutral samarocene. In this restrained conformation the stereocenter is in the metallacycle, where it is better able to exert its influence than in the corresponding species in which the metallacycle has been opened. Therefore, the opening of the metallacyclic resting state, assisted by an incoming MA molecule, is likely to be the one responsible for stereoregularity (or lack thereof), as well as the rate-determining step.

In the ring opening, a MA molecule substitutes the ester ligand of the metallacycle, approaching the metallacycle axially from the same side of the ester group ("frontside attack") or from the opposite axial position ("backside attack") (Figure 8.3). The direction of the incoming MA molecule and the orientation of the enolate group in the metallacycle determine the stereochemistry of the following stereocenter. The possibility of a non-stereoselective two-step mechanism, in which first the ring opens spontaneously, followed by coordination of MA, needs not to be considered, because the energy spent for breaking the Zr···O=C interaction is recovered only in the second step.

With a backside attack of MA to the metallacycle, the enolate coordination site is maintained, as shown in Scheme 8.4. In contrast, frontside attack results in inversion of the enolate coordination site. For the metallacycle obtained from the "down/down" or the "up/up" type of complex, the orientation and coordination site of the enolate is maintained following repeated backside attacks, which leads

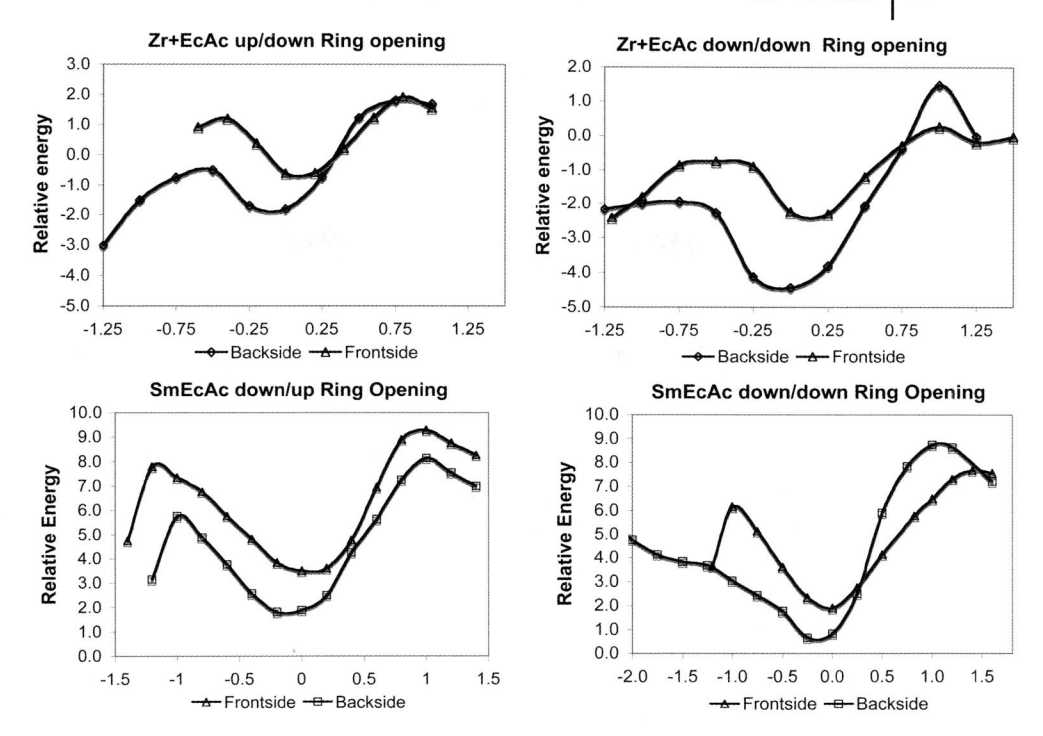

Figure 8.3 Energy profiles for ring-opening reactions involving a monometallic cationic zirconocene or a monometallic neutral samarocene. Scheme 8.3 illustrates the process. Energies correspond to the constrained geometries optimizations along the reaction coordinate. Geometrical parameters and energies for the fully optimized stationary points are shown in Tables 8.6 and 8.7, respectively.

to an isotactic polymer. The opposite result is achieved by repeated frontside attacks: orientation of the enolate is always the same, but its coordination site changes every time, thereby leading to a syndiotactic polymer. The metallacycles obtained from the "down/up" or "up/down" complexes follow the same rules: enolate coordination site is maintained with the backside attack and inverted with the frontside attack. However, the MeO group of the enolate at every step is oriented in the opposite direction compared with the previous one. Therefore, the backside attack in this case leads to the syndio product, whereas the frontside attack leads to the iso product (Scheme 8.4).

The reaction coordinates for the MA-assisted ring openings present two transition states separated by an intermediate for all systems and all direction of the attack (Figure 8.4). The first transition state in each curve is higher in energy than the second one and, therefore, it is the one controlling the rate of the ring opening (Section 8.3.2). As MA approaches the metallacycle, the energy of the system increases because of added steric crowding around the metal center. The

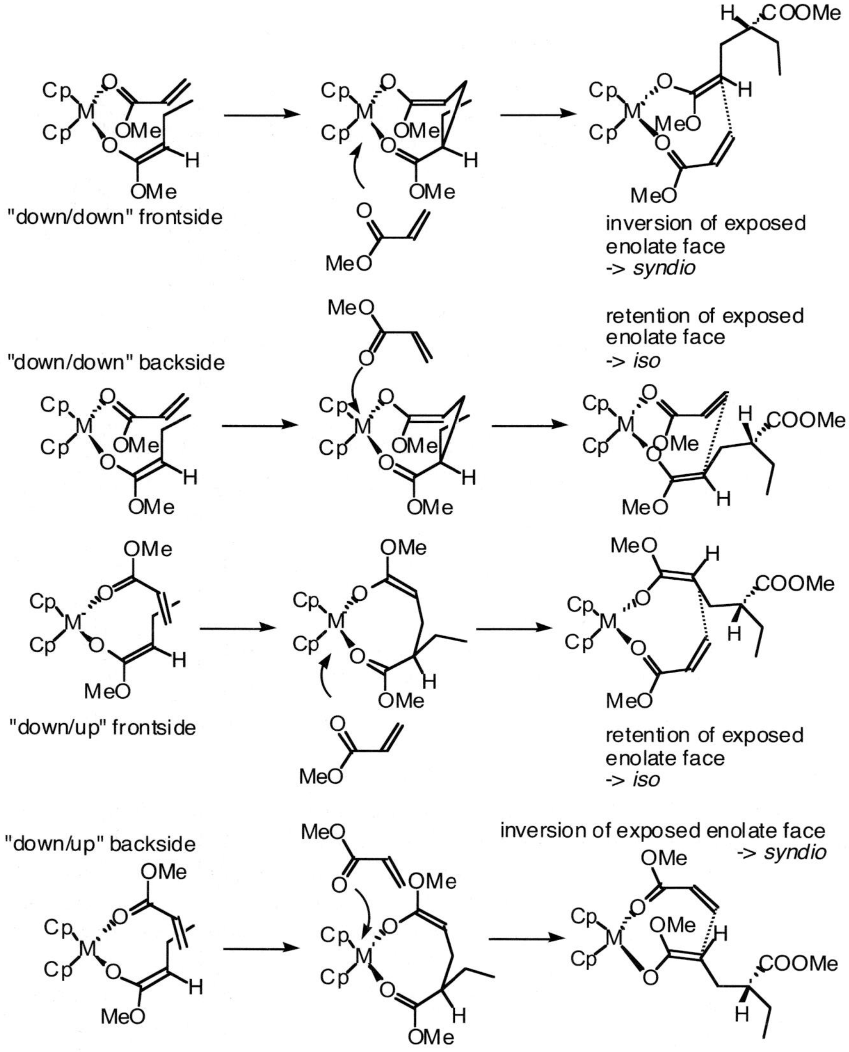

Scheme 8.4 Stereoregularity control in neutral Sm- and cationic Zr-catalyzed processes.

geometric rearrangement forces a smaller O–Metal–O angle in the metallacycle, causing more ring strain (Table 8.6). This provides a qualitative criterion for establishing the relative stability of competing transition states: the transition states in which the O–Metal–O angle is tighter are higher in energy. The energy reaches a maximum when the reaction coordinate ranges approximately from 1.4 to 0.75 Å. Thereafter, the stabilizing electronic interactions between MA and the metal center increases, driving the energy of the system to a local minimum

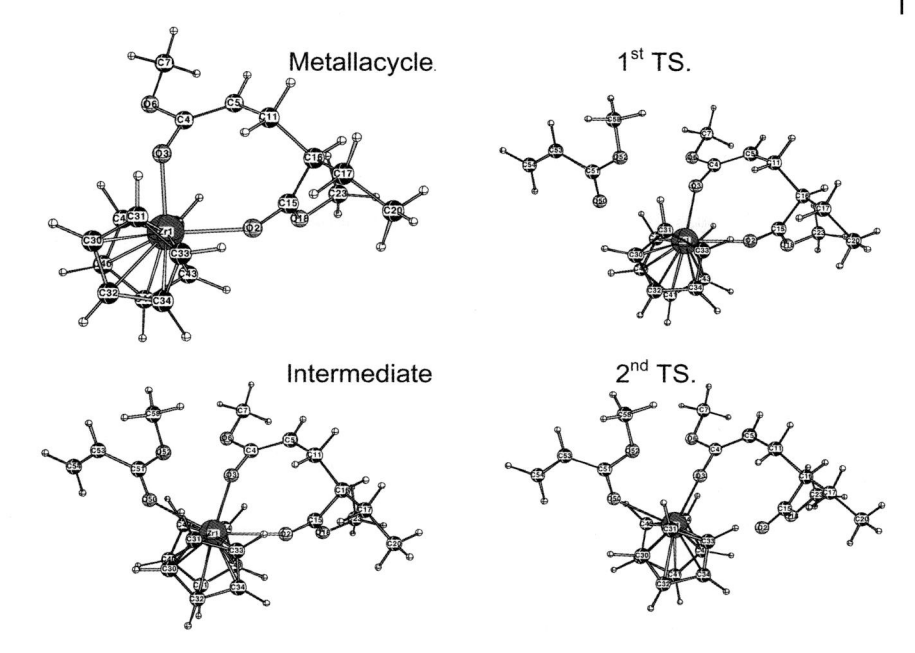

Figure 8.4 Computed structures for the species involved in the backside ring opening of the "down/down" cationic zirconocene metallacycle.

around 0 Å (see Table 8.6) on the reaction coordinate. The ring opening is faster with the Zr-based system than with the Sm-based one. The internal activation energy for the fastest channel of the former system (Zr^+ dd frontside) is −0.6 kcal mol^{-1}, whereas for the latter it is 6.3 kcal mol^{-1} (Sm^+ du backside, see Table 8.7). Also the intermediate is more stable for the cationic zirconocene, although to a lesser extent.

In most cases the transition states leading to the syndio product are around 1 kcal mol^{-1} lower than those leading to the iso product, with the exception of the Zr "up/down" metallacycle, whose opening is not selective. After reaching the ring-opening intermediate, the ester group leaves the coordination site on the metal while the MA-to-metal distance barely changes. The energy rises again, due to the weakening of the ester–metal interaction and to the strain created in the metallacycle being opened. When the ester group is sufficiently far from the metal the metallacycle is completely open and the geometry can relax, recovering the strain energy and generating the metallocene-enolate-MA complex for the next coupling reaction.

Figure 8.5 shows representative structures for the ring-opening process, corresponding to the backside ring opening of the "down/down" cationic zirconocene metallacycle.

Table 8.6 Selected interatomic distances for metallacycles, the corresponding first ring-opening transition states and the ring-opening intermediates.[a]

Metallacycles	MA⋯M[b,c] (Å)	=O⋯M[d] (Å)	−O−M[e] (Å)	O−M−O[f] (°)
Zr$^+$ dd[g]	–	2.21	2.04	90.0
Zr$^+$ ud[h]	–	2.22	2.04	99.5
Sm dd	–	2.46	2.19	83.0
Sm du[i]	–	2.43	2.22	82.0

Transition states	MA⋯M (Å)	=O⋯M (Å)	−O−M (Å)	O−M−O (°)
Zr$^+$ dd backside	3.25	2.25	2.04	81.9
Zr$^+$ dd frontside	3.29	2.24	2.05	83.4
Zr$^+$ ud backside	2.97	2.28	2.01	84.1
Zr$^+$ ud frontside	2.99	2.25	2.05	81.9
Sm dd backside	3.46	2.50	2.21	79.5
Sm dd frontside	3.46	2.49	2.21	78.4
Sm du backside	3.38	2.38	2.27	82.0
Sm du frontside	3.55	2.44	2.21	78.1

Intermediates	MA⋯M (Å)	=O⋯M (Å)	−O−M (Å)	O−M−O (°)
Zr$^+$ dd backside	2.39	2.36	2.05	74.2
Zr$^+$ dd frontside	2.48	2.30	2.08	75.7
Zr$^+$ ud backside	2.41	2.39	2.04	75.9
Zr$^+$ ud frontside	2.44	2.32	2.07	76.8
Sm dd backside	2.43	2.56	2.25	75.1
Sm dd frontside	2.53	2.55	2.23	74.4
Sm du backside	2.43	2.51	2.25	76.9
Sm du frontside	2.55	2.54	2.26	73.2

a) Process is displayed in Scheme 8.3.
b) M = Sm or Zr (as appropriate).
c) MA⋯M = distance between the methyl acrylate incoming group and the central metal.
d) =O⋯M = distance between the ester leaving group and the central metal.
e) −O−M = distance between the enolate and the central metal.
f) O−M−O = angle formed by the oxygen ligands in the metallacycle.
g) dd = "down/down".
h) ud = "up/down".
i) du = "down/up".

Table 8.7 Relative energies of the species involved in the ring-opening reactions.[a]

	1st TS	E_{rel}[b] (kcal mol^{-1}) Intermediate	2nd TS
Zr$^+$ dd backside[c]	−0.4	−5.0	−3.2
Zr$^+$ dd frontside	−0.6	−3.0	−0.3
Zr$^+$ ud backside[d]	0.9	−2.4	−0.8
Zr$^+$ ud frontside	1.0	−1.7	0.6
Sm dd backside	7.7	−0.8	2.4
Sm dd frontside	6.6	1.6	2.4
Sm du backside[e]	6.3	0.0	4.5
Sm du frontside	8.1	2.8	6.2

a) Process is displayed in Scheme 8.3.
b) Reference energy corresponds to the metallacycle + MA at infinite distance.
c) dd = "down/down".
d) ud = "up/down".
e) du = "down/up".

8.3.2
Kinetic Scheme for the Prediction of Stereoregularity

Scheme 8.5 depicts the backside or frontside opening of the metallacycle (A or A′, depending on the conformation) to form a new complex (D, D′, E or E′), whose conformation is not necessarily the same as in the previous cycle. An intermediate (B, B′, C or C′) is formed after the first transition state. Once the second transition state is passed, the metallacycle is fully open and a new enolate-acrylate complex is formed. A rotational movement is necessary to bring together the partners in the coupling reaction, as in the first propagation step, for which no barrier was found on the potential energy surface. At this point, fast interconversion between complexes could result in loss of stereoselectivity. However, the coupling is likely to be faster because the rotation involved in the conformational change is hindered in two ways. Firstly, it results in steric repulsion between the ligands. Secondly, it requires changing the oxygen lone pair involved in the σ-donation, which proceeds through a transition state in which there is no overlap between the orbitals of oxygen and the metal. Formation of a new C–C bond is rapid and irreversible, and the system returns almost immediately to its metallacycle resting state.

Since the coupling reaction proceeds without barrier on the potential energy surface, the first step of the ring opening, which has a higher transition state than the second, is the rate-determining step. Each consecutive pair of stereo-

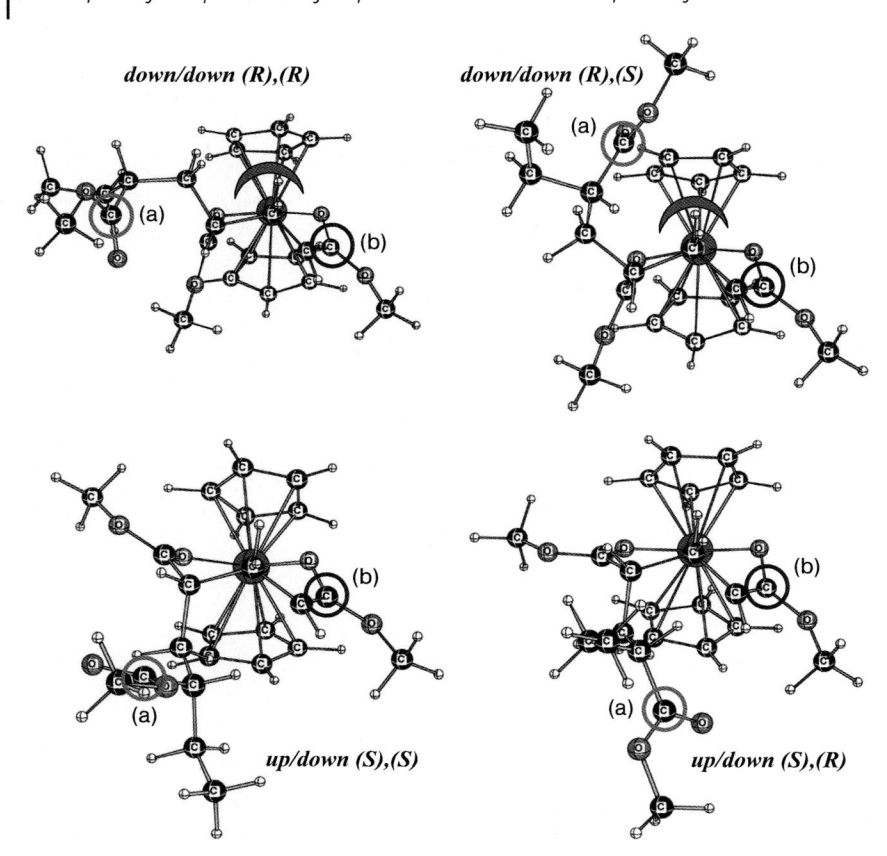

Figure 8.5 Diastereomeric zirconocene metallacycles obtained after the second monomer addition. Encircled carbon atoms labeled (a) are the electrophile in the backbiting reaction, those labeled (b) are the nucleophile. The crescent in the upper two structures indicates that the backbiting reaction cannot occur for steric reasons. See text for details. Energies are shown in Table 8.9.

genic carbons forms either an r syndiotactic dyad, or a m isotactic dyad. The ring opening channel that maintains the enolate enantioface involved in the coupling generates an m dyad, the other generates an r dyad. At each cycle the growing polymer chain can rest either in state A or A′ because crossovers are made possible during the ring-opening step.

Knowing the free energy of activation, rate constants can be computed using transition state theory. Since we ultimately are interested in relative rates between iso and syndio channels, entropic effects can be neglected, as they will be nearly the same in either case.

Detailed derivation of the kinetic equations, which has been reported elsewhere [17], leads to the approximate expression for the syndio/iso dyad composition

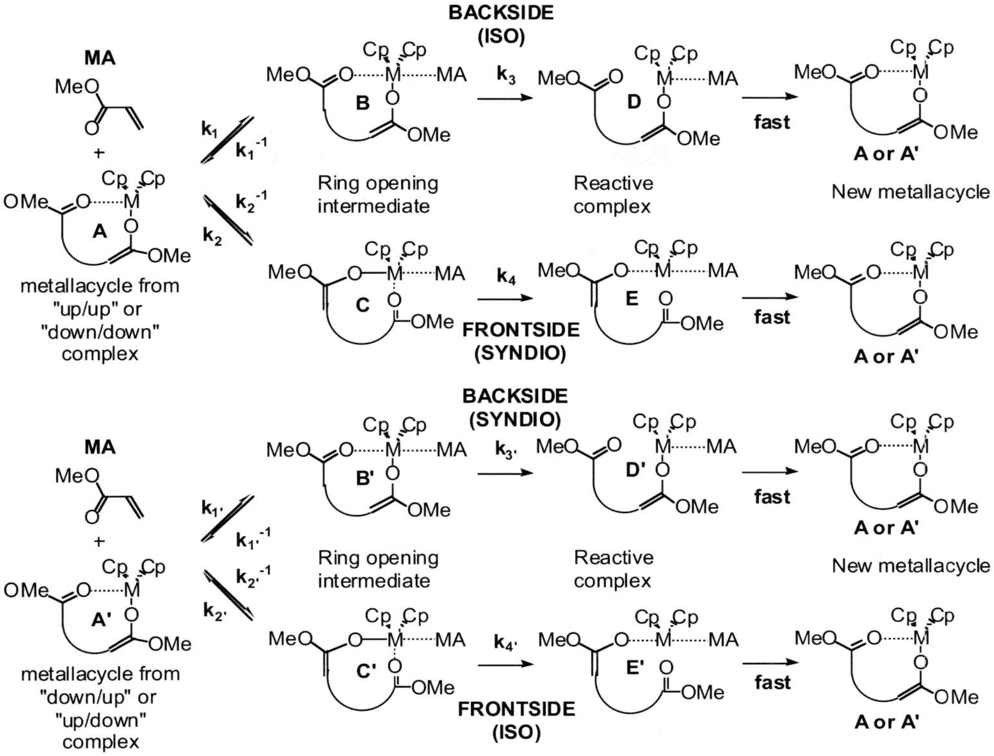

8.3 Discussion

Scheme 8.5 Kinetic scheme for the prediction of stereoregularity.

ratio in the polymer, χ:

$$\chi = \frac{(k_1 + k_2)\dfrac{k_2}{k_1} + (k_{1'} + k_{2'})\dfrac{k_{2'}}{k_{1'}}}{k_1 + k_2 + k_{1'} + k_{2'}} \quad (1)$$

The syndiotactic character in the polymer is predicted as follows:

$$\text{SYNDIO\%} = \frac{\chi}{\chi + 1} \cdot 100 \quad (2)$$

PMA obtained from the neutral samarocene is predicted to have 99% syndiotactic character at $-95\,°C$; syndiotacticity decreases down to 92% at 50 °C (Table 8.8). In the same temperature range, the syndiotactic character in the PMA obtained from the cationic zirconocene is much lower, ranging from 64 to 58% between $-95\,°C$ and 50 °C. Despite the approximations of the kinetic model, there is qualitative agreement with the results obtained by two experimental groups. Höcker

Table 8.8 Predicted percent syndiotacticities for the GTP of methyl acrylate, catalyzed by a neutral zirconocene or a neutral samarocene.

T (°C)	Syndiotacticity (%)	
	Cp_2Zr^+	Cp_2Sm
−95	64	99
−50	62	98
−25	61	96
0	60	95
25	59	94
50	58	92

et al. report 89% syndiotacticity for PMMA produced at −45 °C using an unsymmetrical cationic *ansa*-zirconocene [10], while they have not attempted acrylate polymerizations. Yasuda et al. used a neutral samarocene that gave 95% syndiotactic PMMA at −95 °C (75% at 25 °C) [4] and up to 60% syndiotactic PMA at −78 °C [22]. The tacticities obtained from methacrylate GTPs should be regarded as an upper limit for acrylate polymerizations, because the presence of the methyl group in the α-position of methyl methacrylate is more effective in inducing stereocontrol, compared with the α-H of MA.

8.3.3
Side Reactions Involving Metallacycles

Conclusive experimental and computational evidence shows that the eight-membered metallacycle formed in the coupling of the enolate and acrylate is the resting state in GTP. Therefore, the first and most intuitive approach in the study of side reactions, like the backbiting and the proton transfer of an ester α-H to the enolate moiety, is that they involve an eight-membered metallacycle. The backbiting reaction consists of the nucleophilic attack by the enolate α-carbon on the carbonyl carbon two monomer units away, i.e., the first free ester group on the polymer chain. The proposed mechanism amounts to a Dieckmann condensation (Claisen intramolecular condensation) and after elimination of a MeOH molecule it leads to the formation of an unreactive product, containing a substituted cyclohexanone ring (Scheme 8.6). Backbiting therefore causes termination of the polymerization process.

Similarly to backbiting, proton transfer involves the α-proton of the first free ester group on the polymer chain (Scheme 8.7). The product of proton transfer is an internal enolate, which can potentially continue the polymerization process. To study the two reactions, the size of the system has been increased to include an extra monomer unit outside the ring. The presence of an extra stereocenter in

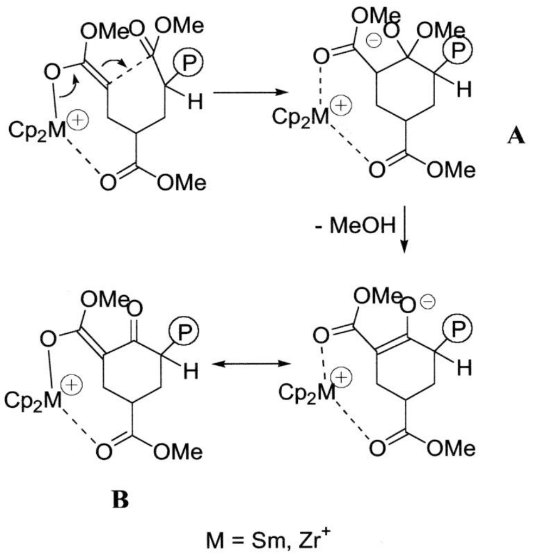

Scheme 8.6 Backbiting reaction mechanism.

the chain doubles the number of metallacycle configurations that need to be considered to give a complete picture of the stereochemistry. Four diastereomeric metallacycles (one from each possible enantiomeric pair) have been optimized for the cationic Zr-based system and two for the neutral Sm-based system. The formation energies relative to the respective THF pre-catalyst complex range from −64.8 and −68.2 kcal mol^{-1} in the case of the Zr-based system to −71.0 and −72.8 kcal mol^{-1} in the case of the Sm-based system (Table 8.9).

Inspection of the complete set of diastereomeric zirconocenium metallacycles systems reveals that only the conformers in which the methoxy groups are on opposite sides of the metallacycle have conformations that permit bringing the ester group of the side chain close enough for the enolate to react, without inflicting excessive strain (Figure 8.5 shows structures of the Zr-based diastereomeric metallocenes). For this reason only the two geometries potentially capable of undergoing the reaction were optimized for the more computationally expen-

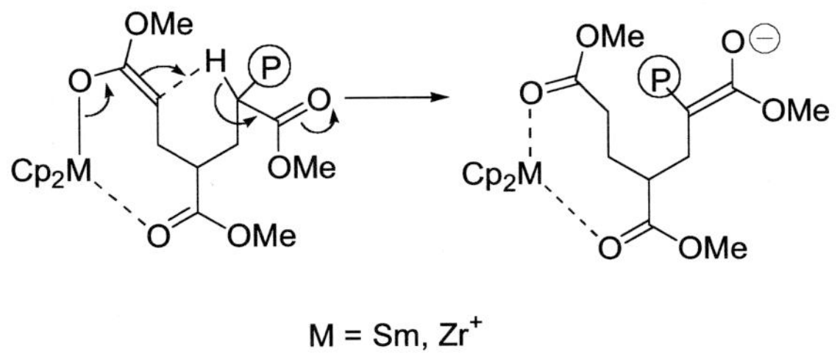

Scheme 8.7 Proton transfer reaction.

Table 8.9 Energies of samarocene and zirconocene metallacycles obtained after the second monomer addition. Reference Zr and Sm species are Cp_2ZrMe^+THF and $Cp_2SmMeTHF$, respectively.

Metallacycle	E_{rel} (kcal mol^{-1})
Zr^+-dd-(R),(R)[a]	−68.2
Zr^+-dd-(R),(S)	−64.8
Zr^+-ud-(S),(R)[b]	−67.5
Zr^+-ud-(S),(S)	−67.3
Sm-du-(R),(R)[c]	−71.0
Sm-du-(R),(S)	−72.8

a) "dd" = "down/down".
b) "ud" = "up/down".
c) "du" = "down/up".

sive Sm-based system and the backbiting and the proton transfer have only been studied for the "up/down" cationic zirconocene metallacycles with (S),(S) and (S),(R) configurations and the equivalent "down/up" neutral samarocene metallacycles having (R),(R) and (R),(S) configurations. Exploring a reaction mechanism for different diastereomers of the same type of metallacycle corresponds to studying that reaction for a syndio-like (alternating chiral centers) or an iso-like polymer (stereocenters of same chirality).

8.3.3.1 Backbiting Reaction in Zr- and Sm-eight-membered Metallacycles

The backbiting reaction (Scheme 8.6) occurs in two steps, first a C–C coupling, in which a six-membered ring is formed (tetrahedral intermediate A), likely followed by MeO$^-$ elimination. The elimination is an unfavorable equilibrium in the classic Claisen condensation, in which the system is negatively charged, so it can be expected to be even more so in a neutral system like the neutral samarocene, or in a positively charged system like the cationic zirconocene. In the Claisen condensation, the equilibrium is driven by deprotonation of the product. The difference between MMA and MA is that only MA has an α-proton, which can possibly be eliminated. In the absence of a strong base, it is possible that MeO$^-$ elimination and deprotonation proceed at least to some extent in a concerted manner.

We have not found a stationary point on the potential energy surface corresponding to the backbiting tetrahedral intermediate A for any of the systems studied, while MeOH elimination products containing the cyclohexanone ring (intermediate B in Scheme 8.6) have been optimized for all the selected configurations of both systems. In all cases they are high energy intermediates, 12.9–13.2 kcal mol^{-1} (Zr) or 10.1–11.0 kcal mol^{-1} (Sm) less stable than the parent metallacycles (Table 8.10). No direct reaction path connecting them to the parent eight-membered metallacycles has been found. The bond lengths (reported in

Table 8.10 Selected interatomic distances and energies (relative to the parent metallacycles) for the backbiting intermediate containing the cyclohexanone moiety (species B in Scheme 8.6).[a]

Species	C=O (ket) (Å)	C–C$_a$ (Å)	C=C (Å)	C–O (Å)	O–Zr (Å)	E$_{rel}$ (kcal mol^{-1})
Zr$^+$-ud-(S),(R)[b]	1.242	1.452	1.389	1.332	2.076	12.9
Zr$^+$-ud-(S),(S)	1.243	1.462	1.422	1.304	2.152	13.2
Sm-du-(R),(R)[c]	1.249	1.435	1.409	1.292	2.257	11.0
Sm-du-(R),(S)	1.248	1.437	1.414	1.289	2.293	10.1

a) Backbiting mechanism is illustrated in Scheme 8.6.
b) "ud" = "up/down".
c) "du" = "down/up".

Table 8.10) show that the ketone C=O double bond and the other C–C and C–O bonds connecting to the metal center through the enolate ligand are intermediate between typical single and double bond distances, indicating resonance between the two canonical forms represented in Scheme 8.6. Although no defined transition state has been characterized, it is quite obvious that transition states and products in the backbiting involving the eight-membered metallacycle are electronically and sterically destabilized, making the process at least very slow and not competitive with the propagation.

8.3.3.2 Proton Transfer in Zr- and Sm-eight-membered Metallacycles

The electrophile involved in the proton transfer is the proton in the α-position to the ester in the side chain. A six-membered transition state, with the α-H being part of such a ring, is expected to be formed (Scheme 8.7). The same metallacycles used for the study on backbiting were used to study of the transfer of the ester α-proton to the enolate α-carbon.

The reaction coordinate for the proton transfer in the up/down, (S),(S)-Zr-metallacycle shows a monotonic increase in energy, while for the other geometry [up/down, (S),(R)] a very shallow local minimum was found at the end point of the constrained linear transit coordinate. In both cases, the product is more than 30 kcal mol^{-1} above the reactant and spontaneously reverts to the starting point when it is optimized freely. Similar results were obtained for the Sm-based system.

As with the backbiting, proton transfer, therefore, does not affect eight-membered metallacycles.

8.3.3.3 Backbiting in Ten-membered Metallocenes

The importance of activating the ester group, i.e., enhancing its electrophilic character, appears to be critical for the feasibility of the backbiting and the proton transfer. The electron-withdrawing effect of a cation coordinated to the ester group, like the metallic center, could activate it sufficiently, but the eight-membered metallacycle is not flexible enough to accommodate the displacement required for the backbiting reaction, while the cumulated six- and four-membered ring system that would result from the reaction would highly strain the product. Furthermore, the proton transfer would impossible since the active α-H is facing outward of the metallacycle.

It is possible, however, especially as the concentration of free MA decreases, that the first ester group displaces the ester group of the metallacycle, forming a larger ten-membered metallacycle, which could better meet the requirements of conformational freedom necessary for the backbiting to occur. The hypothetical backbiting product would benefit thermodynamically from the formation of a bicyclic, cumulated, system of six-membered rings. Conversion of the (S),(S) up/down cationic zirconocene eight-membered metallacycle into a ten-membered one is exothermic by 1.3–5.0 kcal mol^{-1}, depending on the conformation of the product. The more stable conformation can undergo easily (internal activation energy: 12.7 kcal mol^{-1}, or 7.6 kcal mol^{-1} above the reference eight-membered

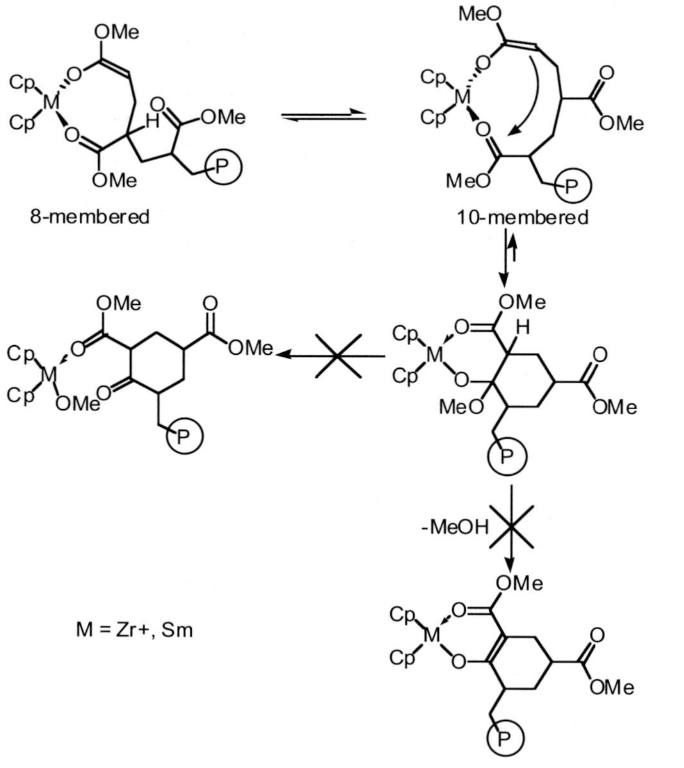

Scheme 8.8 Backbiting and decomposition in ten-membered metallacycles.

metallacycle) and exothermically the first step of the backbiting reaction, forming the tetrahedral intermediate, as represented in the first reaction of Scheme 8.8. Another conformation of the intermediate, more stable by 6.6 kcal mol^{-1} (9.5 kcal mol^{-1} more stable than the eight-membered metallacycle) is accessible from the first one through a conformational equilibrium. The energies of the species along the reaction coordinate are shown in Figure 8.6. The conformation of the more stable tetrahedral intermediate resembles that of the more stable ten-membered metallacycle, although no direct reaction channel connecting the two has been found. The tetrahedral intermediate is potentially capable of undergoing further reactions that irreversibly give stable products. Elimination of MeOH would yield a stable cationic zirconocene acac-type complex. The reaction is favored thermodynamically (5.5 kcal mol^{-1} exothermic from the backbiting intermediate in its more stable conformation and 15.0 kcal mol^{-1} from the eight-membered metallacycle, see Figure 8.6) and the reverse process is less likely because its barrier is higher than that of the forward process and because it is also disfavored by entropy. However, the two mechanisms that have been considered (a concerted elimination and a stepwise elimination, following an E1cB mechanism) are kinetically very unfavorable.

Another possibility is that the methoxy group in the tetrahedral intermediate is transferred to the Zr center. The process should be irreversible due to the strength of the Zr–OMe bond, but also the transfer of MeO$^-$ to Zr has a very high activation energy (>30 kcal mol^{-1}); therefore, the backbiting from the ten-membered metallacycle is relatively easy and exothermic, but the process is not irreversible. Transformation into a ten-membered metallacycle and subsequent backbiting, therefore, are likely to trap the propagating center, greatly slowing down chain propagation and causing it to be not living (as experimentally observed), but do not necessarily lead to termination.

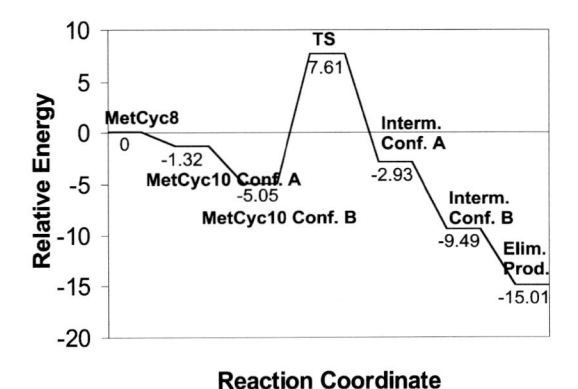

Figure 8.6 Species along the reaction coordinate from the eight-membered metallacycle to the backbiting product after elimination of a MeOH molecule. The energies are in kcal mol^{-1}, relative to the eight-membered metallacycle.

Proton transfer in the ten-membered metallacycle has not been studied. As with eight-membered metallacycles, in both conformations the α-H points out of the ring, making transfer to the enolate α-carbon on the opposite side of the ring not viable.

8.3.4
Polymer Chain Transfer Reactions

Metallacycle ring opening by ester groups on the polymer chain becomes increasingly more competitive with the previously described MA-assisted ring opening as the MA concentration drops during the polymerization, and for groups sufficiently far away, or the on other polymer chains, it can be considered that a new, open type of ester/enolate complex is obtained. This event has been studied using $CH_3CH(COOMe)CH_3$ as a model ligand representing an ester group on a polymer chain. Complexation of the polymer model is more exothermic by 7.3–12.1 kcal mol^{-1} (Zr) and 7.4–12.7 kcal mol^{-1} (Sm) (Table 8.11). The complexes so-obtained have an ester group activated by the metal, capable of reacting intramolecularly with the enolate α-carbon. Since the shortcomings of the backbiting reaction were already known (i.e., difficult formation of an irreversible product because of the high energetic cost for cleavage of the C–OMe bond), the transfer of a proton has been studied instead. As shown in Scheme 8.9, the product of proton transfer is a new enolate/ester metallocene complex. The enolate is internal and therefore likely to be less reactive. Growth of the polymer chain is stopped and the propagation center is transferred to the α-position of an ester group (the new, internal enolate), resulting in chain branching. Indeed, this reaction has been exploited in some cases, in which active hydrogen compounds such as thiols or enolizable ketones have been employed as chain transfer agents [41–43].

Table 8.11 Comparison of formation energies of samarocene and zirconocene enolate-MA complexes with samarocene and zirconocene enolate-polymer model complexes. Energies are relative to the pre-catalyst THF complex.

Species	E_{rel} (kcal mol^{-1})
Zr "down/up" MA complex	−25.4
Zr "down/down" MA complex	−30.2
Zr "down/up" pol. complex	−37.5
Zr "down/down" pol. complex	−37.5
Sm "down/up" MA complex	−38.4
Sm "down/down" MA complex	−39.0
Sm "down/up" pol. complex	−51.1
Sm "down/down" pol. complex	−46.4

Scheme 8.9 Polymer chain transfer reaction mechanism.

Transfer of the α-H to the enolate has been studied in down/down and down/up enolate-ester complexes. Before the transfer, the polymer model ligand rotates to align the C–H bond as perpendicular as possible to the π-plane of the enolate. The process hardly involves any change in energy, and in the Zr-based system the two starting geometries lead to the same transition state. The internal barrier for the process (Figure 8.7 and Table 8.12) is 17.9 kcal mol^{-1}. The product is more stable than the reactants by around 4 kcal mol^{-1}. The chain transfer has a higher activation energy than the C–C coupling in the backbiting reaction, in agreement

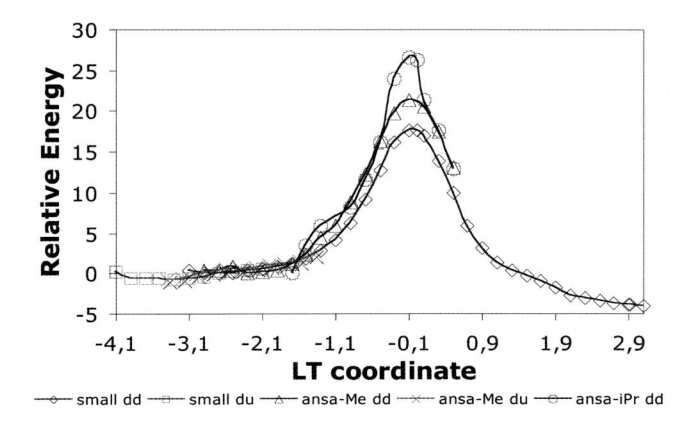

Figure 8.7 Comparison of chain transfer reaction profiles for differently substituted zirconocenes. The lowest points on the left-hand side (reactants) of the "dd" curves are assigned 0 relative energy.

Table 8.12 Effect of catalyst modifications on the energies of local minima in the chain transfer reaction involving samarocene and zirconocene enolate-polymer model complexes. Activation and reaction energies are relative to the ester-enolate metallocenium complexes of the respective geometry (down/down or down/up). The Sm-based "down/up" complex with simple Cp ligands is 4.67 kcal mol^{-1} more stable than the "down/down" complex.

Species	E_{rel} (kcal mol^{-1})	
	TS	Product
Cp$_2$ Zr$^+$ dd[a]	17.9	−4.0
ansa/Me Zr$^+$ dd	21.3	
ansa/i-Pr Zr$^+$ dd	31.2	
Cp$_2$Sm dd	12.7	−4.0
Cp$_2$Sm du[b]	16.8	0.7
Cp*$_2$Sm dd	23.6	

a) dd = "down/down".
b) du = "down/up".

with recent work by Chen and coworkers, according to which backbiting is the sole origin for termination products [26].

For the Sm-based system, the internal barriers (Figure 8.8 and Table 8.12) are 12.7 and 16.8 kcal mol^{-1} (from the down/down and down/up geometries, respectively). In this case, the TSs are different, but very close in geometry and energy. The reaction from the down/down complex is exothermic by 4 kcal mol^{-1}, while the reaction from the down/up complex, which is more stable, is nearly thermoneutral. The better results obtained polymerizing acrylates with Cp*$_2$SmMeTHF,

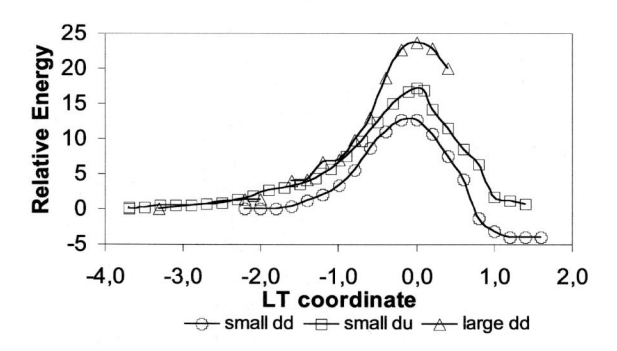

Figure 8.8 Comparison of chain transfer reaction profiles for differently substituted samarocenes. The lowest point in each curve is assigned 0 relative energy.

compared with Zr-based systems, should be interpreted as the effect of the bulky Cp* ligands, more than to the nature of the metal. Yasuda's systems are permethylated, while most of the Zr-based systems, even the ansa ones, do not have as much steric hindrance on the coordination sites for the oxygen-based ligands. Changing the basic zirconocene to a more complex system, with substituents on the front region, should give further elements for the comparison. Such a comparative study could give information on whether it is possible to reduce side reactions through rational modifications of the steric demands of the ligands.

8.3.5
Minimization of Side Reactions

To minimize a side reaction it is necessary to increase its activation energy compared with that of the propagation, which can be achieved through modifications to the catalyst that selectively destabilize the transition states for the backbiting and chain transfer reactions, thus slowing down the side reactions more than the propagation. Modifications to the catalyst should ideally slow down the side reactions, while not affecting adversely the propagation steps.

In the transition states for the chain transfer and the backbiting in the ten-membered metallacycle, the side groups of the enolate and ester ligands are roughly parallel to the plane formed by the metal center and the Cp centroids (representative transition states for the chain transfer are shown in Figure 8.9). Bulkier groups in the positions of the Cp rings that are close to the ligands are likely to increase steric repulsions in the transition state, thus slowing down the side reactions. To prove that, the reaction was explored again with bulkier substituents on the Cp rings.

For the Sm system, the real, permethylated system used by Yasuda has been studied. For the cationic Zr system, a 3,4,3′,4′-tetramethylated $ansa$ system has been selected, bearing a $CH_2–CH_2$ bridge, as in the bis-indenyl system used by Chen and coworkers [23–26], which has been proven to be an effective mononuclear zirconocenium catalyst for the GTP of methyl methacrylate (Figure 8.10). Preliminary full QM calculations have been conducted for the initiation step and the enolate-acrylate complexes. The energies of these species are reported in Table 8.13. The equivalent structures for the systems with simple Cp ligands have been reported in Section 8.3.1.1. Abstraction of THF from the precatalyst complexes in the larger systems is easier by 4.7 kcal mol^{-1} in the case of the neutral samarocene, and 6.3 kcal mol^{-1} for the cationic zirconocene. The weaker interaction between the metal centre and the THF oxygen atom is also witnessed by an increase of the O–M distance (M = Sm, Zr$^+$). Generation of the enolate is also slightly less exothermic (by 3.8 kcal mol^{-1} for Zr, and 0.4 kcal mol^{-1} for Sm). No stable Cp*$_2$Sm-MA-enolate complexes have been found. Instead, the complexes evolved during the calculation to the corresponding eight-membered metallacyclic products. For the Zr-based system, stable acrylate-enolate complexes are still local minima because of the smaller size of the modified ligand, compared with Cp*.

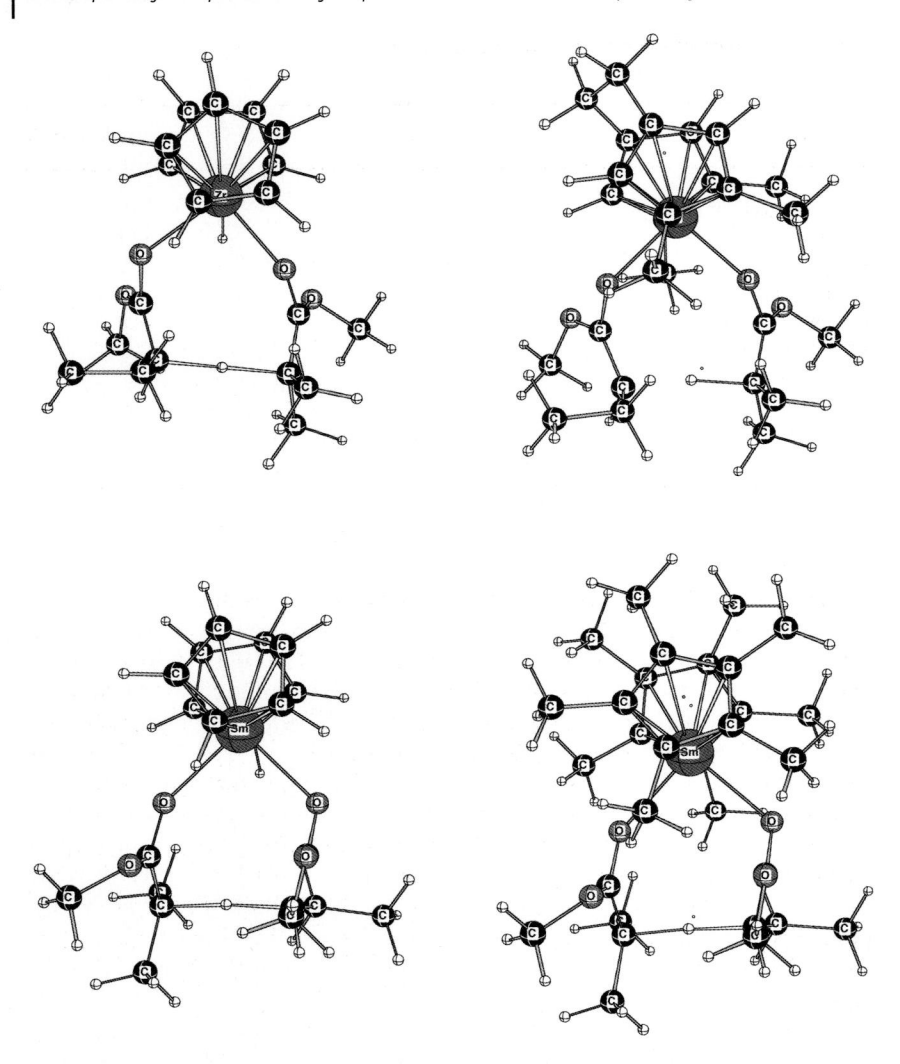

Figure 8.9 Transition states for the chain transfer with the Zr-based system (top) and for the Sm-based system. Small, full QM models are shown on the right, modified, QM/MM models are shown on the left.

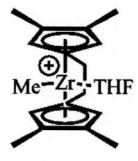

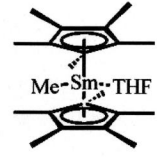

Figure 8.10 Bulkier systems used in the study of the minimization of side reactions.

Table 8.13 Energies of samarocene and zirconocene species (large systems, full QM calculations).

Species	E_{rel} (kcal mol^{-1})
Et-Cp$^\dagger_2$ZrMeTHF$^+$	0
Et-Cp$^\dagger_2$ZrMe$^+$	35.1
Et-Cp$^\dagger_2$ZrMeMA$^+$	1.4
Et-Cp$^\dagger_2$ZrEno$^+$	−12.0
Zr-down/down complex	−24.0
Zr-down/up complex	−28.9
Cp$^*_2$SmMeTHF	0
Cp$^*_2$SmMe	13.8
Cp$^*_2$SmMeMA	3.8
Cp$^*_2$SmEno	−32.8

The effect of the modified ligand systems has been investigated first on the chain transfer reaction. This part of the study has been carried out using hybrid QM/MM calculations. For the Sm system with Cp* ligands, the barrier for the faster chain transfer reaction channel is increased by 10.9 kcal mol^{-1} compared with the system with simple Cp ligands (Figure 8.8 and Table 8.12), giving further evidence that the better results obtained experimentally in the GTP of acrylates catalyzed by Yasuda's neutral samarocene are in fact due to the ligand system.

In the Zr-based system, the internal activation energy increases as bulkier groups are introduced on the 3,4 and 3′,4′ positions. Using methyl groups, the barrier increases by just 3.5 kcal mol^{-1} (Figure 8.7 and Table 8.12), which is not as effective as the neutral samarocene bearing Cp* ligands. Only two methyl groups per Cp ring do not exert an effect as strong as that found with Cp*, as could also be inferred by the fact that the acrylate-enolate complexes are stable species.

Substituting the Me groups with *i*Pr groups increases the internal activation energy for the chain transfer by 13.3 kcal mol^{-1}, bringing it to 31.2 kcal mol^{-1}. The increase of the barrier by 3.5 kcal mol^{-1} with the first modified system is, however, already reasonably effective. Further calculations show that this simple modification works even better for the control of the main side reaction, the back-biting in the ten-membered metallacycle. The effect of the modification is to increase the barrier by almost 10 kcal mol^{-1}, from 12.7 to 22.5 kcal mol^{-1} (Figure 8.11). A further increase in the steric bulk (i.e., substituting the methyl groups with *i*-propyl ones) would probably further increase the barrier, but could affect negatively the propagation steps.

Possible effects of the modification to the Zr-based catalyst on the propagation have been investigated. There are only minor changes in the C–C coupling reaction profile of the "up/down" complex (Figure 8.12). In the frontside ring-open-

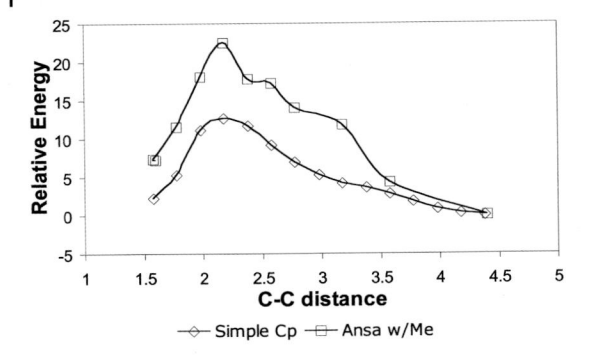

Figure 8.11 Comparison of backbiting reaction profiles for differently substituted ten-membered zirconocenium metallacycles ["up/down" (S),(S) diastereomer]. The ten-membered metallacycle start points are assigned 0 relative energy.

ing reaction of the "down/down" metallacycle the general shape of the energy profile, as well as the relative energies of the approximate transition states, do not change much compared with the smaller unsubstituted system, although the relative stability of the first and the second transition state is inverted (Figure 8.13). This last fact would require that a different and possibly more complex kinetic model be used to predict the stereoselectivity, compared with that presented in Section 8.3.2. This task, however, is beyond the scope and the objectives of this chapter.

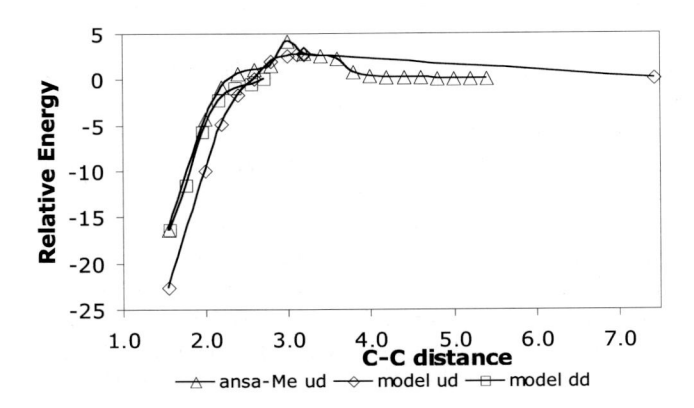

Figure 8.12 Comparison of C–C coupling reaction profiles for differently substituted zirconocenium-enolate-MA complexes ("up/down" geometry). The respective reacting complex start points are assigned 0 relative energy.

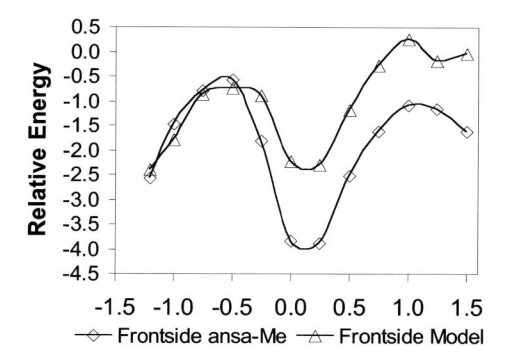

Figure 8.13 Effect of catalyst modification on the frontside MA-assisted ring-opening reaction profile for the cationic zirconocenium system ("down/down" metallacycle). The respective metallacycle start points are assigned 0 relative energy.

8.4 Conclusions

Through this DFT study, we have explored the mechanism of the Group Transfer Polymerization of methyl acrylate, catalyzed by d- or f-block monometallic metallocenes, the possible side reactions that affect the polymerization, and what modifications can be introduced in the catalytic system, to minimize the side reactions without affecting adversely the propagation. The two isoelectronic systems behave similarly in the GTP. For both systems, no energy barrier for the C–C coupling reaction has been encountered on the potential energy surface. The product obtained after C–C bond formation is a stable metallacycle, which is the resting state in the GTP.

The first transition state in the opening of the metallacycle resting state, assisted by an incoming acrylate molecule, is the most important elementary step in determining the microstructure of the polymer. A kinetic model for the prediction of stereoregularity has been developed, which has shown to be qualitatively in agreement with the available experimental data.

Investigations of different side reaction mechanism involving the eight-membered metallacycle resting state strongly suggest that an isomerization to a ten-membered metallacyclic species is necessary for the backbiting reaction to be competitive with propagation. The ten-membered metallacycle easily undergoes backbiting to yield an unreactive, stable tetrahedral intermediate, which is sufficiently stable to trap the propagating centers and bring chain growth to a halt.

Ring opening from ester groups in remote locations of the polymer chain, and possibly on other polymer chains, is increasingly competitive with MA-assisted ring opening at higher monomer conversions and can produce polymer-enolate complexes capable of undergoing chain transfer, thus creating branching and increased polydispersity. The chain transfer reaction is a less likely side reaction than the backbiting, due to its higher activation energy.

The transition states of these two processes can be destabilized by introducing substituents in the front positions of the Cp rings, The modified ansa system [C_2H_4-bis(3,4-diMe-Cp) ligand] can address this need while at the same time not affecting adversely the key steps of propagation.

Acknowledgments

This work has been supported by BASF. T.Z. thanks the Canadian government for a Canada Research Chair in theoretical inorganic chemistry.

An important part of the calculations was performed on the Westgrid cluster of Canada.

The authors are grateful to E.-X. Chen for useful discussions and for disclosing data on acrylate polymerizations still unpublished when the research was ongoing.

References

1 Collins, S., Ward, D. G., *J. Am. Chem. Soc.* 1992, **114**, 5460.
2 Boffa, L. S., Novak, B. M., *Macromolecules* 1994, **27**, 6993.
3 Yasuda, H., Yamamoto, H., Yokota, K., Miyake, S., Nakamura, A., *J. Am. Chem. Soc.* 1992, **114**, 4908.
4 Yasuda, H., Yamamoto, M., Yamashita, H., Yokota, K., Nakamura, A., Miyake, S., Kai, Y., Kanehisa, N., *Macromolecules* 1993, **26**, 7134.
5 Yasuda, H., Furo, M., Yamamoto, H., Nakamura, A., Miyake, S., Kibino, N., *Macromolecules* 1992, **25**, 5115.
6 Collins, S., Ward, D. G., Suddaby, K. H., *Macromolecules* 1994, **27**, 7222.
7 Yasuda, H., *J. Organomet. Chem.* 2002, **647**, 128.
8 Nguyen, H., Jarvis, A. P., Lesley, M. J. G., Kelly, W. M., Reddy, S. S., Taylor, N. J., Collins, S., *Macromolecules* 2000, **33**, 1508.
9 Boffa, L. S., Novak, B. M., *Chem. Rev.* 2000, **100**, 1479.
10 Frauenrath, H., Keul, H., Höcker, H., *Macromolecules* 2001, **34**, 14.
11 Hölscher, M., Keul, H., Höcker, H., *Chem. Eur. J.* 2001, **7**, 5419.
12 Hölscher, M., Keul, H., Höcker, H., *Macromol. Rapid Commun.* 2000, **21**, 1093.
13 Li, Y., Ward, D. G., Reddy, S. S., Collins, S., *Macromolecules* 1997, **30**, 1875.
14 Bandermann, F., Ferenz, M., Sustmann, R., Sicking, W., *Macromol. Symp.* 2001, **174**, 247.
15 Sustmann, R., Sicking, W., Bandermann, F., Ferenz, M., *Macromolecules* 1999, **32**, 4204.
16 Hölscher, M., Keul, H., Höcker, H., *Macromolecules* 2002, **35**, 8194.
17 Tomasi, S., Weiss, H., Ziegler, T., *Organometallics* 2006, **25**, 3619.
18 Stojcevic, G., Kim, H., Taylor, N. J., Marder, T. B., Collins, S., *Angew. Chem. Int. Ed.* 2004, **43**, 5523.
19 Karanikolopoulos, G., Batis, C., Pitsikalis, M., Hadjichristidis, N., *Macromolecules* 2001, **34**, 4697.
20 Rodriguez-Delgado, A., Chen, E. Y.-X., *Macromolecules* 2005, **38**, 2587.
21 Rodriguez-Delgado, A., Mariott, W. R., Chen, E. Y.-X., *Macromolecules* 2004, **37**, 3092.
22 Ihara, E., Morimoto, M., Yasuda, H., *Macromolecules* 1995, **28**, 7886.

23 Chen, E. Y.-X., Cooney, M. J., *J. Am. Chem. Soc.* 2003, **125**, 7150.
24 Bolig, A. D., Chen, E. Y.-X., *J. Am. Chem. Soc.* 2002, **124**, 5612.
25 Rodriguez-Delgado, A., Chen, E. Y.-X., *J. Am. Chem. Soc.* 2004, **126**, 5612.
26 Mariott, W. R., Rodriguez-Delgado, A., Chen, E. Y.-X., *Macromolecules* 2006, **39**, 1318.
27 Tomasi, S., Weiss, H., Ziegler, T., *Organometallics* 2007, **26**, 2157.
28 te Velde, G., Bickelhaupt, F. M., van Gisbergen, S. J. A., Fonseca Guerra, C., Baerends, E. J., Snijders, J. G., Ziegler, T., *J. Comput. Chem.* 2001, **22**, 931.
29 Fonseca Guerra, C., Snijders, J. G., te Velde, G., Baerends, E. J., *Theor. Chem. Acc.* 1998, **99**, 391.
30 ADF2004.01, SCM, *Theoretical Chemistry*, Vrije Universiteit, Amsterdam, The Netherlands (http://www.scm.com).
31 Becke, A. D., *Phys. Rev. A* 1988, **38**, 3098.
32 Perdew, J. P., *Phys. Rev. B* 1986, **33**, 8822.
33 Perdew, J. P., *Phys. Rev. B* 1986, **34**, 7406.
34 Snijders, J. G., Baerends, E. J., Ros, P., *Mol. Phys.* 1979, **38**, 1909.
35 Boerrigter, P. M., Baerends, E. J., Snijders, J. G., *Chem. Phys.* 1988, **122**, 357.
36 Ziegler, T., Tschinke, V., Baerends, E. J., Sijders, J. G., Ravenek, W., *J. Phys. Chem.* 1989, **93**, 3050.
37 Pulay, P., *Chem. Phys. Lett.* 1980, **73**, 393.
38 Morokuma, K., Maseras, F., *J. Comput. Chem.* 1995, **117**, 5179.
39 Woo, T. K., Cavallo, L., Ziegler, T., *Theor. Chem. Acc.* 1998, **100**, 307.
40 Clark, M., Cramer, R. D. I., van Opdenbosch, N., *J. Comput. Chem.* 1989, **10**, 982.
41 Nodono, M., Tokimitsu, T., Tone, S., Mskino, T., Yanagase, A., *Macromol. Chem. Phys.* 2000, **201**, 2282.
42 Lian, B., Toupet, L., Carpentier, J-F., *Chem. Eur. J.* 2004, **10**, 4301.
43 Lian, B., Lehmann, C. W., Navarro, C., Carpentier, J-F., *Organometallics*, 2005, **24**, 2466.

9
Insights into the Mechanism of H_2O_2-based Olefin Epoxidation Catalyzed by the Lacunary $[\gamma\text{-}(SiO_4)\,W_{10}O_{32}H_4]^{4-}$ and di-V-substituted-γ-Keggin $[\gamma\text{-}1,2\text{-}H_2SiV_2W_{10}O_{40}]^{4-}$ Polyoxometalates. A Computational Study

Rajeev Prabhakar, Keiji Morokuma, Yurii V. Geletii, Craig L. Hill, and Djamaladdin G. Musaev

9.1
Introduction

Catalytic epoxidation of olefins by hydrogen peroxide is a long-standing goal of industry and synthetic organic chemistry [1]. Over the years, several environmentally hazardous epoxidation processes that utilize various catalysts and oxidants have been developed, but the use of hydrogen peroxide offers an environmentally and economically attractive alternative [2–6]. The advantages of catalytic epoxidation of olefins with H_2O_2 are derived from the low cost and a high active oxygen content of this oxidant and production of water as the only by-product [7].

In the search for an efficient catalyst for H_2O_2-based epoxidations, numerous materials containing transition metal, including non-heme iron complexes [8–12], metalloporphyrins [13], titanium silicalites [14], methyl-trioxorhenium [15–17], tungsten compounds [2, 18, 19], manganese complexes [20] and polyoxometalates (POMs) [3–6, 21–29], have been used under either homogeneous or heterogeneous conditions, but all these systems demonstrated certain disadvantages. Recently, Mizuno and coworkers have reported a di-lacunary (a cavity-containing or "defect") Keggin-type silicodecatungstate, $[\gamma\text{-}(SiO_4)\,W_{10}O_{32}H_4]^{4-}$ (**1**), that catalyzes the epoxidation of olefins with 99% selectivity, 99% H_2O_2 utilization efficiency and high stereospecificity [26, 30]. At almost the same time, another complex, $[\gamma\text{-}1,2\text{-}H_2SiV_2W_{10}O_{40}]^{4-}$ (**2**), with a $\{VO\text{-}(\mu\text{-}OH)_2\text{-}VO\}$ core, has been characterized by the same group and claimed to be the best catalyst for epoxidation of terminal olefins, with very high yields, $\geq 99\%$ selectivity and $\geq 87\%$ hydrogen peroxide utilization efficiency [31–34].

However, very little is known about the mechanisms of these promising catalyzed H_2O_2-based epoxidation reactions. The experimentally available information on the chemistry of **1** can be summarized as follows: (a) There are four Me_4N^+ counter-cations per lacunary compound. Notably, in the paper by Mizuno et al. [30] the structural analysis of $[\gamma\text{-}(SiO_4)\,W_{10}O_{32}H_4]^{4-}$ was of the tetramethylammonium, $(Me_4N)_4$, salt, while all the epoxidation data provided in Table 9.1 below and discussed in this chapter were obtained using the $(Bu_4N)_4$ salt. (b) The high stereospecificity of alkene oxidation suggests a structurally rigid and non-radical substrate-attacking intermediate. (c) Diepoxide is not produced in the reaction with diene [30].

Relatively more information is available on the (**2**)-catalyzed epoxidation by H_2O_2: (a) Experimental studies show no epoxidation of alkenes in the absence of **2** or when *tert*-butyl hydroperoxide (TBHP) is used in place of H_2O_2 as the oxidant [31]. (b) The silicotungstate $[\gamma\text{-}(SiO_4)W_{12}O_{36}]^{4-}$ compound with a structure similar to **2** has been observed to be far less active, suggesting that the W atoms in **2** are not the active sites. (c) The monovanadium, $[\gamma\text{-}SiVW_{11}O_{40}]^{5-}$, and trivanadium, $[\gamma\text{-}1,2,3\text{-}SiV_3W_9O_{40}]^{7-}$, compounds are far less catalytically active, which consistent with the V–O–W and V=O centers being unimportant in catalysis by **2**. (d) Stereospecificity, diastereoselectivity and regioselectivity of epoxidations catalyzed by **2** are quite different from those observed when **1** is used as the catalyst. This suggests that **1** and **2** involve distinct substrate-oxygenating intermediates [35]. (e) IR and NMR experiments confirm that the catalytically active complexes, $[\gamma\text{-}(SiO_4)W_{10}O_{32}H_4]^{4-}$ and/or $[W_2O_3(O_2)_4(H_2O)_2]^{2-}$, are not generated during catalysis by **1** [30]. (f) Finally, ^{51}V and ^{183}W NMR experiments strongly suggest [31] that, during the epoxidation, H_2O_2 reacts with the {VO-(μ-OH)$_2$-VO} core of **2** to produce a hydroperoxy species, {VO-(μ-OH)(μ-OOH)-VO} (**2a**). Formation of **2a** is also supported by the fact that **2** and methanol react to form a monomethyl ester [33]. Thus, either the hydroperoxy species, {VO-(μ-OH)(μ-OOH)-VO} (**2a**), could be the epoxidizing intermediate or it could rearrange to other species with {VO-(μ-OOH)(μ-OOH)-VO} (**2b**), {VO-(η^2-O$_2$)-VO} (**2c**) cores, or others. Since the substitution of one proton in the initial complex **2** with a {VO-(μ-OH)$_2$-VO} core by *n*-Bu$_4$NOH inhibits both esterification by CH_3OH and epoxidation by H_2O_2, **2c** seems to be the most plausible [31].

Here, we applied computational methods to shed light on the mechanism of olefin epoxidation by H_2O_2, catalyzed by **1** and **2**, to elaborate the nature and structures of intermediates and transitions states and the effect of countercations.

In recent computational studies of the mechanism of epoxidation of C_2H_4 by H_2O_2 catalyzed by the lacunary complex $[\gamma\text{-}(SiO_4)W_{10}O_{32}H_4]^{4-}$ (**1**) we proposed the intermediacy of a W–OOH species [36]. In the "hydroperoxy" mechanism, a terminal hydroxyl-ligand containing (W–OH) site was found to be catalytically active. In this chapter we summarize our previous results on epoxidation of C_2H_4 by H_2O_2 catalyzed by the lacunary complex $[\gamma\text{-}(SiO_4)W_{10}O_{32}H_4]^{4-}$ (**1**), and report our new findings on the "hydroperoxy" mechanism of olefin epoxidation by H_2O_2 catalyzed by the di-V-substituted γ-Keggin polyoxometalate $[\gamma\text{-}1,2\text{-}H_2SiV_2W_{10}O_{40}]^{4-}$ (**2**).

9.2
Computational Details

9.2.1
Methods

All calculations were performed using the *Gaussian 03* program [37]. The geometries of reactants, intermediates, transition states and products were optimized without any symmetry constraints at the B3LYP/Lanl2dz level of theory with additional d polarization functions for the Si atom ($\alpha = 0.55$) and the corresponding Hay–Wadt effective core potential (ECP) for V and W atoms [38, 39]. This approach is referred to subsequently as "B3LYP/[Lanl2dz+d(Si)]". The final energetics of the optimized structures were further improved by performing single point calculations, including additional d and p polarization functions for O ($\alpha = 0.96$) and H ($\alpha = 0.36$) atoms, respectively, in the basis set used for optimization of the geometries. This approach is henceforth referred to as "B3LYP/[Lanl2dz+d(Si,O)+p(H)]".

Hessians, calculated for all the structures and transition states, were confirmed to have one imaginary frequency corresponding to the reaction coordinates. However, due to the large size of models used in this study it was not technically possible to calculate Hessians with the same basis set [Lanl2dz+d(Si)] utilized in the optimizations, and therefore they were calculated without additional polarization functions for Si atoms, i.e., using the Lanl2dz basis set. Dielectric effects from the surrounding environment were estimated using the self-consistent reaction field IEF-PCM method [40] at the B3LYP/[Lanl2dz+d(Si)] level. Based on experimental conditions, solvent effects have been included using acetonitrile solvent [dielectric constant (ε) = 36.64]. Throughout this chapter, we report the B3LYP/[Lanl2dz+d(Si,O)+p(H)] energies including solvent effects using the [Lanl2dz+d(Si)] basis set and zero-point vibrational (un-scaled, using the Lanl2dz basis set), thermal (at 298.15 K and 1 atm) and entropy corrections (at 298.15 K) using the Lanl2dz basis set. The energies without solvent effects are provided in parentheses.

9.2.2
Models

The entire structures of both **1** and **2** have been used as a model in the calculations. Notably, in this study we use the tetrahydroxyl, $[\gamma\text{-}(SiO_4)W_{10}O_{28}(OH)_4]^{4-}$, form of **1**, $[\gamma\text{-}(SiO_4)W_{10}O_{32}H_4]^{4-}$, rather than the experimentally reported [30] bis(oxo) and bis(aquo), $[\gamma\text{-}(SiO_4)W_{10}O_{28}(H_2O)_2]^{4-}$, form. The choice of this model is based on our previous computational study [41], which clearly demonstrated that the $[\gamma\text{-}(SiO_4)W_{10}O_{28}(OH)_4]^{4-}$ form of **1** is energetically more favorable than $[\gamma\text{-}(SiO_4)W_{10}O_{28}(H_2O)_2]^{4-}$, in the gas-phase. Thus, our "full" model has the four hydroxyl (–OH) groups representing the possible reaction sites but we will utilize only one such site to probe the mechanism.

C_2H_4 is used to model olefins, and H_2O_2 is utilized as an oxidant. Complexes **1** and **2** have been isolated as the TBA (n-Bu$_4$N$^+$) salt; TBA is a large, bulky molecule. Since in this study the whole of compounds **1** and **2** have been used in the calculations, it is technically impossible to explicitly include n-Bu$_4$N$^+$ in the calculations. Therefore, tetramethylammonium ion (Me$_4$N$^+$) has been used as a model for the n-Bu$_4$N$^+$ counter-cation. An overall charge of "4−" and a singlet ground electronic state have been chosen.

The modeled overall epoxidation process, Reaction (1), is calculated to be exothermic by 38.2 (37.5) kcal mol^{-1}:

$$C_2H_4 + H_2O_2 \rightarrow C_2H_4O + H_2O \tag{1}$$

9.3
Results and Discussion

9.3.1
Mechanism of the Olefin Epoxidation Catalyzed by the Lacunary POM, $[\gamma\text{-}(SiO_4)W_{10}O_{32}H_4]^{4-}$ (1)

Since there is no information concerning the mechanism of the epoxidation catalyzed by $[\gamma\text{-}(SiO_4)W_{10}O_{28}(OH)_4]^{4-}$, previously we used the "small" model, $OW(OH)_4(H_2O)$, to identify the possible mechanisms. As discussed previously, the formation of peroxo, superoxo and hydroperoxo species during the reaction of the model with H_2O_2 were considered. To generate peroxo species, H_2O_2 has to donate both its protons and there are no obvious proton acceptors in the neighborhood of **1**. Moreover, a peroxo species is highly reactive and will start abstracting protons from the hydroxyl groups bonded to the W atoms and alter their oxidation states. This situation would contradict the experimental proposal that a structurally rigid intermediate is generated during the reaction [30]. In contrast, the generation of superoxide will require the formation of a radical-pair, which is clearly ruled out by experiments, suggesting the creation of a non-radical substrate-attacking intermediate. Furthermore, the barrier for the cleavage of the W=O bond to form epoxide was calculated to be prohibitively high. Based on those detailed studies we have predicted a "hydroperoxy" mechanism, which includes two steps: (a) Formation of a W-hydroperoxy (W−OOH) species and (b) formation of epoxide. This mechanism is discussed in detail below.

9.3.1.1 Formation of W-hydroperoxy (W−OOH) Species
In this step of the reaction, H_2O_2 is bonded to the lacunary compound to form complex **I_1** [**1**(H_2O_2), Figure 9.1].

Here and below we use the notation of **A_B**, where **A** stands for the calculated structure (**A** = **I, TS1, II, TS2,** and **III**, see below), while **B** stands for the catalyst used and can be either **1** or **2**.

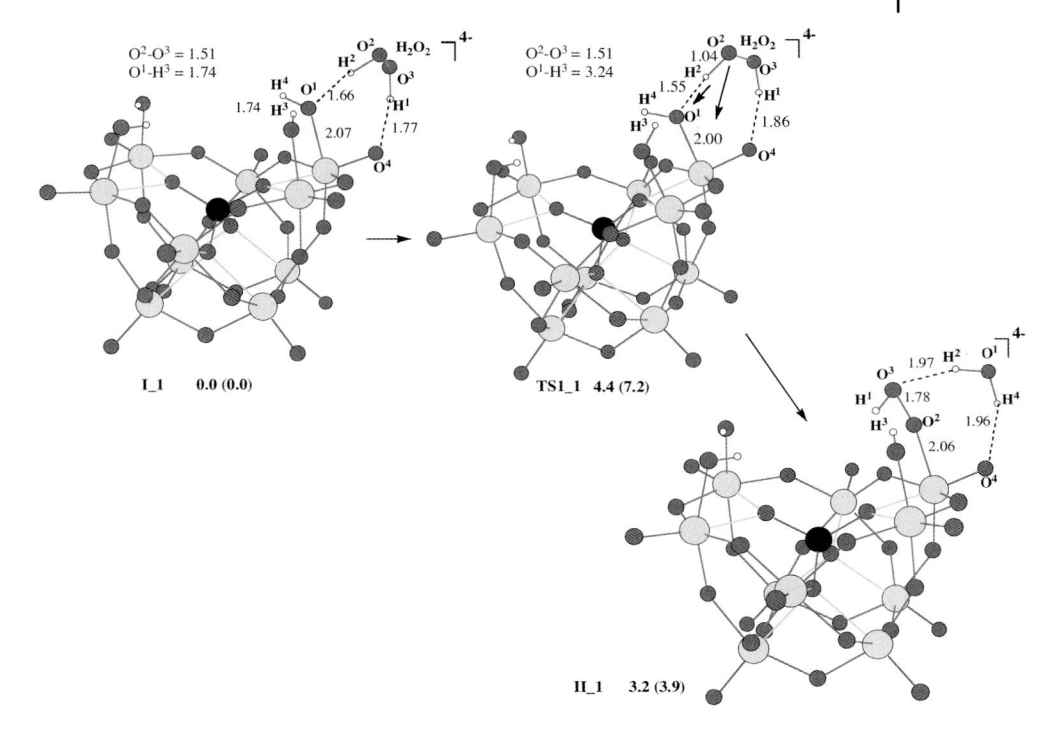

Figure 9.1 Optimized structures (in Å) and energies [with and without (in parenthesis) solvent effects, in kcal mol^{-1}] of the reactant, intermediates, transition states and the product for the reaction of $[\gamma\text{-}(SiO_4)W_{10}O_{28}(OH)_4]^{4-}$ with hydrogen peroxide.

In **I_1**, H_2O_2 forms strong hydrogen bonds with a hydroxyl ($-O^1H$) and terminal oxo (W=O) groups of the lacunary compound. At this point, in a concerted manner, the O^2-H bond of H_2O_2 is broken and the proton is transferred to the distal hydroxyl group ($-O^1H$). This process yields a hydroperoxide ($-O^2O^3H$) and a water molecule. The resulting $-O^2O^3H$ in turn makes a nucleophilic attack at the vacant site (created as the result of water formation and dissociation) on the metal center to generate a W-hydroperoxy ($W-O^2O^3H$) species (structure **II_1**). Figure 9.1 shows the optimized transition state (**TS1_1**) for this process, and this step has a barrier of 4.4 (7.2) kcal mol^{-1}. In **TS1_1**, the O^2-H bond is activated (O^2-H bond distance = 1.04 Å) and the corresponding $W-O^1H$ distance is 2.00 Å, which is slightly longer than the other W-OH (1.95 Å) bonds, indicating the formation of a water molecule. The connections between **TS1_1** and the corresponding minima are confirmed by Intrinsic Reaction Coordinate (IRC) [42] calculations. The formation of the $W-O^2O^3H$ species is a slightly endothermic, by 3.2 (3.9) kcal mol^{-1}. The formation of this species is in line with experimental suggestions that a non-radical oxidant is generated on the lacunary compound and only a water molecule is produced as the by-product [30].

9.3.1.2 Formation of Epoxide

In this most critical step the W–O^{2}O^{3}H intermediate reacts with C$_2$H$_4$ to produce ethylene epoxide (C$_2$H$_4$O^3). Here, in a concerted fashion, the O^3–H bond of the W–O^{2}O^{3}H species is broken and the proton is transferred through a water molecule to the W-bound oxygen atom (W–O^2). In this process, the O^2–O^3 bond of the W–O^{2}O^{3}H species is cleaved and the released oxygen atom (O^3) is immediately abstracted by the C$_2$H$_4$ molecule to form epoxide. The optimized transition state structure (**TS2_1**; Figure 9.2) and the corresponding distances (O^2–O^3 = 1.73 Å and C–O^3 = 2.07 Å) indicate that this process is synchronous.

This step has some similarities with the O–O bond splitting of the Fe(II)–(HOOH) species in the peroxidase pathway of cytochrome P450. In this enzyme, the distal His has been suggested to play the role of an intermediate catalyst by accepting a proton from the proximal oxygen and then donating it to the distal oxygen [43]. In the present study, like the distal His in cytochrome P450, a water molecule has been proposed to perform the similar role of a proton donor and acceptor. Another example of such a process is the O–O bond cleavage of the Cu(II)–OOH species in copper amine oxidase (CAO) [44]. In CAO, indirectly through a water molecule, a proton transfer from a sp^3 carbon of the phenol ring to one of the oxygens of the peroxide splits the O–O bond and creates a Cu(II)–OH species. The transition structure (**TS2_1**) is confirmed to be connected to the corresponding minima by IRC calculations. The formation of epoxide (structure **IV_1**) has a very high barrier of 35.5 (36.1) kcal mol^{-1}. Moreover, this process follows a 3.2 (3.9) kcal mol^{-1} endothermic step, which makes an overall barrier of 38.7 (40.0) kcal mol^{-1}, calculated from **I_1**. This is the rate-limiting step of the entire catalytic cycle. In the product (**IV_1**), epoxide is hydro-

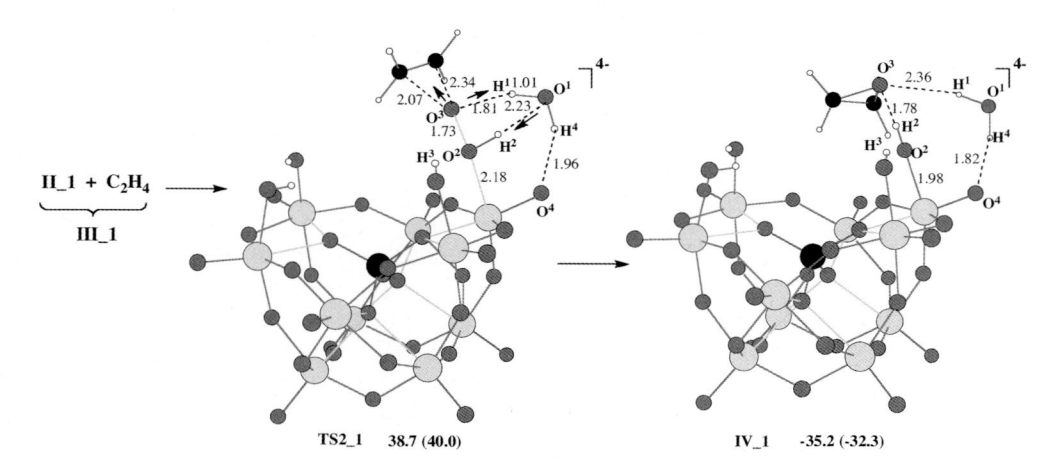

Figure 9.2 Optimized structures (in Å) and energies [with and without (in parenthesis) solvent effects, in kcal mol^{-1}] of the reactant, intermediates, transition states and the product for ethylene epoxidation by the [γ-(SiO$_4$)W$_{10}$O$_{28}$(OH)$_3$(OOH)(H$_2$O)]$^{4-}$ intermediate.

gen bonded to the lacunary complex and the water molecule. The formation of product ethylene epoxide from **II_1** is exothermic by 38.3 (35.2) kcal mol^{-1}.

Thus, the overall barrier, 38.7 (40.0) kcal mol^{-1}, is very high, which indicates a low reactivity for **1**. This number is obtained in the absence of counter-cations, which are always present in catalytic solutions. To elucidate roles of counter-cations in this process at the next stage we have included up to two Me$_4$N$^+$ cations in the calculations.

9.3.1.3 Counter Cation Effect

Four Bu$_4$N$^+$ counter-cations per unit of **1** are present during the catalysis; their effect on the barrier of the rate-limiting process (**I_1** → **TS2_1**) was also investigated using Me$_4$N$^+$ as a model.

These effects have been computed by explicitly incorporating one or two Me$_4$N$^+$ molecules into the models. The reactant in the presence of one and two Me$_4$N$^+$ molecules (structures **I_1$_{1CC}$** and **I_1$_{2CC}$** are not presented in Figure 9.3) are fully optimized, whereas in the transition state structures (**TS2_1$_{1CC}$** and **TS2_1$_{2CC}$** are given in Figure 9.3) only the main reaction coordinates (O^2–O^3, O^1–H, O^2–H, O^3–H and O^3–C bond distances) from the **TS2_1** are kept frozen.

This type of approximation can be justified by considering the fact that the entire [γ-(SiO$_4$)W$_{10}$O$_{28}$(OH)$_4$]$^{4-}$ compound is already included in the calculations. Solvent effects are included in the calculated energetics by performing single-point calculations at the IEF-PCM level. The addition of the first molecule of Me$_4$N$^+$, located inside the cavity, reduces the barrier from **I_1** by 7.6 (8.3) kcal mol^{-1}, and the overall barriers for the formation of epoxide becomes 31.1 (31.7) kcal mol^{-1}. Furthermore, inclusion of the second Me$_4$N$^+$ molecule, positioned outside the cavity, further reduces the barrier from **I_1** by 4.3 (4.3) kcal mol^{-1}, and the overall barriers becomes 26.8 (27.5) kcal mol^{-1}. The reason for the relatively small effect of the second Me$_4$N$^+$ counter-cation is that, in comparison to seven hydrogen bonds formed by the first Me$_4$N$^+$, the second Me$_4$N$^+$ forms only five hydrogen bonds with the lacunary complex. The remaining two counter-cations may have smaller effects but their presence is also expected to reduce the rate-limiting barrier. This pronounced counter-cation effect on **TS2_1** derives, in good measure, from the fact that the –OOH fragment of the W–OOH bears a significant negative charge that is strongly stabilized by the counter-cation. These results clearly indicate that the presence of counter-cations strongly influences the reactivity of **1** and makes it a very efficient catalyst under moderate conditions. They also suggest that ion-pairing effects in the H$_2$O$_2$-based oxidation catalyzed by POMs could be of general importance.

In summary, computational studies of ethylene epoxidation by H$_2$O$_2$ catalyzed by the lacunary POM **1** indicate that **1** effectively acts as a mononuclear W(VI)-complex in activating the oxidant. In the first step of the reaction, the O^2–H bond of H$_2$O$_2$ is broken and the hydroperoxide (W–O^{2}O^{3}H) species and a water molecule are produced in a concerted manner. This process has a barrier of 4.4 (7.2) kcal mol^{-1} and it is slightly endothermic by 3.2 (3.9) kcal mol^{-1} (Table 9.1). The formation of products (W–O^{2}O^{3}H and H$_2$O) in this step is in agreement

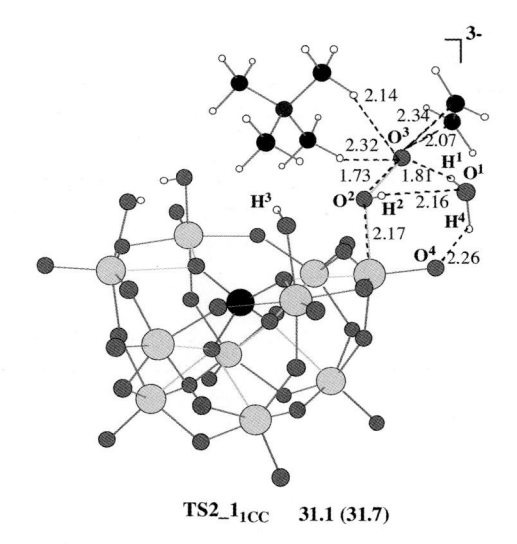

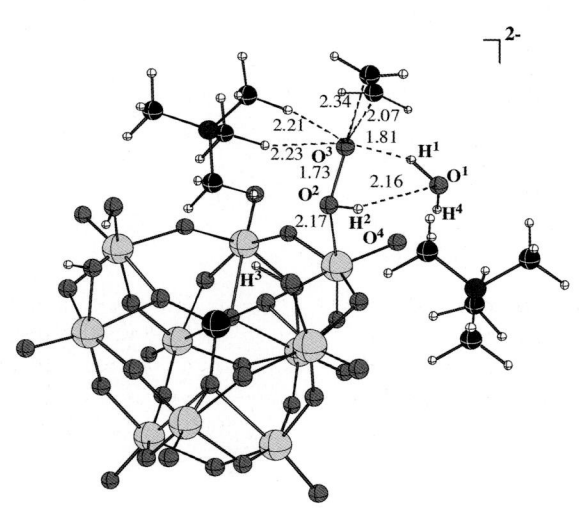

Figure 9.3 Optimized structures (Å) and energies (kcal mol^{-1}) of the transition state (**TS2_1**) for epoxidation in the presence of one (1CC) and two (2CC) Me$_4$N$^+$ counter-cations.

with experimental suggestions that a non-radical oxidant is generated on the lacunary POM and only a water molecule is produced as a by-product [30]. In the second and the final step of the reaction, the O^3–H bond of the W–O^{2}O^{3}H species is cleaved and the proton is transferred via a water molecule to the W-bound oxygen atom (W–O^2), leading to cleavage of the O^2–O^3 bond and formation of an epoxide. This is the rate-limiting step of the entire catalytic cycle

9.3 Results and Discussion

Table 9.1 Calculated energetics[a] for H_2O_2-based C_2H_4 epoxidation catalyzed by the lacunary $[\gamma\text{-}(SiO_4)W_{10}O_{32}H_4]^{4-}$ ($x=1$) and di-V-substituted-γ-Keggin $[\gamma\text{-}1,2\text{-}H_2SiV_2W_{10}O_{40}]^{4-}$ ($x=2$) polyoxometalates.

Structure	$x = 1$				$x = 2$		
	cc = 0[b]		cc = 1	cc = 2	cc = 0		cc = 1
	Energetics (kcal mol^{-1})						
	Gas-phase	Solv.[c]	Solv.	Solv.	Gas-phase	Solv.	Solv.
I_x	0.0	0.0	0.0	0.0	0.0	0.0	0.0
TS1_x	7.2	4.4			6.7	6.3	
II_x	3.9	3.2			0.1	−0.8	
TS2_x	40.0	38.7	31.1	26.8	44.5	27.7	25.6
IV_x	−32.3	−35.2			−29.6	−39.6	

a) Relative to the I_x + C_2H_4 reactants, where I_x = $x\,(H_2O_2)$ complexes.
b) "cc" indicates the number of counter-cations included into the calculations.
c) In the text the numbers including solvent (acetonitrile) effects are presented without parenthesis, while those obtained in the gas phase are given in parenthesis.

and occurs with an overall barrier of 38.7 (40.0) kcal mol^{-1}. Formation of the product ($C_2H_4O^3$) is exothermic by 38.3 (35.3) kcal mol^{-1}. Inclusion of the one and two Me_4N^+ counter-cations reduces the rate-limiting barrier, at the second step, by 7.6 (8.3) and 11.9 (12.6) kcal mol^{-1}, respectively. These results clearly indicate that ion-pairing interactions make **1** a more efficient catalyst for the epoxidation of C_2H_4 by H_2O_2.

9.3.2
Mechanism of the Olefin Epoxidation Catalyzed by the di-V-substituted γ-Keggin POM $[\gamma\text{-}1,2\text{-}H_2SiV_2W_{10}O_{40}]^{4-}$ (2)

As mentioned in the Introduction, despite the wealth of available experimental information, the exact mechanism of the epoxidation of alkenes by H_2O_2 catalyzed by **2**, and the effect of the counter-cations such as TBA on the energetics, are not known. Elucidation of the mechanism of this reaction, i.e.,

$$\{VO\text{-}(\mu\text{-}OH)_2\text{-}VO\} + H_2O_2 + \text{olefin}$$
$$\rightarrow \{VO\text{-}(\mu\text{-}OH)_2\text{-}VO\} + \text{epoxide} + H_2O \qquad (2)$$

requires an understanding of several elementary processes, including but not limited to the reaction of hydrogen peroxide with the initial di-vanadium polyoxotungstate:

$$\{VO\text{-}(\mu\text{-}OH)_2\text{-}VO\} + H_2O_2 \rightarrow \{VO\text{-}(\mu\text{-}OH)(\mu\text{-}OOH)\text{-}VO\} + H_2O \quad (3a)$$

$$\{VO\text{-}(\mu\text{-}OH)(\mu\text{-}OOH)\text{-}VO\} + H_2O_2$$
$$\rightarrow \{VO\text{-}(\mu\text{-}OOH)(\mu\text{-}OOH)\text{-}VO\} + H_2O \quad (3b)$$

epoxidation of the olefin by viable reactive oxygenating intermediates:

$$\{VO\text{-}(\mu\text{-}OH)(\mu\text{-}OOH)\text{-}VO\} + \text{olefin} \rightarrow \{VO\text{-}(\mu\text{-}OH)_2\text{-}VO\} + \text{epoxide} \quad (4a)$$

$$\{VO\text{-}(\mu\text{-}OOH)(\mu\text{-}OOH)\text{-}VO\} + 2\ \text{olefin}$$
$$\rightarrow \{VO\text{-}(\mu\text{-}OH)_2\text{-}VO\} + 2\ \text{epoxide} \quad (4b)$$

$$\{VO\text{-}(\eta^2\text{-}O_2)\text{-}VO\} + \text{olefin} + H_2O \rightarrow \{VO\text{-}(\mu\text{-}OH)_2\text{-}VO\} + \text{epoxide} \quad (4c)$$

and the hydroperoxy-to-peroxo rearrangement:

$$\{VO\text{-}(\mu\text{-}OOH)(\mu\text{-}OOH)\text{-}VO\} + 2H^+ \rightarrow \{VO\text{-}(\eta^2\text{-}O_2)\text{-}VO\} + 2H_2O \quad (5)$$

In the present DFT study, we elucidate the mechanisms of the Reactions (3a) and (4a). The mechanisms of the Reactions (3b) and (4b) are expected to be similar to those for Reactions (3a) and (4a), respectively. In addition, the effect of counter-cations on the energetics of these important reactions has also been investigated. Studies of the mechanisms of the Reactions (4c) and (5) are in progress and will be reported separately.

9.3.2.1 Mechanism of Reaction (3a)

At the starting point of this reaction the oxidant, H_2O_2, is coordinated to compound 2 [structure **I_2 = 2(H_2O_2)** in Figure 9.4]. In **I_2**, H_2O_2 forms strong hydrogen bonds with the $V^1=O^3$ and the V^1–O with the bond distances of O^3–H^2 = 1.80 Å and O–H^1 = 1.85 Å respectively.

From structure **I_2**, in a concerted manner, the O^5–H^1 bond of H_2O_2 is broken and the proton is transferred to the hydroxyl group (–O^1H) of the {VO-(μ-OH)$_2$-VO} core. This process leads to formation of the –$O^5O^4H^2$ species and a water molecule, H_2O^1. The resulting –$O^5O^4H^2$ nucleophile attacks the vacant site (created as the result of the formation and dissociation of a water molecule) on the metal center to generate the experimentally proposed vanadium-hydroperoxy structure $\{V^1O^3\text{-}(\mu\text{-}O^5O^4H^2)(\mu\text{-}OH)\text{-}V^2O\}(H_2O)$ (**II_2**). This process occurs with a barrier of 6.3 (6.7) kcal mol^{-1} at the transition state **TS1_2**. In **TS1_2**, all the corresponding distances (O^5H^1 = 1.06 Å, O^1H^1 = 1.46 Å and V^1–O^1H = 2.11 Å) indicate proton migration from H_2O_2 to O^1H and the formation of a water molecule (H_2O^1). Notably, in comparison to **I_2**, in **TS1_2**, V^1–O^1 and V^2–O^1 bond

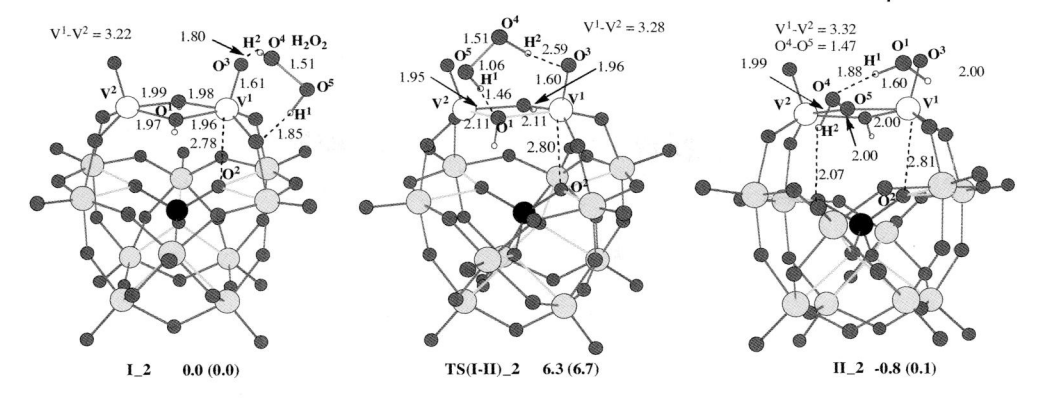

Figure 9.4 Calculated important bond distances (Å) of the reactant, transition state and product of the reaction {VO-(μ-OH)$_2$-VO} + H$_2$O$_2$ → {VO-(μ-OH)(μ-OOH)-VO} + H$_2$O (3a).

distances elongate by 0.15 Å to facilitate the formation and release of the water molecule, and, once this process is completed, both distances, in **II_2**, decrease by 0.11 Å. In **TS1_2**, in comparison with **I_2**, there is a small decrease in the charges on V^1 (0.1e) and O^1 (0.08e). During **I_2** → **TS1_2** → **II_2**, both V^1–V^2 and V^1–O^2 bond distances were found to increase continuously, i.e., 3.22 < 3.28 < 3.32 Å and 2.78 < 2.80 < 2.81 Å, respectively. Formation of the vanadium-hydroperoxy species, {V^{1}O^3-(μ-O^{5}O^{4}H^2)(μ-OH)-V^{2}O}(H$_2$O), **II_2** is slightly exergonic: −0.8 (0.1) kcal mol^{-1}.

As seen from comparison of the energetics presented with (gas phase) and without (solution phase) parenthesis, the inclusion of solvent effects at the IEF-PCM level, using CH$_3$CN as a solvent, does not significantly change the calculated energetics of Reaction (3a).

9.3.2.2 Mechanism of the Epoxidation Reaction (4a)

In this step of the hydroperoxy mechanism, the intermediate {V^{1}O^3-(μ-O^{5}O^{4}H^2)(μ-OH)-V^{2}O}(H$_2$O), **II_2**, reacts with C$_2$H$_4$ to produce ethylene epoxide (C$_2$H$_4$O^4) (Figure 9.5).

The O^4–H^2 bond of the μ-O^{5}O^{4}H^2 species is broken in a concerted fashion and the proton is transferred through a water molecule (produced in Reaction 3a) to the atom bound to the two vanadium atoms (μ-O^5). In this process, the O^4–O^5 bond of the μ-O^{5}O^{4}H^2 species is cleaved and the released oxygen atom (O^4) is immediately abstracted by the C$_2$H$_4$ molecule to form epoxide. Figure 9.5 shows the optimized transition state structure, **TS2_2**; the barrier for this process is 27.7 (44.5) kcal mol^{-1}. All the corresponding bond distances (O^{4}H^2 = 1.07 Å, O^{1}H^2 = 1.80 Å and O^4–O^5 = 1.87 Å) indicate that this is a synchronous process. In **TS2_2**, in comparison to **III_2**, V^1–O^5 and V^2–O^5 bond distances shrink by 0.08 and 0.07 Å, respectively, and the metal–metal (V^1–V^2) bond distance elongates by 0.08 Å. In **TS2_2**, the charge on V^1 reduces by only 0.05e, whereas on

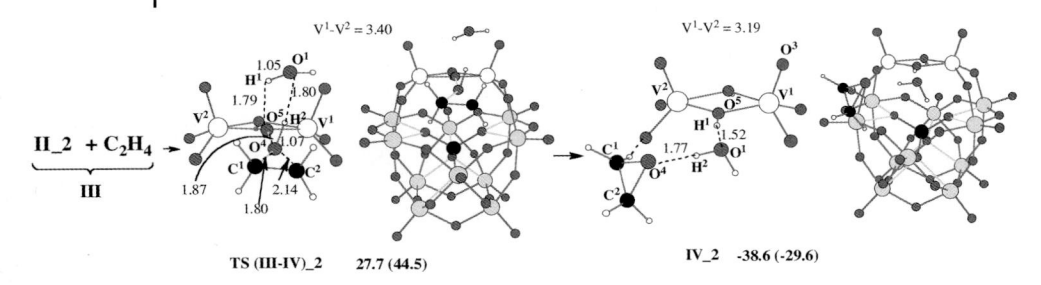

Figure 9.5 Calculated important bond distances (Å) of the reactant, transition state and product of the reaction {VO-(μ-OH)(μ-OOH)-VO} + $C_2H_4 \rightarrow$ {VO-(μ-OH)$_2$-VO} + epoxide.

O^3 it increases by 0.13e. In this process a significant solvent effect that reduces the barrier by 16.8 kcal mol^{-1} is found. A reasonable explanation for this pronounced effect is that **TS2_2** with a small dipole moment 4.5 D is more strongly stabilized by the highly polar solvent acetonitrile than **III_2** with a larger dipole moment of 8.53 D. This is the rate-limiting step of the entire catalytic cycle. In the product (**IV_2**), epoxide is hydrogen bonded to the lacunary complex and the water molecule. In the **III_2 $\rightarrow$ TS2_2 $\rightarrow$ IV_2** process, both V^1-V^2 and V^1-O^2 bond distances were first found to increase in **TS2_2** and then decrease in the product, i.e., 3.32 < 4.40 > 3.19 Å and 2.81 < 2.87 > 2.86 Å. The elongation in V^1-V^2 in **TS2_2** is due to an increase in the single bond character of the V^1-O^5 caused by splitting of the O^4-O^5 bond. The formation of product ethylene epoxide from **IV_2** is found to be exothermic by 39.6 (29.6) kcal mol^{-1}.

As seen from a comparison of the gas- and solution-phase energetics, the inclusion of solvent effects at the IEF-PCM level using CH_3CN as a solvent significantly changes the calculated energetics of Reaction (3a). It reduces the barrier for epoxidation in **TS2_2** from 44.5 to 27.7 kcal mol^{-1} and increases the reaction energy from 29.6 to 39.6 kcal mol^{-1}.

However, as shown above, the presence of one and two $(CH_3)_4N^+$ counter-cations impacts on the mechanism of epoxidation of olefins by H_2O_2 catalyzed by the lacunary complex $[\gamma\text{-}(SiO_4)W_{10}O_{28}(OH)_4]^{4-}$ and reduces the rate-limiting barrier by 7.6 (8.3 and 11.9 (12.6) kcal mol^{-1}, respectively. As a consequence, the impact of counter-cations on the calculated energetics of Reactions (3a) and (4a) has also been investigated.

9.3.2.3 Counter Cation Effect

Since compound **2** was isolated as the tetra-n-butylammonium ion (t-Bu_4N^+) salt, the effect of counter-cation on the barrier of the rate-limiting step (**III_2 $\rightarrow$ TS2_2**) of overall reaction:

$$2 + H_2O_2 + C_2H_4 \rightarrow I\_2 + C_2H_4 \rightarrow TS1\_2 + C_2H_4$$
$$\rightarrow II\_2 + C_2H_4 \rightarrow \rightarrow III\_2 \rightarrow TS2\_2 \rightarrow IV\_2 \quad (6)$$

(CH₃)₄N⁺ V^1-V^2 = 3.35   (CH₃)₄N⁺
O^4-O^5 = 1.47

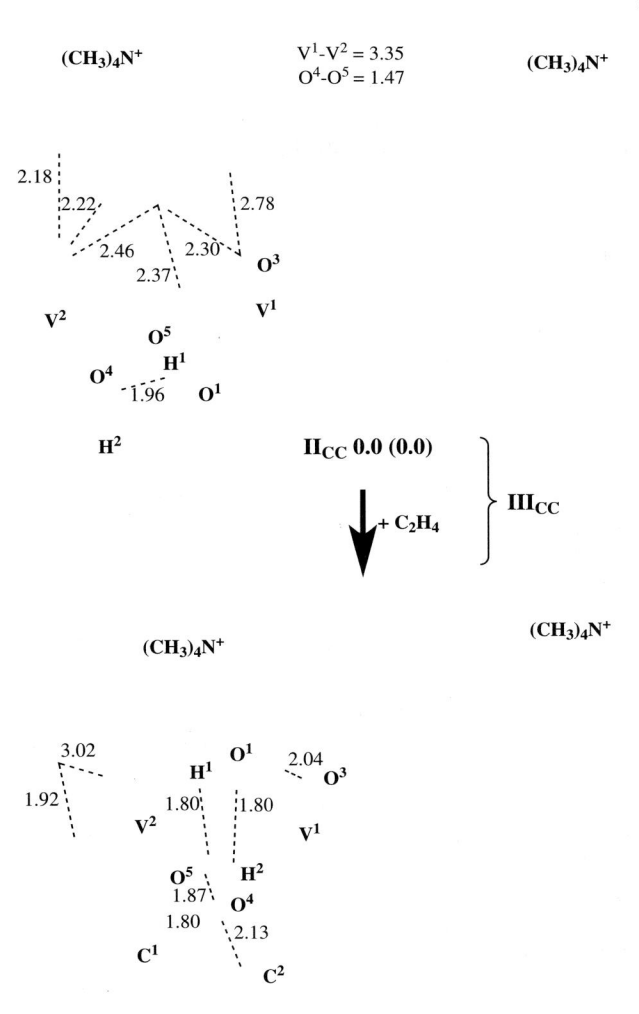

2.18
2.22, 2.78
2.46 2.30
2.37 O^3
V^2 V^1
O^5
O^4 H^1
1.96 O^1
H^2

II$_{CC}$ 0.0 (0.0)

↓ + C₂H₄ } III$_{CC}$

(CH₃)₄N⁺   (CH₃)₄N⁺

3.02
       O^1 2.04
1.92 H^1 O^3
    1.80   1.80
  V^2 V^1
    O^5 H^2
    1.87 O^4
    1.80   2.13
    C^1
        C^2

TS(III-IV)$_{CC}$ 25.6 (34.6)

Figure 9.6 Calculated important bond distances (Å) of the reactant and transition state of the reaction {VO-(μ-OH)(μ-OOH)-VO} + C₂H₄ → {VO-(μ-OH)₂-VO} + epoxide in the presence of one molecule of counter-cation Me₄N⁺.

was investigated. In this study, Me₄N⁺ was used to model the larger but electronically similar t-Bu₄N⁺ counter-cation. Previously, we demonstrated that inclusion of the first counter-cation molecule in the calculations has the most pronounced effect on the calculated energetics and geometries [36]. Therefore, here we only elucidate the effects from one counter-cation. Reactant **III_2** in the presence of one Me₄N⁺ (structures **III_2₁CC**, Figure 9.6) is fully optimized, whereas in the transition state structure (**TS2_2**$_{1CC}$) only the main reaction coordinates (O^4–O^5, O^4–H^2, O^1–H^2, O^1–H^1 and O^4–C^1 bond distances) from the **TS2_2** are kept frozen.

This is a valid approximation because the entire $[\gamma\text{-}1,2\text{-}H_2SiV_2W_{10}O_{40}]^{4-}$ compound is already included in the calculations. Solvent effects are included in the calculated energetics by performing single-point calculations at the IEF-PCM level using CH_3CN as the solvent. Addition of Me_4N^+ proximal to the catalytically active unit, $\{V^1O^5\text{-}(\mu\text{-}O^6O^7H)(\mu\text{-}OH)\text{-}V^2O\}$, reduces the barrier from **III_2** by 2.1 (9.9) kcal mol^{-1}, and the barrier for formation of epoxide: 25.6 (34.6) kcal mol^{-1}.

In summary, notably, the exact mechanism for H_2O_2-based alkene epoxidation catalyzed by **2**, Reaction (2), could be very complex. Elucidation of the full mechanism for this and closely related catalytic oxygenation reactions requires a fairly complete comprehension of the mechanisms of several elementary processes. These include but are not limited to (a) reaction of hydrogen peroxide with initial species, Reactions (3a) and (3b); (b) epoxidation of olefin by various oxygenation-competent intermediates such as hydroperoxy, $\{VO\text{-}(\mu\text{-}OH)(\mu\text{-}OOH)\text{-}VO\}$ and $\{VO\text{-}(\mu\text{-}OOH)_2\text{-}VO\}$, and peroxo, $\{VO\text{-}(\eta^2\text{-}O_2)\text{-}VO\}$, species, i.e., Reactions (4a)–(4c), respectively, and (c) hydroperoxy-to-peroxo rearrangement, Reaction (5).

The density functional studies of the mechanisms of Reactions (3a) and (4a), presented above, show that: (a) Reaction of $\{VO\text{-}(\mu\text{-}OH)_2\text{-}VO\}$, **2**, with H_2O_2 proceeds in a concerted manner, and produces the vanadium-hydroperoxy species $\{V^1O^3\text{-}(\mu\text{-}O^5O^4H^2)(\mu\text{-}OH)\text{-}V^2O\}(H_2O)$, **II_2**. This process has a barrier of 6.3 (6.7) kcal mol^{-1} and is slightly exothermic: -0.8 (0.1) kcal mol^{-1}. Formation of the vanadium-hydroperoxy product, that with the $\{VO\text{-}(\mu\text{-}OH)(\mu\text{-}OOH)\text{-}VO\}$ core, during this reaction is in excellent agreement with the available experiments [31] (b) The water molecule produced in Reaction (3a) plays an important role and is involved the epoxidation of C_2H_4 by **II_2**, Reaction (4a). During this reaction, the O–H bond of the μ-OOH group in **II_2** is broken in a concerted fashion and the proton is transferred through the water molecule to the μ-oxo center bound to two V-atoms. The barrier associated with this process is 27.7 (44.5) kcal mol^{-1}.

9.4
Polyoxometalate-catalyzed Ethylene Epoxidation by Hydrogen Peroxide. Comparison of the Hydroxyl Mechanism involving Complexes 1 and 2

Data presented in Table 9.1 show that the rate-limiting step for the hydroxy mechanism of both the **1** and **2** catalyzed C_2H_4 epoxidation by H_2O_2 is the O-atom transfer from hydroperoxo intermediate **II** (**II_1** and **II_2**, respectively) to olefin. It proceeds with 38.7 (40.0) and 27.7 (44.5) kcal mol^{-1} barriers at the transition states **TS2_1** and **TS2_2**, respectively. The inclusion of one Me_4N^+ counter-cation reduces these values to 31.1 (31.7) and 25.6 (34.6) kcal mol^{-1}, respectively. Comparison of these data shows that this rate-limiting barrier is slightly larger for lacunary POM **1** than that for di-V-substituted-γ-Keggin [γ-1,2-H$_2$SiV$_2$W$_{10}$O$_{40}$]$^{4-}$ (**2**). In other words, epoxidation of olefins by H$_2$O$_2$ catalyzed by **2** is expected to proceed slightly faster than the closely related reaction catalyzed by the lacunary complex, [γ-(SiO$_4$)W$_{10}$O$_{28}$(OH)$_4$]$^{4-}$ (**1**). Furthermore, in this step, the water molecule, produced by reaction of **1** and **2** with H$_2$O$_2$, plays a dual role of proton acceptor and donor.

As seen in Table 9.1, the inclusion of one counter-cation (Me$_4$N$^+$) reduces the rate-limiting barrier by 7.6 (8.3) and 2.1 (9.9) kcal mol^{-1} for the reaction with **1** and **2**, respectively. Comparison of these data shows that the counter-cation has a relatively small effect on the rate-limiting barrier of the reaction C$_2$H$_4$ + H$_2$O$_2$ → OC$_2$H$_4$ + H$_2$O catalyzed by **2** compared with that catalyzed by **1**. This could be explained by the fact that in the intermediate II_1 the active unit (W–OOH), that involved in the O-atom transfer process, is located slightly closer to the surface (better exposed to the counter-cation) than that (V–OOH) in the intermediate II_2.

Finally, however, definitive elaboration of all aspects of the mechanism of ethylene epoxidation by hydrogen peroxide catalyzed by the lacunary [γ-(SiO$_4$)W$_{10}$O$_{32}$H$_4$]$^{4-}$ (**1**) and di-V-substituted-γ-Keggin [γ-1,2-H$_2$SiV$_2$W$_{10}$O$_{40}$]$^{4-}$ (**2**) polyoxometalates requires the elucidation of the mechanisms and energetics of other elementary processes, such as (4c) and (5), which comprise Reaction (2). These studies are under investigation in our laboratory.

Acknowledgment

The present research is supported by a grant (DE-FG02-03ER15461) from the U.S. Department of Energy. Computer resources were provided by the Cherry Emerson Center for Scientific Computation.

References

1 Tullo, A. *Chem. Eng. News* 2004, **82**, 15.
2 Ishii, Y., Yamawaki, K., Ura, T., Yamada, H., Yishida, T., Ogawa, M. *J. Org. Chem.* 1998, **53**, 3587–3593.
3 Neumann, R., Gara, M. *J. Am. Chem. Soc.* 1995, **117**, 5066–5074.
4 Mizuno, N., Nozaki, C., Kiyoto, I., Misono, M. *J. Am. Chem. Soc.* 1998, **120**, 9267–9272.
5 Hill, C. L., Prosser-McCartha, C. M. *Coord. Chem. Rev.* 1995, **143**, 407–455.
6 Kozhevnikov, I. V. *Chem. Rev.* 1998, **98**, 171–198.
7 Lane, B. S., Burgess, K. *Chem. Rev.* 1989, **103**, 2457–2760.
8 White, M. C., Doyle, A. G., Jacobsen, E. N. *J. Am. Chem. Soc.* 2001, **123**, 7194–7195.
9 Que, L. Jr. *Science* 1991, **253**, 273–274.
10 Pestovsky, O., Stoian, S., Bominaar, E. L., Shan, X., Munck, E., Que, L. Jr., Bakac, A. *Angew. Chem. Int. Ed.* 2005, **44**(42), 6871–6874.
11 Bukowski, M. R., Koehntop, K. D., Stubna, A., Bominaar, E. L., Halfen, J. A., Muenck, E., Nam, W., Que, L. Jr. *Science* 2005, **310**(5750), 1000–1002.
12 MacBeth, C. E., Golombek, A. P., Young, V. G., Jr., Yang, C., Kuczera, K., Hendrich, M. P., Borovik, A. S. *Science* 2000, **289**, 938–941.

13 Battioni, P., Renaud, J. P., Bartoli, J. F., Reinaartiles, M., Fort, M., Mansuy, D. *J. Am. Chem. Soc.* 1988, **110**, 8462–8470.
14 Notari, B. *Adv. Catal.* 1996, **41**, 253.
15 Romao, C. C., Kühn, F. E., Herrmann, W. A. *Chem. Rev.* 1997, **97**, 3197–3246.
16 Vasbinder, M. J., Espenson, J. H. *Organometallics* 2004, **23**, 3355–3358.
17 Oesz, K., Espenson, J. H. *Inorg. Chem.* 2003, **42**, 8122–8124.
18 Venturello, C., Alneri, E., Ricci, M. *J. Org. Chem.* 1983, **48**, 3831–3833.
19 Duncan, D. C., Chambers, R. C., Hecht, E., Hill, C. L. *J. Am. Chem. Soc.* 1995, **117**, 681–691.
20 De Vos, D. E., Meinershagen, J. L., Bein, T. *Angew. Chem. Int. Ed. Engl.* 1996, **35**, 2211–2213.
21 Neumann, R. *Prog. Inorg. Chem.* 1998, **47**, 317.
22 (a) Neumann, R., Dahan, M. *Nature* 1997, **388**, 353–355. (b) Neumann, R., Dahan, M. *J. Am. Chem. Soc.* 1998, **120**, 11969–11976.
23 Nishiyama, Y., Nakagawa, Y., Mizuno, N. *Angew. Chem. Int. Ed.* 2001, **19**, 3751–3753.
24 Okun, N. M., Anderson, T. M., Hill, C. L. *J. Am. Chem. Soc.* 2003, **125**, 3194–3195.
25 Weinstock, I. A., Barbuzzi, E. M. G., Wemple, M. W., Cowan, J. J., Reiner, R. S., Sonnen, D. M., Heintz, R. A., Bond, J. S., Hill, C. L. *Nature* 2001, **414**, 191–195.
26 Nakagawa, Y., Kamata, K., Kotani, M., Yamaguchi, K., Mizuno, N. *Angew. Chem. Int. Ed.* 2005, **44**, 5136–5141.
27 Hill, C. L. *Comprehensive Coordination Chemistry II*, ed. A. G. Wedd, Elsevier Science, New York, 2004, Vol. 4, pp. 679–759.
28 Neumann, R. *Prog. Inorg. Chem.* 1998, **47**, 317–370.
29 Neumann, R. *Modern Oxidation Methods*, ed. J. E. Backvall, Wiley VCH, Weinheim, 2004, pp. 233–251.
30 Kamata, K., Yonehara, K., Sumida, Y., Yamahuchi, K., Hikichi, S., Mizuno, N. *Science* 2003, **300**, 964–966.
31 Nakagawa, Y., Kamata, K., Kotani, M., Yamaguchi, K., Mizuno, N. *Angew. Chem. Int. Ed.* 2005, **44**, 5136.
32 Mizuno, N., Nakagawa, Y., Yamaguchi, K. *J. Mol. Catal. A* 2006, **251**, 286.
33 Nakagawa, Y., Uehara, K., Mizuno, N. *Inorg. Chem.* 2005, **44**, 14.
34 Nakagawa, Y., Uehara, K., Mizuno, N. *Inorg. Chem.* 2005, **44**, 9068.
35 (a) Kamata, K., Yonehara, K., Sumida, Y., Yamaguchi, K., Hikichi, S., Mizuno, N. *Science* 2003, **300**, 964. (b) Kamata, K., Kotani, M., Yamaguchi, K., Hikichi, S., Mizuno, N. *Chem. Eur. J.*, 2007, **13**, 639.
36 Prabhakar, R., Morokuma, K., Hill, C. L., Musaev, D. G. *Inorg. Chem.* 2006, **45**, 5703.
37 *Gaussian 03 (Revision C1)*, Frisch, M. J. et al. 2004, Gaussian Inc., Pittsburgh, PA.
38 (a) Becke, A. D. *Phys. Rev. A* 1988, **38**, 3098–3107. (b) Lee, C., Yang, W., Parr, R. G. *Phys. Rev. B* 1988, **37**, 785–789. (c) Becke, A. D. *J. Chem. Phys.* 1993, **98**, 5648.
39 Hay, P. J., Wadt, W. R. *J. Chem. Phys.* 1985, **82**, 270–283.
40 Cances, E., Mennucci, B., Tomasi, J. *J. Chem. Phys.* 1997, **107**, 3032–3041.
41 Musaev, D. G., Morokuma, K., Geletii, Y. V., Hill C. L. *Inorg. Chem.* 2004, **43**, 7702–7708.
42 Gonzalez, C., Schlegel, H. B. *J. Phys. Chem.* 1990, **94**, 7467–7471.
43 Denisov, I. G., Makris, T. M., Sligar, S. G., Schlichting, I. *Chem. Rev.* 2005, **105**, 2253–2278.
44 Prabhakar, R., Siegbahn, P. E. M. *J. Am. Chem. Soc.* 2004, **126**, 3996–4006.

10
C–H Bond Activation by Transition Metal Oxides

Joachim Sauer

10.1
Introduction

The catalytic activation of C–H bonds is of fundamental interest, for it is the crucial step in important technological processes and also in many enzymatic reactions. Examples are the oxidative dehydrogenation (ODH) of lower alkanes to alkenes [1] and of alcohols to aldehydes or ketones [2] by supported transition metal oxide catalysts and the oxygenation of C–H bonds to C–OH groups by cytochrome P450 [3].

For the complex solid catalysts neither the structure of the active phase on the support nor the individual mechanistic steps are known. It is not clear (or controversial) how the activity depends on the size of the active species and which of the different oxygen species is active in the different steps of the reaction. These oxygen species include those on terminal metal oxo (double) bonds O_t, bridging oxygen O_b in polynuclear oxide clusters, species in the active metal–support interphase bond O_i, and peroxo species O_p-O_p formed on re-oxidation of the catalysts by O_2 (Scheme 10.1).

To reduce the complexity of supported metal oxide catalysts, attempts are made to learn from studies on related model systems. Thin oxide films on metal substrates are studied by surface science techniques in ultrahigh vacuum [4], reactions with cationic or anionic oxide clusters are studied in the mass spectrometer [5], and inorganic model compounds are synthesized and used in

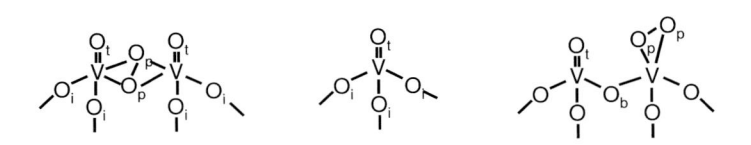

Scheme 10.1

Computational Modeling for Homogeneous and Enzymatic Catalysis.
A Knowledge-Base for Designing Efficient Catalysts. K. Morokuma and D. G. Musaev (Eds.)
Copyright © 2008 WILEY-VCH Verlag GmbH & Co. KGaA, Weinheim
ISBN: 978-3-527-31843-8

"homogeneous" phase with the aim to characterize possible intermediates [6]. In this chapter we focus on vanadium oxide as active phase [7] and compare supported species with bulk oxide surfaces, gas phase cluster ions and model compounds. The reactions of interest are the oxidative dehydrogenation of propane to propene and of methanol to formaldehyde. By comparison with other transition metal oxide catalysts and other substrates we will analyze the unifying aspects of the proposed mechanisms.

10.2
Gas Phase and Surface Species Considered

Figure 10.1 shows the models for monomeric and dimeric V_2O_5 species on silica surfaces that have been used in computational studies [8–11]. The basic silsesquioxane structures have proved to be very good models [12, 13] for the electronic properties and the structural flexibility of silica surfaces [14, 15]. With cyclohexyl ligands instead of H the $O=VSi_7O_{12}H_7$ structure has also been synthesized [16].

Gas-phase studies have used VO_2^+, VO_3^-, $V_3O_7^+$ and $V_4O_{10}^+$ ions [17–20]. While the former three are closed shell V^V systems, the latter is a radical cation derived from a neutral V^V system [9]. Trimethoxyvanadate, $O=V^V(OCH_3)_3$ and its radical cation have also been investigated [10, 21].

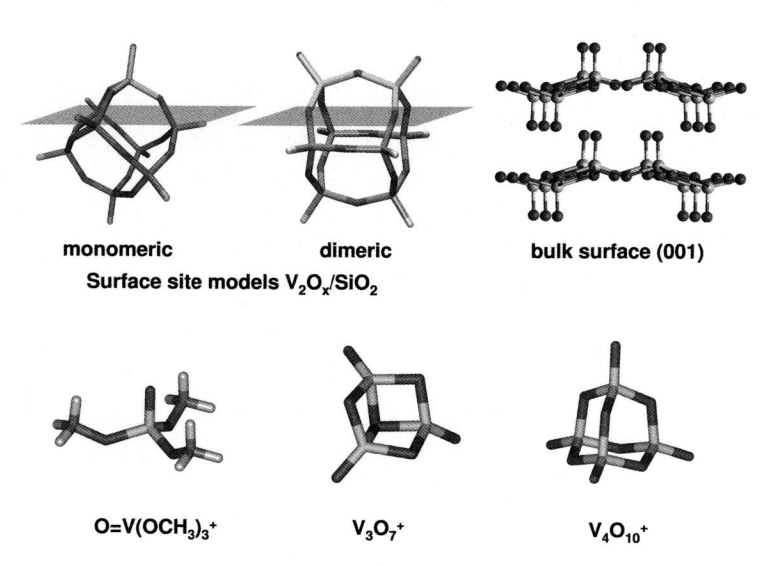

Figure 10.1 Models for monomeric and dimeric V_2O_5 species on silica surfaces, the (001) surface of V_2O_5 crystals (top row) and different gas phase species

Table 10.1 Energies of different modes of methanol binding onto $(O=)_n M(O-)_m$ transition metal oxide sites.

	Reaction	Energy (kJ mol^{-1})
Ia	$O=VCl_3 + CH_3OH \rightarrow HO(HOCH_2)VCl_3$	131 [22]
Ib	$O=VCl_3 + CH_3OH \rightarrow HO(CH_3O)VCl_3$	31 [22]
II	$(O=)_2MoCl_2 + CH_3OH \rightarrow HO(CH_3O)(O=)MoCl_2$	−38 [23]
III	$O=VO_2^- + CH_3OH \rightarrow HO(CH_3O)VO_2^-$	−244 [20]
IV	$O=V(O-)_3 + CH_3OH \rightarrow HO(CH_3O)V(O-)_3$	49 [24]
V	$O=V(O-)_3 + CH_3OH \rightarrow O=V(OCH_3 \cdot HO-)(O-)_2$	−32/−42 [10]

10.3
Oxidation of Methanol to Formaldehyde

The first step of the oxidation of methanol is binding of the methanol substrate onto the catalyst. There are several possibilities for this reaction (Table 10.1, I–V). Addition of the OH group onto the V=O site (Ib) is clearly preferred over CH addition (Ia), as shown for the O=VCl$_3$ molecule [22]. For $(O=)_2$Mo sites OH addition is an exoenergetic process (II) [23], and even more so in the case of the gas phase anion VO$_3^-$ (III) [20]. The reason for the strong binding onto VO$_3^-$ is probably the low coordination of vanadium, which changes from three to four on binding methanol. For O=V(O–)$_3$ sites anchored on silica, OH addition is endoenergetic on the O=V bond (IV, coordination changes from four to five), but exoenergetic onto the V–O(–Si) bond (V) [10].

As result of OH addition onto the V–O(–Si) bond, there is a methoxy group attached to the same V as the vanadyl group (1) and neighbored to a HO–Si group on the surface (Scheme 10.2). Oxidation occurs by H transfer from the C–H bond to the O=V^V group. The transition structure, Ts1/2, is tentatively described by an HO and an OCH$_2^{\bullet}$ group, both attached to a V^{IV}(d^1) site, and has diradical character [10]. The reaction involves decoupling of the electron

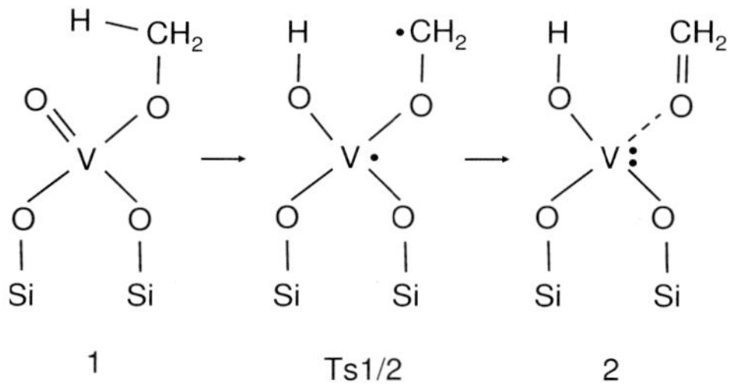

Scheme 10.2

pairs in the C–H bond and in the (V=O)π bond. Its DFT description requires a broken symmetry solution. Closed shell treatments yield much higher barriers, which confirms that the reaction involves H transfer, not hydride transfer. Oxidative dehydrogenation is completed by splitting the V–O(CH$_2$·) bond, which yields a V^{III} site with two electrons in d-orbitals (triplet) and formaldehyde, **2**.

When the two (V–)O–Si bonds in **1**, which anchor the vanadium oxide species on the silica surface, are replaced by methyl groups the trimethoxyvanadate molecule **1a** is obtained, which can be considered a gas-phase model of the surface site (Scheme 10.3).

Indeed, DFT calculations show that the intrinsic barriers for formaldehyde formation differ by a few kJ mol^{-1} only (Table 10.2). Experimentally, this reaction can be observed on photo-ionization of **1a** [21]. Figure 10.2 shows how the reaction energy profile changes from the neutral system to the radical cation. Ionization of **1a** requires much more energy (9.25 eV) than ionization of the product **2a** (6.83 eV) because **2a** features vanadium(III) with two electrons in high lying d-orbitals. This turns the endoenergetic reaction with a late transition structure for the neutral molecule into an exoenergetic one with an early transition structure for the radical cation. Consequently, the energy barrier is much lower (80 kJ mol^{-1}) than in the neutral molecule. This is confirmed by the measured barrier of 52 ± 11 kJ mol^{-1} [21]. For the neutral molecule CCSD(T) calculations have shown that the DFT barriers obtained with B3LYP are still too high (154 compared with 131 kJ mol^{-1}) [10]. The best B3LYP estimate for the radical cation is 72 kJ mol^{-1} (Table 10.2).

Table 10.2 Intrinsic and apparent energy barriers for the oxidative dehydrogenation of methanol by gas phase and surface vanadium oxide species.

System	Energy barrier (kJ mol^{-1})			
	Intrinsic	Best estimate	Apparent	Observed
O=V(OCH$_3$)$_3$	147 [10]	131[a] [10]		52 ± 11 [21]
O=V(OCH$_3$)$_3{}^+$	80 [10]	72 [10]		≤0 [20]
CH$_3$OH + O=VO$_2{}^-$	157 [20]		−87 [20]	132 [25]
CH$_3$OH + O=V(OSi–)$_3$	154 [10]	137 [10]	98 [10]	82 ± 10 [26]

a) CCSD(T) calculations.

10.4 Oxidative Dehydrogenation of Alkanes on Supported Transition Metal Oxide Catalysts

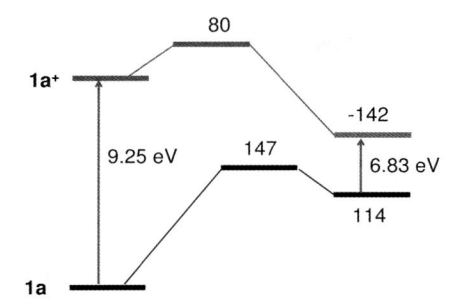

Figure 10.2 Change of the reaction energy profile for formaldehyde formation on ionization of O=V(OCH$_3$)$_3$ (**1a**).

Comparison with experiments for the surface sites should consider both intrinsic barriers and surface binding energies. If we use independent measurements to estimate the heat of adsorption (-50 kJ mol^{-1}, cf. the calculated binding energy of -42 kJ mol^{-1} in Table 10.1) and the apparent barrier (82 ± 10 kJ mol^{-1}), we get an experimental estimate of 137 ± 10 kJ mol^{-1} for the intrinsic barrier [10]. A similar value (132 kJ mol^{-1}) has been observed for the ODH of ethanol on V$_2$O$_5$/SiO$_2$ catalysts [25]. Both are in close agreement with the best B3LYP estimate (Table 10.2) [10].

The reaction of methanol with the O=VO$_2^-$ anion in the gas phase is an example of the large effect of the substrate binding. B3LYP calculations yield -244 kJ mol^{-1} for the OH addition on the V=O bond (Table 10.1, III) resulting in a negative apparent barrier [20]. In the mass spectrometer, besides [VO$_2$(OH)(OCH$_3$)]$^-$, [V, O$_3$, H$_2$]$^-$ (CH$_2$O loss), and [VO$_2$(OCH$_2$)]$^-$ (H$_2$O loss) have been observed [20]. The intrinsic barrier for converting [V^{V}O$_2$(OH)(OCH$_3$)]$^-$ into [V^{III}O(OH)$_2$(OCH$_2$)]$^-$ is 157 kJ mol^{-1} [20], very close to the results for **1** and **1a**. At the end of this chapter we show that the similarities and differences between the reactivity of different vanadium oxide species can be understood by looking at their reducibility, defined by:

$$\frac{1}{2}H_2 + O=M(d^0) \rightarrow HO-M(d^1) \tag{1a}$$

$$\frac{1}{2}H_2 + [O=M(d^0)]^{\cdot +} \rightarrow [HO-M(d^0)]^+ \tag{1b}$$

10.4
Oxidative Dehydrogenation of Alkanes on Supported Transition Metal Oxide Catalysts

In the catalytic literature it has been suggested that propane reacts with O=V(O−)$_3$ surface sites by C−H addition onto the O=V bond [27, 28]. B3LYP calculations show that this process is highly endoenergetic and has activation barriers of over 200 kJ mol^{-1} (Table 10.3) [11]. CH addition on the V–O–(Si) bond is similarly unfavorable. The same is true for the neutral MO$_3$O$_9$ cluster (coordination four) [29]. For the V$_3$O$_7^+$ and VO$_2^+$ gas phase ions the barriers are much lower and the process is exoenergetic [17, 18]. This is ascribed to the low coordination (three and two, respectively) at the reaction site. Owing to a substantial ion–molecule interaction (127 and 191 kJ mol^{-1} for V$_3$O$_7^+$ and VO$_2^+$, respectively) the apparent barrier almost vanishes for V$_3$O$_7^+$ and is negative for VO$_2^+$.

Hydrogen abstraction by the oxygen atom of the vanadyl oxo bond (transition metal oxo bond, in general) is found to be the initial step in the oxidative dehydrogenation of propane by O=V(O–)$_3$ sites on silica surfaces [11]. The diradical intermediate that is reached via the diradicaloid transition structure consists of a propyl radical attached to a HO-V(O–)$_3$ site and has similar energies in the singlet and triplet states (Figure 10.3). Spin-crossing to the triplet state most likely occurs in this step.

For the second hydrogen abstraction, many possible pathways lead to propene formation [11]. First, we consider cases in which a second hydrogen abstraction occurs at the same site as the first one, yielding H$_2$O·V^{III}(O–)$_3$ and propene (Figure 10.3). This product state can also be reached via an isopropanol intermediate attached to a V^{III} surface site. The latter forms by a propyl radical rebound mechanism. Oxygenation of alkanes to alcohols by the ferryl (O=Fe) groups in enzyme cytochrome P450 occurs by a similar rebound mechanisms [3]. In our case, this is not the main channel, in particular not at high temperatures. Moreover, the surface alcohol can not only eliminate propene but also undergo further oxidation to ketones [11].

Second, the i-propyl radical can desorb from the biradical intermediate, C$_3$H$_7$·HOV(O–)$_3$, and re-adsorb at a fresh unreacted O=V(O–)$_3$ site (Figure 10.3). At such a site the second hydrogen abstraction leading to propene has a

Table 10.3 Barriers and reaction energies of propane binding onto transition metal oxide sites. Results in parenthesis refer not to the separated reactants but to the propane–vanadium oxide ion encounter complex.

	Reactant	Product	Barrier (kJ mol^{-1})	Energy (kJ mol^{-1})
1	VO$_2^+$	HO(C$_3$H$_7$)VO$^+$	−76 (115) [17]	−273 (−82) [17]
2	V$_3$O$_7^+$	[HO(i-C$_3$H$_7$)VO$_2$V$_2$O$_4$]$^+$	13 (120)[a]	−166 (−59)[a]
3	Mo$_3$O$_9$	HO(i-C$_3$H$_7$)(O)MoO$_2$Mo$_2$O$_5$	210[b]	89[b]
4a	O=V(O–)$_3$	HO(i-C$_3$H$_7$)V(O–)$_3$	263 [11]	168 [11]
4b	O=V(O–)$_3$	O=V((i-C$_3$H$_7$)·HO–)(O–)$_2$	225 [11]	172 [11]

a) ΔH(0 K) [18].
b) ΔH(688 K) [29].

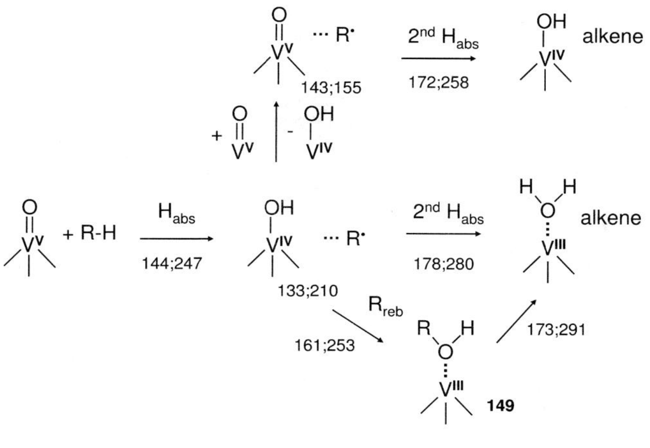

10.4 Oxidative Dehydrogenation of Alkanes on Supported Transition Metal Oxide Catalysts

Figure 10.3 Oxidative dehydrogenation of propane on monomeric vanadium(v) oxide sites supported on silica. The numbers are energies at 0 K followed by the standard Gibbs free energies at 800 K for transition structures and intermediates (in kJ mol^{-1}).

lower barrier, and the difference compared with the second hydrogen abstraction occurring at the same site increases with increasing temperature (the Gibbs free energies differ by 22 kJ mol^{-1}). Desorption of the propyl radical requires 10 kJ mol^{-1} only, and is favored by the entropy gain of the radical when released into the gas phase. The Gibbs free energy change at 800 K is −95 kJ mol^{-1}.

Microkinetic modeling of the full kinetic network with all steps included [11, 30] shows that the surface propyl radical is the central intermediate and that its formation is the rate-determining step. This surprising finding relies on the assumption that reoxidation of HOV^{IV}(O−)$_3$ sites to O=V^V(O−)$_3$ is fast (Mars–van Krevelen mechanism [31]), which is supported by isotopic tracer studies [32]. That the initial hydrogen abstraction is the rate-determining step can be rationalized with a reduced kinetic scheme, which includes the formation of the radical, k_1, and its consumption by recombination, k_{-1}, or by any reaction of the propyl radical with O=V^V(O−)$_3$ and HOV^{IV}(O−)$_3$ sites, yielding propene, k_2 and k_3, respectively:

$$\text{O=V}^V(\text{O-})_3 + \text{C}_3\text{H}_8 \underset{k_{-1}}{\overset{k_1}{\rightleftarrows}} \text{HO-V}^{IV}(\text{O-})_3 + \text{C}_3\text{H}_7^\bullet \quad (2)$$

$$\text{O=V}^V(\text{O-})_3 + \text{C}_3\text{H}_7^\bullet \overset{k_2}{\rightarrow} \text{HO-V}^{IV}(\text{O-})_3 + \text{C}_3\text{H}_6 \quad (3)$$

$$\text{HO-V}^{IV}(\text{O-})_3 + \text{C}_3\text{H}_7^\bullet \overset{k_3}{\rightarrow} \text{H}_2\text{O} \cdot \text{V}^{III}(\text{O-})_3 + \text{C}_3\text{H}_6 \quad (4)$$

For a stationary concentration of propyl radicals in the gas phase we obtain Eq. (5) for propene formation if we assume that $x = [\text{HOV}^{IV}(\text{O-})_3]/[\text{O=V}^V(\text{O-})_3] \ll 0$ (fast reoxidation):

$$\frac{d[C_3H_6]}{dt} = \frac{(k_2 + k_3 x) \cdot k_1[O=V^V(O-)_3][C_3H_8]}{(k_{-1} + k_3)x + k_2} \approx k_1[O=V^V(O-)_3][C_3H_8]$$
(5)

The high energy and Gibbs free energy barriers of different possible elementary processes for the second H abstraction, represented by k_2, and k_3, will only affect the stationary concentration of the propyl radicals, not the rate of the propene formation.

Observed rates and activation barriers are compatible with the conclusion that the first hydrogen abstraction is the rate-determining step. For highly dispersed vanadium oxide on mesoporous silica (MCM-41, 0.5 mol V nm^{-2}) and on amorphous silica (3 mol V nm^{-2}) turnover frequencies for propene formation of 0.01 (748 K) and 0.3×10^{-3} s^{-1} (623 K) have been measured [33, 34], which compare with calculated values of 0.26 (750 K) and 1.7×10^{-3} s^{-1} (600 K) [11]. The observed (apparent) Arrhenius activation energy of 122 ± 20 kJ mol^{-1} [33, 35] is in close agreement with the estimate from B3LYP calculations of 123 ± 5 kJ mol^{-1}, which includes an experimental estimate (37 ± 5 kJ mol^{-1}) [36] for the van der Waals binding of propane onto the catalyst surface.

For the vanadium oxide surface species, hydrogen abstraction is much more favorable than alternative mechanisms such as C–H addition. Hydrogen abstraction by the metal oxo bond seems to be the rate-determining step also for oxidative dehydrogenation of propane on many other supported transition metal oxide catalysts, for which similar energy barriers, between 99 and 126 kJ mol^{-1}, have been reported (vanadium, molybdenum and tungsten oxide on zirconia [37], vanadium oxide on alumina [38]). For example, for MO$_x$/ZrO$_2$ an Arrhenius activation energy of 117 kJ mol^{-1} has been measured [37], while a computational study (B3LYP) of the reaction of propane with Mo$_3$O$_9$ found the lowest activation barrier (135 kJ mol^{-1}, ΔH at 688 K) for hydrogen abstraction from the methylene group among many other possible mechanisms, including C–H addition (Table 10.3) [29]. The kinetic study [32] on MO$_x$/ZrO$_2$ provided further support for the mechanism shown in Figure 10.3, in particular for the assumption of a fast reoxidation of reduced sites, $x = [\text{HOV}^{IV}(O-)_3]/[O=V^V(O-)_3] \ll 0$ (Eq. 5), which makes hydrogen abstraction irreversible (the backward reaction in Eq. 1 is suppressed). If this were not the case, rapid isotope scrambling should be observed when using a 1:1 mixture of propane and fully deuterated propane as fee, but no mixed isotopomers C$_3$H$_n$D$_{8-n}$ have been detected [32].

10.5
C–H Activation of Alkanes by Transition Metal Oxide Species in the Gas Phase

The reaction of propane with vanadium oxide has also been examined in the gas phase using V$_3$O$_7^+$, which is a closed shell ion with vanadium in the oxidation state +v. B3YP calculations predict an intrinsic energy barrier of 102–108 kJ mol^{-1} (Table 10.4) for H abstraction, which results in a vanishing apparent barrier (-5–6 kJ mol^{-1}). Similar results are obtained for C–H addition (see above, see also Table 10.3). While the 2 + 2 C–H addition involves only closed shell singlet species, H abstraction requires decoupling of the electron pair and proceeds via a biradicaloid transition structure, yielding a biradical intermediate: C_3H_8 + $[O=VO_2V_2O_4]^+ \rightarrow C_3H_7{\cdot}[HOVO_2V_2O_4]^+$.

Within DFT, the transition structure can only be localized on the "broken symmetry" potential energy surface, from which the energy of the open shell singlet state can be estimated using the triplet state energy for spin-projection (see Refs. [10, 11] for details). In the mass spectrum, only the mass of the $C_3H_8{\cdot}V_3O_7^+$ encounter complex was seen, whereas masses belonging to $[V_3O_7H_2]^+$ or to $[V_3O_6C_3H_6]^+$, which would indicate oxidative dehydrogenation after loss of C_3H_6 or H_2O, were not found. This came as a surprise in view of the near-zero energy barriers. The explanation is provided by entropy. As indicated by the Gibbs free energies in Table 10.4, dissociation of the encounter complex into the reactants ($+63$ kJ mol^{-1}) is entropically favored compared with crossing either of the transition structures ($+122$ and $+92 \ldots +103$ kJ mol^{-1}).

We may ask if oxidative dehydrogenation would be observed if a more reactive hydrocarbon were used. This is indeed the case. If the experiment is repeated with 1-butene instead of propane, $[V_3O_7H_2]^+$ concomitant with formation of neutral butadiene is observed [18], together with the encounter complex and other minor products.

Table 10.4 Energies and Gibbs free energies (298 K), for the transition structures and intermediates for two reactions of 1-butene and propane with $V_3O_7^+$: CH addition and H abstraction [18].

Label (propane)	E_0 a) (kJ mol^{-1})	G_{298} (kJ mol^{-1})	Label (1-butene)	E_{ZP} (kJ mol^{-1})	G_{298} (kJ mol^{-1})
$C_3H_8 + V_3O_7^+$	0	0	$C_4H_8 + V_3O_7^+$	0	0
$C_3H_8{\cdot}V_3O_7^+$	-107	-63	$C_4H_8{\cdot}V_3O_7^+$	-199	-152
TS (CH addition)	13	59	TS (CH addition)	-108	-56
$[HO(i\text{-}C_3H_7)VO_2V_2O_4]^+$	-166	-119	$[HO(C_4H_7)VO_2V_2O_4]^+$	-210	-161
TS (H abstraction) b)	$6^{bs}, -5^s$	$44^{bs}, 33^s$	TS (H abstraction) b)	$-61^{bs}, -65^s$	$-20^{bs}, -24^s$
$C_3H_7{\cdot}[HOVO_2V_2O_4]^+$	$-70^{bs}, -57^s$ -97^t	$-35^{bs}, -22^s$ -52^t	$C_4H_7{\cdot}[HOVO_2V_2O_4]^+$	-166^t	-121^t

a) Energy at 0 K, including zero point vibrational contributions.
b) Superscripts bs and s refer to the broken symmetry and (spin-projected) open-shell singlet energies, respectively, and t to the triplet state.

The B3LYP data for 1-butene in Table 10.4 are in line with experimental results. The substrate binding is so strong and the apparent energy barriers are so much below zero that they more than outweigh the entropy effect of creating two gas-phase species. The Gibbs free energy required to dissociate the complex (152 kJ mol^{-1}) is much larger than the one to cross the barrier (96 kJ mol^{-1}). The reaction of 1-butene with $V_3O_7^+$ is also a case where C–H addition is more favorable than H abstraction.

Instead of a more reactive hydrocarbon we can also use a more reactive vanadium oxide species. The V_4O_{10} cluster, which has the same cage structure as P_4O_{10}, is a closed shell system with vanadium in the +v oxidation state (empty d-shell). On ionization, the $V_4O_{10}^+$ radical cation (Figure 10.1) is produced, which has been predicted to be particularly reactive for oxygen loss [9]. Recent experiments have shown that it is so reactive that it can abstract H even from the least reactive alkane, CH_4 [19]. Without any barrier, a methyl radical is formed together with $[V_4O_{10}H^+]$, which is detected in the mass spectrometer. On ionization of V_4O_{10} the electron is removed from the O=V π bond, leaving a highly reactive ˙O–V$^{(+)}$ radical site at the terminal oxygen. After H attachment a closed shell system with a very stable (V)OH group is formed:

$$H_3C-H + {}^{\bullet}O-V^{(+)}(O-)_3 \to H_3C^{\bullet} + HO-V^{(+)}(O-)_3 \quad \Delta E_0 = -88 \text{ kJ mol}^{-1}$$

Such highly reactive ˙O–M$^{(+)}$ oxygen radical sites will be only formed on ionization of a transition metal oxide, if it is in its highest oxidation state. If not, there are higher lying occupied d-states from which the electron can be removed and the transition metal oxo bonds remain double bonds.

This explains also why MoO_3^+ abstracts hydrogen from CH_4, while $Mo(d^1)^VO_2^+$ and $Mo(d^3)^{III}O^+$ do not react at all [39], and also why OsO_4^+ abstracts hydrogen from CH_4, but $Os^{VII}O_3^+(d^1)$ does not react (the lower osmium oxides yield different products) [40]. B3LYP calculations of the reaction energies confirm the experimental observations [41]:

$$H_3C-H + O=Mo^{(+)}=O \to H_3C^{\bullet} + HO-Mo^{(+)}=O \quad \Delta E_0 = +70 \text{ kJ mol}^{-1}$$

$$H_3C-H + {}^{\bullet}O-Mo^{(+)}O_2 \to H_3C^{\bullet} + HO-Mo^{(+)}O_2 \quad \Delta E_0 = -81 \text{ kJ mol}^{-1}$$

$$H_3C-H + {}^{\bullet}O_2Os^{(+)}O_2 \to H_3C^{\bullet} + HO-Os^{(+)}O_3 \quad \Delta E_0 = -38 \text{ kJ mol}^{-1}$$

In ˙$O_2Os^{(+)}O_2$ the spin-density is delocalized over two oxygen sites.

In the ionic picture of oxides the ˙O–M$^{(+)}$ active site corresponds to a O˙$^-$ radical anion. This is also the active species in Li-doped MgO that is used as solid catalyst for the oxidative coupling of CH_4 [42]. The reaction is initiated by generation of methyl radicals at the surface according to [43]:

$$H_3C-H + [O^{\bullet-}Li^+]_{MgO} \to H_3C^{\bullet} + [HO^-Li^+]_{MgO}$$

10.6 Conclusions

Above we have presented experimental and computational evidence that hydrogen abstraction by transition metal oxo bonds is the initial step in the C–H activation by oxides:

$$R-H + O=M(d^n) \rightarrow R^{\bullet} + HO-M(d^{n+1}) \qquad (6)$$

with d^n denoting the occupation of non-bonding d-states on the transition metal site. In the oxidative dehydrogenation of alkanes, the alkyl radical can directly form alkenes by losing another hydrogen atom, but it can also form an alcohol by a rebound mechanism, as suggested for enzyme P450, see, for example, Ref. [3] and references therein:

$$R^{\bullet} + HO-M(d^{n+1}) \rightarrow H(R)O\cdot M(d^{n+2}) \qquad (7)$$

In the oxidation of alcohols they bind first as alkoxides to the surface site before the metal oxo bond abstracts a hydrogen atom. Oxidation is completed when the aldehyde is eliminated.

Many other examples for H abstractions by different types of oxide catalysts can be found in the review of Limberg [6]. A specific example is the permanganate oxidation of alkanes by hydrogen abstraction and rebound [44].

The ability of different transition metal oxide species (solid surface, gas phase etc.) can be characterized by the strength of the H–O(M) bond formed in comparison with the strength of a model substrate such as H_2:

$$\frac{1}{2}H_2 + O=M(d^0) \rightarrow HO-M(d^1) \qquad (1a)$$

$$\frac{1}{2}H_2 + [O=M(d^0)]^{\bullet+} \rightarrow [HO-M(d^0)]^{+} \qquad (1b)$$

Table 10.5 shows the data for species discussed in this chapter. The reactivity of vanadyl groups in supported vanadium oxide species, $O=V(O-)_3$, and in the $O=V(OCH_3)_3$ molecules is similar. Calculations for the bulk V_2O_5 surface, $(V_2O_5)_s$, suggest a much higher reactivity than for supported vanadium oxide species. The reason for this is a special relaxation mechanism involving different layers of the crystal (Figure 10.1) [9, 46].

When gas-phase studies are made on active species "isolated" in the gas phase, we need to take into account two important differences: (a) Gas phase experiments need charged species, anions or cations, whose reactivity may differ from the one of neutral oxide species on surfaces. (b) Usually in the gas phase only reaction channels can be observed that have Gibbs free energies below the entrance channel. Because of the entropy loss connected with the binding of the substrate molecule on the oxide ion, usually only reactions can be observed with a negative apparent energy barrier.

With respect to the electronic properties of the active oxide we can again distinguish two cases:

(a) Closed shell ions ($V_3O_7^+$ and VO_3^-) may have barriers comparable to neutral systems in the redox step (Table 10.5), but bind the substrate much more strongly due to ion–molecule interactions and/or lower coordinated transition metal sites. This effect is large enough for VO_3^- to make the oxidation of methanol to formaldehyde possible [20], and for VO_2^+ to enable the reaction with propane [17]. For $V_3O_7^+$ it is sufficient to bring about oxidative dehydrogenation of 1-butene, but not of propane [18]. Ions with a partially filled d-shell show a similar behavior to the closed shell species, for both of them the d-occupation increases on reduction. Examples are MoO_2^+ and MoO^+, which do not react with CH_4 while MoO_3^+ does [39].

(b) Very reactive radical cations are formed on ionization of parent (neutral) oxide species in their highest oxidation state (empty d-shell). The most reactive species, $^\bullet O-M^{(+)}$, are formed when an electron is removed from the transition metal oxo bond, $O=M(d^0)$. Examples are OsO_4^+, $V_4O_{10}^+$ and MoO_3^+ which all are able to activate the C–H bond in CH_4 [19, 39, 40]. Among them $V_4O_{10}^+$ is the only polynuclear oxide species for which this reaction has been observed.

We conclude that neutral transition metal oxide species supported on surfaces and the corresponding ions isolated in the gas phase can show very different abilities to activate a C–H bond. However, these differences can be understood on the basis of the same mechanism, i.e., hydrogen abstraction preceded by the binding of the substrate on the respective catalyst.

Table 10.5 Energies for reduction by hydrogen (Eq. 1) of metal–oxo bonds in different environments [41].

Metal–oxo species	Energy (kJ mol^{-1})	
	B3LYP	PBE
$O=V(OCH_3)_3$	−19	44
$OVSi_7O_{12}H_7$	−48	19
VO_3^-	–	−69
$V_3O_7^+$	−160 [24]	−110
MoO_2^+	−189	–
$[O=V(OCH_3)_3]^+$	−235	−204
$(V_2O_5)_s$	–	−274 [45]
OsO_4^+	−303	–
$V_4O_{10}^+$	−310	−297
MoO_3^+	−344	

Acknowledgment

I thank my co-authors of the original publications on which this chapter is based and all my colleagues with whom I enjoyed cooperation in the Center of Collaborative Research 546 "Transition metal oxide aggregates" supported by the German Research Foundation (DFG).

References

1 J. Haber, in *Handbook of Heterogeneous Catalysis*, eds. G. Ertl, H. Knözinger, J. Weitkamp, VCH, Weinheim, 1997, Vol. 5, p. 2253.
2 M. Muhler, in *Handbook of Heterogeneous Catalysis*, eds. G. Ertl, H. Knözinger, J. Weitkamp, VCH, Weinheim, 1997, Vol. 5, p. 2274.
3 S. Shaik, S. P. d. Visser, F. Ogliaro, H. Schwarz, D. Schröder, *Curr. Opin. Chem. Biol.* 2002, **6**, 556.
4 S. Guimond, M. Abu Haija, S. Kaya, J. Lu, W. J. S. Shaikhutdinov, H. Kuhlenbeck, H.-J. Freund, J. Döbler, J. Sauer, *Top. Catal.* 2006, **38**, 117.
5 D. K. Böhme, H. Schwarz, *Angew. Chem., Int. Ed.* 2005, **44**, 2336.
6 C. Limberg, *Angew. Chem., Int. Ed.* 2003, **42**, 5932.
7 B. M. Weckhuysen, D. E. Keller, *Catal. Today* 2003, **78**, 25.
8 N. Magg, B. Immaraporn, J. B. Giorgi, T. Schroeder, M. Bäumer, J. Döbler, Z. Wu, E. Kondratenko, M. Cherian, M. Baerns, P. C. Stair, J. Sauer, H.-J. Freund, *J. Catal.* 2004, **226**, 88.
9 J. Sauer, J. Döbler, *Dalton Trans.* 2004, **19**, 3116.
10 J. Döbler, M. Pritzsche, J. Sauer, *J. Am. Chem. Soc.* 2005, **127**, 10861.
11 X. Rozanska, R. Fortrie, J. Sauer, *J. Phys. Chem. C* 2007, **111**, 6041.
12 J. Sauer, J.-R. Hill, *Chem. Phys. Lett.* 1994, **218**, 333.
13 B. Civalleri, E. Garrone, P. Ugliengo, *Chem. Phys. Lett.* 1998, **294**, 103.
14 G. Calzaferri, R. Hoffmann, *J. Chem. Soc., Dalton Trans.* 1991, 917.
15 A. M. Bieniok, H.-B. Bürgi, *J. Phys. Chem.* 1994, **98**, 10735.
16 F. J. Feher, J. F. Walzer, *Inorg. Chem.* 1991, **30**, 1689.
17 M. Engeser, M. Schlangen, D. Schröder, H. Schwarz, T. Yumura, K. Yoshizawa, *Organometallics* 2003, **22**, 3933.
18 S. Feyel, D. Schröder, X. Rozanska, J. Sauer, H. Schwarz, *Angew. Chem., Int. Ed.* 2006, **45**, 4677; *Angew. Chem.* **118**, 4793–4797.
19 S. Feyel, J. Döbler, D. Schröder, J. Sauer, H. Schwarz, *Angew. Chem., Int. Ed.*, 2006, **45**, 4681; *Angew. Chem.*, **118**, 4797–4801.
20 T. Waters, A. G. Wedd, R. A. J. O'Hair, *Chem. Eur. J.* 2007, **13**, 8818.
21 D. Schröder, J. Loos, M. Engeser, H. Schwarz, H.-C. Jankowiak, R. Berger, R. Thissen, O. Dutuit, J. Döbler, J. Sauer, *Inorg. Chem.* 2004, **43**, 1976.
22 L. Q. Deng, T. Ziegler, *Organometallics* 1996, **15**, 3011.
23 J. N. Allison, W. A. Goddard III, *J. Catal.* 1985, **92**, 127.
24 X. Rozanska, J. Sauer 2006, unpublished calculations.
25 T. Feng, J. M. Vohs, *J. Phys. Chem. B* 2005, **109**, 2120.
26 G. Deo, I. E. Wachs, *J. Catal.* 1994, **146**, 323.
27 S. T. Oyama, *J. Catal.* 1991, **128**, 210.
28 T. Blasco, J. M. L. Nieto, *Appl. Catal. A* 1997, **157**, 117.
29 G. Fu, X. Xu, X. Lu, H. Wan, *J. Phys. Chem. B* 2005, **109**, 6416.
30 X. Rozanska, R. Fortrie, J. Sauer, 2007, in preparation.
31 P. Mars, D. W. van Krevelen, *Chem. Eng. Sci. Spec. Suppl.* 1954, **3**, 41.
32 K. D. Chen, E. Iglesia, A. T. Bell, *J. Phys. Chem. B* 2001, **105**, 646.
33 E. Kondratenko, M. Cherian, M. Baerns, D. Su, R. Schlögl, X. Wang, I. E. Wachs, *J. Catal.* 2005, **234**, 131.

34 H. Tian, E. I. Ross, I. E. Wachs, *J. Phys. Chem. B* 2006, **110**, 9593.
35 E. V. Kondratenko, M. Baerns, 2006, personal communication.
36 A. Kämper, A. Auroux, M. Baerns, *Phys. Chem. Chem. Phys.* 2000, **2**, 1069.
37 K. Chen, A. T. Bell, E. Iglesia, *J. Phys. Chem. B* 2000, **104**, 1292.
38 M. D. Argyle, K. Chen, A. T. Bell, E. Iglesia, *J. Catal.* 2002, **208**, 139.
39 I. Kretzschmar, A. Fiedler, J. N. Harvey, D. Schroder, H. Schwarz, *J. Phys. Chem. A* 1997, **101**, 6252.
40 K. K. Irikura, J. L. Beauchamp, *J. Am. Chem. Soc.* 1989, **111**, 75.
41 J. Sauer, 2007, default TZVP basis set of the Turbomole library with default effective core potentials for Mo and Os; calculations for this chapter.
42 J. H. Lunsford, *Angew. Chem., Int. Ed.*, 1995, **34**, 970.
43 C. R. A. Catlow, S. A. French, A. A. Sokol, J. M. Thomas, *Philos. Trans. Royal Soc. Ser. A*, 2005, **363**, 913.
44 T. Strassner, K. N. Houk, *J. Am. Chem. Soc.*, 2000, **122**, 7821.
45 M. V. Ganduglia-Pirovano, J. Sauer, 2007, calculations for this chapter.
46 M. V. Ganduglia-Pirovano, J. Sauer, *Phys. Rev. B*, 2004, **70**, 045422.

11
Mechanism of Ru- and Mo-catalyzed Olefin Metathesis

Andrea Correa, Chiara Costabile, Simona Giudice, and Luigi Cavallo

11.1
Introduction

Recent decades have seen olefin metathesis emerging as one of the most useful and versatile tools for the synthesis of C=C bonds [1–6]. Indeed, this reaction allows the synthesis of complicated molecules in fewer and simpler steps relative to classical synthetic approaches, and it is particularly suited for the synthesis of medium-size rings, which are otherwise difficult to close. The first family of effective catalysts has been developed by Schrock and coworkers [3–5]. Schrock's catalysts are Mo alkylidene complexes such as **1** and **2** (Figure 11.1). They present very high reactivity and are particularly useful for asymmetric olefin metathesis. The main drawbacks are the poor tolerance to other functional groups, and air and moisture sensitivity. The second family of effective catalysts has been developed by Grubbs and coworkers [1, 2]. Grubbs' catalysts are Ru carbene complexes such as **3** and **4** (Figure 11.1). They benefit from air and moisture resistance, and

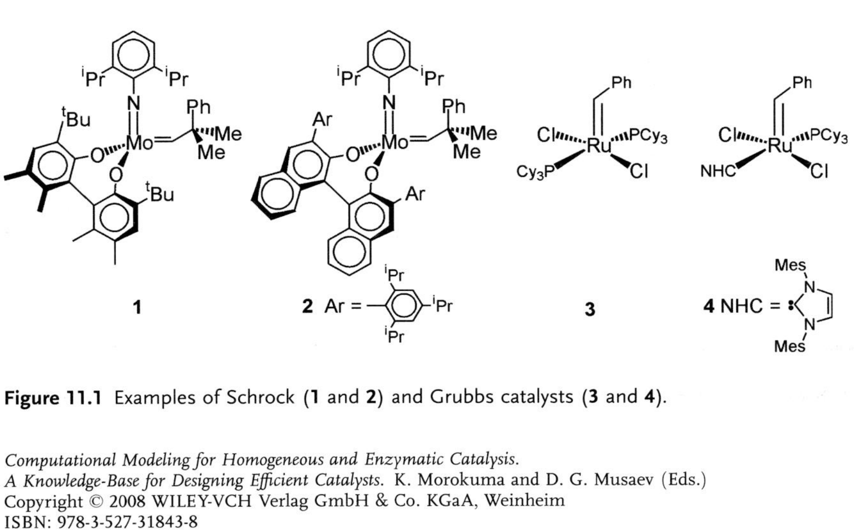

Figure 11.1 Examples of Schrock (**1** and **2**) and Grubbs catalysts (**3** and **4**).

Computational Modeling for Homogeneous and Enzymatic Catalysis.
A Knowledge-Base for Designing Efficient Catalysts. K. Morokuma and D. G. Musaev (Eds.)
Copyright © 2008 WILEY-VCH Verlag GmbH & Co. KGaA, Weinheim
ISBN: 978-3-527-31843-8

are highly tolerant to other functional groups. The main drawbacks are the somewhat low activity (compared with the Mo-based catalysts), and the asymmetric version of this catalyst is not well explored yet. In short, combining the advantages of the two families of catalysts, the organic chemist is offered a large set of choices when a new synthetic strategy has to be devised.

The development of such effective catalysts was considerably accelerated when fundamental mechanistic studies elucidated the real nature of the active species, and offered a more or less detailed picture of the overall mechanism [7–15].

It is now well established that the mechanism of olefin metathesis is that proposed by Chauvin [16]. In the case of ring-closing metathesis of a diene such as 1,6-hexadiene, the main steps are those illustrated in the box of Scheme 11.1. After the first C=C bond has reacted, coordination of the second C=C bond leads to the coordination intermediate, which is followed by the four-center transition state that collapses into the metallacyclobutane intermediate. Similar steps lead to release of the final five-membered ring product.

Although the basics of the mechanism have been clarified, a full understanding of these catalysts has not been achieved yet, and a clear structure–function relationship is not established as well. An understanding of the details of the mechanisms operative with both classes of catalysts is key to designing both better performing catalysts and catalysts with new scope.

In this respect, computational chemistry techniques are a very powerful tool for shedding light on this class of reactions, which can contribute to a detailed understanding of the laws that rule these systems. Thus, unsurprisingly, many theoretical studies have been performed in the field. However, while the Ru-based catalysts have been already investigated at a good level (the different activity of the first and second generation of Grubbs' catalyst has been rationalized, as

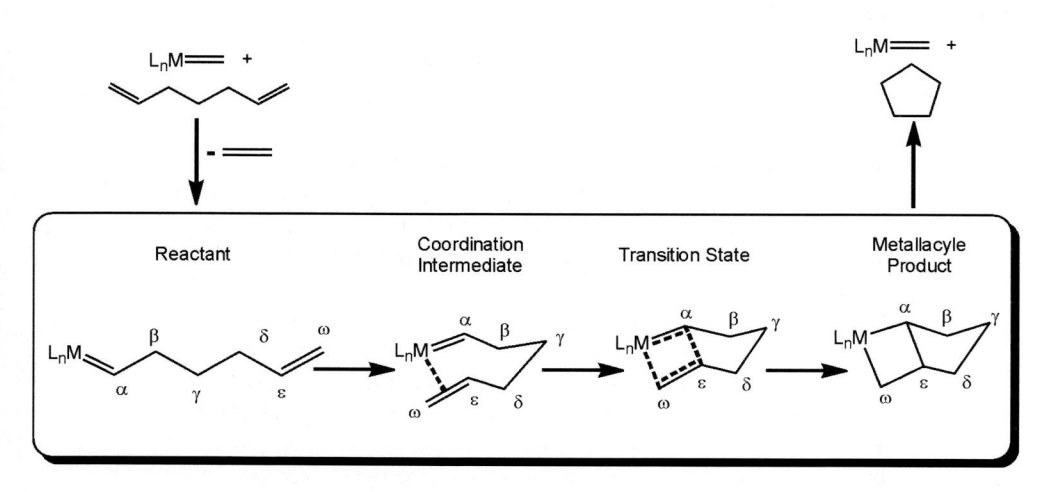

Scheme 11.1 Mechanism of ring-closing metathesis.

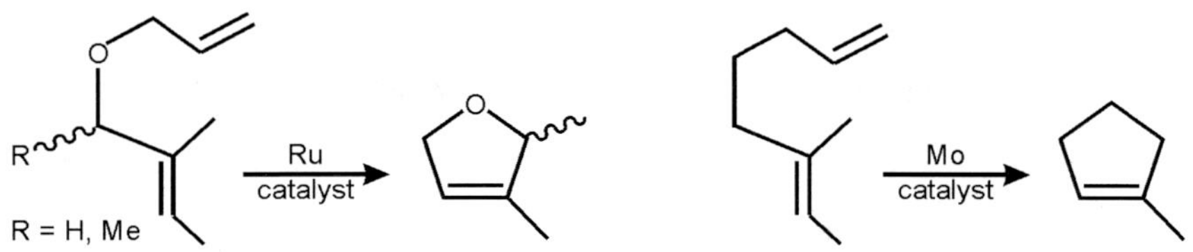

Scheme 11.2 Asymmetric ring-closing metathesis of 1,6-dienes with Ru- and Mo-based catalysts.

well as the elementary steps of the metathesis reaction and of the mechanism of enantioselectivity) [15, 17–25], Schrock's Mo-catalysts have been investigated less thoroughly [26–29]. These studies have been performed at different levels of theory, and to date no systematic comparison of the results obtained with different computational approaches has been performed.

For this reason, and both for Mo and Ru-based catalysts, we present such a comparative study. For Grubbs' catalysts we investigated the asymmetric ring-closing metathesis of 1,6-dienes with Ru-based catalysts (Scheme 11.2) using full QM approaches based on the pure BP86 [30–32] and the hybrid B3LYP [33–35] functionals, and QM/MM approaches [36–38] with different partitioning of the system into QM and MM regions. Further, we compare results obtained with two of the most popular packages, *ADF* [39, 40] and *Gaussian* [41]. For Schrock catalysts we investigated the ring-closing metathesis of 1,6-dienes (Scheme 11.2) with a Mo-based catalyst using a QM/MM approach based on the pure BP86 or the hybrid B3LYP functionals for the QM part.

11.2
Models and Computational Details

11.2.1
Models

For Grubbs' catalysts, the models we considered are derived from (pre)catalyst **4**, in which the phosphine trans to the NHC ligand has been removed, and the SIMes NHC has been substituted by a chiral NHC with two Ph groups on the NHC ring, and with *i*Pr substituted phenyl rings bonded to the N atoms (Figure 11.2). Finally, the =CH–Ph group bonded to Ru in **4** has been replaced with the =CH–CH$_2$–O–CHR–C(CH$_3$)=CH(CH$_3$) group, with R=H or CH$_3$, while the original Cl atoms of **4** have been replaced with I atoms, since higher enantiomeric excesses are obtained experimentally for systems with I as halide [42]. The catalyst chirality is provided by the two C atoms bearing the Ph substituents in the NHC ring.

The Grubbs' catalyst has been treated by full QM calculations using the ADF and *Gaussian 03* (G03) implementations of the pure BP86 functional, and the

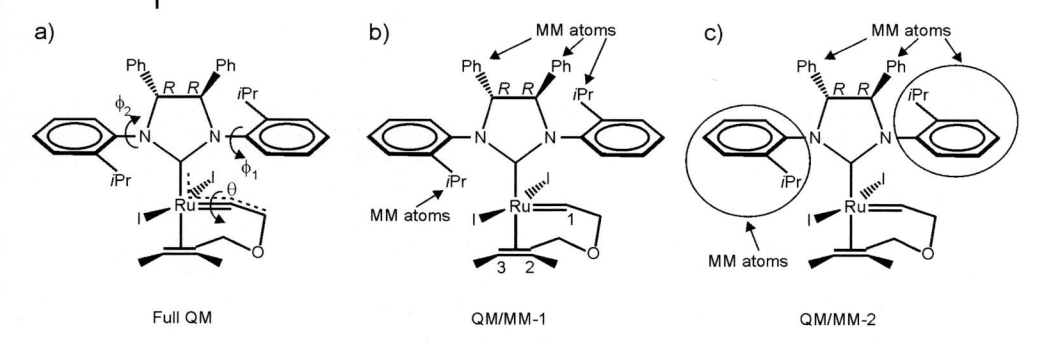

Figure 11.2 Representation of the partitioning of Grubbs systems into QM and MM regions, and definition of the dihedral angles φ_1, φ_2, and θ. The configuration of the chiral C atoms of the NHC ring is also indicated.

G03 implementation of the hybrid B3LYP functional. Further, the system was treated with QM/MM approaches, and two different partitioning of the system into QM and MM parts have been used (Figure 11.2). The QM part was treated again with the ADF and G03 implementations of the pure BP86 functional, while for the MM part we adopted the Amber force field [43], with the Lennard-Jones curves fitted to a purely repulsive potential in the case of the ADF calculations [37, 44], whereas we adopted the UFF force field [45] with standard Lennard-Jones curves for the G03 calculations. QM/MM-2 calculations were performed with the *Gaussian 03* program only. Double-ζ quality basis sets plus one polarization function were used for main group atoms (DZP [39, 40] and SVP [46] in the ADF and G03 calculations, respectively), while a triple-ζ quality basis set plus one polarization function was used for the Ru atom (TZVP [39, 40] and SDD/ECP [47, 48] for the ADF and G03 calculations, respectively).

Figure 11.3 Representation of the partitioning of Schrock systems into QM and MM regions, and definition of the dihedral angle θ. The configuration G$^+$ around the central dihedral angle of the chiral biphen ligand is also indicated.

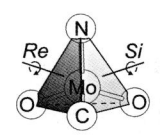

Scheme 11.3 Representation of the two chiral CNO faces of a Mo-based catalyst.

For the Schrock's catalysts, the models we considered are derived from (pre)-catalyst **1**, in which the =CH–C(Me$_2$)Ph group bonded to Mo has been replaced with the =CH–CH$_2$–CH$_2$–CH$_2$–CR=CHR (Figure 11.3). Due to the size of the system, we always used the QM/MM partitioning of Figure 11.3.

Catalyst chirality is provided by the central dihedral angle of the biphen ligand. This key dihedral angle can assume either a G$^+$ or a G$^-$ conformation. In this chapter we always assume a G$^+$ conformation (Figure 11.3). The presence of a chiral ligand, such as the biphen, makes the two diastereotopic CNO faces of the Mo-catalyst no longer equivalent (Scheme 11.3). To distinguish between them we use the *Re/Si* nomenclature originally defined for specifying heterotopic half spaces [49].

The QM part was treated with the G03 implementations of the pure BP86 and of the hybrid B3LYP functionals, while for the MM part we adopted the UFF parameters with standard Lennard-Jones curves. Consistently with the Ru-based catalyst, the SVP basis set was used for main group atoms, while the SDD/ECP basis set was used for the Mo atom.

Finally, and common to olefin metathesis with both catalysts, coordination of a prochiral olefin to a metal atom, such as the substrates of Figure 11.1, gives rise to non-superimposable coordinations [49]. As an example, Scheme 11.4 illustrates the two chiral geometries that originate from coordination of the second C=C double bond of the 1,6-diene in the coordination intermediate that immediately precedes the ring-closing step.

To distinguish between coordination of the two enantiofaces we prefer the nomenclature *re/si* [49], instead of the nomenclature *R/S*, defined for double or triple bonds π-bonded to a metal atom [50, 51]. To avoid confusion, we reserve the symbols *R/S* to define the configuration of the chiral sp^3-C atoms in both reactants and products.

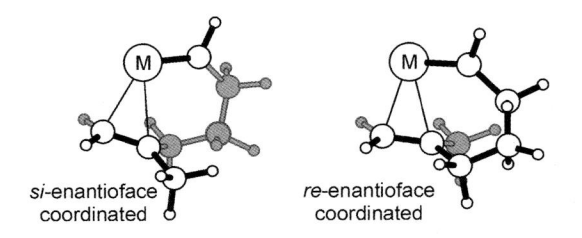

si-enantioface coordinated

re-enantioface coordinated

Scheme 11.4 Coordinations of the two prochiral enantiofaces of the substrate to a M atom.

11.3
Results and Discussion

11.3.1
Nomenclature

Before presenting results, we define the nomenclature that we follow. For the Ru-based system, and for the substrate with R=H (Scheme 11.2), structures discussed hereafter will be labeled, for example, as **CO-S-re-syn**, with the following meaning: (a) the first label, **CO** or **TS**, indicates whether the structure is a coordination intermediate or a ring-closing transition state, respectively; (b) the second label, **B** or **S**, indicates if olefin coordination occurs trans to the NHC ligand (bottom-path in Scheme 11.5) or cis to the NHC ligand (side-path in Scheme 11.5); (c) the third label, **re** or **si** indicates which enantioface of the double bond of the substrate is coordinated to the metal (Scheme 11.4); (d) the fourth label, **syn** or **anti**, holds only for side-coordinated structures, and indicates the relative orientation of the two CH_3 groups on the double bond of the substrate with respect to the NHC ligand. Thus, **CO-S-re-syn** indicates the coordination intermediate for the ring-closing metathesis derived from the side-coordination of the C=C double bond. The coordinated enantioface is *re*, and *syn* is the orientation of the two CH_3 groups of the substrate relative to the NHC ligand. For the substrate with R=Me, (Scheme 11.2), the additional label R or S, to indicate the chirality of C atom of the substrate, will be added at the end, as in **TS-B-re-R**.

For the Mo-based system, structures discussed hereafter will be labeled, for example, as **TS-Re-si**, with the following meaning: (a) the first label, **CO**, **TS** or **MC**, indicates whether the structure is a coordination intermediate, a ring-closing transition state, or the metallacycle product, respectively; (b) the second label, **Re** or **Si** indicates which diastereotopic CNO face of the catalyst is attacked by the substrate (Scheme 11.3); (c) the third label, **re** or **si** indicates which enantioface of the double bond of the substrate is coordinated to the metal (Scheme 11.4). Thus, **TS-Re-si** indicates the transition state for the ring-closing metathesis

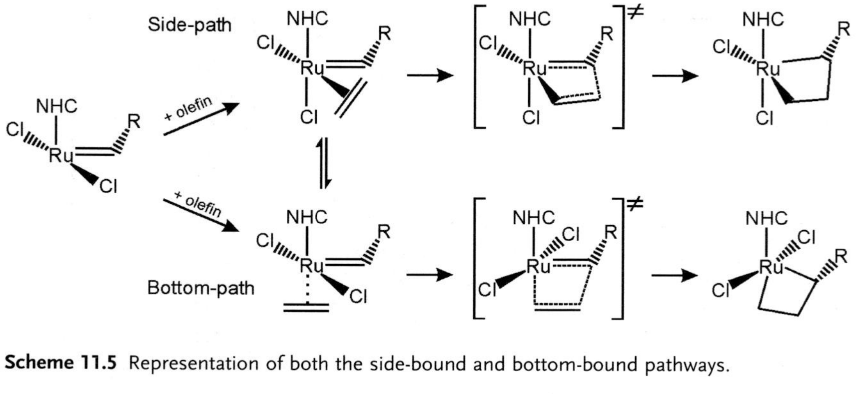

Scheme 11.5 Representation of both the side-bound and bottom-bound pathways.

derived from attack of the *si* enantioface of the substrate to the *Re* CNO face of the catalyst.

11.3.2
Ru-based Catalysts

In this section we compare different functionals and different QM/MM implementations in the metathesis of olefins catalyzed by the chiral Grubbs' catalyst of Figure 11.2.

Focusing on the coordination intermediates, the first global result is that in the gas phase the bottom-coordinated isomers **CO-B-re** and **CO-B-si** are almost isoenergetic, and are clearly more stable than the side-coordinated isomers, independently of the program, QM/MM implementation or functional used (Table 11.1). However, more detailed comparison reveals some differences. The G03 full QM B3LYP calculations indicate that the hybrid functional predicts the most favored **CO-S-*si-syn*** side-coordinated species to be more than 10 kcal mol^{-1} higher in energy relative to the most favored bottom-coordinated species, whereas the BP86 functional (same program and same basis set) predicts the **CO-S-*si-syn*** side-coordinated species to be slightly more than 5 kcal mol^{-1} higher in energy than the most stable bottom-coordinated species. This difference was already underlined in a previous paper [17], and it is not a minor difference, since solvent effects can easily overcome the stereoelectronic preference for the bottom-bound geometries. Indeed, investigation of the preferred pathway (bottom-bound or side-bound, see Scheme 11.5) is still an open field. With bulky substituents, like the

Table 11.1 Energetics of ring-closing metathesis, catalyzed by the chiral 2nd generation Grubbs Ru catalyst of Figure 11.2, and obtained by different computational approaches. Energies in kcal mol^{-1}. QM/MM-1 ADF values are taken from Ref. [18].

	ADF Full QM	ADF QM/MM	G03 bp86 Full QM	G03 bp86 QM/MM-1	G03 bp86 QM/MM-2	G03 b3lyp Full QM
CO-B-re	0	0.1	0.1	1.4	2.5	0
CO-B-si	0.2	0	0	0	0	0.1
CO-S-re-anti	9.1	6.7	6.2	5.8	6.8	11.9
CO-S-re-syn	11.0	15.1	9.0	9.2	5.3	14.0
CO-S-si-anti	10.5	10.5	8.3	9.2	7.3	13.8
CO-S-si-syn	7.5	8.1	5.5	4.6	5.9	10.4
TS-B-re	2.8	2.8	2.4	4.1	6.4	3.8
TS-B-si	2.4	1.4	1.5	3.3	5.3	3.1
TS-B-si-S_{eq}	0	0	0	0	0	0
TS-B-re-R_{eq}	0.3	2.4	1.4	0.9	0.8	0.6
TS-B-si-R_{ax}	6.0	6.6	6.5	6.8	4.2	7.1

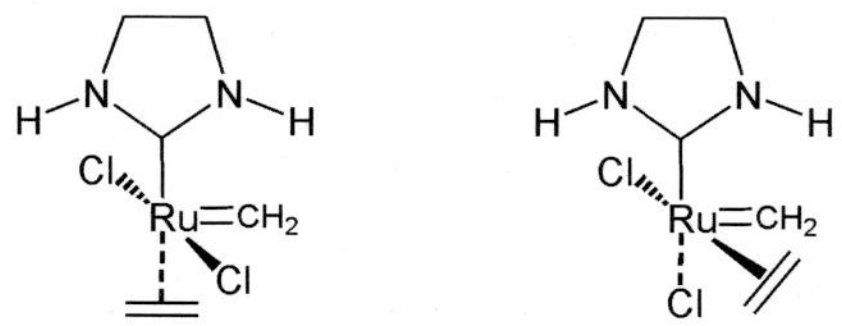

Figure 11.4 A model system for calculations at different levels of theory.

iPr substituted aromatic rings of the system reported in Figure 11.2, or with mesityl groups such as in the very popular IMes and SIMes based catalysts, it has been calculated that solvent effects are not able to revert the order of stability at the level of the transition state, and thus the bottom-bound pathway is followed [17]. However, with less bulky systems, such as systems in which the aromatic substituents on the N atoms is replaced by a simple methyl group, the balance between steric, electronic and solvent effects can be very delicate [52], and the proper behavior of a given functional becomes fundamental to understanding which pathway is actually followed.

For this reason, we performed a series of calculations on the model system of Figure 11.4 at different levels of theory. In particular, we calculated the energy difference between the side- and bottom-coordinated species with the BP86 and B3LYP functionals, and at the MP2 and CCSD(T) level of theory using different basis sets. Table 11.2 gives the results.

The DFT results clearly indicate that the B3LYP functional indeed favors more the bottom-bound geometry than the side-bound geometry (high values of $\Delta E_{\text{side-bottom}}$), and that within a given functional the results do not depend on the basis set used. In contrast, the MP2 results clearly favor the side-bound geometry, and this preference increases as the quality of the basis set used improves.

Table 11.2 Energy difference, $\Delta E_{\text{side-bottom}}$, between the side- and bottom-bound geometries of the system reported in Ref. [52].

Method	Basis set (Ru/other atoms)	$\Delta E_{\text{side-bottom}}$ (kcal mol^{-1})
BP86	SDD/SVP	1.0
BP86	QZVP/QZVP	1.5
B3LYP	SDD/SVP	3.9
B3LYP	QZVP/QZVP	4.1
MP2	SDD/SVP	−0.6
MP2	TZVPP/TZVP	−1.7
MP2	QZVP/QZVP	−1.9
CCSD(T)	SDD/SVP	3.6

Instead, the CCSD(T) calculations strongly favor the bottom-bound geometry, although this estimate is quite unreliable since a limited basis set used. The best estimate for the $\Delta E_{\text{side-bottom}}$ is then calculated as reported in Eq. (1):

$$\Delta E_{\text{side-bottom}}(\text{best}) = \Delta E_{\text{side-bottom}}(\text{MP2/QZVP}) + [\Delta E_{\text{side-bottom}}(\text{CCSD(T)/SVP}) - \Delta E_{\text{side-bottom}}(\text{MP2/SVP})] \quad (1)$$

Combining the $\Delta E_{\text{side-bottom}}$ reported in Table 11.2, our best estimate for the energy difference between the side- and bottom-bound geometries, $\Delta E_{\text{side-bottom}}(\text{best})$, is 1.1 kcal mol^{-1}, which is far closer to the $\Delta E_{\text{side-bottom}}$ calculated with the BP86 functional than to those calculated with the B3LYP functional. In conclusion, although our $\Delta E_{\text{side-bottom}}(\text{best})$ can be improved certainly, the value we calculated suggests that the BP86 functional probably performs better than the B3LYP functional in the prediction of the relative stability of the side- and bottom-bound geometries. Finally, as indicated by previous work, inclusion of solvent effects is absolutely mandatory for any reliable prediction of the $\Delta E_{\text{side-bottom}}$ in real reaction conditions [52].

Returning to full systems of Figure 11.2, and to the energy values of Table 11.1, the small differences between the ADF and G03 BP86 full QM calculations can be easily ascribed to the difference in the basis sets. Differently, the QM/MM-1 ADF results are in good agreement with the full QM BP86 ADF and G03 results, whereas the QM/MM-1 predicts the **CO-B-*si*** to be favored by more than 1 kcal mol^{-1} relative to the **CO-B-*re*** structure, and this preference increases to more than 2 kcal mol^{-1} in the QM/MM-2 model, which has a larger MM contribution. Thus, we ascribe this difference to the different force fields used (AMBER and UFF in the ADF and G03 calculations, respectively) and to the different treatment of the Van der Waals contribution (truncated to the repulsive part and including the long attractive tail in ADF and G03, respectively). In this specific case it seems that the truncated-repulsive approach used in the ADF calculations leads to a slightly better agreement with full QM calculations.

Geometrically (Table 11.3 and Figures 11.5 and 11.6), all the methods result in a very similar folding of the aromatic rings bonded to the N atoms. Indeed, the φ_1 and φ_2 angles are very close to the experimental values found in the X-ray structures of closely related compound ($\varphi_1 = -76°$, $\varphi_2 = -77°$) [42]. This is a very important result, since we ascribed the enantioselectivity in the desymmetrization of achiral trienes to the chiral folding of the catalyst. This chiral folding is imposed by the Ph rings on the chiral C atoms of the NHC ring [18]. With regards to the Ru=C1 carbene bond in the bottom-bound coordination intermediates, all methods based on the BP86 functional predict that this bond is very close to 1.84 Å, while the full G03 B3LYP calculations predict this bond to be slightly shorter (close to 1.83 Å). All methods consistently predict this bond to be slightly longer, by roughly 0.01 Å, in the side-bound coordination intermediates, with the full G03 B3LYP values again slightly shorter than the values from calculations based on the BP86 functional. Further differences are in the distances Ru–C2

Table 11.3 Key geometric parameters of ring-closing metathesis, catalyzed by the 2nd generation Grubbs Ru catalyst of Figure 11.2, and obtained by different computational approaches. QM/MM-1 ADF values are taken from Ref. [18].

Parameter	Structure	ADF Full QM	ADF QM/MM	G0Full 3 bp86 Full QM	G03 bp86 QM/MM-1	G03 bp86 QM/MM-2	G03 b3lyp Full QM
φ_1 (°)	CO-B-re	−67.99	−72.77	−67.99	−72.57	−72.95	−68.90
	CO-B-si	−68.57	−72.74	−68.57	−74.09	−75.73	−68.03
	TS-B-re	−65.12	−65.37	−64.79	−66.40	−63.42	−66.29
	TS-B-si	−65.38	−73.76	−65.36	−66.59	−63.30	−66.32
φ_2 (°)	CO-B-re	−64.30	−73.46	−64.30	−72.50	−72.76	−65.92
	CO-B-si	−72.34	−72.87	−72.35	−75.47	−76.45	−71.50
	TS-B-re	−62.60	−63.71	−62.57	−66.34	−64.15	−64.02
	TS-B-si	−60.84	−66.13	−60.75	−62.19	−58.93	−62.03
Ru–C1 (Å)	CO-B-re	1.841	1.838	1.841	1.839	1.841	1.833
	CO-B-si	1.841	1.841	1.841	1.834	1.832	1.830
	CO-S-re-anti	1.854	1.855	1.854	1.851	1.854	1.840
	CO-S-si-syn	1.847	1.857	1.847	1.846	1.846	1.839
	TS-B-re	1.889	1.864	1.870	1.871	1.945	1.875
	TS-B-si	1.887	1.869	1.872	1.869	1.945	1.877
Ru–C2 (Å)	CO-B-re	2.605	2.528	2.605	2.556	2.559	2.696
	CO-B-si	2.625	2.546	2.625	2.414	2.388	2.900
	CO-S-re-anti	2.361	2.353	2.361	2.357	2.379	2.462
	CO-S-si-syn	2.336	2.360	2.336	2.335	2.386	2.416
	TS-B-re	2.458	2.496	2.482	2.474	2.349	2.447
	TS-B-si	2.460	2.467	2.472	2.471	2.350	2.435
Ru–C3 (Å)	CO-B-re	2.395	2.352	2.395	2.337	2.322	2.487
	CO-B-si	2.385	2.351	2.385	2.249	2.232	2.617
	CO-S-re-anti	2.275	2.347	2.276	2.257	2.289	2.343
	CO-S-si-syn	2.265	2.287	2.265	2.248	2.262	2.337
	TS-B-re	2.220	2.253	2.223	2.214	2.084	2.192
	TS-B-si	2.220	2.212	2.208	2.206	2.082	2.176
C2–C3 (Å)	TS-B-re	1.432	1.420	1.438	1.439	1.559	1.442
	TS-B-si	1.430	1.429	1.442	1.439	1.559	1.447

and Ru–C3, which refer to coordination of the C=C double bond of the substrate. All methods systematically predict a non symmetrical coordination of the C=C double bond, with the Ru–C3 bond sensibly shorter than the Ru–C2 bond. Notably, the two full QM BP86 calculations predict very similar Ru–C2 and Ru–C3 distances, which are longer than the corresponding distances calculated with the

11.3 Results and Discussion | 255

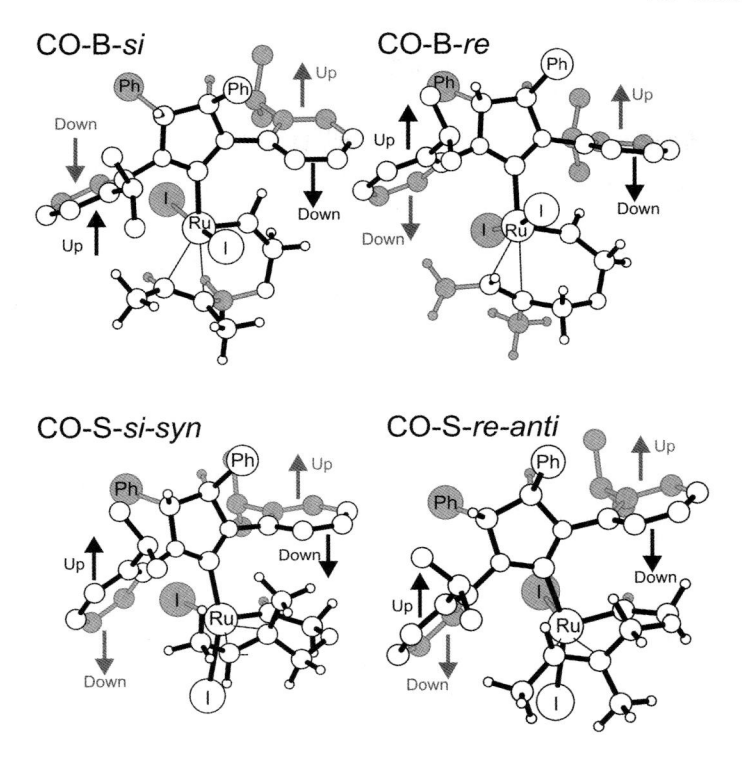

Figure 11.5 Coordination intermediates for the ring-closing metathesis of the substrate with R=H (Scheme 11.2).

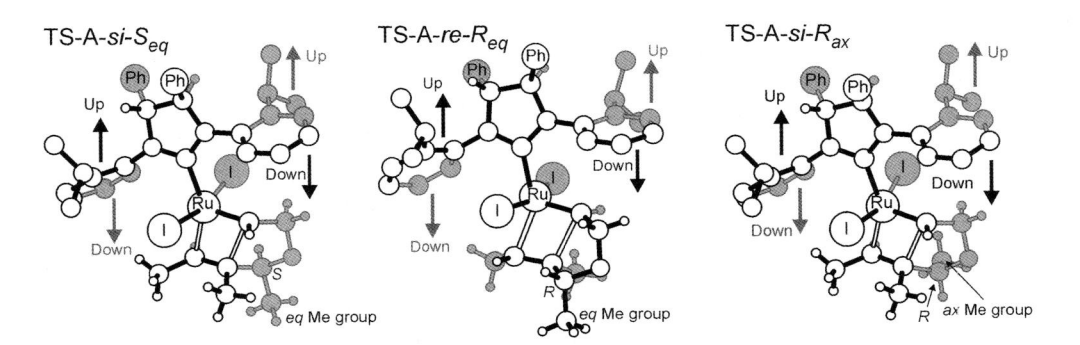

Figure 11.6 Transition states for the ring-closing metathesis of the chiral substrate with R=Me (Scheme 11.2). Also indicated are the configuration of the chiral C atom of the substrate and the pseudo-axial or equatorial position of the additional Me group of the substrate.

QM/MM BP86 based methods. In our opinion, this is a clear effect of the QM/MM partitioning. Unlike from the Ru–C1 distance, the full QM B3LYP calculations predict quite longer distances for C=C coordination relative to the BP86 based calculations. Finally, all methods agree that the C=C bond is more tightly coordinated in the side-bound geometry than in the bottom-bound geometry. Probably, an effect of the strong trans influence of the NHC ring in the bottom-bound coordination intermediates.

Moving to the **TS-B-re** and **TS-B-si** transition states, all methods clearly predict a sensible elongation of the Ru–C1 bond with a simultaneous shortening of the Ru–C3 bond, and all methods predict the transition state to be a very asynchronous and early transition state (Ru–C3 $\gg$ Ru–C1). The forming C–C bond is always close to 1.40 Å, with the exception of the QM/MM-2 value, which clearly indicates that a too drastic QM/MM partitioning has some geometrical consequences.

Energetically, all methods predict that the **TS-B-si** transition state is favored by roughly 0.5–1.5 kcal mol^{-1} relative to the **TS-B-re** transition state. Furthermore, all the BP86 based methods predict the barrier for metallacycle formation to be in the range 1–4 kcal mol^{-1}, with the exception of the QM/MM-2 approach, which predicts this barrier to be in the range 5–6 kcal mol^{-1}. These are key failures of the QM/MM-2 approach, which clearly indicates that a careful partitioning of the system between QM and MM parts has to be devised. Finally, the BP86 functional predicts barriers somewhat lower than those obtained with the B3LYP functional.

Moving to the transition states for the enantioselective step, the two transition states leading to formation of a ring with R chirality, **TS-B-re-R$_{eq}$** and **TS-B-si-R$_{ax}$**, are predicted to be of higher energy with respect to the transition state **TS-B-si-S$_{eq}$**, which leads to the formation of a ring with S chirality (eq and ax stand for pseudo-equatorial and pseudo-axial position of the methyl group in the chair conformation of the substrate, see Figure 11.6). That is, all approaches can reproduce the experimental result that, in the desymmetrization of achiral trienes, catalysts with a (R,R) configuration on the NHC ring preferentially close rings with a S configuration [42, 53]. This is a remarkable result, since the energy differences needed to rationalize the experimental results are pretty small, and even small failures of the QM/MM approaches would lead to a disagreement with the experimental data.

Thus, all methods support the mechanism of enantioselectivity we proposed in previously [18]. Namely, the two Ph groups in positions 4,5 of the imidazolyl ring of the NHC ligand impose a chiral folding to the N-bonded aromatic groups. In particular, the side of the N-bonded aromatic group near the Ph group is bent-down, i.e., bent toward the Ru atom and the substrate. This folding of the catalyst implies a chiral orientation of the dihedral angle θ in the transition state. The preferred orientation is the one that places the R group bonded to the Ru=C carbene atom away from the bent-down side of the N-bonded aromatic group. For an R,R configuration of the catalyst, this implies that values of $\theta \approx -90°$ are favored. This implies that with an R,R catalyst the si enantioface of the substrate is more reactive, which explains the preference for the **TS-B-*si*** transition state. At this point, an additional R group near the reacting C=C double bond of the substrate is rather easily accommodated in a pseudo-equatorial position, independently of the reacting enantioface of the substrate, which explains the preference for the **TS-B-*si*-S$_{eq}$** transition state. Conversely, transition states with the additional group in a pseudo-axial position are of remarkably higher energies.

11.3.3
Mo-based Catalysts

Here we compare different functionals in the ring-closing metathesis of olefins catalyzed by the Schrock's catalyst of Figure 11.3. Using the B3LYP functional within a QM/MM approach, we were able to locate the four coordination intermediates corresponding to attack of the second C=C double bond of the substrate

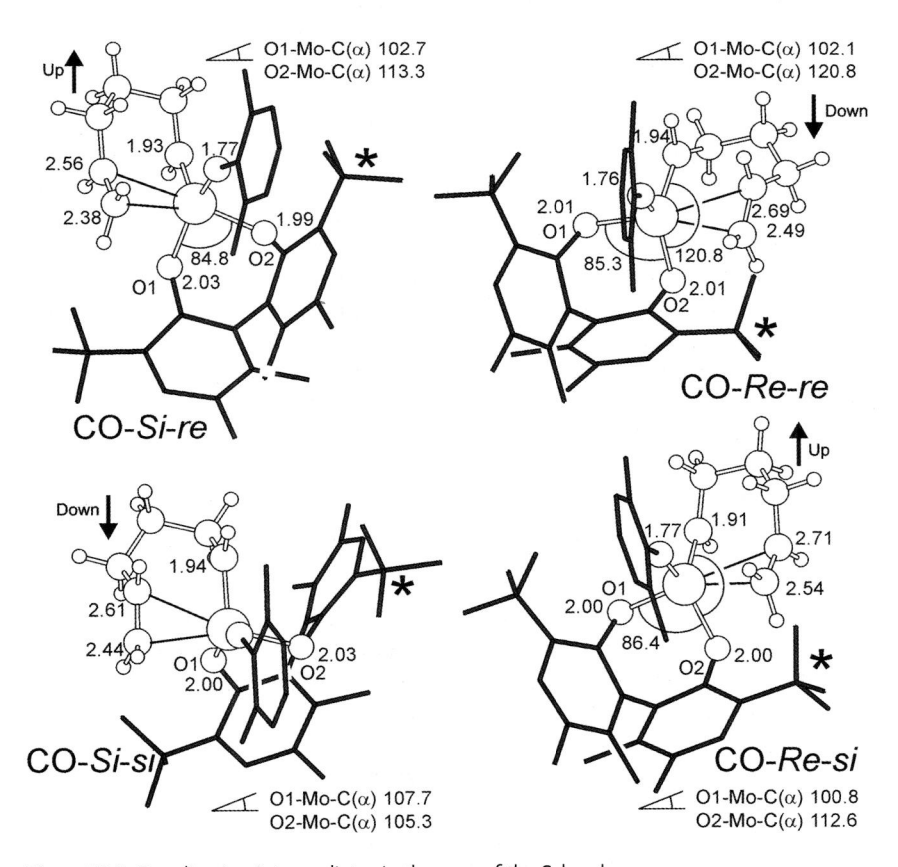

Figure 11.7 Coordination intermediates in the case of the Schrock catalyst of Figure 11.3. Distances are in Å, angles are in degrees.

to the two diastereotopic CNO faces of the Mo-catalyst. These structures are reported in Figure 11.7. In all cases, coordination of the C=C double bond pushes one of the O atoms trans to the N atom.

The overall geometry can be described as square pyramidal. The **CO-Si-re** and **CO-Re-si** geometries present the β, γ and δ sp^3-C atoms oriented toward the N atom, whereas the **CO-Si-si** and **CO-Re-re** geometries present the same C atoms away from the N atom. Coordination of the double bond is highly asymmetric, with the terminal C(ω) atom quite closer to the Mo atom. This is due to the trans effect of the O atom, which is almost perfectly trans to the C(ε) atom [average O–Mo–C(ε) angle 169°], while the average O–Mo–C(ω) angle is 156°. Coordination of the C=C double bond is almost a barrierless process, despite the fact that it requires the noticeable rearrangement of one of the O atoms from almost cis to trans to the N atom. Indeed, test calculations performed on smaller systems (i.e., without the t-Bu and Me groups on the ligand, and using ethene as substrate) invariably led from reactant directly into the metallacycle products, without any barrier both at the coordination steps and at the following ring-closing step. Similar no-barriers behavior was shown by the complete systems when calculated with the BP86 functional. This agrees with previously reported results [27].

In contrast, with the B3LYP functional the coordination intermediates are stable intermediates along the reaction path, as confirmed by frequency calculations. Energetically (Figure 11.8), coordination of the *si* enantioface of the substrate is quite favored relative to coordination of the *re* enantioface, independently of the attacked diastereotopic face of the Mo-catalyst. The reason for this preference is better understood when the geometrical arrangements associated with the ligand deformation upon C=C coordination are examined. The model structures

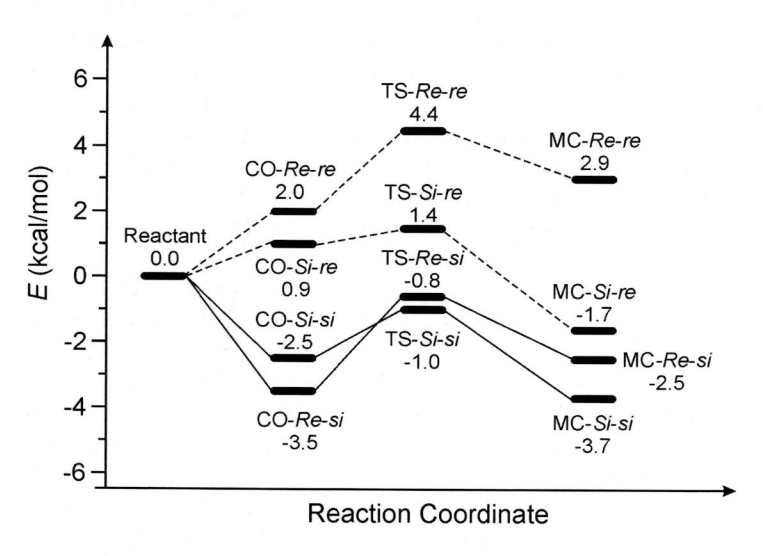

Figure 11.8 Reaction profile for the ring-closing step of 1,6-dienes in the presence of **1**.

of Figure 11.7 indicate that, after attack to the *Si* face of the Mo-catalyst, the *t*-Bu group marked by a star (*t*-Bu*) is pushed up. This makes the **CO-Si-re** geometry of high energy because the "up" C(β) atom of the substrate sterically interacts with the *t*-Bu* group. Through a similar rearrangement, after attack to the *Re* face of the Mo-catalyst the same *t*-Bu group is pushed down. This makes the **CO-Re-re** geometry of high energy because the "down" C(δ) atom of the substrate sterically interacts with the *t*-Bu* group. Instead, the **CO-Si-si** and **CO-Re-si** geometries are of lower energy because in these cases the "down" and "up" oriented C(δ) and C(β)γ atoms are away from the *t*-Bu* group, which is "up" and "down" oriented, respectively, in the *Si* and *Re* geometries. This simple scheme explains the preference for the *si* enantioface of the substrate, whatever diastereotopic face of the Mo-catalyst is attacked, and it is key to understanding the whole mechanistic scenario. Of course, this conclusion holds for Mo-catalyst with a G^+ configuration at the biphen ligand.

Figure 11.9 reports the transition states for the ring-closing step. They are rather similar to the coordination intermediates of Figure 11.7, since minor movements are required to convert the coordination intermediates into the transition states. The emerging C–C bond is around 2.00 Å in all the transition states. Transition states **TS-Si-si** and **TS-Re-si** are of quite lower energy relative to the **TS-Si-re** and **TS-Re-re** transition states (Figure 11.8). This can be easily explained having in mind the points used to rationalize the energy ordering in the case of the coordination intermediates. Finally, the four metallacycles metallacycle products are reached from the corresponding transition states. Since the geometries of the metallacycles are consistent with the those of the coordination intermediates and of the transition states we shall not discuss them in detail.

The overall energy profile of the four different reaction paths just described is shown in Figure 11.8. The more stable coordination intermediates **CO-Si-si** and **CO-Re-si** are of lower energy with respect to the reactant with the uncoordinated C=C double bond. Instead, formation of the coordination intermediates **CO-Si-si** and **CO-Re-si** is slightly endoergonic. Energy barriers lower than 3 kcal mol^{-1} are required to reach the transition states for the ring-closing step. Again, transition states **TS-Si-si** and **TS-Re-si** are of lower energy than the product, which indicates that attack of the *si* enantioface along either the *Si* or *Re* direction should be a very easy process. Differently, attack of the *re* enantioface of the substrate seems to be unlikely whatever is the attacked face of the Mo-catalyst.

Concluding this section, it is worth noting a geometric feature of the forming five-membered ring. The C(δ)H$_2$ group close to the C=C double bond presents the two H atoms in a pseudo-equatorial and a pseudo-axial position. These H atoms are marked as pro-(*R*) and pro-(*S*) in the structures **TS-Re-si** and **TS-Si-si** of Figure 11.9, according to the configuration (*R* or *S*, respectively) that would be assumed by the C atom to which they are bonded when they are replaced with a OR group. This configuration would also correspond to the configuration of the products at the end of the metathesis reaction. Consistently with the case of the Ru-based system of the previous section, this suggests that formation of a substrate of *S* configuration is favored, since coordination of the favored *si* enantio-

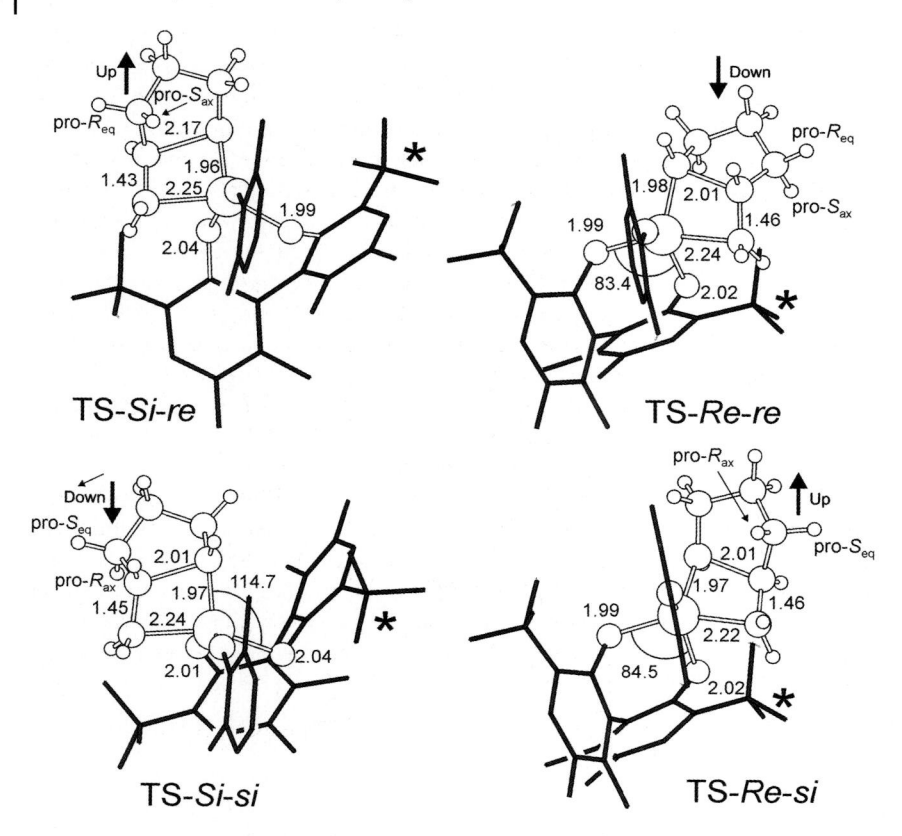

Figure 11.9 Transition states in the case of the Schrock catalyst of Figure 11.3. Distances are in Å, angles are in degrees.

face of the substrate presents a pro-S_{eq} H atom, whereas pro-R_{eq} atoms are present only after coordination of the *re* enantioface.

11.4
Conclusions

In this chapter we have discussed the different performances of different QM and QM/MM computational approaches in the modeling of Ru- and Mo-catalyzed olefin metathesis reactions. With regards to the Ru-based systems, our calculations indicate that the BP86 and the B3LYP functionals substantially perform the same. The main difference is in the different stability predicted for the side- and bottom-bound coordination intermediates. Both functionals predict the side-bound geometry to be of higher energy. However, the BP86 functional predicts a smaller energy difference with respect to the B3LYP functional. The "best" value for this difference, based on CCSD(T) calculations with a reduced basis set, suggests that the BP86 could perform better in this case. We also tested different QM/MM partitioning schemes, and we found that even a large MM part, such as in the QM/MM-2 partitioning scheme, leads to rather reasonable results, although fine details are not reproduced well. This set of calculations suggests that fast screening of catalysts, where fine details are not needed, can be performed even with the QM/MM-2 partitioning scheme, while accurate results require that a smaller MM part, as in the QM/MM-1 partitioning scheme, is used. The QM/MM-1 partitioning scheme leads to quite good agreement with full QM calculations. As for the functional to be used, the pure DFT BP86 is to be preferred, due to the very fast implementation of this functional in many packages.

Very different is the case of the Mo-based Schrock's catalysts. Here the performances of the two functional are totally different. With the BP86 functional we were unable to locate any coordination intermediate and any transition state for metallacycle formation. That is, from the olefin unbound intermediate the BP86 functional predicts a downhill path until the metallacycle is formed. Instead, using the B3LYP functional, coordination intermediates and, consequently, the following transition state for metallacycle formation were located for each of the different attacks we explored. At this point is of course difficult to establish which of the two functionals is correct. Reliable higher level calculations are needed to answer this question.

Acknowledgments

We thank CINECA (Grant INSTM/Supercalcolo) for easy access to computer facilities.

References

1 T. M. Trnka, R. H. Grubbs, *Acc. Chem. Res.* 2001, **34**, 18.
2 R. H. Grubbs, *Handbook of Olefin Metathesis*, Wiley-VCH, Weinheim, 2003.
3 A. H. Hoveyda, R. R. Schrock, *Chem. Eur. J.* 2001, **7**, 945.
4 A. H. Hoveyda, R. R. Schrock, *Comprehensive Asymmetric Catal, Suppl.* 2004, **1**, 207.
5 R. R. Schrock, A. H. Hoveyda, *Angew. Chem., Int. Ed.* 2003, **42**, 4592.
6 A. Fürstner, *Angew. Chem., Int. Ed.* 2000, **39**, 3012.
7 D. S. La, E. S. Sattely, J. G. Ford, R. R. Schrock, A. H. Hoveyda, *J. Am. Chem. Soc.* 2001, **123**, 7767.
8 W. C. P. Tsang, J. A. Jernelius, G. A. Cortez, G. S. Weatherhead, R. R. Schrock, A. H. Hoveyda, *J. Am. Chem. Soc.* 2003, **125**, 2591.
9 E. L. Dias, S. T. Nguyen, R. H. Grubbs, *J. Am. Chem. Soc.* 1997, **119**, 3887.
10 M. Ulman, R. H. Grubbs, *Organometallics* 1998, **17**, 2484.
11 M. S. Sanford, J. A. Love, R. H. Grubbs, *J. Am. Chem. Soc.* 2001, **123**, 6543.

12 J. A. Tallarico, P. J. Bonitatebus Jr., M. L. Snapper, *J. Am. Chem. Soc.* 1997, **119**, 7157.
13 P. E. Romero, W. E. Piers, *J. Am. Chem. Soc.* 2005, **127**, 5032.
14 C. Hinderling, C. Adlhart, H. Baumann, P. Chen, *Angew. Chem., Int. Ed.* 1998, **37**, 2685.
15 C. Adlhart, C. Hinderling, H. Baumann, P. Chen, *J. Am. Chem. Soc.* 2000, **122**, 8204.
16 J.-L. Herisson, Y. Chauvin, *Makromol. Chem.* 1971, **141**, 161.
17 A. Correa, L. Cavallo, *J. Am. Chem. Soc.* 2006, **128**, 13352.
18 C. Costabile, L. Cavallo, *J. Am. Chem. Soc.* 2004, **126**, 9592.
19 L. Cavallo, *J. Am. Chem. Soc.* 2002, **124**, 8965.
20 G. Occhipinti, H. R. Bjorsvik, V. R. Jensen, *J. Am. Chem. Soc.* 2006, **128**, 6952.
21 C. Adlhart, P. Chen, *Angew. Chem., Int. Ed.* 2002, **41**, 4484.
22 C. Adlhart, P. Chen, *J. Am. Chem. Soc.* 2004, **126**, 3496.
23 F. Bernardi, A. Bottoni, G. P. Miscione, *Organometallics* 2000, **19**, 5529.
24 S. F. Vyboishchikov, W. Thiel, *Chem. Eur. J.* 2005, **11**, 3921.
25 S. Fomine, J. V. Ortega, M. A. Tlenkopatchev, *Organometallics* 2005, **24**, 5696.
26 A. Poater, X. Solans-Monfort, E. Clot, C. Coperet, O. Eisenstein, *Dalton Trans.* 2006, 3077.
27 T. P. M. Goumans, A. W. Ehlers, K. Lammertsma, *Organometallics* 2005, **24**, 3200.
28 Y. D. Wu, Z. H. Peng, *J. Am. Chem. Soc.* 1997, **119**, 8043.
29 K. Monteyne, T. Ziegler, *Organometallics* 1988, **17**, 5901.
30 A. Becke, *Phys. Rev. A* 1988, **38**, 3098.
31 J. P. Perdew, *Phys. Rev. B* 1986, **33**, 8822.
32 J. P. Perdew, *Phys. Rev. B* 1986, **34**, 7406.
33 A. D. Becke, *J. Chem. Phys.* 1993, **98**, 5648.
34 P. J. Stephens, F. J. Devlin, C. F. Chabalowski, M. J. Frisch, *J. Phys. Chem.* 1994, **98**, 11623.
35 C. Lee, W. Yang, R. G. Parr, *Phys. Rev. B* 1988, **37**, 785.
36 F. Maseras, K. Morokuma, *J. Comput. Chem.* 1995, **16**, 1170.
37 L. Cavallo, T. K. Woo, T. Ziegler, *Can. J. Chem.* 1998, **76**, 1457.
38 T. K. Woo, L. Cavallo, T. Ziegler, *Theor. Chem. Acc.* 1998, **100**, 307.
39 G. te Velde, F. M. Bickelhaupt, E. J. Baerends, C. Fonseca Guerra, S. J. A. Van Gisbergen, J. G. Snijders, T. Ziegler, *J. Comput. Chem.* 2001, **22**, 931.
40 ADF 2005, Vrije Universiteit Amsterdam, Amsterdam, 2005.
41 *Gaussian 03 Revision C.02*, M. J. Frisch, G. W. Trucks, H. B. Schlegel, G. E. Scuseria, M. A. Robb, J. R. Cheeseman, J. Montgomery, J. A., T. Vreven, K. N. Kudin, J. C. Burant, J. M. Millam, S. S. Iyengar, J. Tomasi, V. Barone, B. Mennucci, M. Cossi, G. Scalmani, N. Rega, G. A. Petersson, H. Nakatsuji, M. Hada, M. Ehara, K. Toyota, R. Fukuda, J. Hasegawa, M. Ishida, T. Nakajima, Y. Honda, O. Kitao, H. Nakai, M. Klene, X. Li, J. E. Knox, H. P. Hratchian, J. B. Cross, C. Adamo, J. Jaramillo, R. Gomperts, R. E. Stratmann, O. Yazyev, A. J. Austin, R. Cammi, C. Pomelli, J. W. Ochterski, P. Y. Ayala, K. Morokuma, G. A. Voth, P. Salvador, J. J. Dannenberg, V. G. Zakrzewski, S. Dapprich, A. D. Daniels, M. C. Strain, O. Farkas, D. K. Malick, A. D. Rabuck, K. Raghavachari, J. B. Foresman, J. V. Ortiz, Q. Cui, A. G. Baboul, S. Clifford, J. Cioslowski, B. B. Stefanov, G. Liu, A. Liashenko, P. Piskorz, I. Komaromi, R. L. Martin, D. J. Fox, T. Keith, M. A. Al-Laham, C. Y. Peng, A. Nanayakkara, M. Challacombe, P. M. W. Gill, B. Johnson, W. Chen, M. W. Wong, C. Gonzalez, J. A. Pople, Gaussian, Inc., Pittsburgh, PA, 2003.
42 T. J. Seiders, D. W. Ward, R. H. Grubbs, *Org. Lett.* 2001, **3**, 3225.
43 W. D. Cornell, P. Cieplak, C. I. Bayly, I. R. Gould, K. M. Merz, Jr., D. M. Ferguson, D. C. Spellmeyer, T. Fox, J. W. Caldwell, P. A. Kollman, *J. Am. Chem. Soc.* 1995, **117**, 5179.
44 T. K. Woo, T. Ziegler, *Inorg. Chem.* 1994, **33**, 1857.
45 A. K. Rappé, C. J. Casewit, K. S. Colwell, W. A. Goddard III, W. M. Shiff, *J. Am. Chem. Soc.* 1992, **114**, 10024.

46 A. Schaefer, H. Horn, R. Ahlrichs, *J. Chem. Phys.* 1992, **97**, 2571.
47 W. Küchle, M. Dolg, H. Stoll, H. Preuss, *J. Chem. Phys.* 1994, **100**, 7535.
48 T. Leininger, A. Nicklass, H. Stoll, M. Dolg, P. Schwerdtfeger, *J. Chem. Phys.* 1996, **105**, 1052.
49 K. R. Hanson, *J. Am. Chem. Soc.* 1966, **88**, 2731.
50 R. S. Cahn, C. Ingold, V. Prelog, *Angew. Chem., Int. Ed. Engl.* 1966, **5**, 385.
51 P. Corradini, G. Paiaro, A. Panunzi, *J. Polym. Sci., Part. C* 1967, **16**, 2906.
52 D. Benitez, W. A. Goddard III, *J. Am. Chem. Soc.* 2005, **127**, 12218.
53 T. W. Funk, J. M. Berlin, R. H. Grubbs, *J. Am. Chem. Soc.* 2006, **128**, 1840.

12
Heterolytic σ-Bond Activation by Transition Metal Complexes

Shigeyoshi Sakaki, Noriaki Ochi, and Yu-ya Ohnishi

12.1
Introduction

σ-Bond activation by transition metal complexes is of considerable importance in modern organometallic and catalytic chemistry and remains a challenging research target. For instance, the C–H σ-bond activations of alkane and aromatics are the first step in introducing the functional group into alkane and aromatic compounds [1], respectively. Theoretical studies of the σ-bond activation are extremely desirable and allow elucidation of the roles of numerous factors (both electronic and steric in origin) in C–H, C–C, etc. bond functionalization.

In general, there are three categories of σ-bond activation reactions: (a) oxidative addition (Reaction 1), (b) metathesis (Reaction 2), and (c) σ-bond addition across to the M=NR bond (Reaction 3).

$$ML_n + A-B \rightarrow (A)ML_n(B) \qquad (1)$$

$$XML_n + A-B \rightarrow (B)ML_n + A-X \qquad (2)$$

$$L_nM=NR + A-B \rightarrow (B)ML_n\{NR(A)\} \qquad (3)$$

Oxidative addition is often observed in the late transition metal complexes and has been extensively investigated using theoretical approaches [2, 3]. Metathesis is observed for the both early and late transition metal complexes. Notably, when X is an anion in Reaction (2) the oxidation state of M does not change in metathesis and the σ-bond activation occurs heterolytically. σ-Bond addition to M=N bond is observed in both early and middle transition metal complexes. This reaction (Reaction 3) is considered to take place in a heterolytic manner, because N is electronegative and M is electropositive.

Here, we present several examples of σ-bond activation reactions. First, we mention the heterolytic σ-bond activation reaction reported by Fujiwara and Moritani [4], where a C–H bond of benzene is activated by $Pd(OAc)_2$ and stylene is produced. Similar metathesis-like heterolytic σ-bond activation has been

Computational Modeling for Homogeneous and Enzymatic Catalysis.
A Knowledge-Base for Designing Efficient Catalysts. K. Morokuma and D. G. Musaev (Eds.)
Copyright © 2008 WILEY-VCH Verlag GmbH & Co. KGaA, Weinheim
ISBN: 978-3-527-31843-8

applied to interesting synthetic reactions, recently [5]. Periana et al. [6] have reported the catalytic conversion of methane into methanol, with catalytica [i.e., dichloro(η^2-{2,2'-bipyrimidyl})platinum(II), Pt(Bpym)Cl$_2$] and cis-diamminedichloroplatinum(II) [i.e., (NH$_3$)$_2$PtCl$_2$, cisplatin] as catalysts. Although oxidative addition of the C–H bond of methane cannot be neglected, it is likely that heterolytic σ-bond activation occurs during these reactions. Metal-promoted hydrogen transfer reactions between secondary alcohols and ketones, which take place through C–H and O–H σ-bond activations [7], as well as H–H bond activation in Ru-catalyzed hydrogenation of ketones [8] are the best examples of heterolytic σ-bond activation.

In this chapter we present theoretical studies on the mechanisms of heterolytic σ-bond activation that occurs via either metathesis or addition across M=N and M≡C bonds.

12.2
Characteristic Features of Heterolytic σ-Bond Activation

Previously, many theoretical chemists have extensively studied [3, 9, 10] σ-bond activation occurring via an oxidative addition pathway, and have proposed a molecular orbital picture of this process. However, little is known about the details of orbital interactions in the heterolytic σ-bond activation process. Therefore, here, we report two pioneering theoretical works on heterolytic σ-bond activation.

12.2.1
Theoretical Study of the Shilov Reaction

In the Shilov reaction, methane is converted into methanol using a Pt(II) catalyst. The first step of this process (Reactions 4 and 5) is C–H σ-bond activation, which has been theoretically investigated by Siegbahn and Crabtree at the DFT(B3LYP) and PCI-80 levels of theory [11].

$$Pt(II) + CH_4 \rightarrow H_3C-Pt(IV)-H \qquad (4)$$

$$Pt(II) + CH_4 \rightarrow H_3C-Pt(II)^- + H^+ \qquad (5)$$

In their studies, the authors used PtCl$_2$(H$_2$O)$_2$ as a model catalyst and included several water molecules into calculations to mimic microsolvation effects. The bulk solvation effect was considered with the SCRF method. The authors demonstrated that the C–H σ-bond activation in this reaction takes place via a metathesis mechanism. During the reaction, Pt–CH$_3$ and H–Cl bonds are formed, and C–H and Pt–Cl bonds are broken. The transition state associated with this mechanism involves Pt, C (from the substrate CH$_4$), H (of activated C–H bond) and the Cl ligand of catalyst. The calculated Pt–CH$_3$, C–H (activated), H–Cl and Pt–Cl distances at the transition state are 2.10, 1.81, 1.46, and 2.63 Å, respectively. These geometry features clearly show that it is a product-like transition state. This conclusion is consistent with the calculated activation energy of 27.0 kcal mol^{-1}: the intermediate connected with this transition state is about 10 kcal mol^{-1} higher in energy than reactants. The calculated atomic charge of the H atom of the broken C–H bond is +0.23 and +0.26e at the η^1-coordinated methane complex and transition state, respectively. It becomes proton-like after the transition state. Charges on the other H atoms of CH$_4$-substrate are +0.19–0.22e. Thus, the broken C–H bond is more polarized at the transition state than at reactant complex, though the polarization is not very significant.

12.2.2
Fujiwara–Moritani Reaction: Driving Force and Orbital Interaction

In the Fujiwara–Moritani reaction [4], Pd(II)-acetate was used as catalyst to activate the C–H bond. Sakaki and coworkers have used Pd(II) and Pt(II) formate complexes as a model to elucidate methane and benzene C–H activations at the DFT, MP2, MP4(SDQ), and CCSD(T) levels of theory [12]. As shown in Figure 12.1, the first step of the reaction is benzene coordination to the metal center to

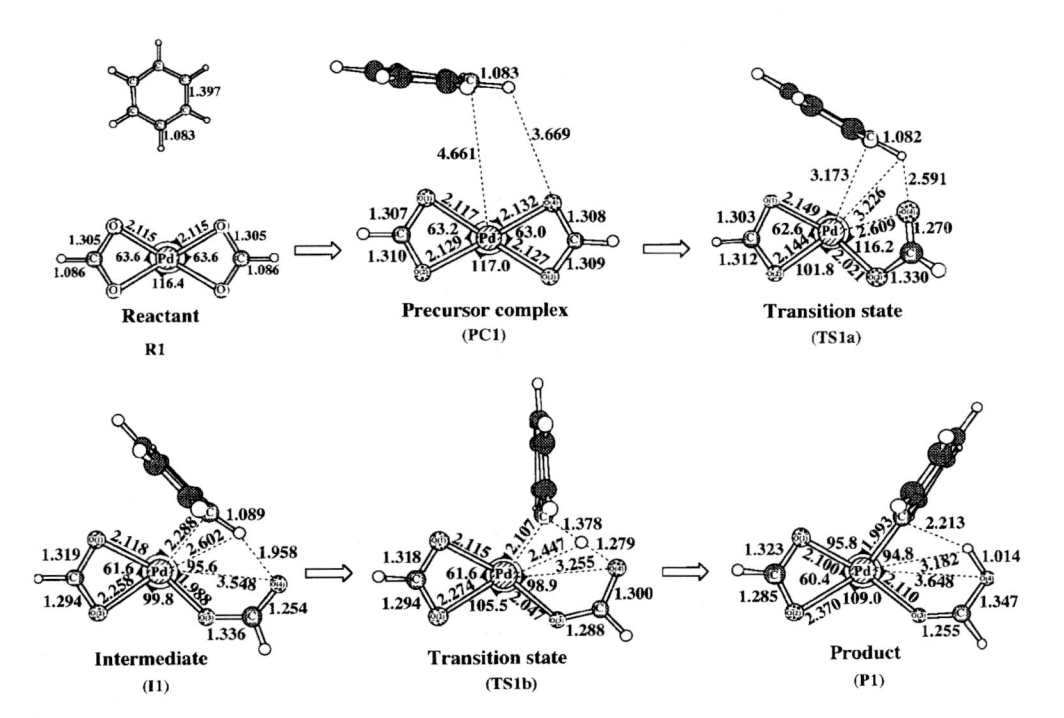

Figure 12.1 Geometry changes in the C–H σ-bond activation of benzene by Pd(η^2-O$_2$CH)$_2$. Bond lengths are in Å and bond angles are in degrees. (From Ref. [12] with permission of the American Chemical Society.)

form benzene adduct **I1**. In the next step, C–H σ-bond activation takes place through the transition state (**TS1b**), to afford Pd(η^2-O$_2$CH)(Ph)(HCOOH) (**P1**). In the transition state **TS1b**, the C–H bond to be broken is elongated to 1.378 Å, and the O–H bond formed shortens to 1.279 Å. The H atom of the broken C–H bond is located in midway between the Ph and OCOH groups. The Ph group is changing its orientation, which is almost intermediate between **I1** and **P1**.

As seen in Table 12.1, the DFT method considerably underestimates the binding energy and the activation barrier, while the MP4(SDQ) and CCSD(T) methods present similar binding energy, activation barrier, and reaction energy. Below, we discuss only the MP4(SDQ) calculated energetics.

Table 12.1 shows that the C–H activation of benzene and methane by Pd(η^2-OCOH)$_2$ occurs with an activation barrier of 15.7 and 21.7 kcal mol^{-1}, respectively. The corresponding values for the platinum catalyst, Pt(η^2-OCOH)$_2$, are 20.9 and 17.3 kcal mol^{-1}. Thus, the Pd-catalyst activates benzene C–H bond with a lower barrier than for methane C–H bond. Conversely, the analogous Pt-catalyst activates the methane C–H bond with a lower barrier than the benzene

Table 12.1 Binding energy (BE), activation barrier (E_a), and reaction energy (ΔE) of the C–H bond activations of benzene and methane by M(η^2-O$_2$CH)$_2$ (M = Pd or Pt). (From Ref. [11].)

Method	Pd(η^2-O$_2$CH)$_2$ + CH$_4$[a]			Pd(η^2-O$_2$CH)$_2$ + C$_6$H$_6$[b]		
	BE[c] (kcal mol^{-1})	E_a[d] (kcal mol^{-1})	ΔE[e] (kcal mol^{-1})	BE[c] (kcal mol^{-1})	E_a[f] (kcal mol^{-1})	ΔE[e] (kcal mol^{-1})
MP2	−1.2	17.5	−12.8	−0.9	11.5	−24.0
MP3	−1.2	19.8	−12.8	−0.4	15.8	−20.4
MP4DQ	−1.2	21.1	−12.0	−0.4	16.3	−19.5
MP4(SDQ)	−1.3	21.5	−8.3	−0.7	15.7	−17.2
CCSD(T)	−1.5	20.5	−6.1	−1.0	14.1	−17.5
DFT(B3LYP)	−0.6	13.9	−4.9	−0.3	9.9	−12.4

	Pt(η^2-O$_2$CH)$_2$ + CH$_4$			Pt(η^2-O$_2$CH)$_2$ + C$_6$H$_6$		
MP4(SDQ)	−2.6	17.3	−13.3	−1.9	20.9	−24.1
CCSD(T)	−3.0	17.7	−12.6	−2.5	20.0	−24.7
DFT(B3LYP	−0.8	11.3	−10.5	−0.4	10.7	−19.3

a) BS-II was used (see Ref. [11]).
b) BS-III was used (see Ref. [11]).
c) BE = E_t(precursor complex) − E_t(sum of reactants).
d) $E_a = E_t$(**TSnb**) − E_t(**PCn**).
e) $\Delta E = E_t$(product) − E_t(sum of reactants).
f) $E_a = E_t$(**TSnB**) − E_t(intermediate).

12.2 Characteristic Features of Heterolytic σ-Bond Activation

C–H bond. The differences in barrier heights is a result of difference in stability of pre-reaction complexes. Indeed, theoretical calculations show that the Pt(II)-benzene complex is more stable than the Pd(II)-benzene complex, as well as the Pt(II)-methane and Pd(II)-methane complexes. The characteristic features of this reaction found are in electron population presented in Figure 12.2(A). Indeed, as seen from this figure, the atomic charge of H (from the broken C–H bond) decreases but the electron populations of methyl and phenyl groups increase during the reaction. In the transition state, the C–H bond to be activated is greatly polarized. These results clearly show that the σ-bond activation takes place heterolytically. It is also noted that the atomic populations of the Pd and Pt considerably increase during the reaction, which is consistent with the experimental proposal that C–H activation is achieved by electrophilic attack of the metal center. In contrast, a completely different trend for the atomic charges was observed in the oxidative addition of the C–H bond to $Pd(PH_3)_2$ and $Pt(PH_3)_2$ complexes (Figure 12.2B). Both the H and phenyl electron populations increase upon reaction, while the Pd and Pt atomic populations decrease considerably.

What orbital interaction induces these population changes in the heterolytic σ-bond activation? Theoretical studies show that the broken C–H bond strongly interacts with orbitals of the $OCOH^-$ ligand via donation and back-donation mechanisms. This orbital picture is completely different from orbital mixing during the oxidative addition process, which involves metal doubly occupied $b_2(d\pi)$ orbitals.

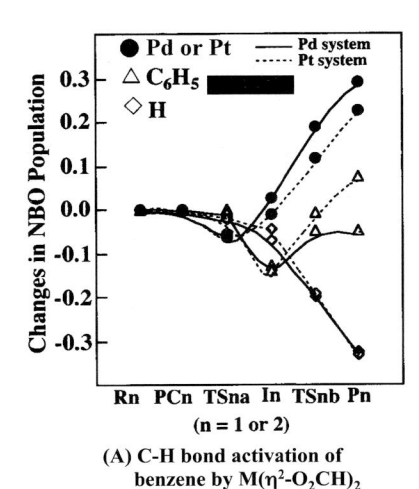

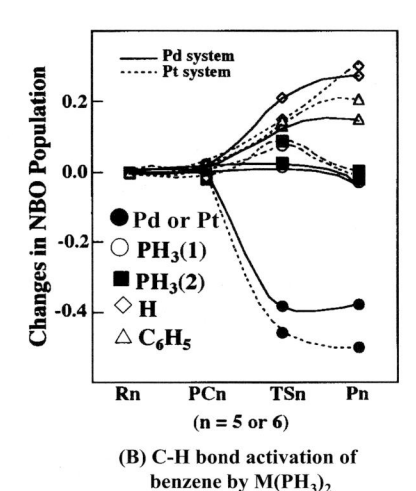

Figure 12.2 Population changes in the C–H σ-bond activations of benzene (A) by $M(\eta^2\text{-}O_2CH)_2$ and (B) by $M(PH_3)_2$ (M = Pd or Pt). A positive value represents an increase in population (and vice versa). Solid lines represent population change for M = Pd, and dotted lines represent population changes for M = Pt. (From Ref. [11] with permission of the American Chemical Society.)

The other important issue is to elucidate why heterolytic σ-bond activation can take place in the Pd(II) complex but not the Pd(0) complex. This is easily interpreted in terms of bond energies. Indeed, in the oxidative addition process a C–H bond (with a 109.0 kcal mol^{-1} bond energy) is broken and the Pd–H (49.5 kcal mol^{-1}) and Pd–Ph (51.0 kcal mol^{-1}) bonds are formed. In heterolytic σ-bond activation, the C–H (with a 109.0 kcal mol^{-1} bond energy) and Pd–O (23.2 kcal mol^{-1}) bonds are cleaved, but the Pd–Ph (51.0 kcal mol^{-1}) and H–OCOH (114.4 kcal mol^{-1}) bonds are formed. Apparently the energy balance (−33.2 kcal mol^{-1}) of Reaction (6) is larger than that (8.5 kcal mol^{-1}) for Reaction (7) because of strong H–OCOH bond formation, heterolytic σ-bond activation occurs much easier for the Pd(II) complex than for the Pd(0)-complex.

$$Pd(II)(\eta^2\text{-}O_2CH)_2 + C_6H_6 \rightarrow (C_6H_5)Pd(II)(\eta^2\text{-}O_2CH)_2(HCOOH) \quad (6)$$

$$Pd(0)(PH_3)_2 + C_6H_6 \rightarrow (C_6H_5)Pd(II)(H)(PH_3)_2 \quad (7)$$

In other words, the formation of a strong H–OCOH bond is driving force of the heterolytic σ-bond activation in this Pd-complex.

12.3
Heterolytic C–H σ-Bond Activation of Methane

Here, we report several important theoretical works on methane C–H bond activation by Pt(II) complexes, metal oxide, and early transition metal complexes.

12.3.1
Methane C–H Bond Activation by Pt(II) Complexes

As mentioned above, Periana and co-workers have experimentally reported the oxidation of methane by Pt(II) complexes in concentrated sulfuric acid [6]. The catalysts are dichloro(η^2-{2,2'-bipyrimidyl})platinum(II), abbreviated as Pt(Bpym)Cl$_2$, and cis-diamminedichloroplatinum(II), which are called catalytica and cisplatin, respectively; methyl bisulfate CH$_3$OSO$_3$H is a direct product of the reaction, which later undergoes hydrolysis to methanol. In the literature there are several theoretical studies on the mechanism of this reaction [13–16]. Because the Pt(II) species are catalysts, both oxidative addition and heterolytic σ-bond activation should be investigated. Ziegler and his colleagues [14] have investigated this reaction for two specific cases: in one case, Cl ligands of the Pt-center remain the same, as proposed by experimentalists, but in the other case Cl ligands were substituted by sulfate anion or OSO$_3$H$^-$. Protonation of bipyrimidine is also considered. When [Pt(Bpym)Cl]$^+$ is an active species, C–H bond activation proceeds via an oxidative addition pathway (Figure 12.3).

However, when [Pt(Bpym)(OSO$_3$H)] used as catalysts it proceeds via a metathesis pathway. Notably, the transition state was not optimized in the case of

12.3 Heterolytic C–H σ-Bond Activation of Methane

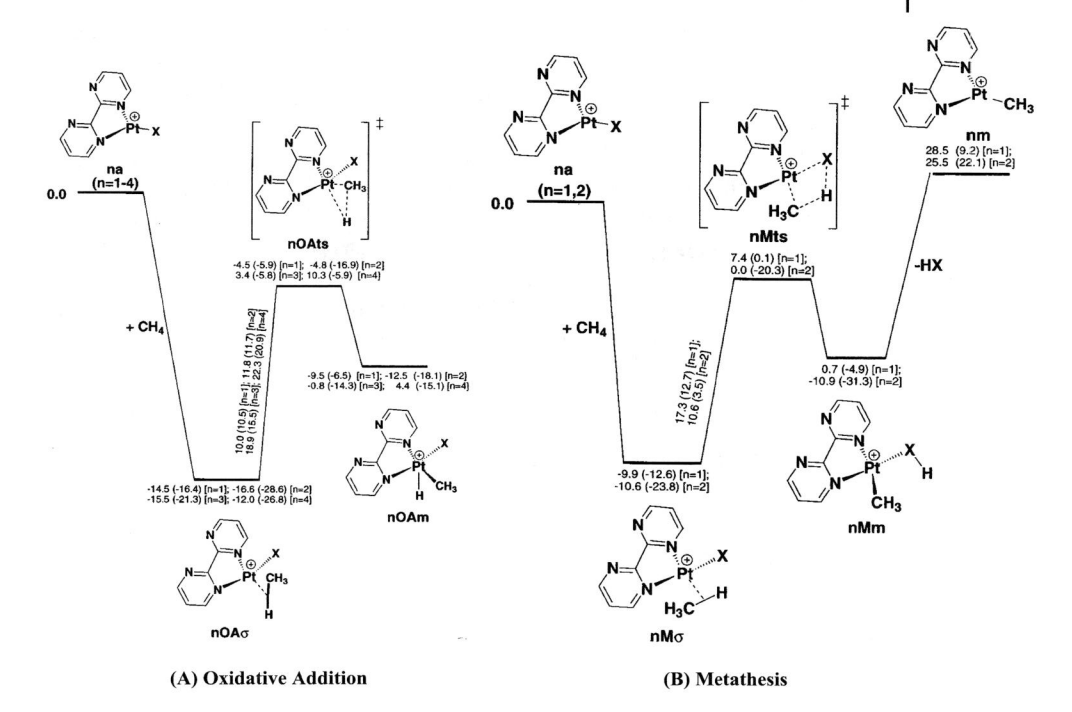

Figure 12.3 Energy changes upon the oxidative addition of methane by Pt(II) and the metathesis processes of the platinum chloride catalysts. (A) The integer *n* in the labels refers to derivatives of (bipyrimidine)PtCl$^+$ ($n=1$), (bipyrimidineH$_2$)PtCl$_3^+$ ($n=2$), (bipyrimidine)Pt(OSO$_3$H)$^+$ ($n=3$), and (bipyrimidineH$_2$)Pt(OSO$_3$H)$_3^+$ ($n=4$), respectively. (B) The integer *n* in the labels refers to derivatives of (bipyrimidine)PtCl$^+$ ($n=1$) and (bipyrimidineH$_2$)PtCl$_3^+$ ($n=2$). The relative energies (kcal mol^{-1}) in the gas phase are shown in parentheses and the corresponding numbers in solution without parentheses. Energies are relative to **na**. (From Ref. [14a] with permission of the American Chemical Society.)

metathesis with [Pt(Bpym)(OSO$_3$H)]$^+$. Goddard and his colleagues also have shown that the reaction mechanisms are different for catalytica and cisplatin [15]. The C–H bond activation of methane proceeds via oxidative addition by *cis*-PtCl$_2$(NH$_3$)$_2$ (Figure 12.4).

In Figure 12.4, **T1** is the transition state of substitution of Cl with methane, and **T2** is the transition state for the heterolytic σ-bond activation. **T2b** is the transition state for oxidative addition leading to species **D**. Apparently, the oxidative addition is slightly easier process than the heterolytic σ-bond activation for this catalyst. Conversely, for PtCl$_2$(Bpym) catalyst the metathesis mechanism occurring via heterolytic σ-bond activation pathway is slightly more favorable than oxidative addition pathway; the transition state for oxidative addition (**T2b**) lies about 9 kcal mol^{-1} higher in energy than that for heterolytic σ-bond activation (**T2**) (Figure 12.5).

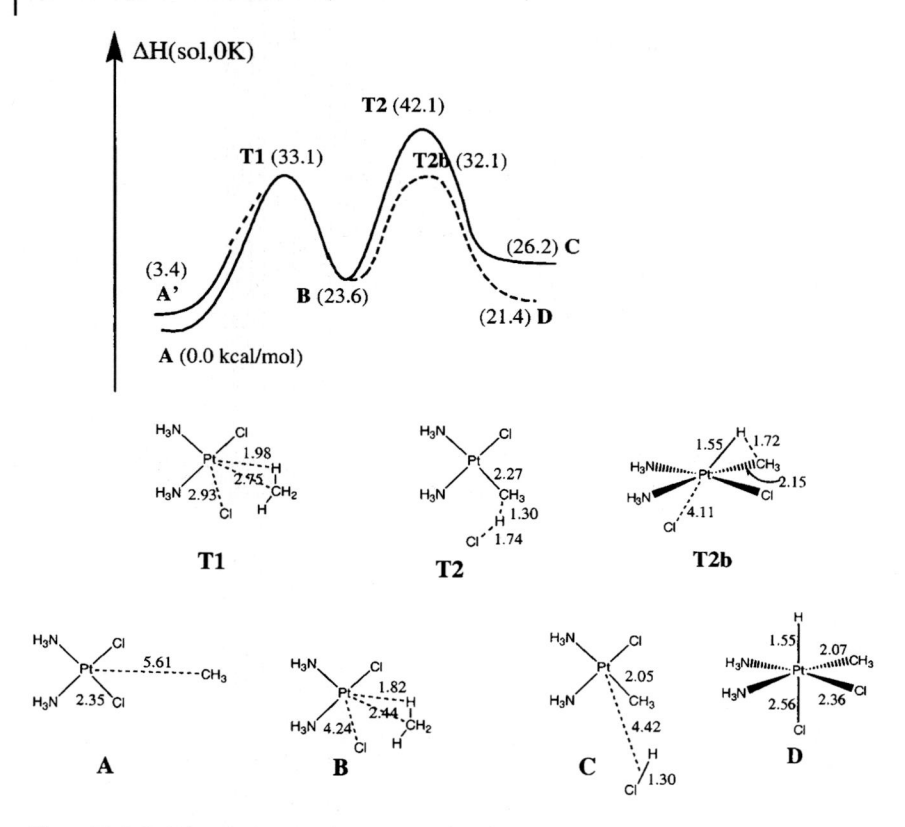

Figure 12.4 C–H bond activation by $(NH_3)_2PtCl_2$. **A'** shows the relative energy for $(NH_3)_2Pt(OSO_3H)Cl$. (From Ref. [15a] with permission of the American Chemical Society.)

The authors discussed the heterolytic C–H σ-bond activation in terms of electrophilic substitution.

It is of considerable interest to clarify the reason why cis-$PtCl_2(NH_3)_2$ favors oxidative addition but $PtCl_2(Bpym)$ favors heterolytic σ-bond activation. The difference between the two complexes is in N-containing ligands coordinated to the Pt-center: NH_3 in cisplatin and bipyrimidine in catalytica. Though the comparison was not made, we expect that the bipyrimidine ligand (which is a conjugated system) tends to keep the +2 oxidation state for the Pt-center.

C–H activation by $Pt(OSO_3H)_2(Bpym)$ and $PtCl_2(Bpym)$ has also been studied [16]. It was shown that in the transition state, Cl^- and OSO_3H^- ligands of these complexes are leaving the coordination sphere of Pt. Therefore, this reaction can be denoted as electrophilic attack of Pt(II) on methane. The Cl system is more reactive than the Bpym system, though the calculated difference is a small. Protonation of Bpym decreases the activation barrier. This would be one of the key points of the reaction, because sulfuric acid was used as solvent.

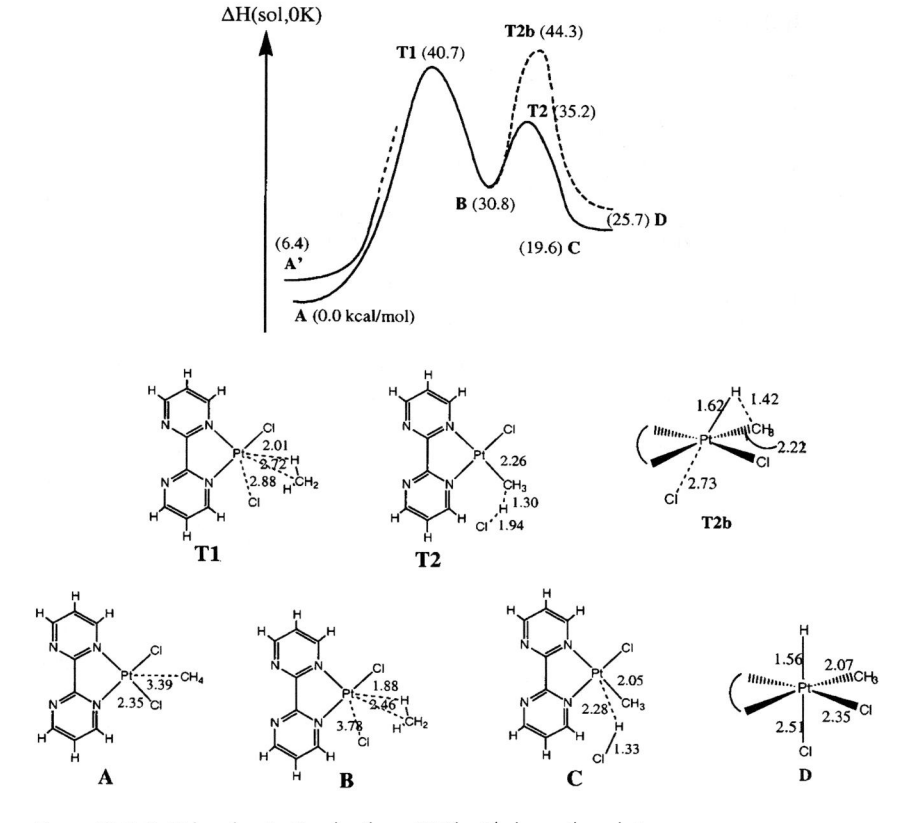

Figure 12.5 C–H bond activation by (bpym)PtCl$_2$. **A'** shows the relative energy for (bpym)Pt(OSO$_3$H)Cl. (From Ref. [15a] with permission of the American Chemical Society.)

12.3.2
Methane C–H bond Activation by Late Transition Metal Catalysts

The conversion of methane into methanol and acetic acid using the [RhI$_2$(CO)$_2$]$^-$ catalyst was experimentally reported by Sen and his colleagues [17]. As could be expected, methane C–H bond activation is one of the important steps of this interesting reaction. Ziegler and Hristov have compared oxidative addition and σ-bond metathesis mechanisms for this reaction. Oxidative addition occurs more favorably than the σ-bond metathesis here [18].

The conversion of methane into methanol by molecular metal oxide has also been reported [19–23] and is discussed in detail by Yoshizawa in Chapter 14.

12.3.3
Heterolytic Methane C–H Bond Activation by Early Transition Metal Systems

Methane C–H bond activation has also been achieved with lanthanocenes, such as $Cp*_2LuCH_3$ ($Cp*$ = pentamethylcyclopentadienyl anion) [24–26]. Eisenstein and coworkers have made a comprehensive theoretical study of the reaction between Cp_2M-CH_3 and methane (M = Y, Sc, and Ln) [27]:

$$Cp_2M-CH_3 + C*H_4 \rightarrow Cp_2M-C*H_3 + CH_4 \tag{8}$$

This reaction proceeds via a metathesis transition state, where the H of the broken C–H bond is located at the middle position between two C atoms (from CH_3 ligand and coordinated CH_4 molecule). Another general feature of this transition state is that the calculated C–H–C angle is close to 180°. This geometrical feature is characteristic of a proton transfer between two negatively charged methyl groups, and indicates that the methane C–H bond activation in these systems occurs in a heterolytic manner. Table 12.2 summarizes the calculated energy barriers of these reactions.

Though the highest values for the activation energies are obtained for the metal with the smallest ionic radii, no clear relationship between the ionic radius and the activation energy is observed. Interestingly, the low activation energies are associated with the large polarization of the C–H bond at the transition state. This large polarization induces large electrostatic interaction between the Cp_2M, CH_3, and $H-CH_3$ moieties, which leads to low activation energy. The early to

Table 12.2 Ionic radius, energy barriers without and with ZPE corrections, enthalpy barriers and activation energies for Reaction (8) $\Delta q(Cp2M)$: difference in the charges on the Cp_2M fragment between the reactant and the transition state, $\Delta q(H-CH_3)$: difference in charges between $H^{(+\delta)}$ and $CH_3^{(-\delta)}$ in the polarized methane at the transition state. (From Ref. [27d].)

M	Ionic radius (Å)	$\Delta E^\ddagger$ (kcal mol^{-1})	$\Delta(E + ZPE)^\ddagger$ (kcal mol^{-1})	$\Delta H^\ddagger$ (kcal mol^{-1})	$\Delta G^\ddagger$ (kcal mol^{-1})	$\Delta q(Cp_2M)$	$\Delta q(H-CH_3)$
Sc	0.87	19.8	20.1	18.6	32.5	−0.20	−0.04
Y	1.019	18.1	17.8	16.6	29.7	−0.10	−0.18
La	1.16	19.0	18.1	17.3	28.5	0.06	−0.32
Ce	1.143	18.7	17.8	17.0	28.0	0.05	−0.32
Sm	1.079	18.1	17.5	16.5	28.4	0.03	−0.30
Ho	1.015	18.9	18.5	17.4	31.1	0.03	−0.20
Yb	0.985	19.8	19.3	18.2	31.9	0.04	−0.30
Lu	0.977	20.3	19.8	18.0	33.3	−0.09	−0.18

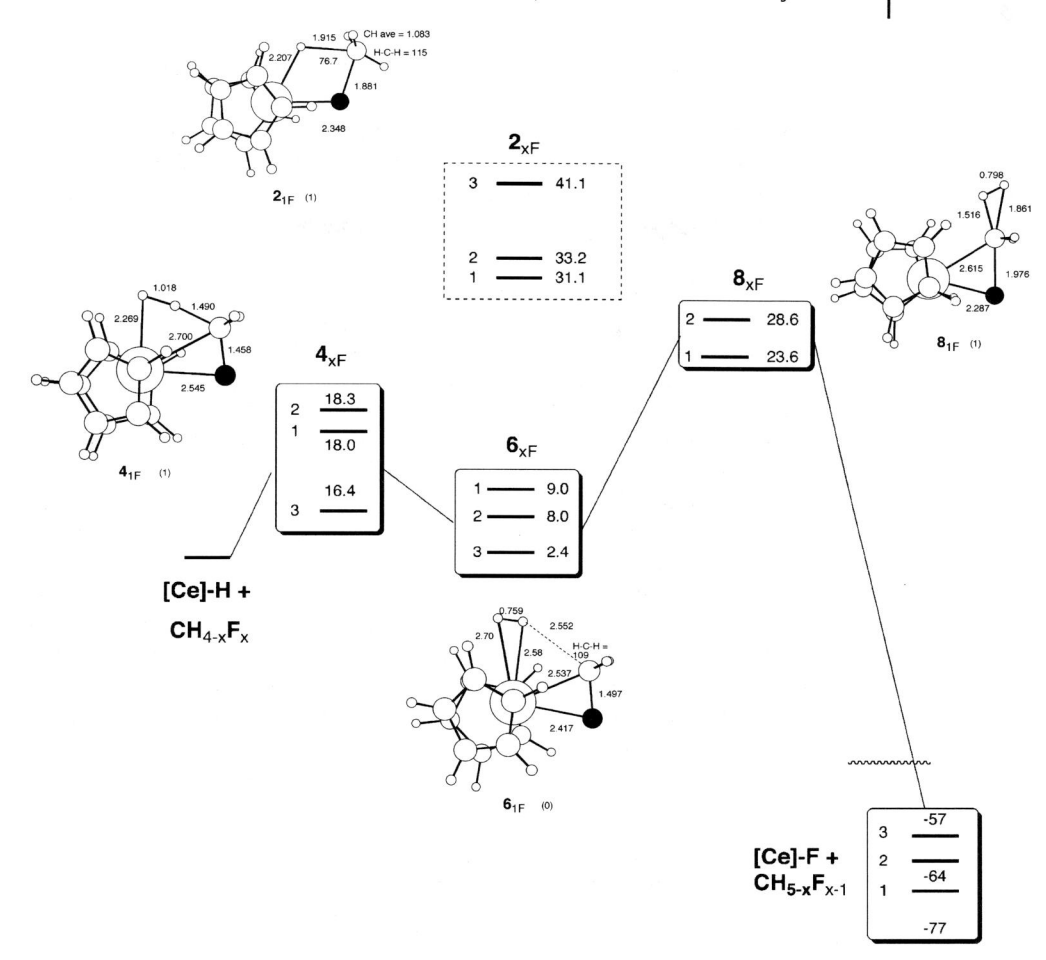

12.3 Heterolytic C–H σ-Bond Activation of Methane | 275

Figure 12.6 Free energy profile in kcal mol^{-1} for the reactions of Cp$_2$CeH and CH$_{4-x}$F$_x$ (x). The value of x is shown on the left at each energy level. Energies in solid boxes correspond to the C–H bond activation via transition state 4$_{xF}$, followed by insertion of the carbene into H$_2$ via transition state 8$_{xF}$ (8$_{3F}$ could not be located). Energies in the dashed box correspond to direct H/F exchange via transition state 2$_{xF}$. (From Ref. [28a] with permission of the American Chemical Society.)

middle lanthanide metals are the most electropositive and, therefore, stronger polarization of C–H bond is expected for these metals. As a result, the heterolytic σ-bond cleavage should occur easily in complexes of these metals.

Exchange of hydrogen by fluorine in CH$_{4-x}$F$_x$ promoted by Cp'$_2$CeH has also been reported [28a]. C–F bond breaking in this reaction is concerted with C–H bond formation, which requires a large activation barrier (see 2$_{1F}$ in Figure 12.6). Thus, direct exchange between F and H should not be considered.

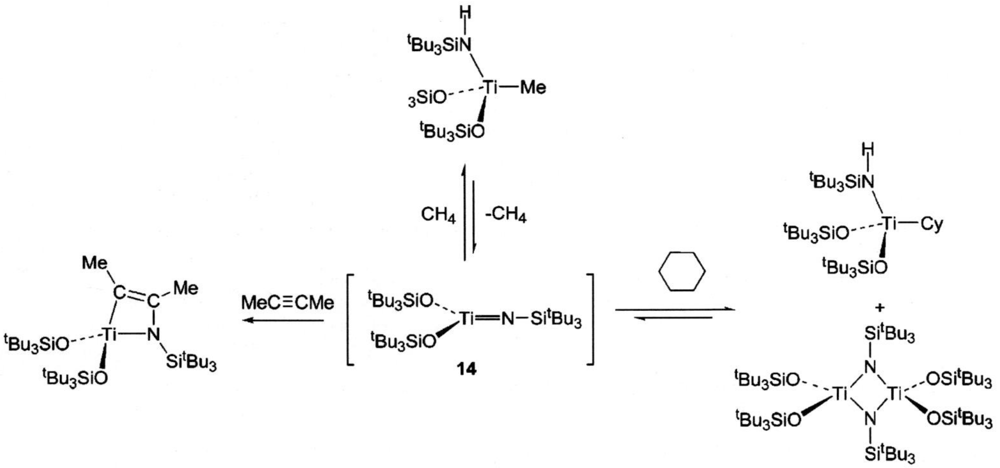

Scheme 12.1 Alkane C–H bond activation by Ti- (and Zr-)imido complexes.

Thus this reaction, most likely, will proceed via C–H bond activation at the transition state 4_{XF}, leading to formation of dihydrogen complex 6_{XF}. At the next stage, the formed H_2 ligand attacks the carbene moiety to form the final product. The later step is shown to be the rate-limiting step of the entire reaction and occurs via transition state 8_{XF}. In the transition state 4_{XF} corresponding to C–H activation, the H–H–CH_2F moiety is collinear, implying that this process is a proton transfer between an anionic alkyl and a hydride ligands. This situation is essentially the same as that of H exchange between $Cp_2M(CH_3)$ and CH_4, discussed above. The relative activation barrier decreases in the order $CH_3F > CH_2F_2 > CHF_3$. Thus, the electron-withdrawing group enhances the proton transfer nature. This means that C–H bond activation in this system takes place heterolytically.

The H-to-F exchange in C_6F_6 and C_6F_5H by the same Cp'_2CeH complex has also been reported [27b]. This reaction takes place via a metathesis pathway including heterolytic σ-bond activation.

Alkane C–H bond activation by Ti- and Zr-imido complexes has been reported experimentally (Scheme 12.1) [29], and theoretically elucidated [30, 31]. This reaction occurs through alkane adduct formation, M(R)(NHR′), and C–H bond activation via a metathesis mechanism. The orbital interaction scheme has been presented clearly based on the component analysis of molecular orbitals of fragments (see Scheme 12.2) [31]. The most important point is mixing of the C–H σ-bonding and σ*-antibonding orbital into the Ti-N d_π-p_π bonding orbital, which leads to the decrease of the H atomic population. The similar C–H bond activation by the Ti-alkylidyne complexes has been experimentally and theoretically investigated by Mindiola et al. [32].

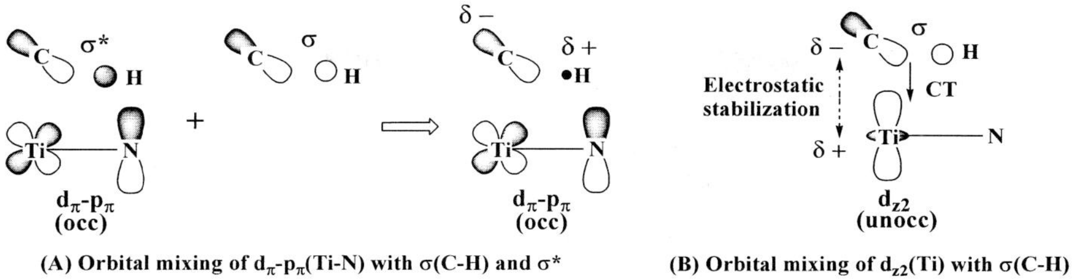

Scheme 12.2 Important orbital interactions in heterolytic C–H bond activation.

(A) Orbital mixing of d_π-p_π(Ti-N) with σ(C-H) and σ^*
(B) Orbital mixing of d_{z^2}(Ti) with σ(C-H)

12.4 Heterolytic σ-Bond Activation of Dihydrogen and Alcohol Molecules

Transition metal catalyzed H–H bond activation is another important process. Through H–H activation via oxidative addition is well known, its heterolytic activation is not well known, but is proposed recently to be important in many reactions. For instance, heterolytic H–H activation has been proposed in the transition metal catalyzed hydrogenation of carbon dioxide. Hydrogen transfer catalysis by Ru complexes, reported by Noyori et al. [33], is also considered to take place via heterolytic σ-bond activation. Here, we report theoretical studies of these two reactions.

12.4.1
Heterolytic H–H Bond Activation

Transition metal catalyzed hydrogenation of carbon dioxide is one of challenging reactions of organometallic chemistry:

$$CO_2 + H_2 \rightarrow HCOOH \tag{9}$$

Rh(I)-catalyzed hydrogenation of carbon dioxide has been investigated at the MP2 and QCISD levels of theory [34]. Rh(H)(PH$_3$)$_2$ was used as a model active species. The first step of this reaction is CO$_2$ insertion into the Rh–H bond. Subsequent reaction with H$_2$ can go either via oxidative addition or metathesis transition states. The latter pathway is more favorable (Figure 12.7).

CO$_2$ insertion into Rh–H bond occurs with a 4.3 kcal mol^{-1} barrier at the MP2 level, while this barrier disappears at the QCISD(T) level. The oxidative addition of H$_2$ to Rh(η^1-OCOH)(PH$_3$)$_2$ is a barrierless process and leads to complex **5d**. However, subsequent reductive elimination from **5d** has large, 24.7 kcal mol^{-1}, barrier at the QCISD(T) level. Addition of H$_2$ to Rh(η^1-OCOH)(PH$_3$)$_2$ via the metathesis pathway is also a barrierless process, which leads to the thermodynamically less (compared with intermediate **5d**) stable intermediate **8**. In **9** (TS) one H (hydride) of H$_2$ molecule starts to be bound with Rh and another one (proton) bound with formate; in other words, the H–H cleavage occurs heterolytically. This metathesis reductive elimination from **8** meets a relatively small

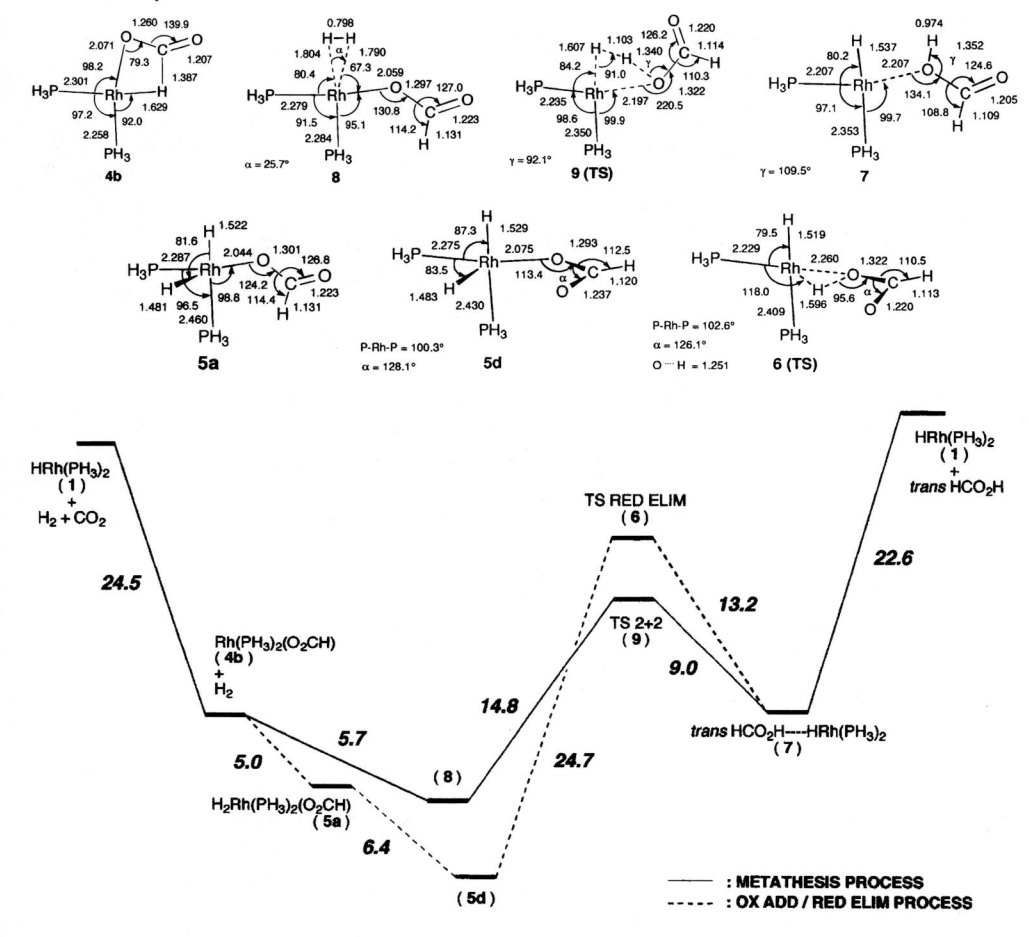

Figure 12.7 (a) Proposed mechanism for the hydrogenation of CO_2 to trans-HCO_2H, catalyzed by $RhH(PH_3)_2$. Energies shown [kcal mol^{-1}; at the QCISD(T)//MP2 level] are relative to the [$(PH_3)_2RhH + CO_2 + H_2$] model system. (From Ref. [34] with permission of the American Chemical Society.)

(only 14.8 kcal mol^{-1}) barrier and, therefore, we expect that the reaction will follow a metathesis pathway through heterolytic H–H activation.

Ru(II)-catalyzed hydrogenation of carbon dioxide has also been reported experimentally [35], and studied theoretically at the DFT, MP2 and MP4(SDQ) levels of theory [36]. The complex cis-$Ru(H)_2(PH_3)_3$ was employed as a model of the active species in earlier theoretical work [36a]. It was found that the first step of the reaction is CO_2 insertion into the Ru–H bond, to form cis-$Ru(H)(\eta^1$-$OCOH)(PH_3)_3$. The next step is either reductive elimination of formic acid or σ-bond metathesis. In the reductive elimination, three- and five-center transition

12.4 Heterolytic σ-Bond Activation of Dihydrogen and Alcohol Molecules

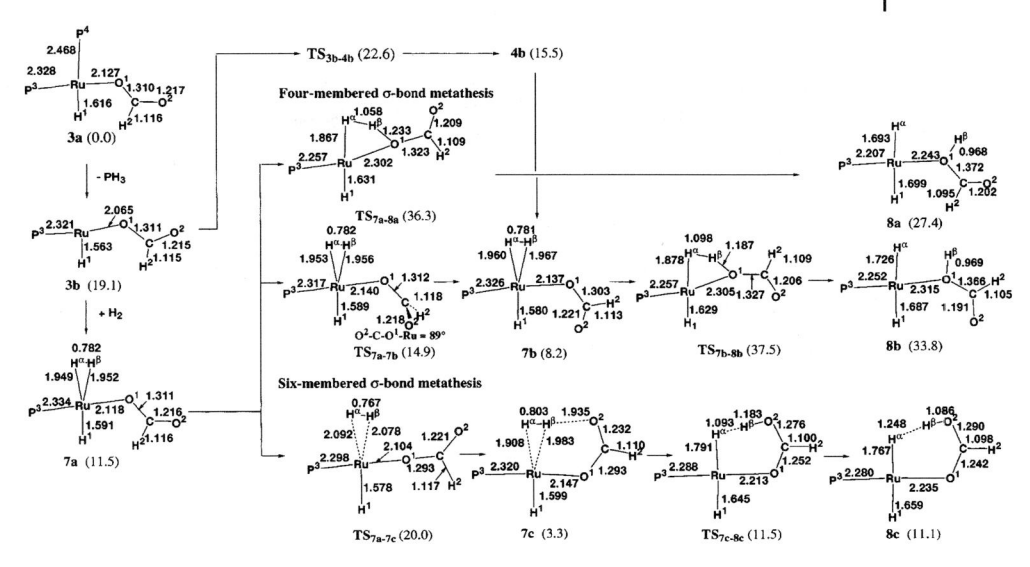

Figure 12.8 Geometry changes in the σ-bond metathesis of $RuH(\eta^1\text{-}OCOH)(PH_3)_3$ with H_2, in which H_2 is trans to H^1. Bond distances are in Å and bond angles in degrees. In parentheses are the energy differences from **3a** [kcal mol^{-1}; DFT(B3LYP)/BS-II//DFT(B3LYP)/BS-I calculation]. The PH_3 ligands perpendicular to the P^3-Rh-H^1 plane are omitted for brevity. (From Ref. [36a] with permission of the American Chemical Society.)

states were investigated. In the metathesis, four- and six-center transition states were investigated. Reductive elimination is not discussed here because it is less favorable than σ-bond metathesis. Figure 12.8 gives important intermediates and transition states along with their geometry parameters.

As seen from Figure 12.8, the four-center transition state is less stable than the six-center transition state. This difference in stability can be explained in terms of better overlap of the O lone pair of formate and the H 1s orbital. A six-center transition state is more favorable than a four-center transition state in many cases.

Later, Sakaki and coworkers have applied a larger real complex, cis-$Ru(H)_2(PMe_3)_3$, to study this reaction [34b], which has provided essentially the same conclusion as the smaller model cis-$Ru(H)_2(PH_3)_3$.

They also have investigated the reason of why metathesis is more favorable than reductive elimination [37]. Theoretical study has shown that in $Ru(H)(\eta^1\text{-}OCOH)(PH_3)_3$ the Ru–H bond is strong, suppressing reductive elimination. In the metathesis, the strong Ru–H bond is not consumed but, rather, an additional Ru–H bond is formed. Thus, the σ-bond metathesis occurs more favorably than the reductive elimination because of the strong Ru–H bond. If the M–H bond is weak, the reductive elimination becomes more favorable. For instance, reductive elimination easily takes place in $[Rh(H)(\eta^1\text{-OCOH})(PH_3)_3]^+$, because the Rh(III)–H bond is not strong.

12.4.2
Heterolytic σ-Bond Activation in Hydrogen Transfer Reactions

Hydrogen transfer from alcohol to ketone is very interesting because this reaction permits the hydrogenation of a ketone without dihydrogen gas. The catalytic system that was theoretically examined consists of $[RuCl_2(\eta^6\text{-benzene})]_2$, N-tosylethylenediamine (or ethanolamine), and KOH [38]. Scheme 12.3 shows the active species **15** formed from these compounds.

The base is necessary to extract X from the Ru center to afford coordinatively unsaturated species **15**. Methanol approaches **15** to form hydrogen bonding with the N atom of the amine moiety. Then, C–H activation is achieved to form **18**. In the transition state **17**, one H atom transfers from the O atom of alcohol to the amine moiety as a proton and the other H atom transfers from the C atom of

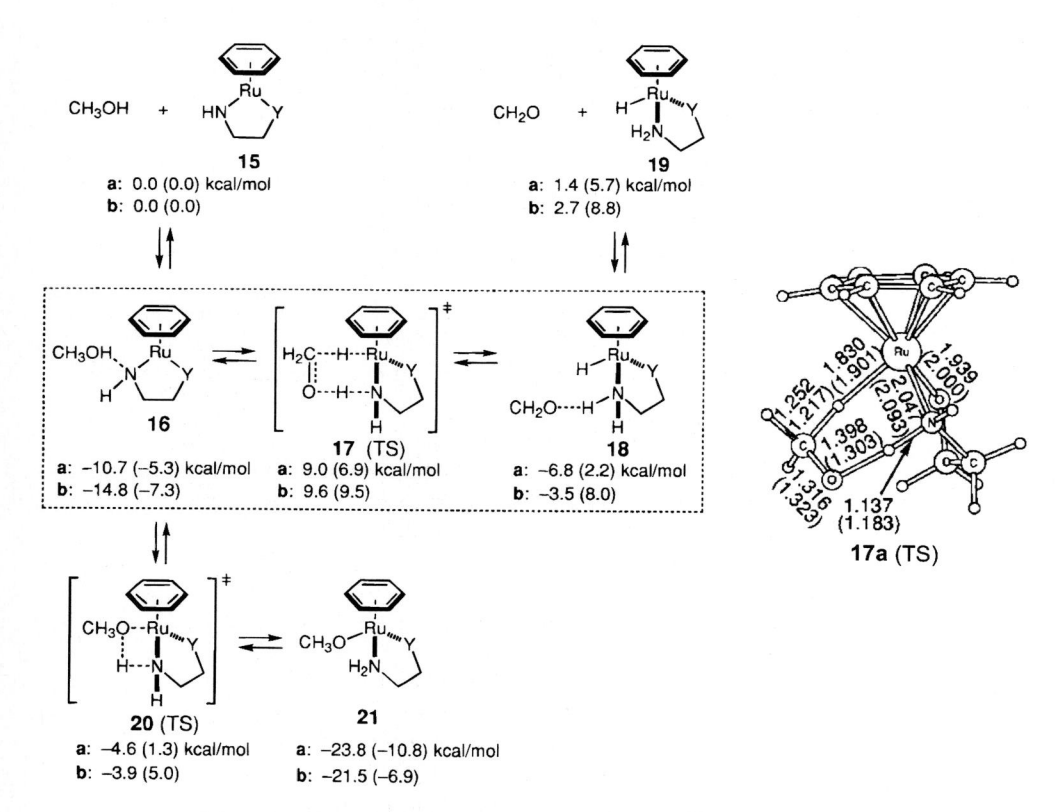

Scheme 12.3 Calculated mechanism for ruthenium(II)-catalyzed hydrogen transfer between methanol and formaldehyde. Relative energies calculated at the B3LYP level are given in parentheses. (From Ref. [38] with permission of the American Chemical Society.)

12.4 Heterolytic σ-Bond Activation of Dihydrogen and Alcohol Molecules

alcohol to the Ru center as hydride. In other words, O–H and C–H bonds are broken simultaneously. This σ-bond activation is completely different from oxidative addition. If formaldehyde dissociates from the catalyst and new ketone reacts with **18**, the transfer hydrogenation of ketone can be completed, which is the reverse reaction of the C–H and O–H σ-bond activations. This reaction course is called a pericyclic mechanism. The other mechanism was also investigated, in which methoxide interacts with the Ru center to form the intermediate **21**, and then β-H abstraction takes place from **21** to form the ruthenium hydride complex **24** (Scheme 12.4).

The intermediate **21** is very stable, and β-H elimination from it has to overcome a large barrier. Also, the elimination of formaldehyde from the Ru center exhibits a considerably large activation barrier. Thus, the β-hydrogen abstraction course is much less favorable than the pericyclic reaction course. It is unclear why the β-hydrogen abstraction is difficult. Usually, β-hydrogen abstraction takes place easily, and the transition state (**23**) is not unusual. Though the details are not clear, the transition state of β-hydrogen abstraction is four-centered, but the transition state of the pericyclic mechanism is six-centered. A smaller strain energy is one reason for the low activation barrier of the pericyclic mechanism; remember six-center transition state is more favorable than four-center one in the heterolytic H–H bond activation by Ru(II)-formate complex [36].

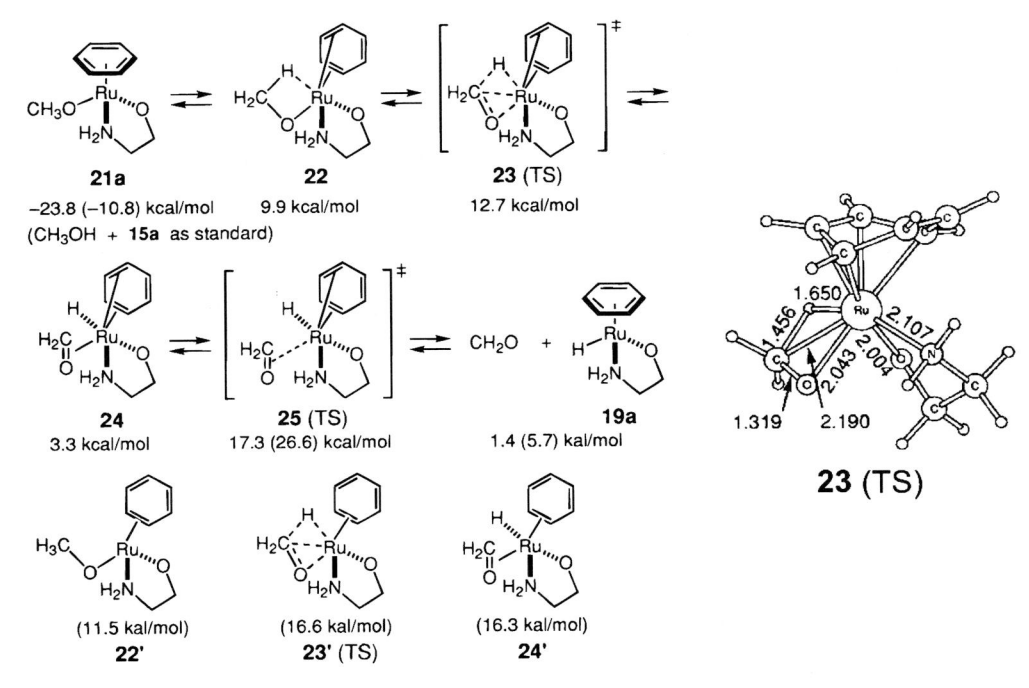

Scheme 12.4 Calculated β-elimination mechanism. Relative energies calculated at the B3LYP level are given in parentheses. (From Ref. [38] with permission of the American Chemical Society.)

12.5
Summary

Various heterolytic σ-bond activation reactions are observed in the chemistry of transition metal complexes. Theoretical studies have succeeded in presenting detailed knowledge of such heterolytic σ-bond activation. Important results are summarized as follows: In many heterolytic σ-bond activation reactions, the alkyl or aryl group starts to interact with the metal center and the H atom starts to interact with another (X) ligand. The σ-bond to be activated is considerably polarized at the transition state. Though only the metal center plays important role in oxidative addition reaction, the metal and ligand cooperatively participate in heterolytic σ-bond activation. Actually, one important driving force is the formation of the X–H bond in the case of heterolytic σ-bond activation. Detailed knowledge of the X–H bond energy is useful to construct the reaction system for heterolytic σ-bond activation. The other important issue of the heterolytic σ-bond activation is the orbital interaction diagram. The doubly occupied orbital of the X ligand undergoes anti-bonding overlap with the C–H σ-bonding orbital and bonding overlap with the C–H σ*-anti-bonding orbital, which greatly decreases the H atomic population and induces the polarization of the C–H bond.

Though we know what metal is reactive for oxidative addition and why, such general knowledge is not enough for the heterolytic σ-bond activation, and so more theoretical studies are needed.

References

1 Recent reviews: (a) B. A. Arndtsen, R. G. Bergman, T. A. Mobley, T. H. Peterson, *Acc. Chem. Res.* 1995, **28**, 154. (b) A. E. Shilov, G. B. Shul'pin, *Chem. Rev.* 1997, **97**, 2879. (c) W. D. Jones, *Top. Organomet. Chem.* 1999, **3**, 9. (d) A. Sen, *Top. Organomet. Chem.* 1999, **3**, 81. (e) R. H. Crabtree, *J. Chem. Soc., Dalton Trans.* 2001, 2437. (f) J. A. Labingerand, J. E. Bercaw, *Nature*, 2002, **417**, 507.
2 A. Dedieu, *Chem. Rev.* 2000, **100**, 543.
3 S. Sakaki, *Top. Organomet. Chem.* 2005, **12**, 31.
4 I. Moritani, Y. Fujiwara, *Tetrahedron Lett.* 1967, 1119; Y. Fujiwara, K. Takagi, Y. Taniguchi, *Synlett* 1996, 591, and references therein.
5 E. Hennessy, S. L. Buchwald, *J. Am. Chem. Soc.* 2003, **125**, 12084; W. C. P. Tsang, N. Zheng, S. L. Buchwald, *J. Am. Chem. Soc.* 2007, **127**, 14560.
6 R. A. Periana, D. J. Taube, E. R. Evitt, D. G. Loffler, P. R. Wentrcek, G. Voss, T. Masuda, *Science* 1993, **259**, 340; R. A. Periana, D. J. Taube, S. Gamble, H. Taube, T. Satoh, F. Fujii, *Science* 1998, **280**, 560.
7 See reviews: G. Zssinovich, G. Mestroni, *Chem. Rev.* 1992, **92**, 1051; C. F. de Graauw, J. A. Peters, H. van Bekkum, J. Huskens, *Synthesis* 1994, 1007.
8 R. Noyori, T. Ohkuma, *Pure Appl. Chem.* 1999, **71**, 1493; R. Noyori, T. Ohkuma, *Angew. Chem., Int. Ed. Engl.* 2001, **40**, 40; R. Noyori, M. Koizumi, D. Ishii, T. Ohkuma, *Pure Appl. Chem.* 2001, **73**, 227; R. Noyori, *Angew. Chem., Int. Ed. Engl.* 2002, **41**, 2008; R. Noyori, *Adv. Synth. Catal.* 2003, **345**, 15.
9 (a) R. J. McKinney, D. L. Thorn, R. Hoffmann, A. Stockis, *J. Am. Chem. Soc.* 1981, **103**, 2595. (b) K. Tatsumi, R. Hoffmann, A. Yamamoto, J. K. Stille, *Bull. Chem. Soc. Jpn.* 1981, **54**, 1857. (c) J. Y. Saillard, R. Hoffmann, *J. Am. Chem. Soc.* 1984, **106**, 2006.
10 J. J. Low, W. A. Goddard III., *J. Am. Chem. Soc.* 1984, **106**, 6928.

References

11 P. E. M. Siegbahn, R. H. Crabtree, *J. Am. Chem. Soc.* 1996, **118**, 4442.

12 B. Biswas, M. Sugimoto, S. Sakaki, *Organometallics* 2000, **19**, 3895.

13 (a) K. Mylavaganam, G. B. Backsay, N. S. Hush, *J. Am. Chem. Soc.* 1999, **121**, 4633. (b) K. Mylavaganam, G. B. Bacskay, N. S. Hush, *J. Am. Chem. Soc.* 2000, **122**, 2041.

14 (a) T. M. Gilbert, I. Hristov, T. Ziegler, *Organometallics* 2001, **20**, 1183. (b) I. H. Hristov, T. Ziegler, *Organometallics*, 2003, **22**, 1668.

15 (a) J. Kua, X. Xu, R. A. Periana, W. A. Goddard, *Organometallics* 2002, **21**, 511. (b) X. Xu, J. Kua, R. A. Periana, W. A. Goddard, *Organometallics* 2003, **22**, 2057.

16 A. Paul, C. B. Musgrave, *Organometallics* 2007, **26**, 791.

17 M. Lin, A. Sen, *Nature* 1994, **368**, 613. (b) M. Lin, T. E. Hogan, A. Sen, *J. Am. Chem. Soc.* 1996, **118**, 4574.

18 I. H. Hristov, T. Ziegler, *Organometallics* 2003, **22**, 3513.

19 (a) D. Schröder, H. Schwartz, *Angew. Chem., Int. Ed. Engl.* 1990, **29**, 1433. (b) D. Schröder, A. Fiedler, J. Hrusak, H. Schwartz, *J. Am. Chem. Soc.* 1992, **114**, 1215. (c) D. Schröder, H. Shwartz, *Angew. Chem., Int. Ed. Engl.* 1995, **34**, 1973. (d) H. Schwartz, D. Schröder, *Pure Appl. Chem.* 2000, **72**, 2319.

20 Y.-M. Chen, D. E. Clemmer, P. B. Armentrout, *J. Am. Chem. Soc.* 1994, **116**, 7815.

21 K. Yoshizawa, *Acc. Chem. Res.* 2006, **39**, 375 and references therein.

22 (a) K. Yoshizawa, Y. Shiota, T. Yamabe, *Chem. Eur. J.* 1997, **3**, 1160. (b) K. Yoshizawa, Y. Shiota, T. Yamabe, *J. Am. Chem. Soc.* 1998, **120**, 564. (c) K. Yoshizawa, Y. Shiota, T. Yamabe, *J. Chem. Phys.* 1999, **111**, 538. (d) Y. Shiota, K. Yoshizawa, *J. Chem. Phys.* 2003, **118**, 5872.

23 Y. Shiota, K. Yoshizawa, *J. Am. Chem. Soc.* 2000, **122**, 12317.

24 P. L. Watson, G. W. Parshall, *Acc. Chem. Res.* 1985, **18**, 51.

25 M. E. Thompson, S. M. Baxter, A. R. Bulls, B. J. Burger, M. C. Nolan, B. D. Santarseiro, W. P. Schaefgr, J. E. Bercaw, *J. Am. Chem. Soc.* 1987, **109**, 203. R. L. Jordan, D. F. Taylor, *J. Am. Chem. Soc.* 1989, **111**, 778.

26 A. D. Sadow, T. D. Tiley, *Angew. Chem., Int. Ed. Engl.* 2003, **42**, 803.

27 (a) L. Maron, O. Eisenstein, *J. Am. Chem. Soc.* 2001, **123**, 1036. (b) L. Maron, L. Perrin, O. Eisenstein, *J. Chem. Soc., Dalton Trans.* 2002, 534. (c) L. Perrin, L. Maron, O. Eisenstein, *Inorg. Chem.* 2002, **41**, 4355. (d) N. Barros, O. Eisenstein, L. Maron, *J. Chem. Soc., Dalton Trans.* 2006, 3052.

28 (a) E. V. Werkema, E. Messines, L. Perin, L. Maron, O. Eisenstein, R. A. Andersen, *J. Am. Chem. Soc.* 2005, **127**, 7781. (b) L. Maron, F. V. Werkema, L. Perin, O. Eisenstein, R. A. Andersen, *J. Am. Chem. Soc.* 2005, **127**, 279.

29 N. Hazari, P. Mountford, *Acc. Chem. Res.* 2005, **38**, 839.

30 (a) T. R. Cundari, *J. Am. Chem. Soc.* 1992, **114**, 10557. (b) T. R. Cundari, *Organometallics* 1993, **12**, 1998. (c) T. R. Cundari, *Organometallics* 1993, **12**, 4971. (d) T. R. Cundari, *J. Am. Chem. Soc.* 1994, **116**, 340. (e) M. T. Benson, T. R. Cundari, E. W. Moody, *J. Organomet. Chem.* 1995, **504**, 1. (f) T. R. Cundari, N. Matsunaga, E. W. Moody, *J. Phys. Chem.* 1996, **100**, 6475. (g) T. R. Cundari, T. R. Klinchkman, P. T. Wolczanski, *J. Am. Chem. Soc.* 2002, **124**, 1481.

31 N. Ochi, Y. Nakao, H. Sato, S. Sakaki, *J. Am. Chem. Soc.* 2007, **129**, 8615.

32 B. C. Bailey, H. Fan, J. C. Huffman, M. M. Baik, D. J. Mindiola, *J. Am. Chem. Soc.* 2007, **129**, 8781.

33 R. Noyori, S. Hashiguchi, *Acc. Chem. Res.* 1997, **30**, 97 and references therein.

34 (a) F. Hutschka, A. Dedieu, W. Leitner, *Angew. Chem., Int. Ed. Engl.* 1995, **34**, 1742. (b) F. Hutschka, A. Dedieu, M. Eichberger, R. Fornika, W. Leitner, *J. Am. Chem. Soc.* 1997, **119**, 4432.

35 (a) P. G. Jessop, T. Ikariya, R. Noyori, *Nature* 1994, **368**, 231. (b) P. G. Jessop, Y. Hsiano, T. Ikariya, R. Noyori, *J. Am. Chem. Soc.* 1994, **116**, 8851. (c) P. G. Jessop, T. Ikariya, R. Noyori, *J. Am. Chem. Soc.* 1996, **118**, 344.

36 (a) Y. Musashi, S. Sakaki, *J. Am. Chem. Soc.* 2000, **122**, 3867. (b) Y. Ohnishi, T. Matsunaga, Y. Nakao, H. Sato, S. Sakaki, *J. Am. Chem. Soc.* 2005, **127**, 4021.

37 Y. Musashi, S. Sakaki, *J. Am. Chem. Soc.* 2002, **124**, 7588.

38 M. Yamakawa, H. Ito, R. Noyori, *J. Am. Chem. Soc.* 2000, **122**, 1466.

13
Hydrosilylation Reactions Discovered in the Last Decade: Combined Experimental and Computational Studies on the New Mechanisms

Yun-Dong Wu, Lung Wa Chung, and Xin-Hao Zhang

13.1
Introduction

Hydrosilylation, which involves the addition of a Si–H bond of hydrosilane across an unsaturated multiple bond of alkene, alkyne or carbonyl compounds, provides the most straightforward and atom-economical method for the generation of versatile silicon-containing intermediates (Scheme 13.1) [1–4]. It plays an important role in organic synthesis, dendrimer and polymer chemistry [1], and can take place under different conditions, including high temperature (>300 °C), UV-irradiation γ-irradiation, electric charge or radical initiator [1]. Milder reaction conditions and high regio-, enantio-, diastereo- and chemo-selectivities can be obtained by using organometallic catalysts [1, 5]. Since the first mechanism, the Chalk–Harrod mechanism, was proposed in 1965 (Scheme 13.2), several more mechanistic pathways for hydrosilylation reactions have been proposed or demonstrated [1]. The reader should refer to excellent reviews on the topic [1]. Several new developments, which implicate new reaction mechanisms, have been reported recently. In this chapter, we highlight these developments. In particular, experimental studies and quantum mechanics calculations that shed light on mechanistic pathways are discussed.

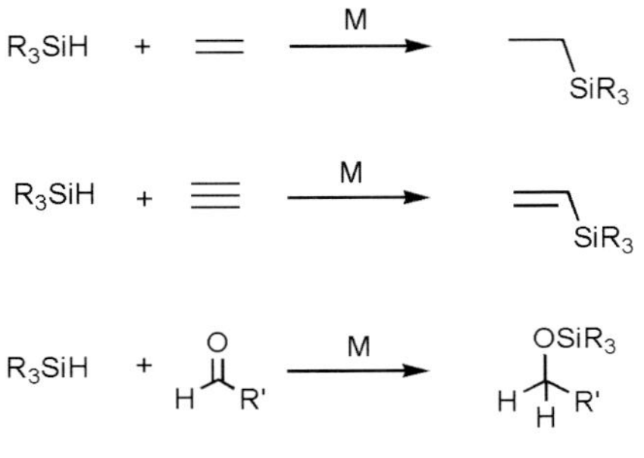

Scheme 13.1 Hydrosilylation of alkenes, alkynes and carbonyl compounds.

Computational Modeling for Homogeneous and Enzymatic Catalysis.
A Knowledge-Base for Designing Efficient Catalysts. K. Morokuma and D. G. Musaev (Eds.)
Copyright © 2008 WILEY-VCH Verlag GmbH & Co. KGaA, Weinheim
ISBN: 978-3-527-31843-8

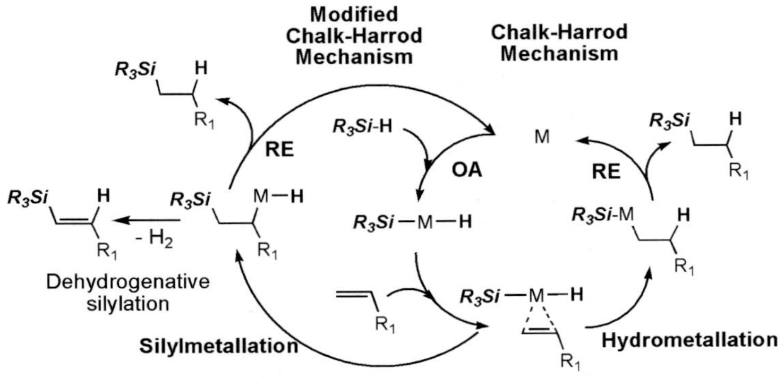

Scheme 13.2 Chalk–Harrod and modified Chalk–Harrod mechanisms.

13.2
General Mechanistic Pathways in the 20th Century

13.2.1
Chalk–Harrod and Modified Chalk–Harrod Mechanisms

Before we advance the latest proposed mechanistic pathways, it is beneficial to briefly present several previously proposed mechanisms [1]. Platinum complexes such as Speier's and Karstedt's catalysts have been widely used as efficient catalysts for alkene hydrosilylation [6, 7]. Chalk and Harrod proposed catalytic cycles for the platinum-catalyzed hydrosilylation of alkene (Scheme 13.2) [8]. This mechanism starts with oxidative addition (**OA**) of the H–Si bond of the silane to the low-valent metal center to generate a metal-hydrido-silyl intermediate. This is followed by insertion of the alkene into the M–H bond (hydrometallation), to give a σ-vinyl intermediate, and final reductive elimination (**RE**) step. The Chalk–Harrod mechanism successfully explains the cis-addition stereochemistry and anti-Markovnikov regiochemistry of platinum-catalyzed hydrosilylation.

The reaction mechanism of hydrosilylation with different silanes (H_4Si, $HSiMe_3$ or $HSiCl_3$) catalyzed by a model catalyst, $Pt(PH_3)_2$, was theoretically studied by Sakaki's group [9]. Isomerization of a $PtEt(SiR_3)(PH_3)$ intermediate formed from hydrometallation was shown to be the rate-determining step in the Chalk–Harrod mechanism with an activation barrier of 22.4–25.7 kcal mol^{-1} at the MP4SDQ level. The final Si–C reductive elimination step proceeds with a low barrier of 9.2–13.8 kcal mol^{-1}, when one π-acid alkene ligand coordinates to the Pt metal. However, the calculated barriers of the Si–C reductive elimination are underestimated, compared with its experimental value for the complex $PtMe(SiPh_3)(PhC\equiv CPh)(PMe_2Ph)$ ($\Delta H^{\ddagger} = 20.8$ kcal mol^{-1} and $\Delta G^{\ddagger} = 22.8$ kcal mol^{-1}) [10]. The discrepancy between the theory and experiment may be resulted from the simplification of the catalyst used in the calculations, since the calcu-

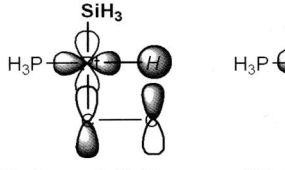

Hydrometallation   **Silylmetallation**

Scheme 13.3 Schematic presentation of molecular orbitals of hydrometallation and silylmetallation steps.

lated barrier of reductive elimination with PtR(SiPh$_3$)(PhC≡CPh)(PMe$_3$) (R = Me or ^{i}Pr) complexes is about 22.0–28.9 kcal mol^{-1} at the B3LYP level [11]. In contrast, a silylmetallation step is the rate-determining step in the modified Chalk–Harrod mechanism, with a barrier of 41–60 kcal mol^{-1}, due to poor orbital overlapping between the Pt and Si groups to form the Pt–Si bond (Scheme 13.3). Therefore, the Pt(0)-catalyzed hydrosilylation of alkenes is expected to proceed via the Chalk–Harrod mechanism.

The reaction mechanism of the Pt(0)-catalyzed hydrosilylation of alkyne has been theoretically studied by Tsipis et al. [12]. The calculated barriers (at the CCSD(T)//B3LYP level) for the oxidative addition of SiH$_4$ to Pt(PH$_3$)(HC≡CH) complex, the subsequent hydrometallation and reductive elimination are 7.1, 11.6 and 5.1 kcal mol^{-1}, respectively. Therefore, hydrometallation step was supposed to be the rate-determining step for hydrosilylation of alkyne.

Roy and Taylor have characterized a platinum(II) disilyl intermediate (**1**) ([Pt(SiR$_3$)$_2$(COD)]), by multinuclear NMR spectroscopy (^{13}C, ^{29}Si and ^{195}Pt) and X-ray crystallography, as the active species for hydrosilylation reaction when a precatalyst [PtCl$_2$(COD)] was used [13]. The Chalk–Harrod mechanism involving Pt(II)/Pt(IV) redox couples was proposed (Scheme 13.4). Concurrently, an X-ray crystal structure of a Pt(IV) silyl complex, [κ^2-((Hpz*)BHpz*$_2$)Pt(H)$_2$(SiEt$_3$)][BAr′$_4$], was reported [14]. The feasibility of the classical Chalk–Harrod mechanism involving the Pt(II)/Pt(IV) redox couple was also implicated by calculations. A computational study on the activation of a model precatalyst, (COD)Pt(C$_2$H$_5$)$_2$, conducted by Thiel et al. [15] indicated that the barrier of oxidative addition of the Si–H bond of HSiMe$_3$ to (COD)Pt^{II}(C$_2$H$_5$)$_2$ is about 27 kcal mol^{-1} (at the BP86 level), which is close to the activation energies experimentally reported for three different precatalysts (25.4–29.5 kcal mol^{-1}). This computational result suggests that the first catalytic step (**1** → **2**) in Scheme 13.4 might have a relatively high barrier.

Whether hydrosilation reactions catalyzed by other late transition metal complexes also undergo via the Chalk–Harrod mechanism has been questioned, owing to the appearance of dehydrogenative silylation products in the reaction of alkenes and alkynes, and trans-addition products in the reaction of alkynes. These unexpected products cannot be envisioned from the Chalk–Harrod mechanism, implying alternative mechanisms. For iron, cobalt and rhodium com-

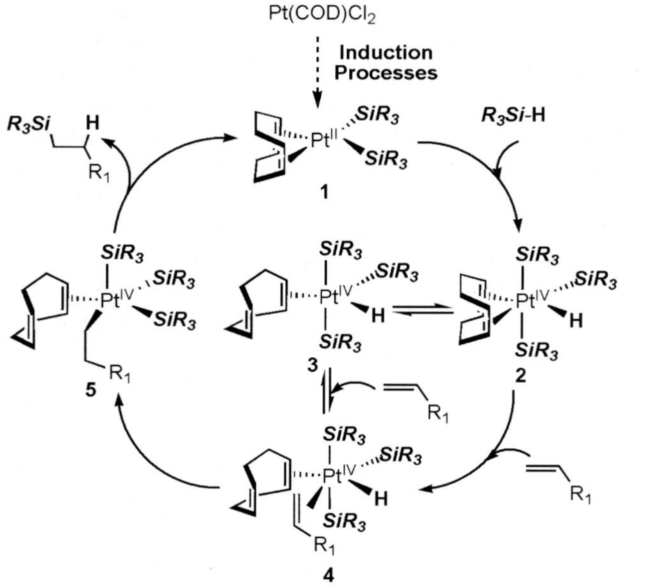

Scheme 13.4 Proposed Chalk–Harrod mechanism involving Pt(II)/Pt(IV) redox couples.

plexes, insertion of alkene into a M–Si bond (silylmetallation) was observed [1, 16]. The modified Chalk–Harrod mechanism was then proposed (Scheme 13.2), in which silylmetallation from the oxidative addition intermediate takes place followed by a C–H reductive elimination. The modified Chalk–Harrod mechanism can successfully explain the formation of the dehydrogenative silylation product. Although an alternative mechanism for formation of vinylsilane that includes an oxidative addition of C–H bond of the alkene, followed by either C–Si reductive elimination or σ-bond metathesis reaction with another silane (Scheme 13.5), was proposed [17], more evidence in support of these pathways is needed.

A subsequent computational study on the reaction mechanism of the hydrosilylation of alkene catalyzed by the model rhodium catalyst, $RhCl(PH_3)_3$ was

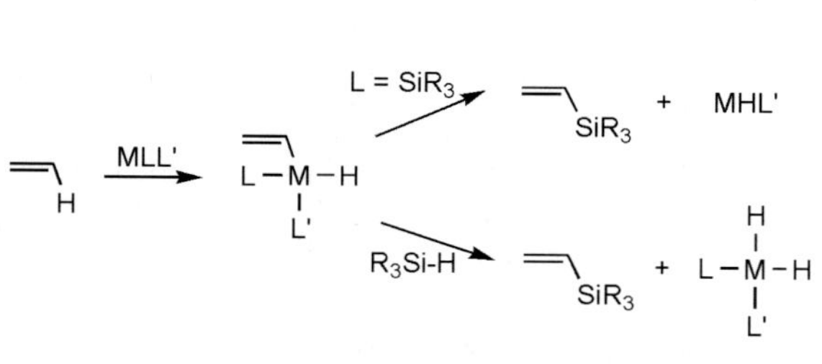

Scheme 13.5 Alternative mechanisms to account for the formation of vinylsilanes.

reported by Sakaki's group [18]. The final Si–C reductive elimination step was found to be the rate-determining step of the Chalk–Harrod mechanism, with an activation barrier of about 27.4–28.8 kcal mol^{-1} at both the DFT and MP4SDQ levels of theory. The rate-determining step of the modified Chalk–Harrod mechanism is either silylmetallation ($E_a = 13.5$–16.9 kcal mol^{-1}) or oxidative addition ($E_a = 11.3$–15.7 kcal mol^{-1}) steps, depending on the computational methods used. Accordingly, they concluded that the hydrosilylation of alkenes catalyzed by rhodium complexes proceed via the modified Chalk–Harrod mechanism. They attributed the observed difference between the Rh and Pt catalysts to a difference in the number of d electrons between the two metals (Rh: d^6/d^8; Pt: d^8/d^{10}), which influences the stability of the intermediates and transition states. However, the major reason for a much lower barrier in the final Si–C reductive elimination step in the Chalk–Harrod mechanism for the Pt(II) complex, compared with the rhodium(III) complex, remains elusive.

13.2.2
Stereochemistry

To account for the trans-addition of Si–H bond to alkynes, Crabtree and Ojima independently combined the modified Chalk–Harrod mechanism with cis-trans isomerization process through a metallacyclopropene or zwitterionic metal carbene intermediate (Scheme 13.6) [19–21]. The proposed mechanistic pathway agrees with experimental observations: neutral rhodium and ruthenium com-

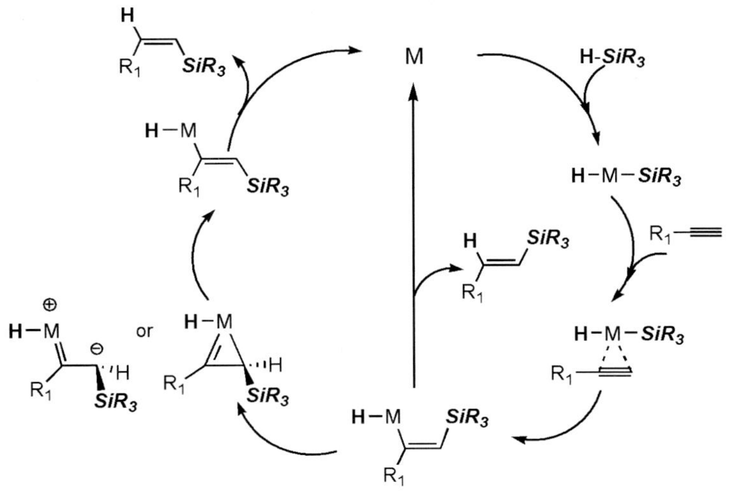

Scheme 13.6 Modified Chalk–Harrod mechanism involving cis-trans isomerization for trans-addition of Si–H bond to alkynes.

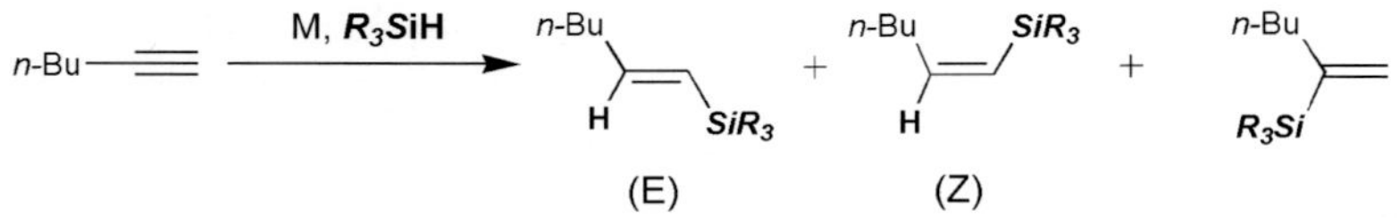

Scheme 13.7

n-Bu—≡ →[M, R₃SiH]

Products:
- (E) anti-Markovnikov: n-Bu-CH=CH-SiR₃
- (Z) anti-Markovnikov: n-Bu/SiR₃ with H
- Markovnikov: n-Bu with R₃Si

Catalyst/Silane	anti-Markovnikov (E)	anti-Markovnikov (Z)	Markovnikov
H_2PtCl_6 /i-PrOH, $HSiCl_3$	78%	0%	22%
[Rh(COD)₂]BF₄, HSiEt₃	99%	1%	0%
RhCl(PPh₃)₃, HSiEt₃	3%	94%	3%
[RuCl₂(p-cymene)]₂, HSiEt₃	4%	96%	0%

Scheme 13.7 Experimentally observed regioselectivity.

plexes predominantly give trans-addition products. Electron-poor cationic rhodium complexes, which are supposed to have a faster reaction rate of reductive elimination and lower stability of the metallacyclopropene intermediate, predominantly give cis-addition products (Scheme 13.7) [1, 20, 22]. Also, the formation of the cis-addition product becomes favorable when 1-alkynes bearing a bulky substituent are used [19].

The hydrosilylation of alkenes catalyzed by organolanthanide complexes, Cp*₂LnCH(SiMe₃)₂ (Ln = La, Nd, Sm, Lu) and Me₂SiCp″₂SmCH(SiMe₃)₂ (Me₂SiCp″₂ = Me₄C₅SiMe₂C₅Me₄), has been reported by Marks and coworkers

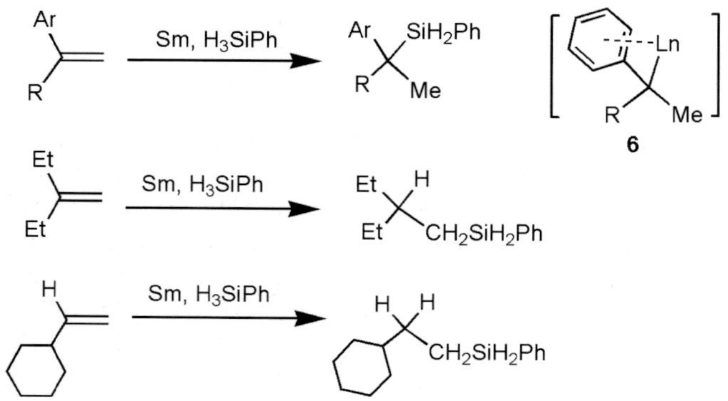

Scheme 13.8 Hydrosilylation of alkenes catalyzed by organolanthanide complexes.

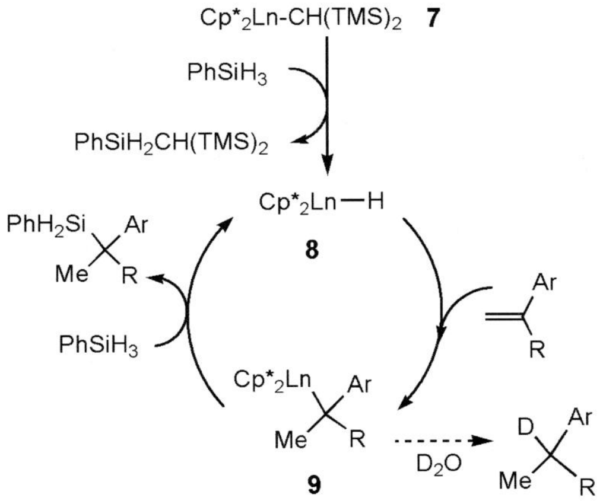

Scheme 13.9 Proposed mechanism for organolanthanide-catalyzed hydrosilylation.

[23]. Regiospecific Markovnikov hydrosilylation was observed from the reaction of vinylarenes, presumably via an interaction of the electrophilic lanthanide and the arene (**6** in Scheme 13.8). In contrast, using 2-ethyl-1-butene or vinylcyclohexene as a substrate exclusively led to anti-Markovnikov products, possibly due to steric repulsion between the alkene substituent and the ancillary ligand. The formation of an exo-dig cyclization product for intramolecular hydrosilylation and the measured kinetic rate law ($k \sim [\text{Ln}]^1[\text{H–Si}]^1[\text{alkene}]^0$) are more consistent with the Chalk–Harrod-type pathway (Scheme 13.9) [23]. Unlike the late transition metal complexes, the first step involves a σ-bond metathesis reaction of the pre-catalyst **7** with the silane as an induction step to give an active metal-hydride intermediate **8**. This is followed by hydrometallation, and a heterolytic σ-bond metathesis reaction with another silane to complete the catalytic cycle.

13.2.3
Carbonyl Compounds

Hydrosilylation of carbonyl compounds plays an important role in organic synthesis, as it provides an efficient method for one-step reduction and protection. Metal complexes, such as Wilkinson's catalyst, titanocene complexes and alkali metal salts (e.g., KF in DMF), have been used to catalyze the hydrosilylation reaction [1, 24–26]. The proposed mechanisms for carbonyl hydrosilylation catalyzed by the titanium and rhodium complexes are analogous to the Chalk–Harrod and modified Chalk–Harrod mechanisms, respectively (Scheme 13.10) [24, 25]. The reaction mechanism of hydrosilylation catalyzed by KF involves the formation of a pentavalent anion $[\text{HSiR}_3\text{F}]^-$ (**10**) as an active hydride source, followed by the rate-determining hydride transfer (reduction via **11**) [26].

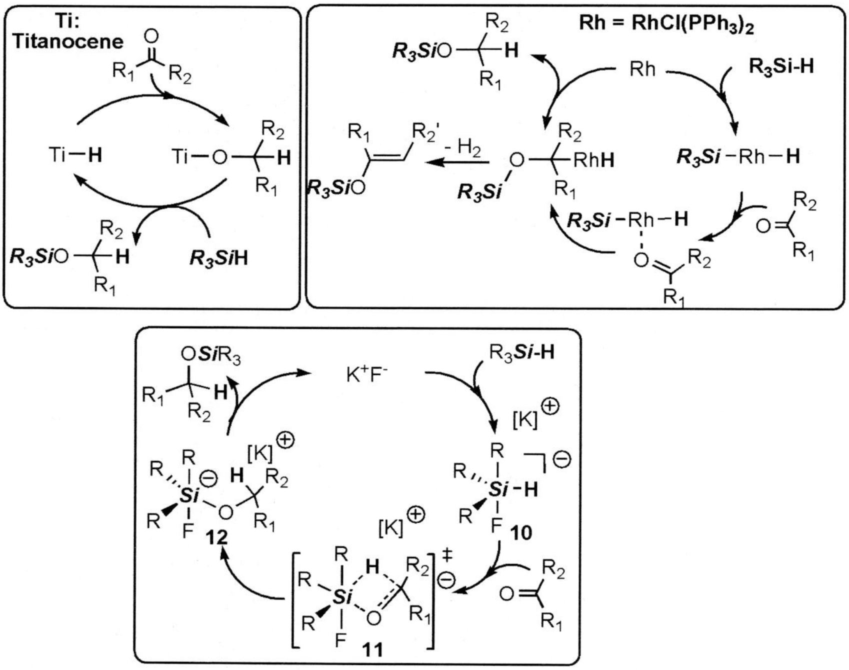

Scheme 13.10 Proposed mechanisms of hydrosilylation of carbonyl compounds.

13.3
New Mechanistic Pathways Discovered at the Beginning of the 21st Century

13.3.1
Main Group Metal Complexes (K, Ca, Sr and B)

Hydrosilylation is often catalyzed by transition metal complexes [1]. Only a few main group metal complexes are used to catalyze the hydrosilylation reaction. In 2006, Harder's group reported the first catalytic hydrosilylation of conjugated alkenes based on four alkali and alkaline-earth metal catalysts (**13–16** in Scheme 13.11) [27]. The reactions show high regioselectivity. In addition, the regioselectivity of hydrosilylation with 1,1-diphenylethylene can be changed by the choice of the metal or/and solvent molecule (Scheme 13.11). The Markovnikov product was exclusively obtained by using calcium (**13** or **14**) and strontium (**15**) complexes under solvent-free conditions. In contrast, only anti-Markovnikov product was obtained when using potassium complex (**16**) under solvent-free conditions or in THF, or using complex **14** in THF, or using complex **15** in THF and Et$_2$O.

Scheme 13.12 shows the proposed mechanism for this novel catalytic reaction. The initiation step involves a σ-bond metathesis reaction of the catalyst **17** with the silane to give a metal hydride intermediate (**18**, Scheme 13.12). With calcium

13.3 New Mechanistic Pathways Discovered at the Beginning of the 21st Century | 293

Scheme 13.11 Alkali and alkaline-earth metal catalysts and regioselectivity of the hydrosilylation reaction.

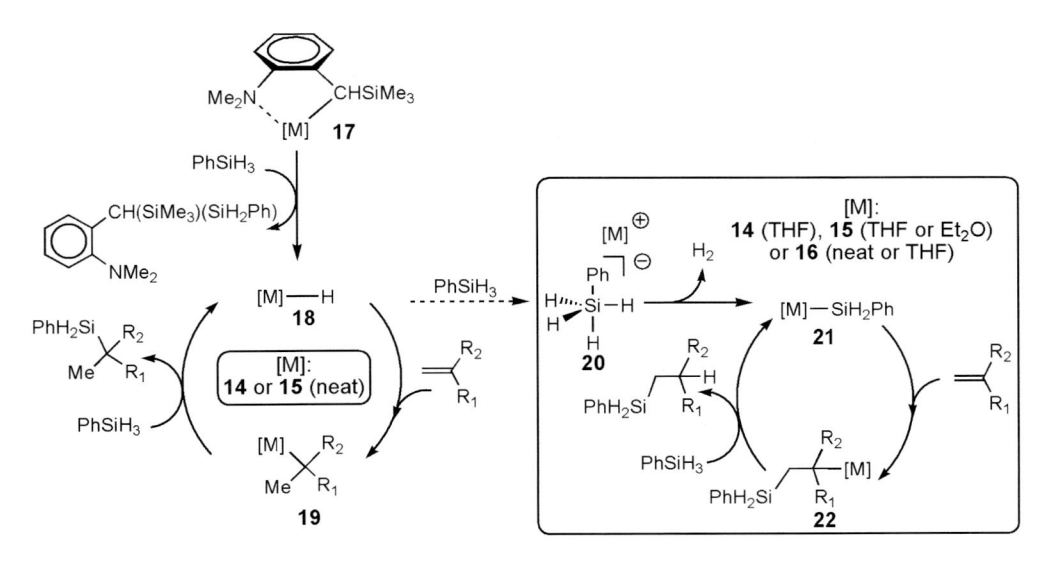

Scheme 13.12 Proposed mechanisms for hydrosilylation catalyzed by alkali and alkaline-earth metal complexes.

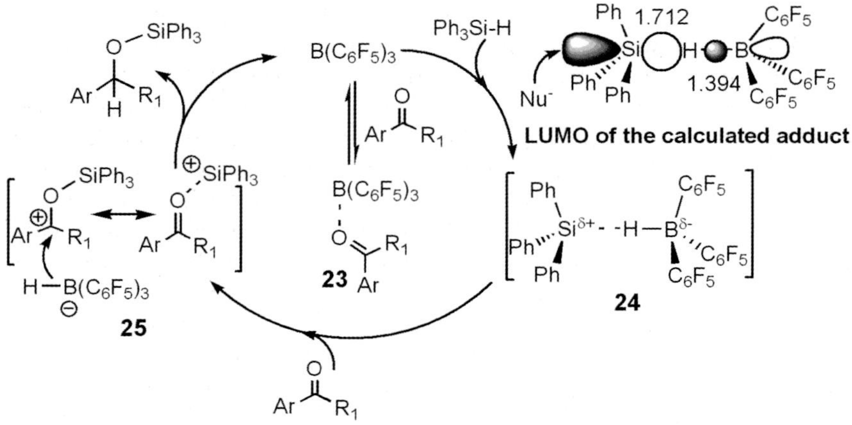

Scheme 13.13 Proposed mechanism for hydrosilylation catalyzed by $B(C_6F_5)_3$.

and strontium complexes under solvent-free condition, a hydrometallation process occurs preferentially, (the Chalk–Harrod-type mechanism), followed by a σ-bond metathesis to give Markovnikov product. Alternatively, the metal hydride intermediate **18** was proposed to react with another silane to form a hypervalent silicon intermediate $[PhSiH_4]^-[M]^+$ (**20**, Scheme 13.12). The formation of **20** is suggested to be favorable with large metal cations and polar solvents. Loss of H_2 from **20** then occurs to give a metal silanide intermediate **21**. This is followed by silylmetallation (the modified Chalk–Harrod-type mechanism) to give a stabilized intermediate $[PhSiH_2CH_2CPh_2]^-[M]^+$ (**22**). A subsequent σ-bond metathesis reaction affords the anti-Markovnikov product.

The reaction mechanism of hydrosilylation of carbonyl compounds catalyzed by the strong Lewis acid $B(C_6F_5)_3$ has been studied by Piers and coworkers [28]. Kinetic studies showed that the reaction rate decreases with increasing basicity or increasing concentration of the carbonyl compound. Therefore, it is suggested that the reaction is initiated by the dissociation of borane from carbonyl-borane complex **23** (Scheme 13.13). The free borane then abstracts a hydride from the hydrosilane to form a silylium/hydridoborate ion pair. The nature of the adduct $B(C_6F_5)_3/R_3SiH$ (R = Et or Ph), **24**, was studied by AM1 calculations, which indicated an incomplete hydride transfer [28]. The carbonyl substrate is suggested to be activated by coordinating with the incipient silylium species (i.e., **25**), and it is reduced by the $[HB(C_6F_5)_3]^-$ counter-anion (Scheme 13.13).

13.3.2
Early Transition Metal Complexes (Zr and Ta)

13.3.2.1 Alkenes
Group IV metallocene complexes Cp_2M (M = Zr, Hf, Ti) are active catalysts for the hydrosilylation of 1-alkenes [17a, 29]. The "olefin-first" mechanism invol-

ving σ-bond metathesis of η^2-alkene-metal and HSiR$_3$ was proposed by Waymouth's and Corey's groups. Several subsequent pathways are possible to give the hydrosilylation product and side-products (Scheme 13.14) [17a, 29c].

A theoretical study on the reaction mechanism of the Cp$_2$Zr (**26**)-catalyzed ethylene hydrosilylation was reported by Sakaki's group [30]. Two coupling reactions of Cp$_2$Zr(C$_2$H$_4$) (**27**) with a Si–H σ-bond of SiH$_4$, analogous to σ-bond metathesis, are found to have activation barriers of about 0.3–5.0 kcal mol^{-1} (in electronic energy) to give Cp$_2$Zr(H)(CH$_2$CH$_2$SiH$_3$) (**30**) and Cp$_2$ZrEt(SiH$_3$) intermediates. Strong π-back-donation between the Zr center and the alkene is suggested to be important in the coupling reaction. Finally, the catalytic cycle is finished with an alkene-assisted C–H reductive elimination step involving **32** from intermediate **31** as the most favorable pathway, with a very small barrier ($\Delta E^{\pm} = 5.0$ kcal mol^{-1}) (Scheme 13.15), rather than direct C–H and Si–C reductive elimination from Cp$_2$Zr(H)(CH$_2$CH$_2$SiH$_3$) and Cp$_2$ZrEt(SiH$_3$) intermediates ($\Delta E^{\pm} = 25.5$–41.8 kcal mol^{-1}). The σ-bond metathesis of **31** or Cp$_2$ZrEt(SiH$_3$)

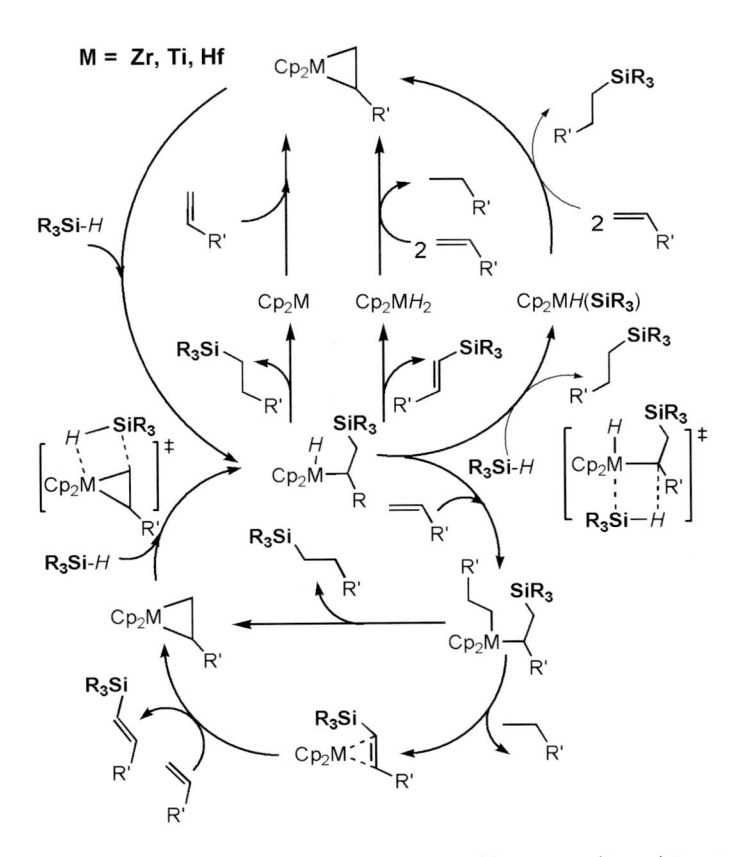

Scheme 13.14 "Olefin-first" mechanism proposed by Waymouth's and Corey's groups.

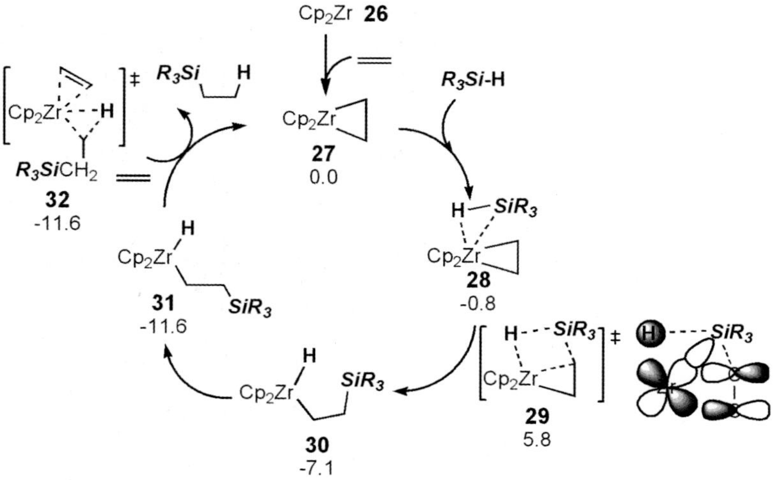

Scheme 13.15 The most favorable pathway for hydrosilylation catalyzed by the Cp$_2$Zr complex [30]. Calculated relative free energies (kcal mol^{-1}) are given.

with one SiH$_4$ should also be overridden because of the higher barriers ($\Delta E^{\pm} = 18.4$–36.8 kcal mol^{-1}).

13.3.2.2 Dinitrogen

Activation and functionalization of the nitrogen molecule by transition metal complexes is of industrial and biological importance (such as the Haber–Bosch process and nitrogenase). Compared with the hydrogenation of N$_2$ [31, 32a], its hydrosilylation is thermodynamically more favourable [31b], due to a weaker Si–H bond and the formation of a stronger N–Si bond. In this regard, Fryzuk and coworkers have reported hydrosilylation with a dizirconium dinitrogen complex {[P$_2$N$_2$]Zr}$_2$(μ-η^2-N$_2$) (33), where [P$_2$N$_2$] is [PhP(CH$_2$SiMe$_2$NSiMe$_2$CH$_2$)$_2$PPh], and a ditantalum dinitrogen complex ([NPN]Ta)$_2$(μ-H)$_2$(μ-η^1:η^2-N$_2$) (35) where [NPN] is (PhNSiMe$_2$CH$_2$)$_2$PPh (Schemes 13.16 and 13.17) [32].

When the ditantalum dinitrogen complex 35 reacted with 2 equivalents of butylsilane (^{n}BuSiH$_3$), a ditantalum disilylimide complex 40 was cleanly obtained (Scheme 13.17) [32b, c]. Two important intermediates (36 and 38) and the disily-

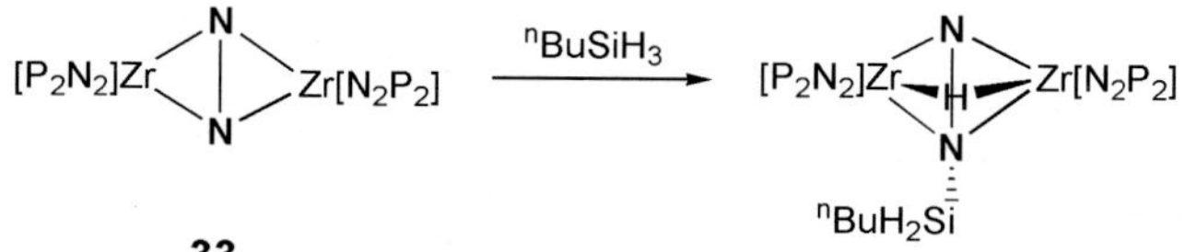

Scheme 13.16 Hydrosilylation of the dizirconium dinitrogen complex.

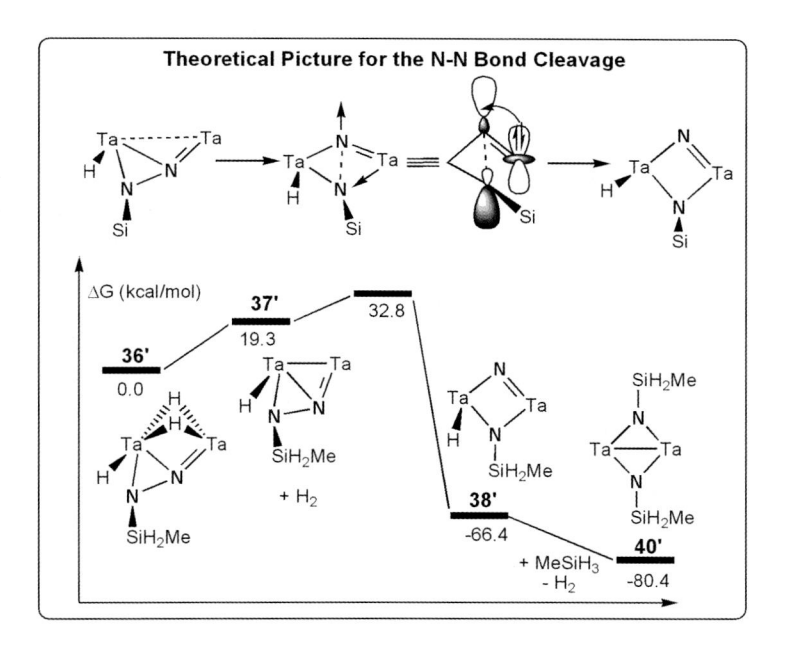

Scheme 13.17 Mechanism for hydrosilylation of the ditantalum dinitrogen complex, proposed by Fryzuk and coworkers [31b and 33].

limide product **40** were observed by NMR. They were also isolated and determined by X-ray crystallography. Later, it was found that excess silane can promote further the transformation of **40**. Unfortunately, the silane cannot further functionalize the two nitrogen atoms to release the silylated amine, rather a cyclometalated product (**41**) was obtained.

Based on these experimental observations, the following mechanism for the hydrosilylation was proposed: The reaction begins with the addition of a Si–H bond across the π-bond of the ditantalum dinitrogen complex **35** to give an intermediate **36**. After that, dinuclear reductive elimination of one hydrogen molecule proceeds to provide two electrons for the cleavage of the N–N bond. Finally, the addition of a Si–H bond of the second silane across the tantalum-nitrogen π-bond of **38** and the second dinuclear reductive elimination occur to give **40**.

Tuczek and Fryzuk reported a theoretical study on the key N–N bond cleavage step with a simplified model using the B3LYP method (Scheme 13.17) [33]. The computational results show that, after the release of H_2, the two tantalum metals gain two electrons (residing in d_{xy}) to form a metal–metal bond. The computational finding of such a transient intermediate (i.e., **37′**) has not yet been observed experimentally. Moreover, the barrier for the subsequent N–N bond cleavage is calculated to be about 32.8 kcal mol^{-1} (relative to the first addition product). The calculations also suggest that the N–N bond cleavage can be regarded as an electron transfer from d_{z2} of the tantalum metal to the $\sigma^*(N-N)$ orbital followed by a subsequent nucleophilic attack of the μ-imido ligand on the tantalum metal. However, it is still unclear how the dinuclear reductive elimination and the transformation of **40** into **41** occur. Further detailed computation studies are necessary to provide more insightful information for the whole reaction process and to help further development of the catalytic reaction.

13.3.3
Middle Transition Metal Complexes (Mo, W and Re)

The hydrosilylation of aldehydes, ketones, imines and nitriles provide an efficient method for reduction. Several novel methods have been reported recently. Bullock's group have found two cationic complexes ($[CpM(CO)_2(IMes)]^+[B(C_6F_5)_4]^-$ (M = Mo or W) as catalysts for solvent-free hydrosilylation of carbonyl compounds under mild conditions [34a]. The tungsten complex is more reactive than the molybdenum complex. Remarkably, the tungsten catalyst precipitates and is separated from the non-polar hydrosilylated product at the end of the reaction, especially when aliphatic substrates are used. Three intermediates, $[CpW(CO)_2(IMes)(Et_2C=O)]^+[B(C_6F_5)_4]^-$ (**42**), $[CpW(CO)_2(IMes)(H)_2]^+[B(C_6F_5)_4]^-$ (**43**) and $[CpW(CO)_2(IMes)(SiEt_3)(H)]^+[B(C_6F_5)_4]^-$ (**44**), were observed in the experiment. In the proposed ionic mechanism [34] (Scheme 13.18) the silylium ion is transferred from the cationic metal hydrido silyl intermediate **44** to the ketone to give the carbocation intermediate **45**. The carbocation intermediate then abstracts a hydride from the silane or from the neutral metal hydride intermediate to afford an alkoxysilane.

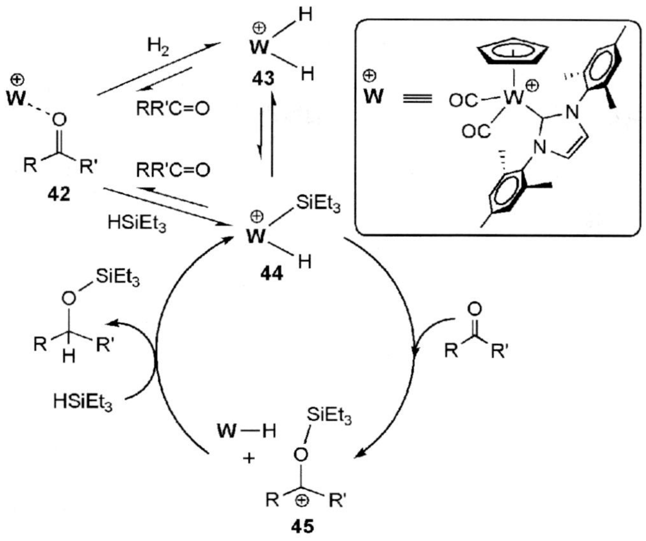

Scheme 13.18 Proposed ionic hydrosilylation mechanism [34].

Another tungsten complex involving the hydrosilylation reaction has been reported by Tobita and coworkers (Scheme 13.19) [35]. With a bulky substituent protection, a neutral base-free silylene complex, $Cp^*(CO)_2(H)W=SiH[C(SiMe_3)_3]$, was synthesized and fully characterized by NMR and X-ray crystallography. This tungsten complex has a hydrido ligand that bears significant interligand interaction with the silylene [36]. The complex was found to react stoichiometrically with acetone and two less reactive nitriles [1a] to produce hydrosilylation pro-

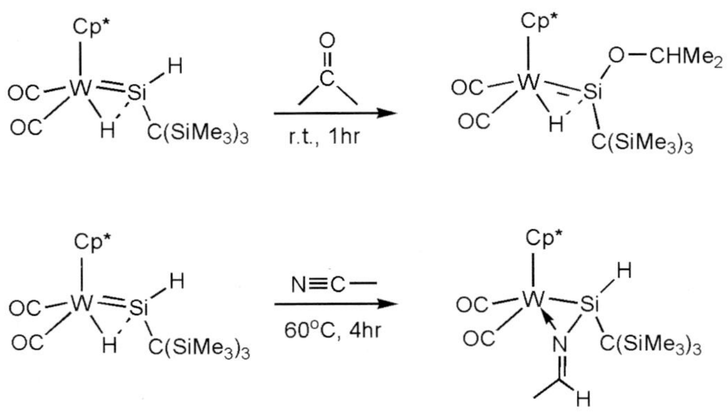

Scheme 13.19 Hydrosilylation of acetone and acetonitrile by the tungsten-silylene complex reported by Tobita and coworkers.

ducts. ^{1}H and ^{29}Si NMR spectroscopy studies revealed that the acetone hydrosilylation product is a tungsten silylene complex with a bridging hydrido ligand as well, while acetonitrile hydrosilylation product is more likely to be a tungsten silyl complex forming a dative bond between the W and N centers. X-ray crystallography shows that acetonitrile is hydrosilylated via a cis addition.

Based on experimental observation, Tobita and coworkers proposed a new mechanism for these hydrosilylation reactions (Path I in Scheme 13.20). According to this mechanism, substrate initially coordinates to the Si atom of the silylene complex, losing the interaction between the hydrido and silylene ligands. Then, the hydride from the W center migrates to the carbon of the substrate to give the product, the tungsten silyl complex, which subsequently isomerizes to a more stable form.

To understand the mechanism of this hydrosilylation, Wu and coworkers performed a computational study [37]. Four possible reaction pathways were calculated with the B3LYP method. The two preferable pathways are illustrated in Scheme 13.20. As seen from this scheme, NCMe preferentially coordinates to the metal center, forming a tungsten silyl complex (47) (Path II). Then, the nitrile inserts into the W–Si bond to give the silyl migration intermediate (48). Subsequent reaction steps are an oxidative addition of the Si–H bond to form a W(IV) hydride intermediate (49) and reductive elimination to form a C–H bond. These two steps complete a hydride transfer from silicon center to carbon, and lead to the experimentally observed product 50.

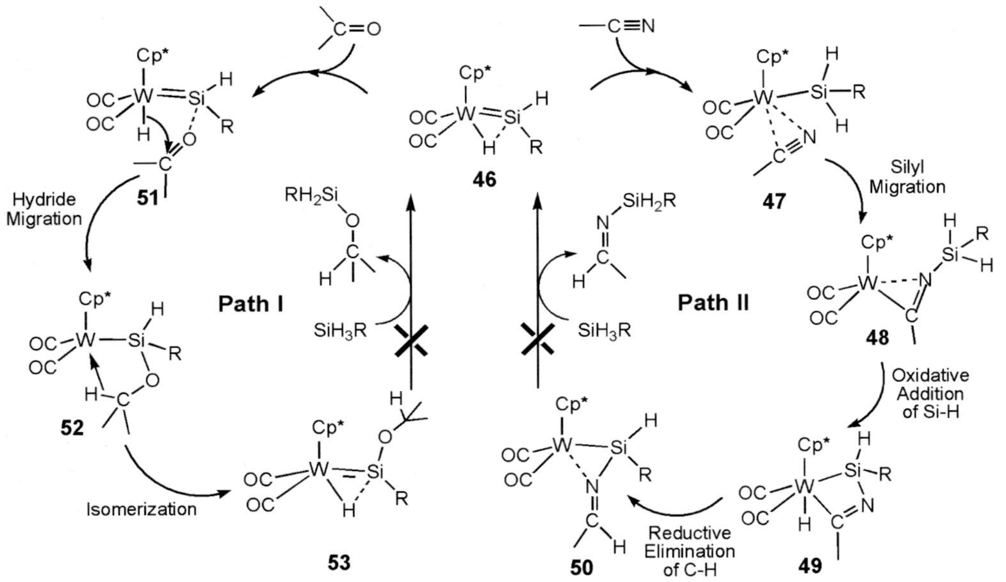

Scheme 13.20 Proposed mechanism of the tungsten-silylene catalyzed hydrosilylation of acetone and acetonitrile.

Acetone as a substrate was also examined. Acetone reluctantly coordinates to the W center due to steric repulsion between the Cp* and the methyl of acetone. Such a coordination intermediate is calculated to be about 22.1 kcal mol^{-1} higher than the reactants in free energy. The subsequent silyl migration barrier is pushed up to about 31.7 kcal mol^{-1}. Therefore, acetone hydrosilylation preferentially proceeds via Tobita mechanism (Path I) with a barrier of 23.1 kcal mol^{-1} (Scheme 13.20).

An unconventional hydrosilylation of aldehydes and ketones catalyzed by a high-valent rhenium(v)-di-oxo complex [ReO$_2$I(PPh$_3$)$_2$] (**54**) was reported by Toste's group [38]. In this connection, several other high-valent metal oxo complexes were reported to be capable of catalyzing the hydrosilylation of carbonyl compounds (**54–61** in Scheme 13.21) [39, 40]. Conventionally, high-valent rhenium-oxo and molybdenum-oxo complexes are usually used for oxidation reactions [41]. These reports of the reduction of carbonyl compounds represent a reverse in the reactivity of these metal complexes as oxidation catalysts.

Toste and coworkers proposed a novel mechanistic pathway for Re(v)-catalyzed hydrosilylation [38a]. It consists of three major steps: (a) [2+2] addition of the Si–H bond across a Re=O bond (e.g., **63**, see Scheme 13.23 below) to generate a rhenium(v) hydrido siloxy intermediate (e.g., **64**); (b) Reduction of the carbonyl from the high-valent rhenium(v) hydrido siloxy intermediate (carbonyl insertion), and (c) retro-[2+2] addition (e.g., **67**) to afford a silyl ether product and to regenerate the active catalyst. In contrast, Abu-Omar and coworkers [40a] found the cationic Re(v)-mono-oxo complex **55** as an efficient catalyst for hydrosilylation at room temperature. They ruled out the [2+2] addition and ionic hydrosilylation mechanisms for the cationic Re(v)-mono-oxo complex **55**. Instead, they proposed a novel metathesis-like pathway (Scheme 13.22) [40a]. Moreover, the Mo(vi) di-oxo complex **61** was proposed to undergo the [2+2] addition across

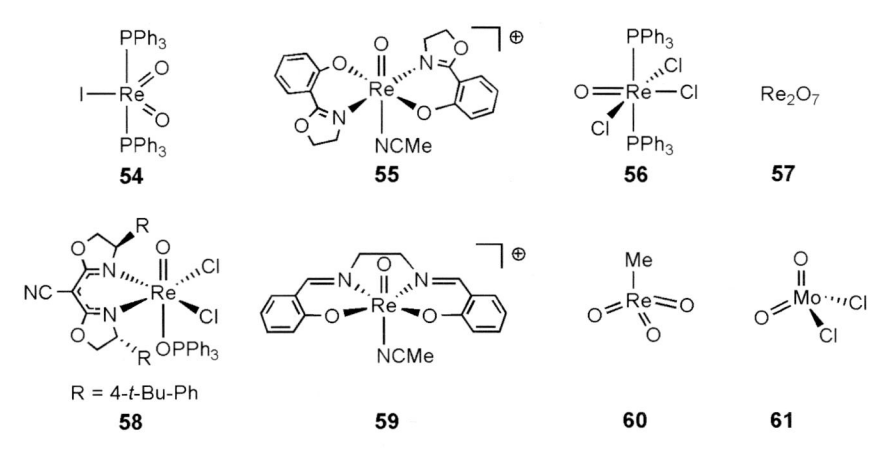

Scheme 13.21 Representative high-valent metal-oxo catalysts for the hydrosilylation of carbonyl compounds.

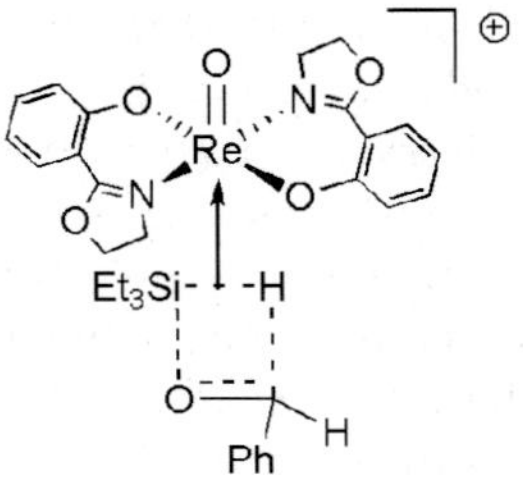

Scheme 13.22 Proposed metathesis-like approach for hydrosilylation catalyzed by complex **55**.

the Mo=O bond to give a molybdenum-hydrido-siloxy intermediate [39a, b, d]. The activation energy for the formation of the molybdenum-hydrido-siloxy intermediate is calculated to be about 42 kcal mol^{-1} at the DFT level [39e]. The reduction step is proposed to proceed via a radical mechanism, since the addition of radical scavengers slows down the reaction.

Wu and coworkers have conducted a computational study on the first unusual hydrosilylation catalyzed by the high-valent Re(v)-di-oxo complex **54** to elucidate the reaction mechanism [42]. Scheme 13.23 gives the key computational results. The B3LYP calculations support the proposed [2+2] addition pathway responsible for the Si–H bond activation. The calculated barrier of the preferable disso-

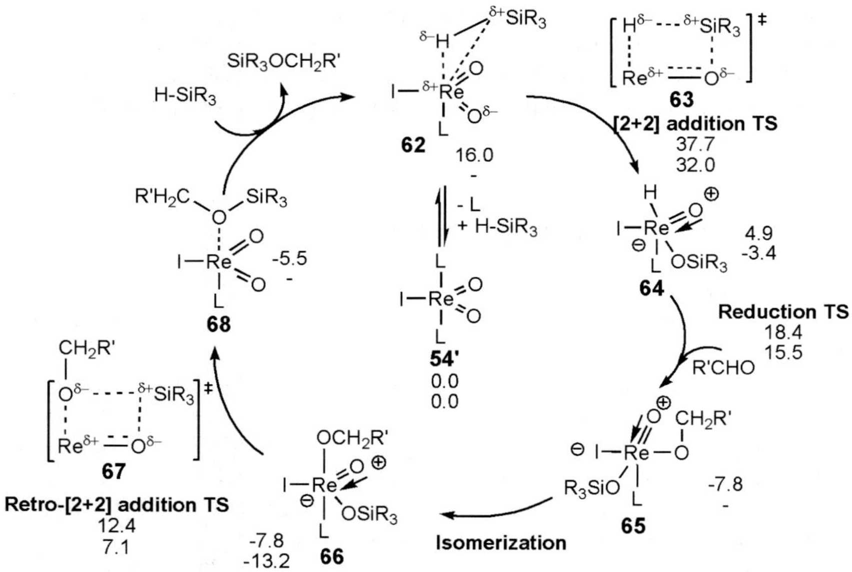

Scheme 13.23 Modified Toste mechanism for the Re(v)-catalyzed hydrosilylation [L = PMe$_3$ (top) and PPh$_3$ (bottom)]. Calculated relative free energies are given (kcal mol^{-1}, at the B3LYP level).

ciative [2+2] addition is about 32.0 and 27.2 kcal mol^{-1} in the gas phase and benzene, respectively, when catalyst **54** is employed. The other processes, including oxidative addition pathway, a direct hydride transfer from the silane to the coordinated carbonyl carbon, metathesis-like pathway and electrophilic Si–H bond cleavage by the carbonyl oxygen, are either higher in energy or are not located with the B3LYP method. The energetic preference towards the proposed [2+2] addition can be ascribed to the favorable electronic matching of the H$^{\delta-}$–Si$^{\delta+}$ and Re$^{\delta+}$=O$^{\delta-}$ bonds and the formation of a strong O–Si bond. In this regard, activation of other small molecules (H$_2$: $\Delta G^{\pm} = 46.1$ and $\Delta G_{rxn} = 10.6$ kcal mol^{-1}; CH$_4$: $\Delta G^{\pm} = 63.2$ and $\Delta G_{rxn} = 32.2$ kcal mol^{-1}; CH$_2$=CH$_2$: $\Delta G^{\pm} = 40.6$ and $\Delta G_{rxn} = 18.4$ kcal mol^{-1}) by the Re(v)-oxo π bond of the model catalyst [ReO$_2$I(PMe$_3$)$_2$] are predicted to be kinetically and thermodynamically inferior than the silane ($\Delta G^{\pm} = 36.4$ and $\Delta G_{rxn} = -7.4$ kcal mol^{-1}) [43]. However, it is difficult to distinguish the [2+2] addition pathway and [3+2] addition pathway experimentally, if the reductive [3+2] addition readily proceeds with a facile tautomeric rearrangement [44]. The B3LYP calculations show that [3+2] pathway has high barriers (81.6–85.5 kcal mol^{-1}). Such high barriers are attributed to the lower stability of the d^4 Re(III) addition product and, particularly, the difficulty in reducing the O=Re^V=O angle suitable for the [3+2] addition because d^2 metal-di-oxo complexes generally prefer to have a very large O=M=O angle to alleviate repulsion between the metal d electrons and oxo p$_\pi$ electrons [45]. After the [2+2] addition, reduction from the Re(v) hydrido siloxy intermediates was calculated to have a modest barrier (13.5–18.9 kcal mol^{-1}). The strong trans-influence of the phosphine ligand was suggested to facilitate the reduction process from the high-valent metal hydrido intermediate **64** [46]. Finally, an intramolecular nucleophilic alkoxy attack on the silicon center (dissociative retro-[2+2] addition) completes the catalytic cycle and affords a silyl ether (Scheme 13.23).

13.3.4
Late Transition Metal Complexes (Ru and Os)

13.3.4.1 Alkynes
In general, hydrosilylation of alkenes and alkynes predominantly leads to anti-Markovnikov products (Scheme 13.7). Methods for Markovnikov hydrosilylation are very limited [1, 22c, 47, 48]. Recently, Trost and Ball reported the first general method of Markovnikov hydrosilylation of terminal alkynes catalyzed by the two cationic Ru(II) complexes, [CpRu(NCMe)$_3$]$^+$PF$_6^-$ (**69**) and [Cp*Ru(NCMe)$_3$]$^+$PF$_6^-$ (**70**) (Scheme 13.24) [49a, b]. Unusual complete trans-addition of the H–Si bond catalyzed by **70** was observed in the reaction of terminal and internal alkynes (Scheme 13.24) [49, 50]. Moreover, intramolecular hydrosilylation of homopropargylic and bis-homopropargylic silyl ethers catalyzed by **70** unexpectedly and solely gave endo-dig cyclization products with trans stereochemistry [50]. Overall, hydrosilylation of alkynes catalyzed by the two cationic Ru(II) complexes (**69** and **70**) lead to very distinct regio- and stereochemical outcomes [1, 47, 51].

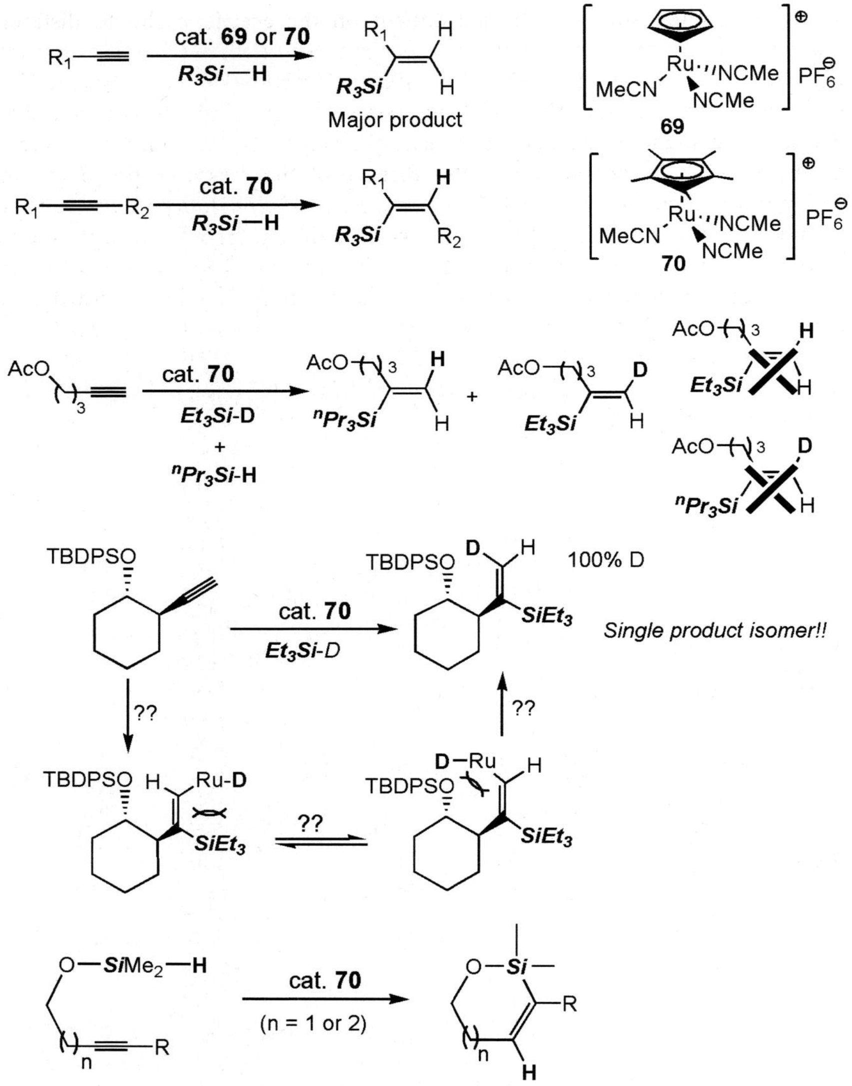

Scheme 13.24 Inter- and intramolecular hydrosilylation of alkynes catalyzed by complexes **69** and **70**.

The Chalk–Harrod and modified Chalk–Harrod-type mechanisms for late transition metals (Schemes 13.2–13.6) could not accommodate the regio- and stereochemistry observed by Trost and Ball (Scheme 13.24), which are inconsistent with the widely-accepted mechanisms (Schemes 13.4–13.7), because steric repulsion between the silyl and the substituent on the acetylene should disfavor Markovnikov silylmetallation [19, 49b, 50]. Moreover, such mechanisms cannot explain the exceptional formation of the six- and seven-membered-ring endo-dig cyclization products [52]. Although a mechanistic proposal involving a back-side attack of the silane to the Lewis-acid activated alkyne can explain trans stereochemistry [53], it cannot explain the absence of the crossover products and Markovnikov regioselectivity [50, 53]. The absence of crossover products further excludes electron-transfer or radical process, as well as a route involving two discrete Ru complexes [50]. Therefore, all available mechanisms cannot explain the unique regio- and stereochemistry of inter- and intramolecular hydrosilylation catalyzed by **69** and **70**. Thus, a new reaction mechanism is needed, which has been proposed by Wu and coworkers (Scheme 13.25). In their B3LYP studies, model catalyst, $[CpRu(NCH)_3]^+$, terminal alkynes, $RC\equiv CH$ for $R = H$ or Me, and $HSiMe_3$ were used [54].

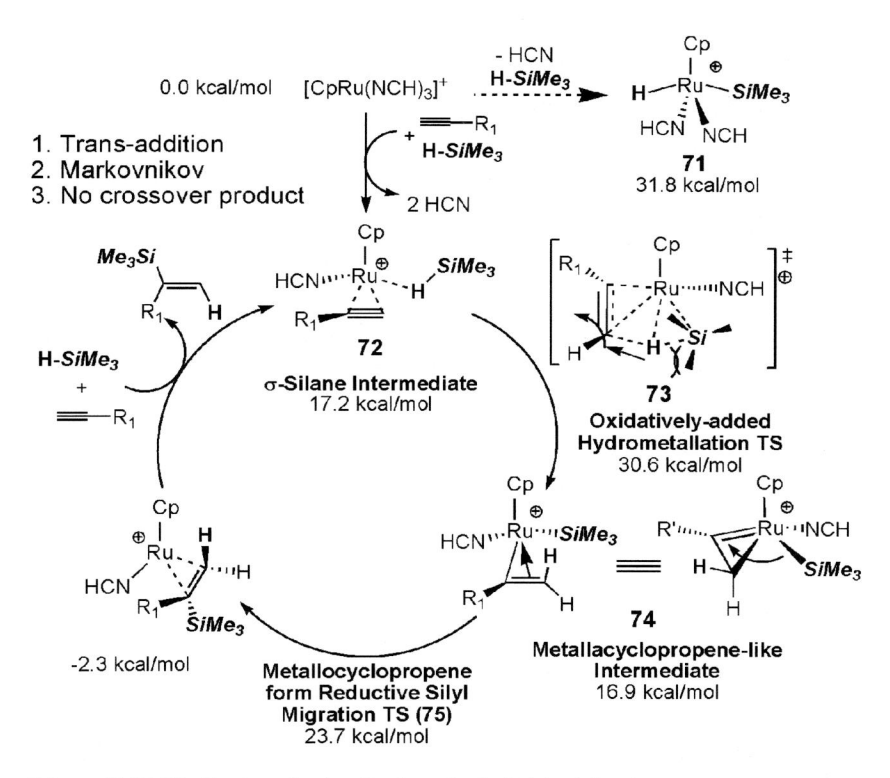

Scheme 13.25 Wu–Trost mechanism for the cationic Ru(II)-catalyzed alkyne hydrosilylation. Relative free energies are given (kcal mol^{-1}, at the B3LYP level of theory).

Oxidatively-added Hydrometallation The oxidative-addition intermediate cis-[CpRu(H)(SiMe$_3$)(NCH)$_2$]$^+$ cannot be obtained with the B3LYP method, but the non-classical Ru(II) σ-silane intermediate [CpRu(η^1-HSiMe$_3$)(NCH)$_2$]$^+$ (Ru–H = 1.817 Å; Ru–Si = 3.361 Å; H–Si = 1.551 Å; Ru–H–Si = 172.6°) was obtained [43]. The formation of an oxidative addition tautomer [2b], trans-[CpRu(H)(SiMe$_3$)(NCH)$_2$]$^+$ **71**, via a very high-barrier oxidative addition ($\Delta G^{\neq}$ = 50.3 kcal mol^{-1}) is less stable than cis-[CpRu(η^1-HSiMe$_3$)(NCH)$_2$]$^+$ by about 21.7 kcal mol^{-1}. The calculations show that oxidative addition to the electronic deficient Ru(II) complexes is unfavorable. The most favorable catalytic pathway starts with the rate-determining oxidative-addition concerted with hydrometallation (oxidatively-added hydrometallation, **73**) from σ-silane intermediates, [CpRu(RC≡CH)(η^1-HSiMe$_3$)(NCH)]$^+$ (**72**, R = Me or H) [55]. Such a concerted process, presumably, stems from the high instability of the oxidative-addition cationic Ru(IV) intermediate and a strong thermodynamic driving force for the formation of a new C–H bond. For the reaction of propyne, the Markovnikov oxidatively-added hydrometallation pathway is kinetically and thermodynamically more favorable than the anti-Markovnikov counterpart by 0.6 and 2.9 kcal mol^{-1}, respectively [43, 54].

For the intramolecular reaction, an interaction between the silyl and hydrido ligands has to be completely lost in exo-hydrometallation transition states (Scheme 13.26) [36, 43, 54]. As discussed before, complete oxidative-addition to the cationic Ru(II) complexes is energetically unfavorable. As a result, unusual endo-dig hydrometallation pathways, which have some interligand interactions (H–Si = 2.188–2.649 Å), are calculated to be lower in energy than the exo-dig hy-

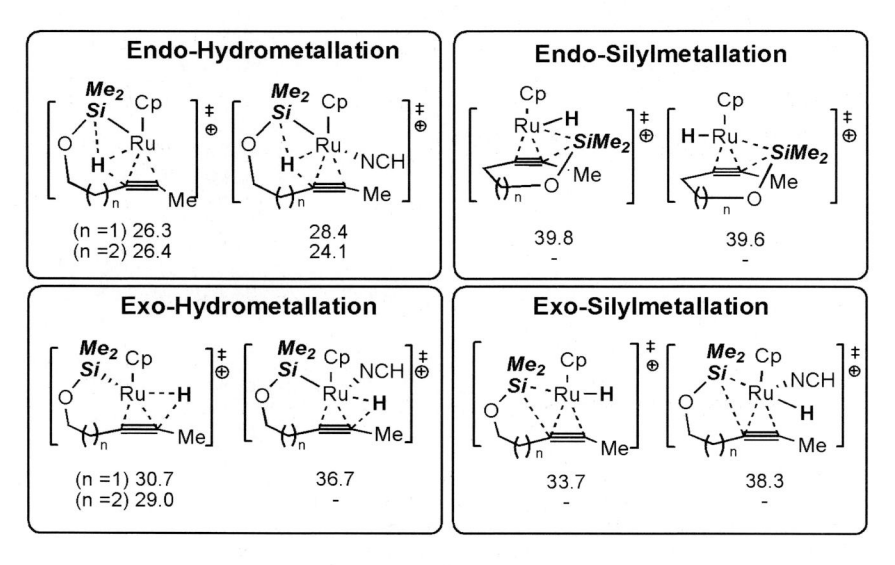

Scheme 13.26 Relative free energies (kcal mol^{-1}) of transition states for intramolecular hydrosilylation, calculated at the B3LYP level.

drometallation pathways (Scheme 13.26) [50, 54]. For the less favorable silylmetallation pathway (the modified Chalk–Harrod type), the observed Markovnikov regioselectivity for the intermolecular reaction and endo-dig regiochemistry for the intramolecular reaction are highly unfavorable, due to steric repulsion between the silyl and methyl group on the alkyne in the former case and a larger ring strain in the latter.

Surprisingly, no σ-vinyl intermediates, $[CpRu(TMS)(NCH)(CRCH_2)]^+$ (R = H or Me), were found. The σ-vinyl form structure for the propene case was characterized as an isomerization transition state (**78**) (C_α–C_β bond coupled with Ru–C_α

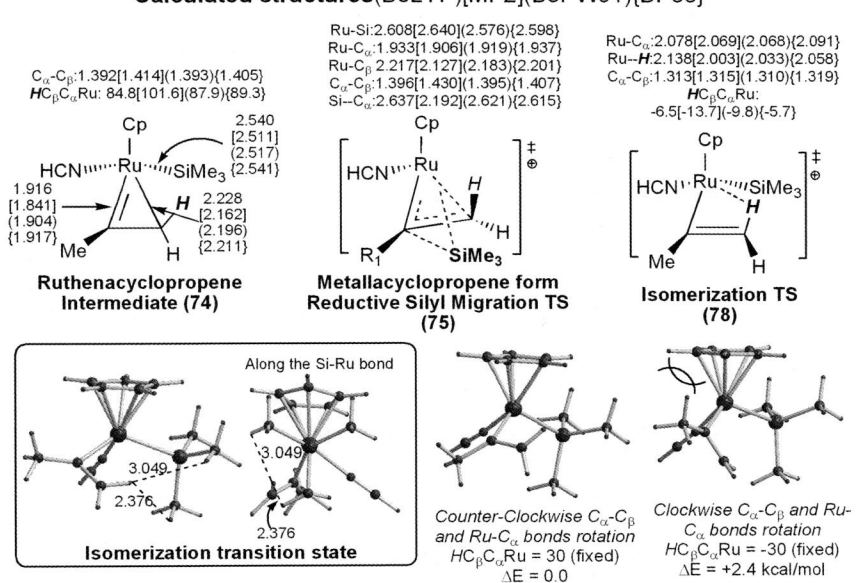

Scheme 13.27 X-ray structures of rhenacyclopropene (**76**) and ruthenacyclopropene (**77**) complexes, and calculated key geometrical parameters of intermediate **74**, and transition states **75** and **78**. Calculated structures of transition state **78** and two constraint structures for counter-clockwise and clockwise C_α–C_β bond rotation are also presented.

bond rotation, Scheme 13.27). Instead, the calculations showed that syn oxidatively-added hydrometallation directly and stereoselectively affords an 18-electron highly-twisted η^2-vinyl Ru(IV) or ruthenacyclopropene intermediate (74) [21]. The transferring hydrogen (in bold) is favored to be on the opposite face of the metallacyclopropene plane with the silyl group in intermediates 74 (Scheme 13.25). Similar stereoselective and direct formation of rhenacyclopropene complex 76 was observed by Casey's group (Scheme 13.27) [56]. The key to the salient trans-addition hydrosilylation is the stereoselective formation of intermediates 74 followed by a facile metallacyclopropene form reductive silyl migration (75) to the carbene center (C_α) (Scheme 13.25) [54]. On the other hand, the normal σ-vinyl from reductive elimination was not found. The unique nature of the metallacyclopropene-like intermediate 74, metallacyclopropene from reductive α-silyl migration and σ-vinyl from isomerization transition states (75 and 79), can also be obtained and characterized by using B3PW91, BP86 and MP2 methods (Scheme 13.27).

Thus, most of the isolated metallacyclopropene complexes are found to contain low-spin d^4 metals [Mo(II), W(II) and Re(III)] [21]. Metallacyclopropene complexes of late transition metals have not been realized, except for two d^4-metals [57]. An X-ray study of metallacyclopropene fragment of Ru(IV) has been reported by Jia and coworkers (Scheme 13.27) [58]. A frontier molecular orbital analysis of this complex elucidates three bonding interactions between the d^4 metal and the 4e η^2-vinyl ligand (Scheme 13.28) [21].

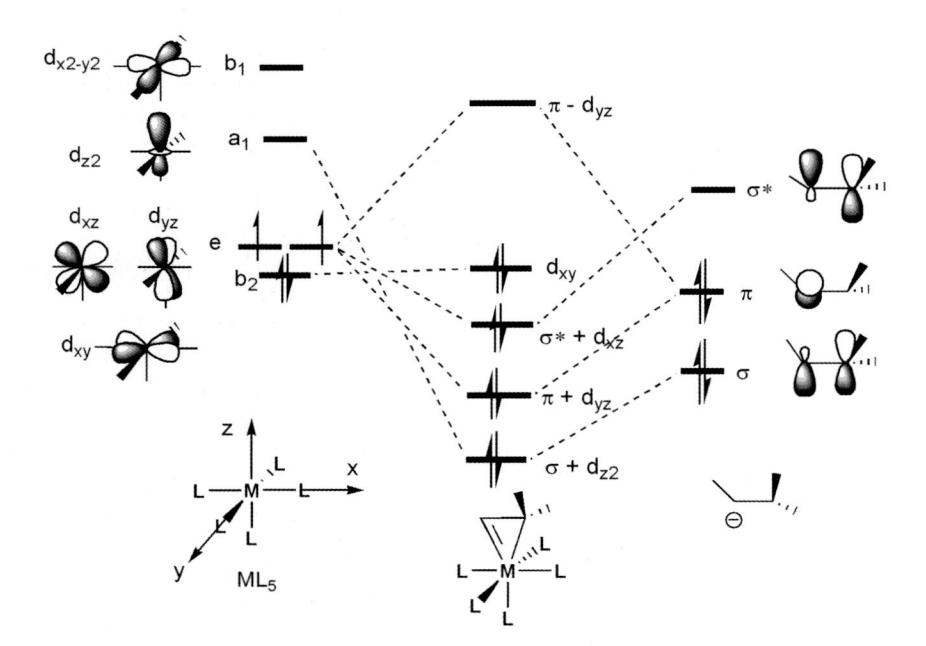

Scheme 13.28 Frontier molecular orbitals of metallacyclopropene complexes.

Intrinsic reaction coordinate calculations also indicated that, once the C–H bond is formed, counterclockwise rotation of the C_α–C_β bond coupled with Ru–C_α bond rotation occurs and results in the stereoselective formation of intermediate 74 [54]. The σ-vinyl isomerization transition state 78 resembles a point where the rotation starts. Viewed from the calculated isomerization transition state shown in Scheme 13.27, two methyl groups of the silyl fragment are somewhat "eclipsed" with a Cp-Ru cone and the remaining methyl group has a very close contact with the transferring hydrogen (H–H = 2.38 Å). Therefore, the transferring hydrogen preferentially moves away from the silyl group and turns towards the Cp ring (Scheme 13.27) [54]. The energy difference between structures with the fixed dihedral angle of $HC_\beta C_\alpha Ru = 30°$ and $-30°$ is about 2.4 kcal mol^{-1}. In addition, when the clockwise C_α–C_β and Ru–C_α bond rotations proceed, the substituent on the C_α position comes closer to the Cp ligand. Therefore, using bulky Cp* ligand further increases the stereoselective formation of the [Cp*Ru(TMS)(NCH)(η^2-CMeCH$_2$)]$^+$ intermediate by 7.4 kcal mol^{-1} [43].

This novel mechanism involving oxidatively-added hydrometallation (73), metallacyclopropene-like intermediate (74) followed by metallacyclopropene 75 formation is, mostly, consistent with the unique observations of regio- and stereochemistry. Moreover, this mechanistic data may imply a new reaction pathway for alkyne insertion catalyzed by middle or late transition metals (Scheme 13.25 vs. Schemes 13.2 and 13.14). Such a novel mechanistic pathway may also extend to other similar reactions [56, 59].

13.3.4.2 Alkenes

Complex 70 (Scheme 13.24) cannot, however, catalyze alkene hydrosilylation [50]. B3LYP calculations showed that the high instability of the hydrometallation product, β-agostic intermediate [CpRu(Et)(SiMe$_3$)(NCH)]$^+$ and its α-rotamer (79 and 80), and high-energy of reductive elimination transition state 81 result in a failure of alkene hydrosilylation (Scheme 13.29).

At the same time, Glaser and Tilley synthesized and characterized Os(IV) and Ru(IV) silylene complexes, [Cp*(P^{i}Pr$_3$)(H)$_2$Os=Si(H)trip][B(C$_6$F$_5$)$_4$], trip = 2,4,6-^{i}Pr$_3$C$_6$H$_2$, and [Cp*(P^{i}Pr$_3$)(H)$_2$Ru=Si(H)Ph•Et$_2$O][B(C$_6$F$_5$)$_4$] (82) [60]. These two metal silylene complexes can stoichiometrically react with 1-hexene to give base-free [Cp*(P^{i}Pr$_3$)(H)$_2$Os=Si(Hex)trip]$^+$ and [Cp*(P^{i}Pr$_3$)(H)$_2$Ru=Si(Hex)Ph]$^+$ complexes (Scheme 13.30). A positive charge of the Os-silylene complex was later shown to be essential for alkene insertion [61]. Catalytic hydrosilylation of alkenes operates when the labile Ru-complex 82 is used. The reaction proceeds selectively with primary silanes, exclusively giving anti-Markovnikov and syn-addition products [60]. No deuterium scrambling is observed when deuterated ethene is used. To account for all experimental observations, Glaser and Tilley proposed a new mechanism via a [2$_\sigma$ + 2$_\pi$] addition transition state (83, Scheme 13.31). An interesting report from Tilley's group demonstrated that the metal silylene complex can also be an active catalyst for hydrosilylation [62, 63].

Although the [2$_\sigma$ + 2$_\pi$] addition pathway proposed by Glaser and Tilley provides the most straightforward explanation to understand the selectivity of pri-

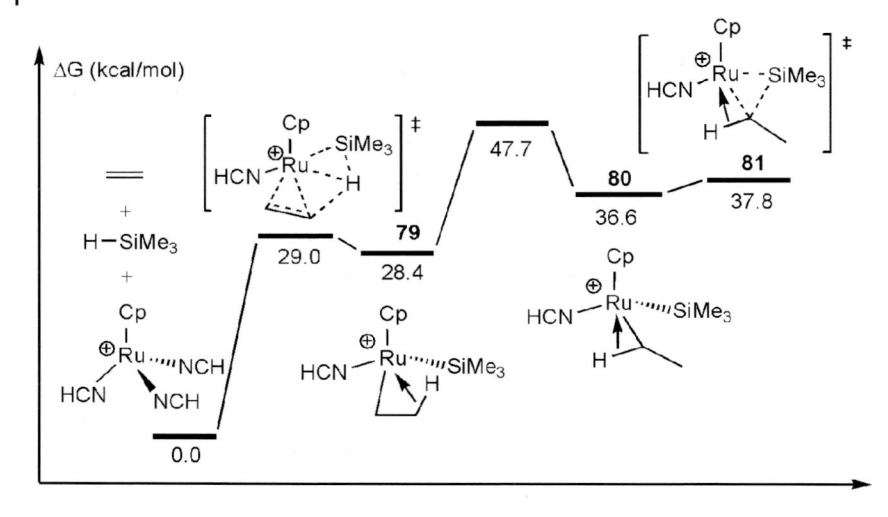

Scheme 13.29 Potential energy surface for hydrosilylation of alkene catalyzed by the [CpRu(NCH)$_3$]$^+$ complex **70**. The relative free energies are given (kcal mol^{-1}, at the B3LYP level).

mary silanes, exclusive anti-Markovnikov regiochemistry and the absence of deuterium scrambling, the selectivity of primary silanes and anti-Markovnikov regiochemistry induced by steric repulsion were also observed in other catalysts via other mechanistic proposals (Schemes 13.8, 13.9 and 13.17) [23, 32c]. Recent

Scheme 13.30 Hydrosilylation of alkenes mediated by the cationic ruthenium- and osmium-silylene complexes.

Scheme 13.31 Glaser–Tilley mechanism for hydrosilylation catalyzed by Ru-silylene complexes.

computational studies have evaluated the feasibility of the Glaser–Tilley mechanistic proposal [43, 64]. Simplified model catalysts, $[Cp(PH_3)(H)_2Ru=SiH_2]^+$ and $[Cp(PH_3)(H)_2Ru=SiPhH]^+$, were used in the B3LYP studies of Hall et al. and Böhme et al., respectively. The unique structure of the simplified and realistic base-free Ru-silylene complexes were found to contain two bridging hydrido ligands (3c–2e interactions, see Scheme 13.32) [43, 64b]. They interact with both the Ru and Si centers; in contrast, the electron-rich Os-silylene complex has two classical hydrido ligands (2c–2e interactions).

When the simplified Ru-silylene complexes were used in the calculations, the $[2_\sigma + 2_\pi]$ addition transition states were lower in energy than the rate-determining step in the Chalk–Harrod-type mechanism (Scheme 13.33). Therefore, the $[2_\sigma + 2_\pi]$ addition is energetically more favorable than the other pathways. These calculations also have shown that oxidatively-added hydrometallation in the Chalk–Harrod-type mechanism can occur and compete with the $[2_\sigma + 2_\pi]$ addition. Reversible alkene insertion/elimination would render the unobserved deuterium scrambling, since in-plane rotation (C_α–C_β bond rotation) could proceed readily [65].

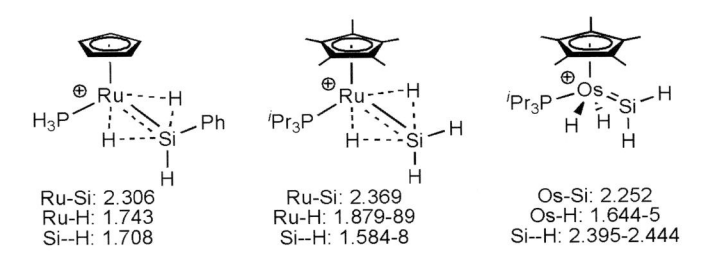

Ru-Si: 2.306
Ru-H: 1.743
Si--H: 1.708

Ru-Si: 2.369
Ru-H: 1.879-89
Si--H: 1.584-8

Os-Si: 2.252
Os-H: 1.644-5
Si--H: 2.395-2.444

Scheme 13.32 B3LYP calculated structures of metal silylene complexes.

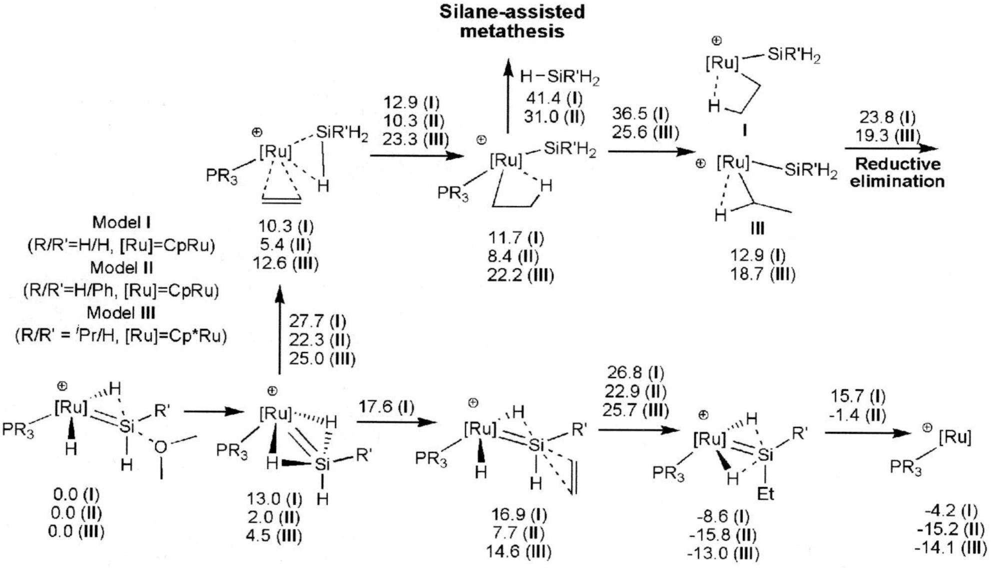

Scheme 13.33 Calculated (at the B3LYP level) relative free energies (kcal mol^{-1}) of hydrosilylation reaction via the Chalk–Harrod-type (top) and the [$2_\sigma + 2_\pi$] addition (bottom) pathways.

The effects of bulky ligands of the Ru-silylene complexes on the reaction mechanisms are quite remarkable (Scheme 13.33) [43]. Steric crowding of the realistic catalyst, [Cp*(P^{i}Pr$_3$)(H$_2$)Ru=(SiH$_2$)]$^+$, promotes formation of the α-agostic intermediate, [Cp*(P^{i}Pr$_3$)Ru(Et)H(SiH$_3$)]$^+$, and reductive elimination in the Chalk–Harrod-type mechanism, which are consistent with available experimental observations [66, 67]. Also, the in-plane rotation is disfavored by steric repulsion with the cis bulky phosphine ligand (P^{i}Pr$_3$) and has a higher barrier (by 4.0 kcal mol^{-1}) than a Ru–C$_\alpha$ bond rotation leading to the α-agostic intermediate. When a more realistic bulky catalyst is considered, the [$2_\sigma + 2_\pi$] addition pathway is more favorable than the Chalk–Harrod-type pathway.

13.4
Conclusions

This chapter summarizes recent advances in hydrosilylation reactions and their new mechanistic pathways, including both experimental and computational studies. Diversified reaction pathways for hydrosilylation reactions can be understood by different bonding and structures of stable silicon intermediates (Scheme 13.34), depending on the catalysts used. These recent novel mechanistic pathways may also operate for other similar reactions, such as hydrogenation and

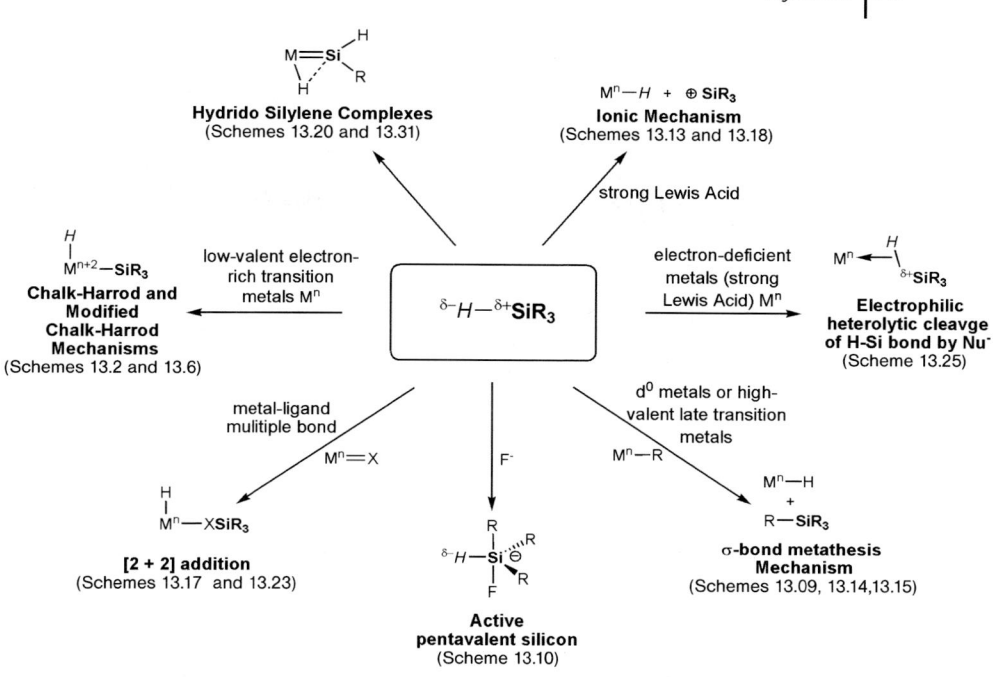

Scheme 13.34 Structures, bonding and mechanistic pathways of the hydrosilylation reaction.

hydroboration reactions. Although most of the new mechanisms were first implicated by experimental observations, computational studies play an important role in characterizing transient intermediates and in evaluating the feasibility of the proposed mechanisms. More computational studies are still needed to understand many remaining unsolved mechanistic issues.

References

1. (a) I. Ojima, in *The Chemistry of Organic Silicon Compounds*, eds. S. Patai, Z. Rappoport, John Wiley & Sons, Chichester, 1989, pp. 1479–1526. (b) I. Ojima, Z. Li, J. Zhu, in *The Chemistry of Organic Silicon Compounds*, eds. Z. Rappoport, Y. Apeloig, John Wiley & Sons, Chichester, 1998, Vol. 2, pp. 1687–1792. (c) T. Hiyama, T. Kusumoto, in *Comprehensive Organic Synthesis*, eds. B. M. Trost, I. Fleming, Pergmon Press, Oxford, 1991, Vol. 8, pp. 763–792.
2. (a) R. H. Crabtree, *Angew. Chem., Int. Ed. Engl.* 1993, **32**, 789–805. (b) G. J. Kubas, *Metal Dihydrogen and σ-Bond Complexes*, Kluwer Academic/Plenum Publishers, New York, 2001. (c) J. J. Shneider, *Angew. Chem., Int. Ed. Engl.* 1996, **35**, 1068–1075.
3. B. M. Trost, *Acc. Chem. Res.* 2002, **35**, 695–705.
4. E. W. Colvin, *Silicon Reagents in Organic Synthesis*, Academic Press, London, 1988.
5. (a) H. Nishiyama, K. Itoh, in *Catalytic Asymmetric Synthesis*, 2nd. edn., ed. I. Ojima, Wiley-VCH, New York, 2000, pp. 111–143. (b) J. Tang, T. Hayashi, in *Catalytic Heterofunctionalization*, eds.

A. Togni, H. Grützmacher, Wiley-VCH, Weinheim, 2001, pp. 73–89. (c) H. Nishiyama, in *Comprehensive Asymmetric Catalysis*, eds. E. N. Jacobsen, A. Pfaltz, H. Yamamoto, Springer, Berlin, 1999, 2000, pp. 267–288. (d) T. Hayashi, in *Comprehensive Asymmetric Catalysis*, eds. E. N. Jacobsen, A. Pfaltz, H. Yamamoto, Springer, Berlin, 1999, pp. 319–333.

6 (a) J. L. Speier, J. A. Webster, G. H. Barnes, *J. Am. Chem. Soc.* 1957, **79**, 974–979. (b) J. L. Speier, *Adv. Organomet. Chem.* 1979, **17**, 407.

7 (a) B. D. Karstedt, (General Electric), US Patent 3 715 334, 1973. (b) P. B. Hitchcock, M. F. Lappert, N. J. W. Warhurst, *Angew. Chem., Int. Ed. Engl.* 1991, **30**, 438–440. (c) J. Stein, L. N. Lewis, Y. Gao, R. A. Scott, *J. Am. Chem. Soc.* 1999, **121**, 3693–3703.

8 A. J. Chalk, J. F. Harrod, *J. Am. Chem. Soc.* 1965, **87**, 16–21.

9 S. Sakaki, N. Mizoe, M. Sugimoto, *Organometallics* 1998, **17**, 2510–2523.

10 F. Ozawa, T. Tani, H. Katayama, *Organometallics* 2005, **24**, 2511–2515.

11 H. Sakurai, M. Sugimoto, *J. Organomet. Chem.* 2004, **689**, 2236–2241.

12 C. A. Tsipis, C. E. Kefalidis, *Organometallics* 2006, **25**, 1699–1706.

13 A. K. Roy, R. B. Taylor, *J. Am. Chem. Soc.* 2002, **124**, 9510–9524.

14 S. Reinartz, P. S. White, M. Brookhart, J. L. Templeton, *J. Am. Chem. Soc.* 2001, **123**, 6425–6426.

15 M. N. Jagadeesh, W. Thiel, J. Köhler, A. Fehn, *Organometallics* 2002, **21**, 2076–2087.

16 (a) M. A. Schroeder, M. S. Wrighton, *J. Organomet. Chem.* 1977, **128**, 345–358. (b) F. Seitz, M. S. Wrighton, *Angew. Chem., Int. Ed. Engl.* 1988, **27**, 289–291. (c) M. Brookhart, B. E. Grant, *J. Am. Chem. Soc.* 1993, **115**, 2151–2156. (d) S. B. Duckett, R. N. Perutz, *Organometallics* 1992, **11**, 90–98.

17 (a) M. R. Kesti, R. M. Waymouth, *Organometallics* 1992, **11**, 1095–1103. (b) J. Ruiz, P. O. Bentz, B. E. Mann, C. M. Spencer, B. F. Taylor, P. M. Maitlis, *J. Chem. Soc., Dalton Trans.* 1987, 2709–2713.

18 S. Sakaki, M. Sumimoto, M. Fukuhara, M. Sugimoto, H. Fujimoto, S. Matsuzaki, *Organometallics* 2002, **21**, 3788–3802.

19 (a) R. S. Tanke, R. H. Crabtree, *J. Am. Chem. Soc.* 1990, **112**, 7984–7989. (b) C.-H. Jun, R. H. Crabtree, *J. Organomet. Chem.* 1993, **447**, 177–187.

20 I. Ojima, N. Clos, R. J. Donovan, P. Ingallina, *Organometallics* 1990, **9**, 3127–3133.

21 D. S. Frohnapfel, J. L. Templeton, *Coord. Chem. Rev.* 2000, **206–7**, 199–235.

22 (a) R. A. Benkeser, R. F. Cunico, S. Dunny, P. R. Jones, P. G. Nerlekar, *J. Org. Chem.* 1967, **32**, 2634–2636. (b) R. Takeuchi, S. Nitta, D. Watanabe, *J. Org. Chem.* 1995, **60**, 3045–3051. (c) Y. Na, S. Chang, *Org. Lett.* 2000, **2**, 1887–1889. (e) J. W. Faller, D. G. D'Alliessi, *Organometallics* 2002, **21**, 1743–1746.

23 P.-F. Fu, L. Brard, Y. Li, T. J. Marks, *J. Am. Chem. Soc.* 1995, **117**, 7157–7168.

24 (a) I. Ojima, M. Nihonyanagi, Y. Nagai, *J. Chem. Soc., Chem. Commun.* 1972, 938. (b) I. Ojima, M. Nihonyanagi, T. Kogure, M. Kumagai, S. Horiuchi, K. Nakatsugawa, *J. Organomet. Chem.* 1975, **94**, 449–461.

25 (a) T. Nakano, Y. Nagai, *Chem. Lett.* 1988, 481–484. (b) J. Yun, S. L. Buchwald, *J. Am. Chem. Soc.* 1999, **121**, 5640–5644.

26 (a) J. Boyer, R. J. P. Corriu, R. Perz, C. Reye, *Tetrahedron* 1981, **37**, 2165–2171. (b) M. Fujita, T. Hiyama, *Tetrahedron Lett.* 1987, **28**, 2263–2264. (c) R. J. P. Corriu, J. C. Young, in *The Chemistry of Organic Silicon Compounds*, eds. S. Patai, Z. Rappoport, John Wiley & Sons, Chichester, 1989, pp. 1241–1288.

27 F. Buch, J. Brettar, S. Harder, *Angew. Chem., Int. Ed.* 2006, **45**, 2741–2745.

28 D. J. Parks, J. M. Blackwell, W. E. Piers, *J. Org. Chem.* 2000, **65**, 3090–3098.

29 (a) M. R. Kesti, M. Abdulrahman, R. M. Waymouth, *J. Organomet. Chem.* 1991, **417**, C12–C15. (b) T. Takahashi, M. Hasegawa, N. Suzuki, M. Saburi, C. J. Rousset, P. E. Fanwick, E.-I. Negishi, *J. Am. Chem. Soc.* 1991, **113**, 8564–8566. (c) J. Y. Corey, X.-H. Zhu, *Organometallics* 1992, **11**, 672–683. (d) J. F. Harrod, S. S. Yun, *Organometallics* 1987, **6**, 1381–1387.

30 S. Sakaki, T. Takayama, M. Sumimoto, M. Sugimoto, *J. Am. Chem. Soc.* 2004, **126**, 3332–3348.

31 (a) J. A. Pool, E. Lobkovsky, P. J. Chirik, *Nature* 2004, **427**, 527–530. (b) H. Basch, D. G. Musaev, K. Morokuma, M. D. Fryzuk, J. B. Love, W. W. Seidel, A. Albinati, T. F. Koetzle, W. T. Klooster, S. A. Mason, J. Eckert, *J. Am. Chem. Soc.* 1999, **121**, 523–528. (c) W. H. Bernskoetter, E. Lobkovsky, P. J. Chirik, *J. Am. Chem. Soc.* 2005, **127**, 14051–14061.

32 (a) M. D. Fryzuk, J. B. Love, S. J. Rettig, V. G. Young, *Science* 1997, **275**, 1445–1447. (b) M. D. Fryzuk, B. A. MacKay, B. O. Patrick, *J. Am. Chem. Soc.* 2003, **125**, 3234–3235. (c) B. A. MacKay, R. F. Munha, M. D. Fryzuk, *J. Am. Chem. Soc.* 2006, **128**, 9472–9483.

33 F. Studt, B. A. MacKay, M. D. Fryzuk, F. Tuczek, *Dalton Trans.* 2006, 1137–1140.

34 (a) V. K. Dioumaev, R. M. Bullock, *Nature* 2003, **424**, 530–532. (b) R. M. Bullock, *Chem. Eur. J.* 2004, **10**, 2366–2374.

35 (a) T. Watanabe, H. Hashimoto, H. Tobita, *Angew. Chem., Int. Ed.* 2004, **43**, 218–221. (b) T. Watanabe, H. Hashimoto, H. Tobita, *J. Am. Chem. Soc.* 2006, **128**, 2176. (c) M. Okazaki, K. A. Jung, K. Satoh, H. Okada, J. Naito, T. Akagi, H. Tobita, H. Ogino, *J. Am. Chem. Soc.* 2004, **126**, 5060–5061. (d) Silyl migration mechanism is implicated by Tobita's recent work: E. Suzuki, T. Komuro, M. Okazaki, H. Tobita, *Organometallics* 2007, **26**, 4379–4382.

36 (a) J. Y. Corey, J. Braddock-Wilking, *Chem. Rev.* 1999, **99**, 175–292. (b) G. I. Nikonov, *J. Organomet. Chem.* 2001, **635**, 24–36. (c) Z. Lin, *Chem. Soc. Rev.* 2002, **31**, 239–245.

37 X.-H. Zhang, L. W. Chung, Z. Lin, Y.-D. Wu, *J. Org. Chem.* (in revision).

38 (a) J. J. Kennedy-Smith, K. A. Nolin, H. P. Gunterman, F. D. Toste, *J. Am. Chem. Soc.* 2003, **125**, 4056–4057. (b) Enantioselective reduction of imines: K. A. Nolin, R. W. Ahn, F. D. Toste, *J. Am. Chem. Soc.* 2005, **127**, 12462–12463.

39 (a) A. C. Fernandes, R. Fernandes, C. C. Romão, B. Royo, *Chem. Commun.* 2005, 213–214. (b) A. C. Fernandes, C. C. Romão, *Tetrahedron Lett.* 2005, **46**, 8881–8883. (c) B. Royo, C. C. Romão, *J. Mol. Catal. A: Chem.* 2005, **236**, 107–112. (d) P. M. Reis, C. C. Romão, B. Royo, *Dalton Trans.* 2006, 1842–1846. (e) P. J. Costa, C. C. Romão, A. C. Fernandes, B. Royo, P. M. Reis, M. J. Calhorda, *Chem. Eur. J.* 2007, **13**, 3934–3941.

40 (a) E. A. Ison, E. R. Trivedi, R. A. Corbin, M. M. Abu-Omar, *J. Am. Chem. Soc.* 2005, **127**, 15374–15375. (b) E. A. Ison, R. A. Corbin, M. M. Abu-Omar, *J. Am. Chem. Soc.* 2005, **127**, 11938–11939. (c) G. Du, M. M. Abu-Omar, *Organometallics* 2006, **25**, 4920–4923.

41 (a) S. D. Burke, R. L. Danheiser, *Handbook of Reagents for Organic Synthesis: Oxidizing and Reducing Agents*, John Wiley and Sons, New York, 1999. (b) W. A. Nugent, J. M. Mayer, *Metal–Ligand Multiple Bonds*, Wiley, New York, 1988. (c) R. H. Holm, *Chem. Rev.* 1987, **87**, 1401.

42 L. W. Chung, H. G. Lee, Z. Li, Y.-D. Wu, *J. Org. Chem.* 2006, **71**, 6000–6009.

43 L. W. Chung, PhD thesis, The Hong Kong University of Science and Technology (HK), 2006.

44 S. K. Tahmassebi, R. R. Conry, J. M. Mayer, *J. Am. Chem. Soc.* 1993, **115**, 7553–7554.

45 (a) K. Tatsumi, R. Hoffman, *Inorg. Chem.* 1980, **19**, 2656–2658. (b) Z. Lin, M. B. Hall, *Coord. Chem. Rev.* 1993, **123**, 149–167.

46 (a) Y. Matano, T. O. Northcutt, J. Brugman, B. K. Bennett, S. Lovell, J. M. Mayer, *Organometallics* 2000, **19**, 2781–2790. (b) T.-Y. Cheng, B. S. Brunschwig, R. M. Bullock, *J. Am. Chem. Soc.* 1998, **120**, 13121–13137.

47 B. M. Trost, Z. T. Ball, *Synthesis-Stuttgart* 2005, **6**, 853–887.

48 Recent example: (a) Y. Kawanami, Y. Sonoda, T. Mori, K. Yamamoto, *Org. Lett.* 2002, **4**, 2825–2827.

49 (a) B. M. Trost, Z. T. Ball, *J. Am. Chem. Soc.* 2001, **123**, 12726–12727. (b) B. M. Trost, Z. T. Ball, *J. Am. Chem. Soc.* 2005, **127**, 17644–17655. (c) B. M. Trost, Z. T. Ball, *J. Am. Chem. Soc.* 2004, **126**, 13942–13944.

50 B. M. Trost, Z. T. Ball, *J. Am. Chem. Soc.* 2003, **125**, 30–31.

51 (a) S. E. Denmark, W. Pan, *Org. Lett.* 2002, **4**, 4163–4166. (b) T. Sudo, N. Asao, Y. Yamamoto, *J. Org. Chem.* 2000, **65**, 8919–8923.

52 R. H. Crabtree, *New J. Chem.* 2003, **27**, 771–772.

53 (a) T. Sudo, N. Asao, V. Gevorgyan, Y. Yamamoto, *J. Org. Chem.* 1999, **64**, 2494–2499. (b) AlCl₃-catalyzed hydrosilylation of alkynes by CPMD simulations: F. Zipoli, M. Bernasconi, A. Laio, *ChemPhysChem* 2005, **6**, 1772–1775.

54 L. W. Chung, Y.-D. Wu, B. M. Trost, Z. T. Ball, *J. Am. Chem. Soc.* 2003, **125**, 11578–11582.

55 Oxidatively-added σ-bond metathesis or hydrogen migration: (a) W. H. Lam, G. Jia, Z. Lin, C. P. Lau, O. Eisenstein, *Chem. Eur. J.* 2003, **9**, 2775–2782. (b) J. Oxgaard, W. A. Goddard, III *J. Am. Chem. Soc.* 2004, **126**, 442–443.

56 C. P. Casey, J. T. Brady, T. M. Boller, F. Weinhold, R. K. Hayashi, *J. Am. Chem. Soc.* 1998, **120**, 12500–12511.

57 M. L. Buil, O. Eisenstein, M. A. Esteruelas, C. Garcia-Yebra, E. Gutiérrez-Puebla, M. Oliván, E. Oñate, N. Ruiz, M. A. Tajada, *Organometallics* 1999, **18**, 4949–4959.

58 S. H. Liu, W. S. Ng, H. S. Chu, T. B. Wen, H. Xia, Z. Y. Zhou, C. P. Lau, G. Jia, *Angew. Chem., Int. Ed.* 2002, **41**, 1589–1591.

59 (a) B. M. Trost, F. D. Toste, *J. Am. Chem. Soc.* 1999, **121**, 9728–9729. (b) E. C. Hansen, D. Lee, *J. Am. Chem. Soc.* 2005, **127**, 3252–3253.

60 P. B. Glaser, T. D. Tilley, *J. Am. Chem. Soc.* 2003, **125**, 13640–13641.

61 P. G. Hayes, C. Beddie, M. B. Hall, R. Waterman, T. D. Tilley, *J. Am. Chem. Soc.* 2006, **128**, 428–429.

62 M. Okazaki, H. Tobita, H. Ogino, *Dalton Trans.* 2003, 493–506.

63 (a) T. D. Tilley, in *The Chemistry of Organic Silicon Compounds*, eds. S. Patai, Z. Rappoport, Wiley & Sons, New York, 1989, pp. 1415–1477. (b) T. D. Tilley, in *The Silicon–Heteroatom Bond*, eds. S. Patai, Z. Rappoport, Wiley & Sons, New York, 1991, p. 245. (c) W. Petz, *Chem. Rev.* 1986, **86**, 1019–1047. (d) H. K. Sharma, K. H. Pannell, *Chem. Rev.* 1995, **95**, 1351–1374. (e) M. S. Eisen, in *The Chemistry of Organic Silicon Compounds*, eds. Z. Rappoport, Y. Apeloig, Wiley & Sons, New York, 1998, Vol. 2, pp. 2037–2128. (f) M. A. Brook, *Silicon in Organic, Organometallic and Polymer Chemistry*, Wiley-Interscience, New York, 2000.

64 (a) C. Beddie, M. B. Hall, *J. Am. Chem. Soc.* 2004, **126**, 13564–13565. (b) U. Böhme, *J. Organomet. Chem.* 2006, **691**, 4400–4410.

65 (a) M. L. H. Green, L.-L. Wong, *J. Chem. Soc., Chem. Commun.* 1988, 677–679. (b) C. P. Casey, C. S. Yi, *Organometallics* 1991, **10**, 33–35. (c) D. J. Tempel, M. Brookhart, *Organometallics* 1998, **17**, 2290–2296.

66 (a) C. P. Casey, J. A. Tunge, T. Y. Lee, M. A. Fagan, *J. Am. Chem. Soc.* 2003, **125**, 2641–2651. (b) J. Jaffart, M. Etienne, F. Maseras, J. E. McGrady, O. Eisenstein, *J. Am. Chem. Soc.* 2001, **123**, 6000–6013. (c) M. Etienne, *Organometallics* 1994, **13**, 410–412. (d) Z. Guo, D. C. Swenson, R. F. Jordan, *Organometallics* 1994, **13**, 1424–1432.

67 (a) C. M. Frech, D. Milstein, *J. Am. Chem. Soc.* 2006, **128**, 12434–12435. (b) D. A. Culkin, J. F. Hartwig, *Acc. Chem. Res.* 2003, **36**, 234–245. (b) J. A. Pool, E. Lobkovsky, P. J. Chirik, *Organometallics* 2003, **22**, 2797–2805. (c) T. Ishiyama, J. F. Hartwig, *J. Am. Chem. Soc.* 2001, **123**, 1232–1233. (d) G. Mann, C. Incarvito, A. L. Rheingold, J. F. Hartwig, *J. Am. Chem. Soc.* 1999, **121**, 3224–3225. (e) J. F. Hartwig, *Acc. Chem. Res.* 1998, **31**, 852–860.

14
Methane Hydroxylation by First Row Transition Metal Oxides

Kazunari Yoshizawa

14.1
Introduction

The direct conversion of methane into methanol under mild conditions is one of the top ten subjects in catalytic chemistry [1]. The activity of the bare first row transition-metal oxides (MO^+) towards methane is probably a key to a better understanding of the energetic and mechanistic aspects of direct methane hydroxylation by soluble methane monooxygenase [2] and Fe-ZSM-5 zeolite [3]. Schwarz and coworkers [4, 5] have investigated the gas-phase reactions of the first-row MO^+ complexes and methane and demonstrated that the late MO^+ complexes after Mn can activate methane (Scheme 14.1). The reaction efficiency and the methanol branching ratio are significantly dependent on the metal. For example, MnO^+ reacts with methane very efficiently, but the branching ratio to methanol is less than 1%. FeO^+ efficiently reacts with methane, forming methanol in 41% yield. CoO^+ exhibits low reactivity toward methane, but the branching ratio to

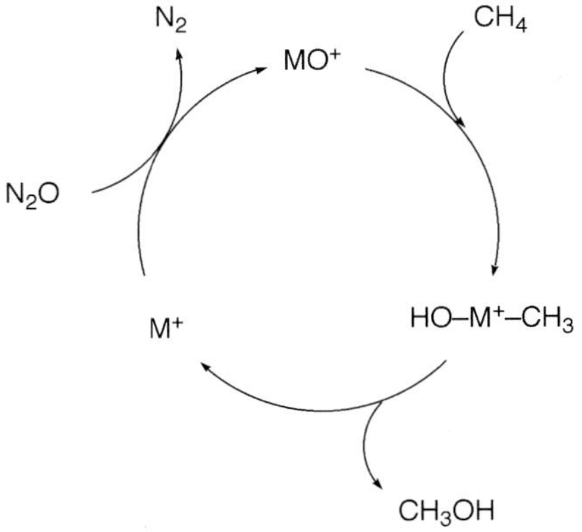

Scheme 14.1

Computational Modeling for Homogeneous and Enzymatic Catalysis.
A Knowledge-Base for Designing Efficient Catalysts. K. Morokuma and D. G. Musaev (Eds.)
Copyright © 2008 WILEY-VCH Verlag GmbH & Co. KGaA, Weinheim
ISBN: 978-3-527-31843-8

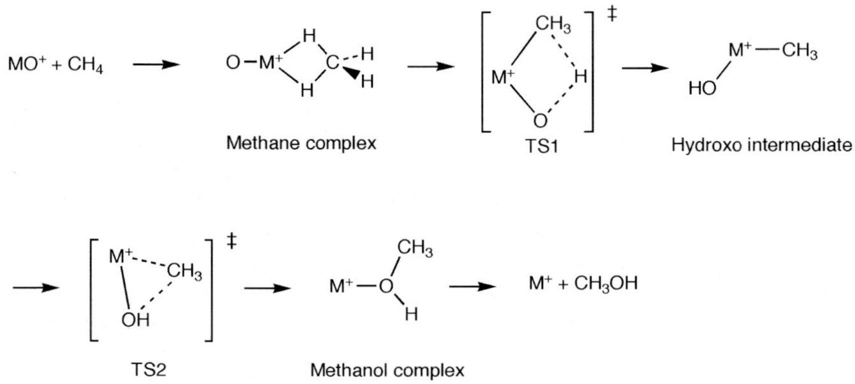

14 Methane Hydroxylation by First Row Transition Metal Oxides

$MO^+ + CH_4 \rightarrow$ O–M⁺···CH₄ (Methane complex) → [TS1]‡ → M⁺–CH₃ with HO (Hydroxo intermediate) → [TS2]‡ → M⁺–O–CH₃ (Methanol complex) → $M^+ + CH_3OH$

Scheme 14.2

methanol is 100%. Both reactivity and methanol branching ratio are excellent in NiO^+. In contrast, the early MO^+ complexes such as ScO^+, TiO^+, and VO^+ have no reactivity towards alkanes and alkenes, due to their strong metal–oxo bonds.

We have studied the reaction pathway and its detailed energetics for the methane–methanol conversion by FeO^+ at the B3LYP level of density functional theory (DFT) [6]. Scheme 14.2 outlines the proposed mechanism of this reaction, which includes two transition states (TS1 and TS2). The nature of these transition states and the HO–Fe^+–CH_3 intermediate were confirmed by intrinsic reaction coordinate (IRC) [7] and femtosecond dynamics [8] calculations. The direct benzene–phenol conversion by FeO^+ in the gas phase [9] and by an iron-oxo species over Fe-ZSM-5 zeolite [10] takes place in a similar non-radical manner. Moreover, we have extended this non-radical mechanism to methane hydroxylation by intermediate **Q** of soluble methane monooxygenase (sMMO) [11], in which a dinuclear iron-oxo complex plays an essential role as an active species [2]. The question as to the reaction pathway and energetics for the methane into methanol conversion by the first-row MO^+ complexes is of particular interest in this chapter.

Another interesting aspect in the methane into methanol conversion by metal-oxo species is the spin inversion, which can occur via the crossing of two potential energy surfaces [12]. In contrast to organic reactions, which proceed on a single potential energy surface in most cases, reactions mediated by organometallic systems can proceed on more than one potential energy surface. For example, the reactions of FeO^+ with dihydrogen [12a] and methane [13]. Thus, besides the classical factors such as barrier heights, spin–orbit coupling (SOC) plays an important role in the reaction of organometallic systems [14]. This phenomenon called "two-state reactivity" is a key feature of the reactions of transition metal systems.

In this chapter we analyze the mechanism and energetics of the reactions of MO^+ (where M is the first-row transition metal atoms, Sc–Cu) and methane.

14.2
Reactivity of the MO⁺ Species

The metal–oxo bond and its catalytic function significantly depend on the number of d-electrons. Figure 14.1 shows the molecular orbitals of MO$^+$, which can be partitioned into bonding (2σ and 1π), nonbonding (1σ and 1δ), and anti-bonding (2π and 3σ) block orbitals. The early transition-metal ions such as Sc$^+$, Ti$^+$, and V$^+$ form a strong triple bond with an oxygen atom, whereas the late transition metals form a weak, reactive double M–O bond. As could be expected, the M–O bond dissociation energy decreases with an increase in the number of d-electrons. Indeed, as seen in Table 14.1, the B3LYP/6-311G** calculated Sc$^+$–O bonding energy is 156 kcal mol^{-1} relative to the dissociation limit of the $^3\Delta$ state of Sc$^+$ and the $^3\Pi$ state of O. This large dissociation energy is a direct consequence of the three fully occupied bonding orbitals. The electronic features of TiO$^+$ and VO$^+$ are very similar to that of ScO$^+$ because the partially occupied nonbonding 1δ set has no direct effect on the metal–oxo bond. Therefore Ti–O$^+$ and V–O$^+$ bonding energies, 155 and 137 kcal mol^{-1}, respectively, are comparable to that for Sc$^+$–O. Since the M–O bond has a very strong triple bond character for the early transition metal atoms, the reactivity of these complexes toward inert alkanes is very low.

In contrast to the early MO$^+$ complexes, those for late transition metal systems have high-spin ground electronic states with singly occupied anti-bonding π

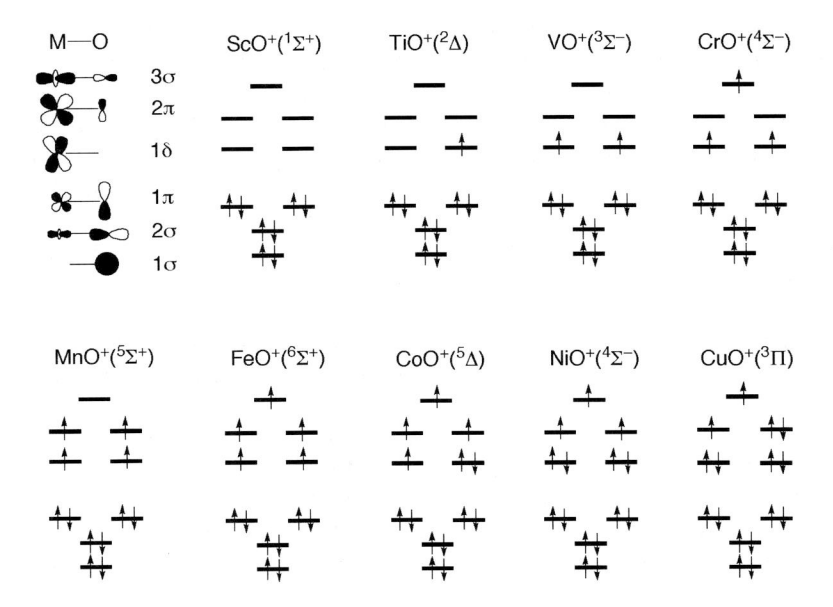

Figure 14.1 Schematic presentation of molecular orbitals and electronic configurations of first row transition-metal oxides.

Table 14.1 Computed bond dissociation energies (BDE), atomic spin densities for the MO$^+$ complexes, and overall heats of reaction (ΔE) for the conversion of methane into methanol calculated at the B3LYP/6-311G** level of theory.

MO$^+$	State	BDE (kcal mol^{-1})	Atomic spin density M	Atomic spin density O	ΔE (kcal mol^{-1})
ScO$^+$	$^1\Sigma^+$	156.1	0.00	0.00	73.5
TiO$^+$	$^2\Delta$	155.1	1.14	−0.14	72.4
VO$^+$	$^3\Sigma^-$	137.2	2.33	−0.33	54.5
CrO$^+$	$^4\Sigma^-$	81.3	3.65	−0.65	−1.3
MnO$^+$	$^5\Sigma^+$	56.4	4.75	−0.75	−26.2
FeO$^+$	$^6\Sigma^+$	75.2	3.86	1.14	−12.6
	$^4\Delta$	69.4	3.62	−0.63	—
CoO$^+$	$^5\Delta$	73.3	2.68	1.32	−25.6
	$^3\Pi$	49.9	2.61	−0.61	—
NiO$^+$	$^4\Sigma^-$	69.3	1.53	1.47	−26.5
	$^2\Sigma^-$	57.9	−0.23	1.23	—
CuO$^+$	$^3\Pi$	37.6	0.47	1.47	−50.0
	$^1\Sigma^+$		0.00	0.00	—

orbitals [16]. A computed dissociation energy for the $^6\Sigma^+$ state of FeO$^+$ (75 kcal mol^{-1}) at the B3LYP/6-311G** level is in good agreement with an experimental value of 81 ± 2 kcal mol^{-1}, while that for the $^5\Delta$ state of CoO$^+$ (73 kcal mol^{-1}) is fully consistent with an experimental value of 77 ± 2 kcal mol^{-1}. For the $^4\Sigma^-$ state of NiO$^+$ the calculated bonding energy is 69 kcal mol^{-1}, which also is in a good agreement with its experimental value (63 ± 3 kcal mol^{-1}) [4, 5]. A computed bond dissociation energy for the $^3\Pi$ ground state of CuO$^+$ is 38 kcal mol^{-1} is also in excellent agreement with an experimental value (37 kcal mol^{-1}).

14.3
Energy Profile for Methane Hydroxylation

Figure 14.2 presents computed energy diagrams for methane hydroxylation by the late transition-metal oxides [6d]. This non-radical reaction pathway for the methane–methanol conversion by MO$^+$ includes two transition states: The first, TS1, corresponds to H-atom abstraction, while the second, TS2, corresponds to a recombination process (Scheme 14.2). TS1 is related to the most important elec-

14.3 Energy Profile for Methane Hydroxylation | 321

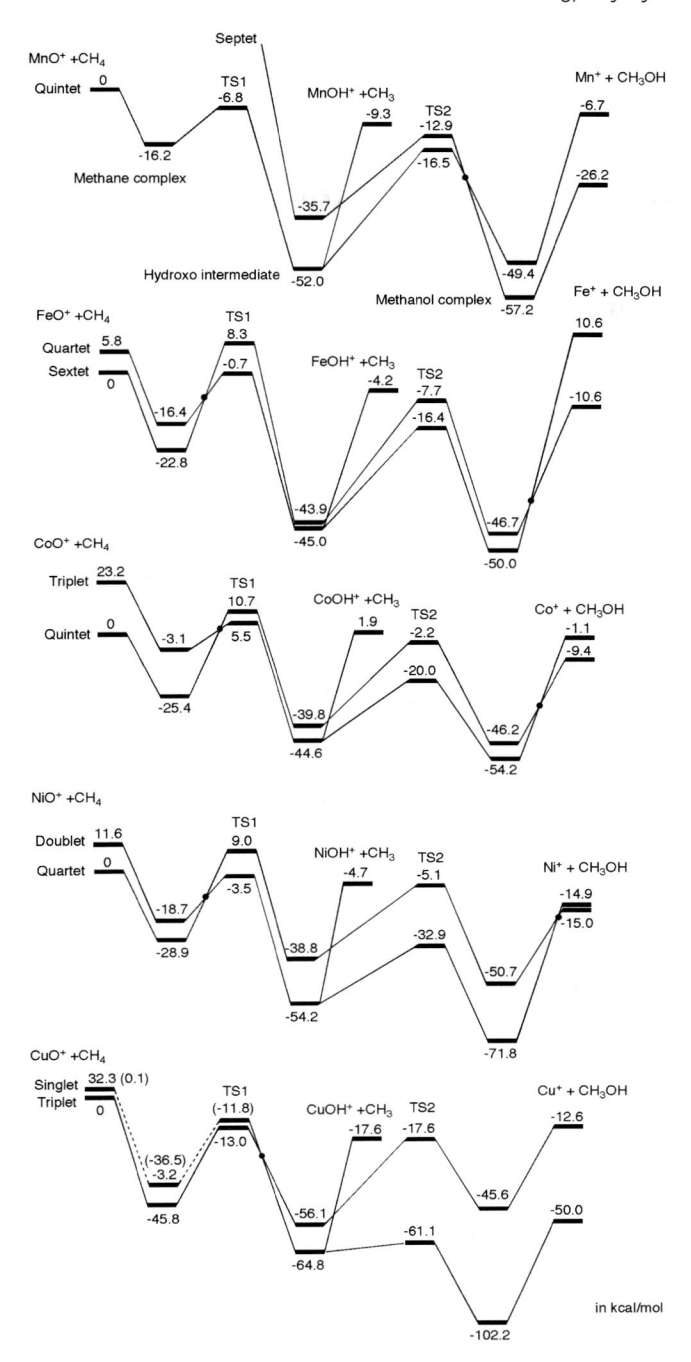

Figure 14.2 Potential energy surfaces for the methane–methanol conversion by the late transition-metal oxides calculated at the B3LYP/6-311G** level of theory.

tronic process that is responsible for the C–H bond dissociation of methane. The reactions of all MO$^+$ with methane are exothermic. As seen in Figure 14.2, for the reaction of FeO$^+$, spin inversion between the sextet and quartet states occurs twice: once in the entrance channel near TS1, and a second time at the exit channel to release methanol. If the reacting system stays only on the sextet potential surface, the reaction will not take place because of large energy barrier at the TS1. However, the quartet electronic state of TS1 is energetically lower than its sextet state, and lies even lower than the ground state dissociation limit. Thus, this reaction can occur in the gas phase under adiabatic conditions. A similar picture can also be seen in the other late transition-metal oxides.

Figure 14.3 shows histograms of computed activation energies for the H-atom abstraction of methane in the MO$^+$/CH$_4$ systems. The presented relative energies (ΔE) are measured from the dissociation limit of the ground state MO$^+$ + CH$_4$. For the reaction of ScO$^+$, TiO$^+$ and VO$^+$, ΔE is a substantial. However, the computed ΔE are negative for the low-spin states of CrO$^+$, MnO$^+$, FeO$^+$, and NiO$^+$ and for both high- and low-spin states of CuO$^+$. These results are fully consistent with the experimental observations indicating that the late MO$^+$ complexes efficiently activate methane [4, 5]. Based on these analysis, we predict that CuO$^+$ should exhibit very high reactivity toward methane molecule. However, the reaction of CuO$^+$ with methane has not yet been observed because CuO$^+$ itself is not stable under ICR conditions, due to the weak Cu–O bond.

As listed in Table 14.2, the observed efficiencies of the reaction of MnO$^+$, FeO$^+$, CoO$^+$, and NiO$^+$ with methane molecule are 40, 20, 0.5, and 20%, respectively [5]. This trend is consistent with that reported from the DFT calculations. Indeed, the computed ΔE (observed reaction efficiencies) for MnO$^+$, FeO$^+$, CoO$^+$, and NiO$^+$ are −6.8 (40%), −0.7 (20%), 5.5 (0.5%), and −3.5 kcal mol^{-1} (20%), respectively.

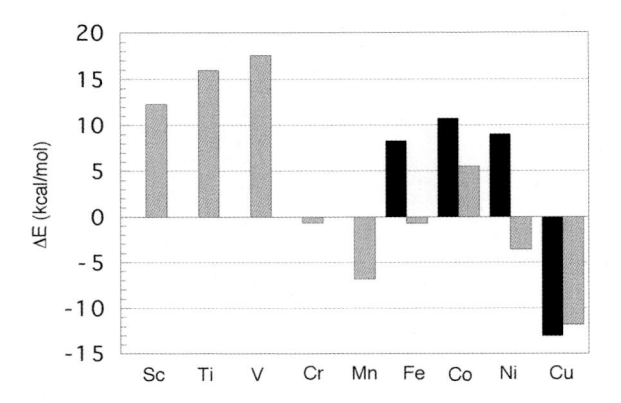

Figure 14.3 Computed activation energies for H-atom abstraction from methane by the MO$^+$ system. The activation energy is measured from the ground state dissociation limit of MO$^+$ + CH$_4$. The black and shaded histograms show high- and low-spin states, respectively.

Table 14.2 Measured reaction efficiencies (φ) and product branching ratios for the reaction of MO$^+$ with methane.

MO$^+$	φ (%)	MOH$^+$ + CH$_3$	MCH$_2^+$ + H$_2$O	M$^+$ + CH$_3$OH
MnO$^+$	40	100	0	<1
FeO$^+$	20	57	2	41
CoO$^+$	0.5	0	0	100
NiO$^+$	20	0	0	100

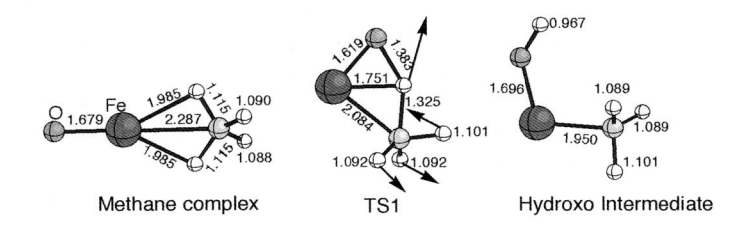

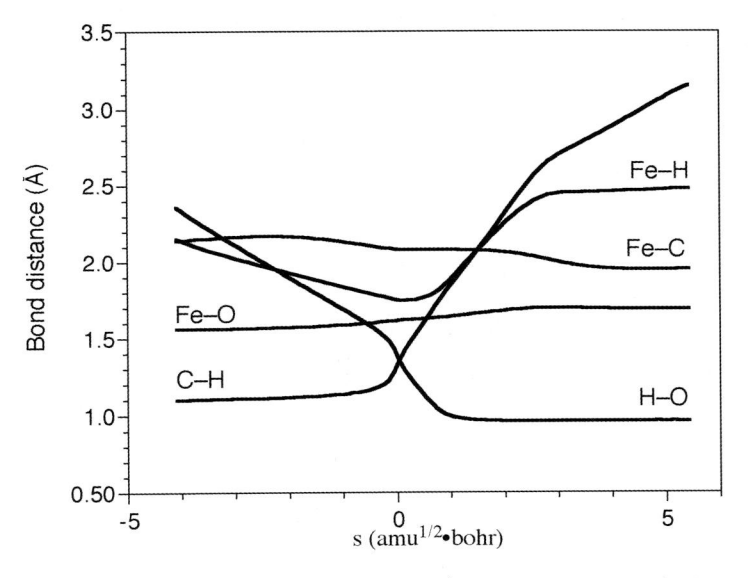

Figure 14.4 Changes of some bond distances (Å) along the IRC initiated from the C–H bond activation transition state of the quartet state calculated at the B3LYP/6-311G** level of theory.

14.4
Intrinsic Reaction Coordinate (IRC) Analysis

As seen from Figure 14.2, the first half of the reaction of FeO^+ and CH_4 is the C–H bond activation which occurs at the TS1 and leads to the hydroxo intermediate. Figure 14.4 gives a detailed description of TS1, in the quartet state. The IRC analysis was traced from TS1 ($s=0$), which exhibits a C_s structure, towards both reactant ($s<0$) and product ($s>0$) directions. As seen in Figure 14.4, the H-atom being abstracted interacts not only with the C and the O atoms but also with the Fe atom.

Let us next look at the second half of the reaction, in which the formation of the methanol complex occurs in a concerted manner via the combination of the OH and CH_3 ligands of the hydroxo intermediate (Figure 14.5). The IRC was again traced from TS2 ($s=0$) towards both reactants ($s<0$) and product ($s>0$) directions.

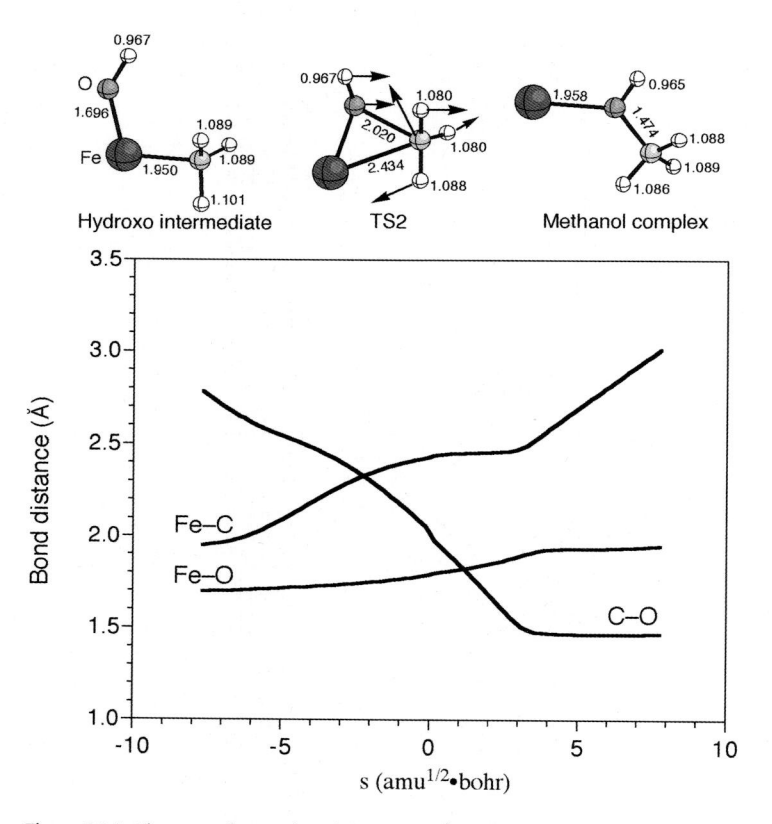

Figure 14.5 Changes of some bond distances (Å) along the IRC initiated from the OH and CH_3 recombination transition state of the quartet state calculated at the B3LYP/6-311G** level of theory.

14.4 Intrinsic Reaction Coordinate (IRC) Analysis

Seam of crossing and spin-inversion are also the subject of our discussion. As discussed above, the transition of spin multiplicity is expected to occur from the ground sextet state to the excited quartet state in the region prior to TS1. Spin inversion is a nonadiabatic process, and its detailed analysis is very complex. Indeed, the total energy of the system that consists of seven atoms ($FeO^+ + CH_4$) is a function of internal degrees of freedom of dimension 15. A crossing seam between the two potential energy surfaces is therefore a "line" of dimension 14, and it is difficult to perform a detailed inspection of the crossing seam. Therefore, we performed single-point calculations of the sextet state of the system as a function of the structural change along the IRC of the quartet state and vice versa; such analyses is expected to provide useful information on the crossing points of energy minimum and maximum.

The solid and dotted lines in Figure 14.6 indicate the computed energy profiles of the quartet and sextet states, respectively, along the quartet IRC. The relative energies are measured from the total energy of the reactant complex of the sextet state as a standard. Seam of crossing point **SI** is located at $s = -2.8$ with a relative energy of 14 kcal mol^{-1}. It can be defined as the crossing point between the quartet and sextet potential energy surfaces. After passing point **SI**, the reaction proceeds on the quartet potential energy surface in the product direction.

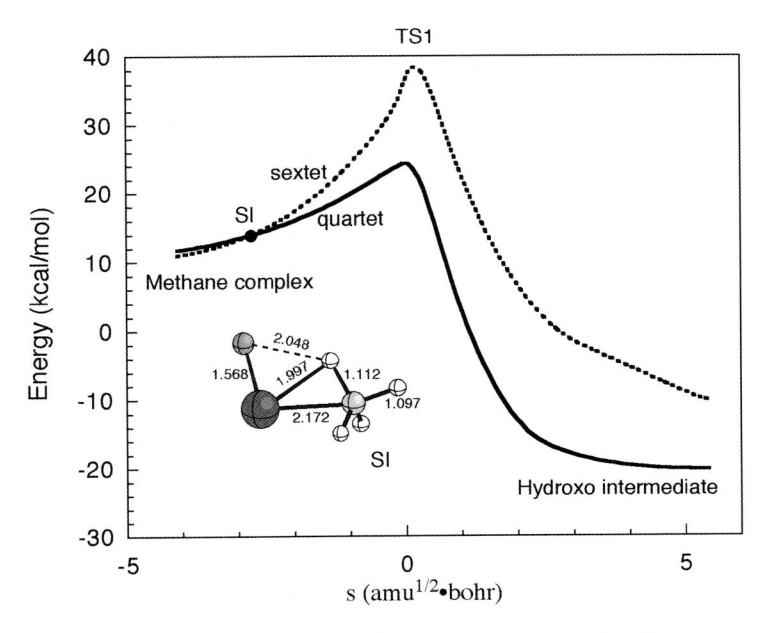

Figure 14.6 Energetics along the IRC of the quartet state connecting the methane complex to the hydroxo intermediate.

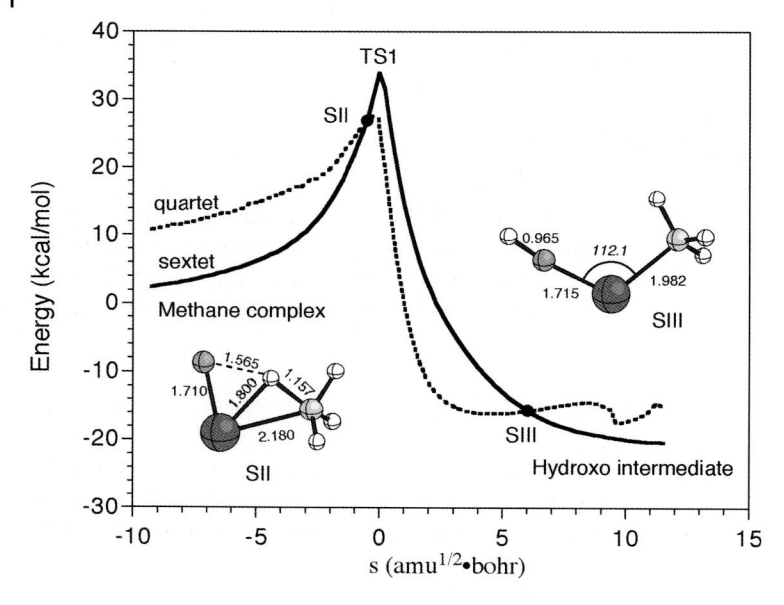

Figure 14.7 Energetics along the IRC of the sextet state connecting the hydroxo intermediate with the methanol complex.

In Figure 14.7, the solid and dotted lines indicate computed energy profiles of the sextet and quartet states, respectively, along the sextet IRC. We find the second seam of crossing point **SII** before TS1.

As indicated in Figure 14.2, there is a third crossing seam between the two potential energy surfaces in the exit channel. This seam of crossing will lead to the spin inversion from the quartet state to the sextet state.

14.5
Spin–Orbit Coupling (SOC) in Methane Hydroxylation

Here, we discuss the SOC between the quartet and sextet states for the reaction of FeO^+ and CH_4 [14]. For this purpose, we have carried out CASSCF/6-311G** calculations along the IRCs of the quartet and sextet states, and we then computed the SOC matrix elements using the SOC-CI method [17]. Since the orbital sets of the quartet and sextet states must share a common set of frozen core orbitals, to calculate the SOC matrix elements we employed the converged CASSCF wavefunction of the quartet state also as a reference state for the sextet CI wavefunction. We used the one-electron effective spin–orbit operator of Eq. (1):

14.5 Spin–Orbit Coupling (SOC) in Methane Hydroxylation

$$\mathbf{H}_{so} = \frac{\alpha^2}{2}\sum_i\sum_k\left(\frac{Z_k^*}{r_{ik}^3}\right)\mathbf{S}_i\cdot\mathbf{L}_{ik} = \sum_i h_i(Z^*) \quad (1)$$

$$\frac{\alpha^2}{2} = \frac{e^2 h}{4\pi m_e^2 c^2} \quad (\alpha^{-1} = 137.036)$$

where L_{ik} and S_i are, respectively, the orbital and spin angular momentum operators for electron i in the framework of nuclei k. The effective nuclear charge Z_k^* is an empirical parameter in the one-electron spin–orbit Hamiltonian. The SOC value is the matrix element that expresses the coupling of the quartet and sextet states by the operator of Eq. (2):

$$\langle\mathbf{H}_{so}\rangle_{s,s'} = \langle^6\Psi_1(M_s)|\mathbf{H}_{so}|^4\Psi_2(M_{s'})\rangle \quad (2)$$

Here $^6\Psi_1$ ($^4\Psi_2$) is the M_s ($M_{s'}$) component of the many-body sextet-state (quartet-state) wavefunction. Considering the generated spin sublevels M_s, a reasonable measure of the SOC-induced quartet–sextet interaction is the root-mean-square coupling constant of Eq. (3):

$$\mathrm{SOC} = \left[\sum_{s,s'}\langle\mathbf{H}_{so}\rangle_{s,s'}^2\right]^{1/2} \quad (3)$$

Figure 14.8 shows computed potential energy surfaces of several low-lying quartet states and SOC values between the quartet and sextet states as a function of reaction coordinate s. The reaction coordinate corresponds to the methane complex at $s=0$, TS1 at $s=4.0$, and the hydroxo intermediate at $s=10.0$. As seen from Figure 14.8, the energy gap between the low-lying quartet and ground sextet states decreases to zero at the first seam of crossing point located at $s=1.3$. The calculated relative energies are from the $^4A_1$ state of methane complex. In the low-lying $^4A_1$ state the C–H bond activation energy at TS1 is 23 kcal mol^{-1} and is 17 kcal mol^{-1} exothermic. Although the energy of the $^4A_2$ state is close to that for the $^4A_1$ before TS1, it becomes unstable after passing TS1. The $^4\Pi_1$ and $^4\Pi_2$ states in the reactant complex lie 30 kcal mol^{-1} above the $^4A_1$ state. Moreover, the activation barriers at TS1 for both Π states are very large (about 90 kcal mol^{-1}). Thus, the $^4A_2$, $^4\Pi_1$, and $^4\Pi_2$ states can be neglected in the spin-crossover analysis reaction after TS1.

Figure 14.8(b) shows the SOC values between the four quartet states and the sextet state. The $^4\Delta$–$^6\Sigma^+$ SOC is a key to understanding the spin-forbidden transition that takes place near the first seam of crossing. Starting from 133 cm^{-1} at $s=0$, $^4A_1$–$^6\Sigma^+$ SOC increases to 169 cm^{-1} at $s=4.0$. After TS1, $^4A_1$–$^6\Sigma^+$ SOC is relatively small: it is 40 cm^{-1} at $s=5.0$ and continues to decrease to 4 cm^{-1} at $s=7.0$, and then increases to 21 cm^{-1} in the hydroxo intermediate. Thus, this SOC analysis clearly demonstrates that the spin inversion between the $^4\Delta$ and $^6\Sigma^+$ states occurs near the first seam of crossing ($s=1.3$).

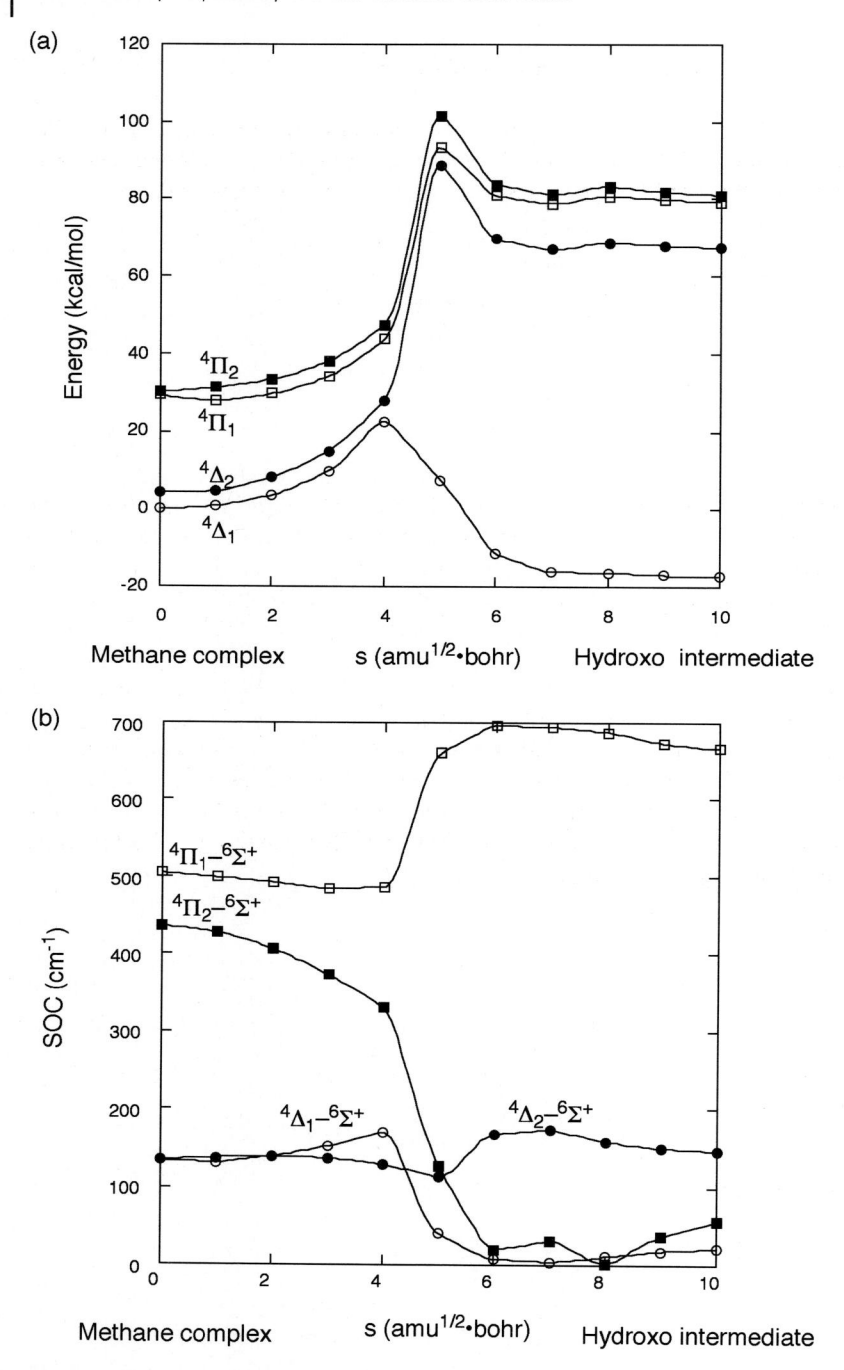

Figure 14.8 (a) Energies of the quartet states at the CASSCF/6-311G** level and (b) spin–orbit coupling (SOC) values between the four quartet states and the ground sextet state along the IRC.

14.6
Kinetic Isotope Effect (KIE) for H-atom Abstraction

Here we discuss the isotope effects for H-atom abstraction via the radical (TSd) and non-radical (TS1) pathways (Scheme 14.3). The KIE is calculated with transition state theory [18]. The values of k_H/k_D were obtained from Eq. (4):

$$\frac{k_{CH_4}}{k_{CD_4}} = \left(\frac{m^R_{CD_4} m^{\#}_{CH_4}}{m^R_{CH_4} m^{\#}_{CD_4}}\right)^{3/2} \left(\frac{I_{CD_4}}{I_{CH_4}}\right)^{3/2} \left(\frac{I^{\#}_{xCH_4} I^{\#}_{yCH_4} I^{\#}_{zCH_4}}{I^{\#}_{xCD_4} I^{\#}_{yCD_4} I^{\#}_{zCD_4}}\right)^{1/2}$$

$$\times \frac{q^R_{vCD_4} q^{\#}_{vCH_4}}{q^R_{vCH_4} q^{\#}_{vCD_4}} \exp\left(-\frac{E^{\#}_{CH_4} - E^{\#}_{CD_4}}{RT}\right) \quad (4)$$

Here superscripts R and # specify the substrate (CH_4) and the transition state, respectively, and m, I, q, and E stand for molecular mass, moment of inertia, vibrational partition function, and activation energy, respectively. The numerator in the last exponential term comes from the fact that the C–H dissociation has a lower activation energy than the C–D dissociation because the former has a larger zero-point vibrational energy.

Table 14.3 summarizes computed values of k_H/k_D for the non-radical abstraction via TS1 for the sextet and quartet states, as well as for the radical-type abstraction via TSd for the sextet state of $FeO^+/CH(D)_4$ system as a function of temperature [19]. At 300 K, the value for the non-radical abstraction is 9.72 and 9.37 for the sextet and quartet states, respectively. The value for the radical-type abstraction in the sextet state is 16.07. Thus, the calculated isotope effect is larger in the radical-type abstraction.

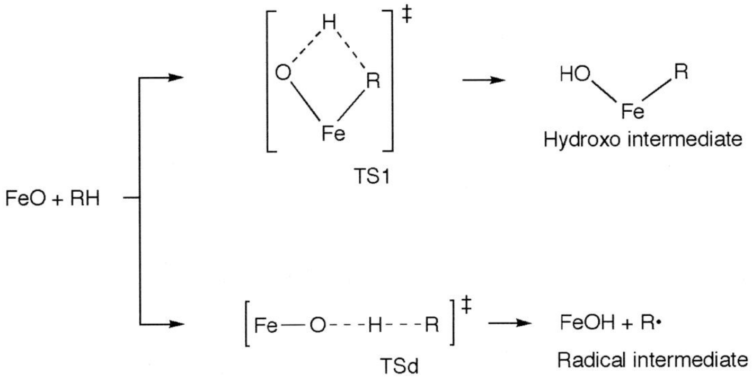

Scheme 14.3

Table 14.3 Kinetic isotope effect (KIE) (k_H/k_D) values in the C–H bond dissociation of methane by the bare FeO$^+$ complex. Values in parentheses include Wigner's tunneling correction.

T (K)	TS1 (^{6}A)	TS1 (^{4}A)	TSd (^{6}A)
200	19.70 (33.95)	19.11 (29.78)	40.52 (67.02)
250	13.08 (21.53)	12.61 (18.54)	23.57 (37.55)
300	9.72 (15.29)	9.37 (13.06)	16.05 (24.62)
350	7.70 (11.57)	7.44 (9.90)	11.9 (17.63)
400	6.36 (9.17)	6.16 (7.88)	9.40 (13.39)

We also have applied the non-radical mechanism to methane hydroxylation by intermediate **Q** of sMMO, the active site of which is a bis(μ-oxo)Fe$^{IV}_2$ complex. Geometry-optimized TS1 in sMMO is essentially identical to that in the gas-phase reaction of FeO$^+$ and CH$_4$ (Scheme 14.4) [11d]. Let us evaluate the non-radical mechanism from the KIE point of view. For sMMO, Lipscomb and coworkers [20] measured from product-distribution analyses after a single turnover KIEs of 4–19 at 277 K; e.g., 19 $\pm$ 3.9 for 1(CH$_4$):1(CD$_4$), 12 $\pm$ 1 for CD$_3$H, 9 $\pm$ 0.5 for CD$_2$H$_2$, and 3.9 $\pm$ 1 for CDH$_3$. Table 14.4 lists KIEs for the H-atom abstraction via TS1, where the values in parentheses include a tunneling correction. As expected, the KIEs significantly decrease with temperature; at 277 K the KIE for the H/D-atom abstraction from CH$_4$/CD$_2$H$_2$ is 9.7 after a tunneling correction. This is in excellent agreement with the value obtained from product distribution analyses at 277 K (9.3 $\pm$ 0.5). Agreement between experiment and theory is also very good for CH$_4$/CD$_4$ and CH$_4$/CD$_3$H.

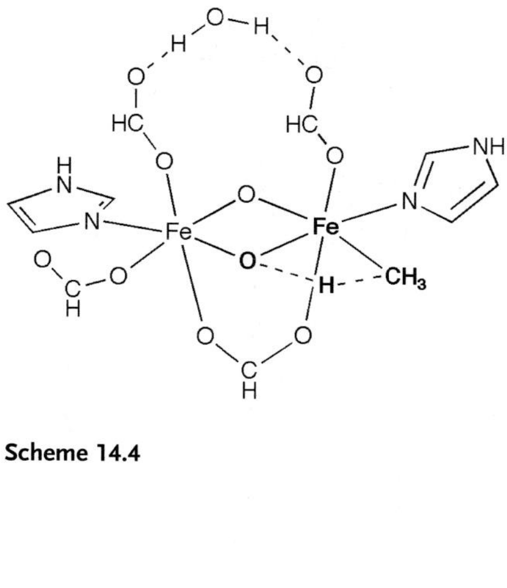

Scheme 14.4

Table 14.4 Computed k_H/k_D values in the H-atom abstraction from methane by intermediate **Q** of sMMO via the transition state (TS1) of the broken-symmetry singlet state. Values in parentheses include Wigner's tunneling correction.

T (K)	CD$_4$	CD$_3$H	CD$_2$H$_2$	CDH$_3$
200	14.8 (25.2)	11.8 (19.7)	9.7 (16.1)	8.3 (13.6)
250	10.8 (17.7)	8.7 (13.9)	7.1 (11.3)	6.0 (9.4)
277	9.5 (15.1)	6.9 (10.7)	6.3 (9.7)	5.3 (8.1)
300	8.6 (13.3)	5.8 (8.5)	5.7 (8.7)	4.8 (7.2)
350	7.1 (10.5)	5.0 (7.1)	4.8 (7.1)	4.1 (5.9)

14.7
Regioselectivity in Alkane Hydroxylation

In this section we discuss primary (1°), secondary (2°), and tertiary (3°) C–H bond activation. According to a textbook [21], the relative reactivity of alkanes in biological hydroxylations generally declines in the order 2° > 3° > 1°, which is different from a general trend we expect from a radical mechanism (3° > 2° > 1°): the calculated C–H bond dissociation energies are 98 and 93 kcal mol^{-1} for primary and tertiary positions. As an example we choose 2-methylbutane (C$_5$H$_{12}$) (Scheme 14.5), and analyze the regioselectivity of its hydroxylation of by FeO$^+$ and FeO^{2+} [22].

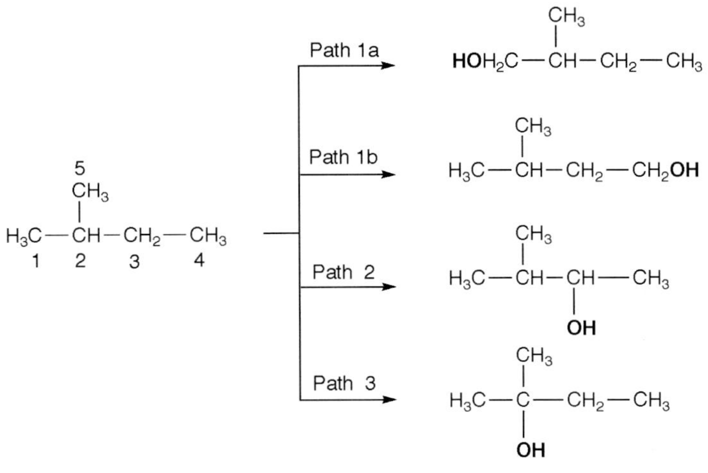

Scheme 14.5

Computed C–H bond dissociation energy on the C1, C2, C3, and C4 atoms of 2-methylbutane reduces in the order primary C1–H (98.8 kcal mol^{-1}) > primary C4–H (98.1 kcal mol^{-1}) > secondary C3–H (95.0 kcal mol^{-1}) > tertiary C2–H (92.8 kcal mol^{-1}). In the initial stages, reactants form ion-molecular complex OFe$^+$(C$_5$H$_{12}$). An H-atom abstraction via TS1 leads to a (HO)Fe(C$_5$H$_{11}$) intermediate, which may have different isomers. This intermediate transforms into the final product Fe$^+$(C$_5$H$_{11}$OH) via a three-centered transition state (TS2).

Figure 14.9 shows a computed energy diagram for the conversion of 2-methylbutane into the secondary alcohol via Path 2 (Scheme 14.5) in the quartet state. The calculated activation energy for the C3–H bond at the TS1-2 is smaller than that at TSd. This makes the non-radical mechanism the energetically more favorable. The hydroxo intermediate in Path 2 yields the corresponding secondary alcohol via TS2-2. The barrier height at TS2-2 is 37.3 kcal mol^{-1} relative to the hydroxo intermediate.

Figure 14.10 summarizes the computed barrier heights for the H-atom abstractions by FeO$^+$ and FeO^{2+} via TS1s and TSds.

In the non-radical mechanism (TS1) the FeO$^+$($^4\Delta$) and FeO^{2+}($^5\Sigma^+$) species prefer the secondary C3–H bond activation and exhibit the following orders of

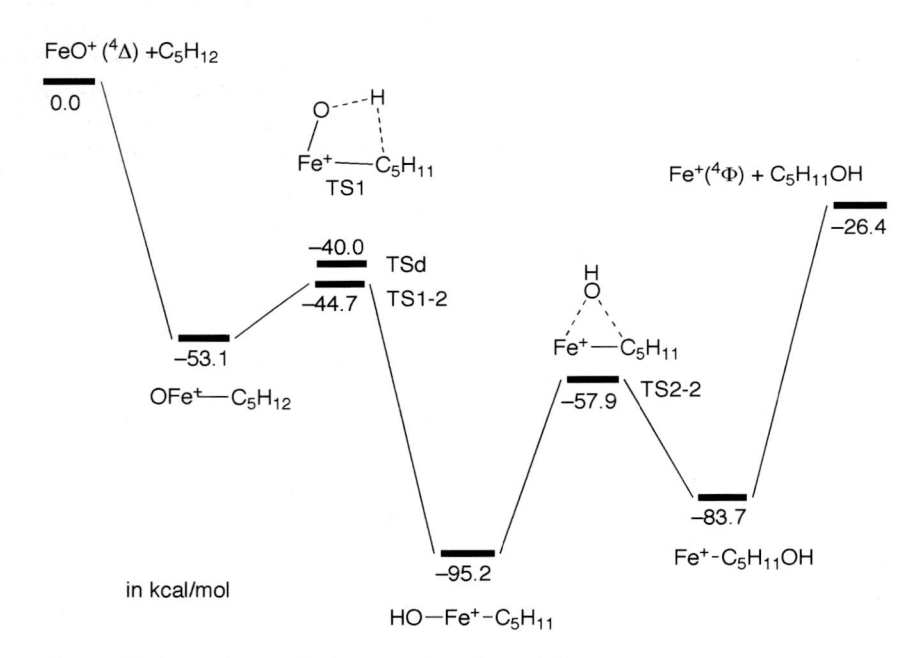

Figure 14.9 Energy diagram for the conversion of 2-methylbutane into 3-methyl-2-butanol by FeO$^+$ (Path 2 in Scheme 14.5) in the quartet state at the B3LYP/(Wachters + D95**) level of theory. Note that the reaction proceeds in two steps, via the TS1-2 and TS2-2 transition states. Relative energies, which include zero-point vibrational energy corrections, are in kcal mol^{-1}.

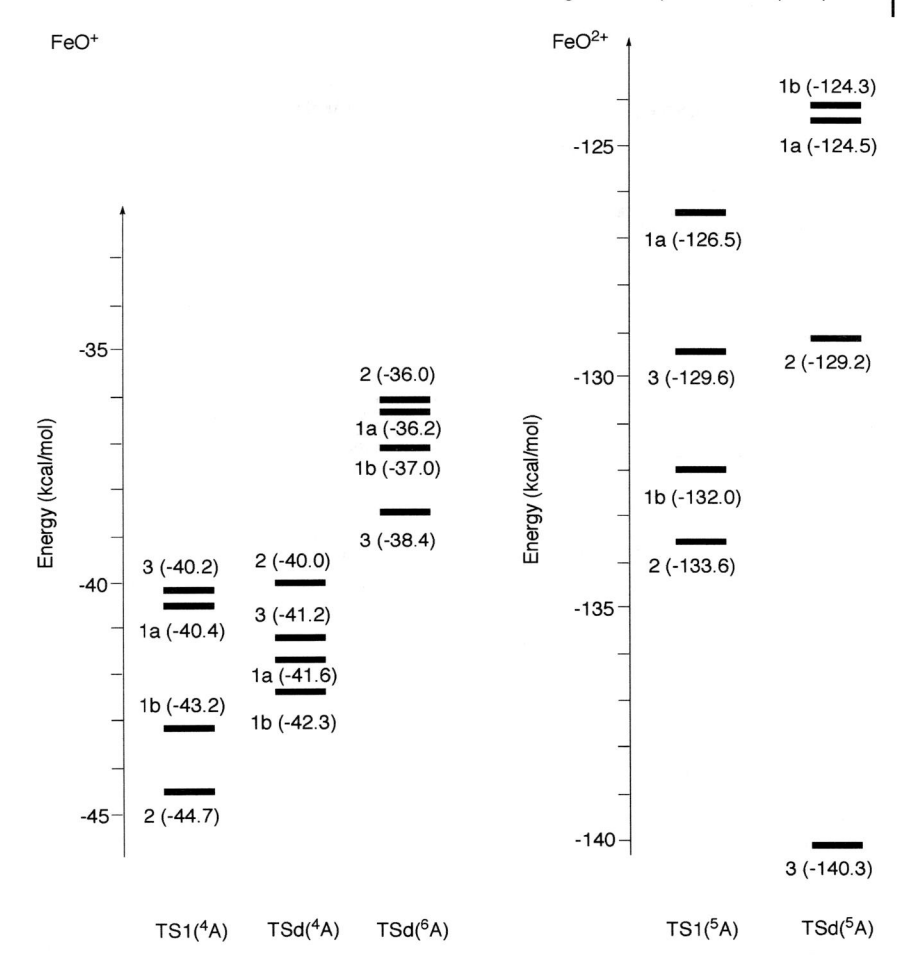

Figure 14.10 Computed C–H bond activation energies of 2-methylbutane by FeO$^+$ and FeO^{2+} via TS1 and TSd measured from the dissociation limits (the $^4\Delta$ state of FeO$^+$ and the $^5\Sigma^+$ state of FeO^{2+}). Values in kcal mol^{-1}.

regioselectivity, respectively:

2° (Path 2) > 1° (Path 1b) > 1° (Path 1a) > 3° (Path 3) for FeO$^+$($^4\Delta$)

2° (Path 2) > 1° (Path 1b) > 3° (Path 3) > 1° (Path 1a) for FeO^{2+}($^5\Sigma^+$)

In remarkable contrast to the non-radical reactions, a strong preference for the tertiary C2–H bond is obtained in the H-atom abstractions in the radical mechanism. FeO$^+$($^6\Sigma^+$) and FeO^{2+}($^5\Sigma^+$) exhibit the following orders of regioselectivity in the C–H bond dissociation of 2-methylbutane via TSds, respectively:

3° (Path 3) > 1° (Path 1b) > 1° (Path 1a) > 2° (Path 2) for FeO$^+$($^6\Sigma^+$)

3° (Path 3) > 2° (Path 2) > 1° (Path 1a) > 1° (Path 1b) for FeO^{2+}($^5\Sigma^+$)

Thus, regioselectivity in the hydrocarbon hydroxylation via TSd is determined by the C–H bond strength of 2-methylbutane. In contrast, regioselectivity in the non-radical mechanism can be explained from the energetical stability of the four-centered transition state structure of TS1.

14.8
Concluding Remarks

Methane hydroxylation occurs in a non-radical, stepwise manner with the bare transition-metal oxide ions as catalyst. It starts with an interaction of a methane molecule with a metal center, forming a weakly bound ion–molecule complex in the initial stage of the reaction. Subsequent C–H bond cleavage leads to the intermediate with the HO–M–CH$_3$ moiety. A combination of the resultant OH and CH$_3$ ligands at the HO–M–CH$_3$ intermediate occurs in a non-radical manner and leads to a final methanol complex. DFT results and experimental observations are in excellent agreement with respect to the reaction efficiency in the C–H bond activation in the MO$^+$/CH$_4$ systems. Spin inversion between the high- and low-spin potential energy surfaces is an essential feature of the methane hydroxylation by the bare transition-metal oxide ions. The final ligand coupling that takes place at the metal center is essentially identical to the mechanism of *Gif* chemistry given by Barton [23]. This mechanism is reasonably extended to methane hydroxylation by sMMO and Fe-ZSM-5 zeolite if their iron centers are coordinatively unsaturated [15]. We are now interested in whether this non-radical mechanism can work at the mononuclear and dinuclear copper sites of particulate methane monooxygenase (pMMO) [24].

Acknowledgments

K.Y. thanks the collaborators in his group, especially Drs. Yoshihito Shiota, Takehiro Ohta, and Takashi Yumura, for their computational efforts on FeO$^+$, sMMO, and Fe-ZSM-5 zeolite. He acknowledges Grants-in-Aid (No. 18350088, 18G0207, and 18066013) for Scientific Research from the Japan Society for the Promotion of Science and the Ministry of Education, Culture, Sports, Science and Technology of Japan (MEXT), the Joint Project of Chemical Synthesis Core Research institution of MEXT, and CREST of Japan Science and Technology Corporation for their support of this work.

References

1. (a) *Chem. Eng. News* 1993, May 31. (b) A. E. Shilov, *The Activation of Saturated Hydrocarbons by Transitions Metal Complexes*, Riedel Publishing, Dordrecht, 1984. (c) A. E. Shilov, G. B. Shul'pin, *Chem. Rev.* 1997, **97**, 2879. (d) A. E. Shilov, A. A. Shteinman, *Acc. Chem. Res.* 1999, **32**, 763. (e) C. L. Hill (ed.), *Activation and Functionalization of Alkanes*, Wiley, New York, 1989. (f) J. A. Davies, P. L. Watson, J. F. Liebman, A. Greenberg, *Selective Hydrocarbon Activation*, VCH, New York, 1990. (g) R. H. Crabtree, *The Organometallic Chemistry of the Transition Metals*, Wiley, New York, 1990. (h) R. H. Crabtree, *Chem. Rev.* 1985, **85**, 245. (i) R. H. Crabtree, *Chem. Rev.* 1995, **95**, 987. (j) B. A. Arndtsen, R. G. Bergman, T. A. Mobley, T. H. Peterson, *Acc. Chem. Res.* 1995, **28**, 154.
2. (a) A. L. Feig, S. J. Lippard, *Chem. Rev.* 1994, **94**, 759–805. (b) B. J. Wallar, J. D. Lipscomb, *Chem. Rev.* 1996, **96**, 2625.
3. (a) J. R. Anderson, P. Tsai, *J. Chem. Soc., Chem. Commun.* 1987, 1435. (b) V. I. Sobolev, K. A. Dubkov, O. V. Panna, G. I. Panov, *Catal. Today* 1995, **24**, 251.
4. (a) D. Schröder, H. Schwarz, *Angew. Chem., Int. Ed.* 1990, **29**, 1433. (b) D. Schröder, A. Fiedler, J. Hrušák, H. Schwarz, *J. Am. Chem. Soc.* 1992, **114**, 1215.
5. D. Schröder, H. Schwarz, *Angew. Chem., Int. Ed.* 1995, **34**, 1973.
6. (a) K. Yoshizawa, Y. Shiota, T. Yamabe, *Chem. Eur. J.* 1997, **3**, 1160. (b) K. Yoshizawa, Y. Shiota, T. Yamabe, *J. Am. Chem. Soc.* 1998, **120**, 564. (c) K. Yoshizawa, Y. Shiota, T. Yamabe, *Organometallics* 1998, **17**, 2825. (d) Y. Shiota, K. Yoshizawa, *J. Am. Chem. Soc.* 2000, **122**, 12317.
7. K. Yoshizawa, Y. Shiota, T. Yamabe, *J. Chem. Phys.* 1999, **111**, 538.
8. K. Yoshizawa, Y. Shiota, T. Kagawa, T. Yamabe, *J. Phys. Chem. A* 2000, **104**, 2552.
9. K. Yoshizawa, Y. Shiota, T. Yamabe, *J. Am. Chem. Soc.* 1999, **121**, 147.
10. (a) K. Yoshizawa, Y. Shiota, T. Kamachi, *J. Phys. Chem. B* 2003, **107**, 11404. (b) Y. Shiota, K. Suzuki, K. Yoshizawa, *Organometallics* 2005, **24**, 3532. (c) Y. Shiota, K. Suzuki, K. Yoshizawa, *Organometallics* 2006, **25**, 3118.
11. (a) K. Yoshizawa, R. Hoffmann, *New J. Chem.* 1997, **21**, 151. (b) K. Yoshizawa, T. Ohta, R. Hoffmann, *J. Am. Chem. Soc.* 1997, **119**, 12311. (c) K. Yoshizawa, *J. Biol. Inorg. Chem.* 1998, **3**, 318. (d) K. Yoshizawa, T. Yumura, *Chem. Eur. J.* 2003, **9**, 2347.
12. (a) S. Shaik, D. Danovich, A. Fiedler, D. Schröder, H. Schwarz, *Helv. Chim. Acta* 1995, **78**, 1393. (b) D. Schröder, S. Shaik, H. Schwarz, *Acc. Chem. Res.* 2000, **33**, 139.
13. K. Yoshizawa, Y. Shiota, T. Yamabe, *J. Chem. Phys.* 1999, **111**, 538.
14. Y. Shiota, K. Yoshizawa, *J. Chem. Phys.* 2003, **118**, 5872.
15. K. Yoshizawa, *Acc. Chem. Res.* 2006, **39**, 375.
16. E. A. Carter, W. A. Goddard III, *J. Phys. Chem.* 1988, **92**, 2109.
17. N. Matsunaga, S. Koseki, M. S. Gordon, *J. Chem. Phys.* 1989, **91**, 1062.
18. D. A. McQuarrie, *Statistical Thermodynamics*, University Science Books, Mill Valley, 1973.
19. K. Yoshizawa, *Coord. Chem. Rev.* 2002, **226**, 251.
20. J. C. Nesheim, J. D. Lipscomb, *Biochemistry* 1996, **35**, 10240.
21. K. Faber, *Biotransformations in Organic Chemistry* 3rd ed., Springer-Verlag, Berlin, 1997.
22. T. Yumura, K. Yoshizawa, *Organometallics* 2001, **20**, 1397.
23. D. H. R. Barton, D. Doller, *Acc. Chem. Res.* 1992, **25**, 504.
24. K. Yoshizawa, Y. Shiota, *J. Am. Chem. Soc.* 2006, **128**, 9873.

15
Two State Reactivity Paradigm in Catalysis. The Example of X–H (X = O, N, C) and C–C Bonds Activation Mediated by Transition Metal Compounds

Maria del Carmen Michelini, Ivan Rivalta, Nina Russo, and Emilia Sicilia

15.1
Introduction

Catalysis has an enormous importance in modern sciences both from fundamental and applied points of view. The understanding of the mechanism and controlling factors of many catalytic reactions are extremely important for industry, environment and life sciences. Despite the enormous progress in theoretical methodologies and experimental techniques achieved in recent decades, the level of understanding of catalytic processes is still limited and remains the subject of intensive investigations. The situation is more complex for processes involving transition metal atoms or ions, because many of these systems posses several lower-lying electronic states in a narrow range of energy, and are open-shell systems with several un-paired electrons in valence space. Consequently, during the reaction system may change its spin multiplicity many times. Such spin non-conserving reactions are often referred to as "spin-forbidden", and require the inclusion of spin–orbital coupling into the calculations. When very strong spin–orbit coupling between the different states is involved, the reaction will behave like any other.

The role of spin flip in organometallic chemistry has been underlined by the introduction of the so-called Two State Reactivity (TSR) paradigm [1]. According to this definition: A thermal reaction that involves spin crossover along the reaction coordinate from reactants to products should be described in terms of two state reactivity.

The idea of state-selective reactivity was first introduced by Armentrout and coworkers [2, 3], who have also pointed out the possible effects that may control this process.

Despite the importance of spin-flip in organometallic chemistry, only recently it has received proper attention [4–11].

Many investigated examples of spin-forbidden processes involve only a single crossing between two surfaces of different multiplicity. Nevertheless, notably,

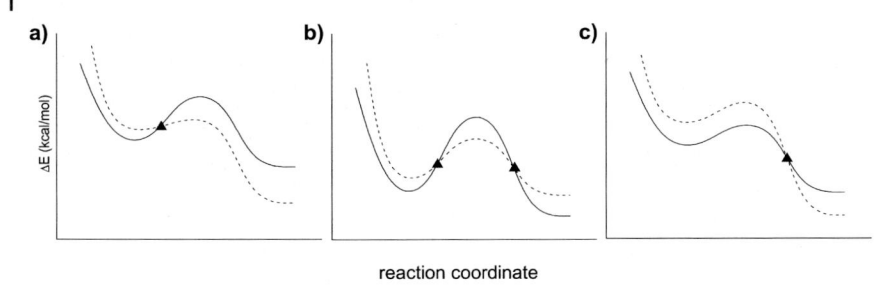

Scheme 15.1

there are also many examples with a double spin crossing between two surfaces of different multiplicity. These two situations are qualitatively depicted in Scheme 15.1, (a) and (b), respectively. In the third situation, described in Scheme 15.1(c), the crossing occurs after the transition state. The third type of spin crossover can occur for systems both with excess energy and with electronically excited states. Therefore, it cannot be considered an example of TSR because, during the spin crossing, it leads to the same state of the product.

The latest progress in this field is due to the success of gradient-corrected Density Functional Theory (DFT), which has become one of the most frequently used tools in quantum chemistry. Experience has shown that structures, frequencies, and many other properties of the transition metal systems can be accurately modeled by hybrid and generalized gradient approximation (GGA) functionals [12].

The present chapter discusses the performance of modern density functional methods in the study of potential energy surfaces (PES) of the reactions of bare transition metal atoms and cations with small molecules. We pay a special attention to reactions involving several lower-lying electronic states of the transition metal centers. We also perform a topological analysis of the electron localization function (ELF) [13, 14] that is useful in rationalizing reaction pathways and in elucidating the features of chemical bonds.

15.2
General Methods

In this section we discuss the computational tools used to explore potential energy surfaces. More computational details are given in specific sections, if required. DFT is the main tool in our investigations of catalysis by transition-metal compounds. The so-called hybrid DFT methods [15, 16] are considered to be the method of choice in many studies. In particular, the B3LYP functional [16, 17–26] is often considered to be the standard approach to study many problems involving transition metals [18]. The optimization of geometries and frequency of the reactants, transition states, intermediates and products of the studied reactions were performed by utilizing the double-ζ quality, DZVP [27], basis sets for the first row transition metals and triple-ζ, TZVP [28], quality basis sets for the main groups elements within the *Gaussian03*/DFT package [29].

For several cases, the energetics of the systems were improved by performing single-point energy calculations using a better quality basis set for the transition metals: TZVP+G(3df,2p) [30] basis set supplemented with a diffuse *s* function (with an exponent that is 0.33 times the most diffuse *s* function on the original set), two sets of *p* functions [31], one set of diffuse *d* function [32], and three sets of uncontracted *f* functions [33].

As it is known, one of the drawbacks of density functional approaches is the incorrect prediction of the energy order of the lower-lying states of atoms and cations. For example, for Fe^+, the B3LYP method puts the ground $^6D(s^1d^6)$ state above the $^4F(d^7)$ excited state. In this connection, several correction schemes have been proposed [18]. However, these correction schemes are not applicable for other cations of the same series [34]. To overcome this irregularity of the B3LYP approach, we have introduced new basis sets of double-ζ quality [35], which will be referred as $DZVP_{opt}$ basis sets.

In many cases we also performed single-point CCSD(T) calculations at the B3LYP optimized geometries. In these calculations we used the TZVP+G(3df,2p) basis sets for the metal atoms. The applicability of the CCSD(T) approach to these systems was checked through the T1 diagnostic [36]. For simplicity, below the B3LYP/TZVP+G(3df,2p)//B3LYP/DZVP and CCSD(T)/TZVP+G(3df,2p)//B3LYP/DZVP approaches will be referred as B3LYP//B3LYP and CCSD(T)//B3LYP, respectively.

For the cyclotrimerization reaction of acetylene by bare second-row transition atoms, the Hay–Wadt effective core potential, denoted as LANL2DZ [37], for transition metals and 6-311+G** basis sets for carbon and hydrogen atoms has been used.

In all calculations no symmetry restrictions were imposed. For each optimized structure, vibrational analysis was performed to determine the nature (minimum or saddle point) of these structures. The nature of all minima connected by a given transition state was confirmed by intrinsic reaction coordinate (IRC) [38] analysis.

For all the studied open-shell species, we checked the $\langle S^2 \rangle$ values. In all cases the calculated values of $\langle S^2 \rangle$ differ from the $S(S+1)$ exact value by less than 10%.

15.3
Activation of X–H (X = O, N, C) Bonds by First-row Transition Metal Cations

Over recent decades, a great body of studies of gas-phase bimolecular reactions of bare transition metal cations with small molecules containing N–H, C–H, O–H bonds has been performed [39–49]. The following reactions are detected during the interaction of first-row transition metal cations, M^+, with water, ammonia and methane:

$$M^+ + XH_n \longrightarrow \begin{cases} MXH_{n-2}^+ + H_2 & (1) \\ MH^+ + XH_{n-1} \quad X = O, N, C & (2) \\ MXH_{n-1}^+ + H & (3) \end{cases}$$

Several factors come into play to determine the final outcome of the process, such as spin, electron configuration, kinetic and electronic energy, and spin–orbit coupling.

Several examples of water, ammonia and methane activation widely studied, both experimentally and theoretically, are discussed in other chapters.

15.3.1
Reaction of the First-row Transition Metal Ions with a Water Molecule

The dehydrogenation of H_2O by first-row transition metal cations and the reverse reaction $MO^+ + H_2 \rightarrow M^+ + H_2O$ have been intensively studied in recent decades [50–57]. Interest in these reactions was mainly due to the inverse reactivity pattern of the M^+/MO^+ couple: high reactivity of M^+ coupled with a sluggish reactivity of MO^+ and vice versa. The early transition metal cations (Sc^+, V^+, Ti^+) appear to be more reactive than their oxides, while the late transition metal cations (Cr^+, Mn^+, Fe^+) are less reactive than their oxides. More intriguingly, the primary product of the reaction $MO^+ + H_2 \rightarrow M^+ + H_2O$ can be formed preferentially on the excited state surface. The latter observation indicates spin-flip during the reaction.

Theoretical studies could be helpful to understand the many aspects of mechanism of the water activation by the first-row transition metal cations. Below, we summarize the mechanism of the reaction of Sc^+ with H_2O, which was reported previously [58].

According to the experiments [53], the reaction $Sc^+ + H_2O \rightarrow ScO^+ + H_2$ is exothermic by 46.8 ± 1.4 kcal mol^{-1}. Two different mechanism of this reaction have been hypothesized. The first is oxidative addition of the H–O bond to Sc^+ to form an H–Sc^+–OH intermediate. The second is direct H-abstraction from H_2O. The first process is thermodynamically more favorable. Dehydrogenation products from the H–Sc^+–OH intermediate are formed by a concerted four-center H_2-elimination transition state. An alternative mechanism is an α-H migration to Sc^+ to form H_2-Sc^+–O followed by H_2 elimination. This path is thermodynamically unfavorable. Another suggested pathway involves the 1,1-dehydrogenation from the initial ion–molecule complex Sc^+–OH_2. The latter process has a high barrier and is not feasible. Scheme 15.2 gives a representation of the structures of the intermediates and transition states involved in the most probable mechanism of this reaction. This scheme is general and is also helpful for the discussion of the analogous reaction with ammonia and methane molecules.

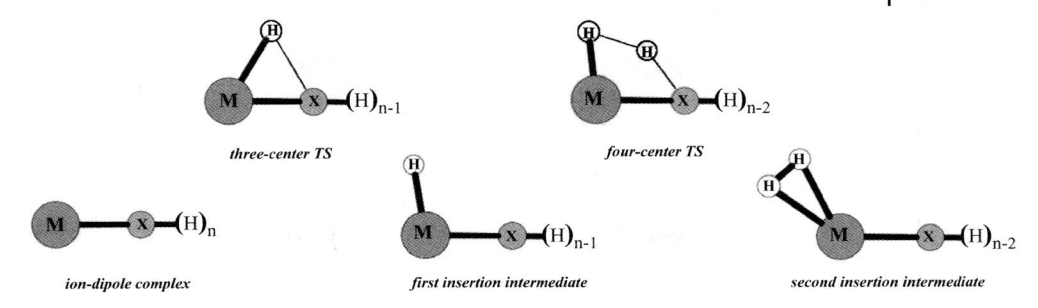

15.3 Activation of X–H (X=O, N, C) Bonds by First-row Transition Metal Cations | 341

three-center TS

four-center TS

ion-dipole complex

first insertion intermediate

second insertion intermediate

Scheme 15.2

Figure 15.1 gives the calculated potential energy surface of the reaction $Sc^+ + H_2O \rightarrow ScO^+ + H_2$.

As outlined above, two possible pathways can be considered: (a) the 1,2-dehydrogenation (Pathway 1) from the intermediate H–Sc$^+$–OH, and (b) the 1,1-dehydrogenation (Pathway 2) from the ion–molecule complex Sc$^+$–OH$_2$.

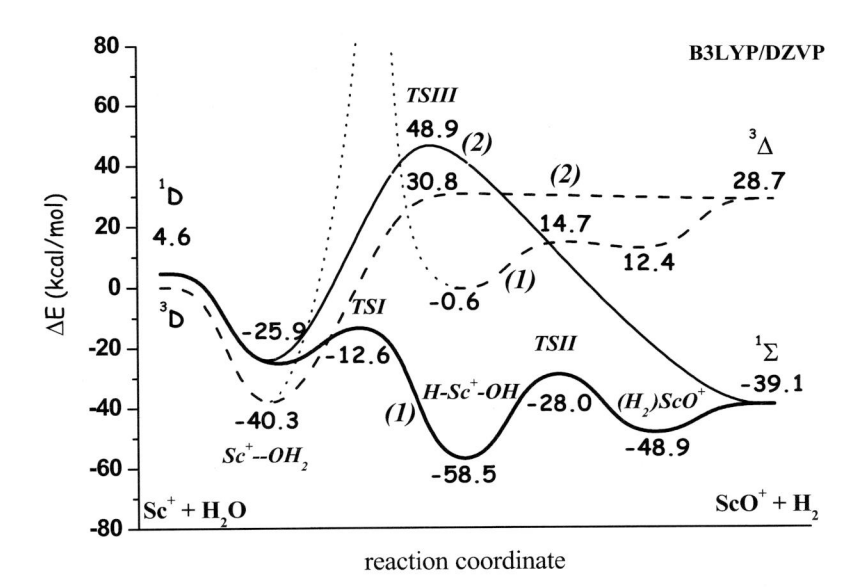

Figure 15.1 B3LYP/DZVP singlet and triplet potential energy surfaces for the reaction of Sc$^+$ with H$_2$O. Explored paths corresponding to 1,2-dehydrogenation (*1*) from the insertion intermediate H–Sc$^+$–OH and 1,1-dehydrogenation (*2*) from the ion–molecule complex Sc$^+$–OH$_2$ are both shown. Energies are in kcal mol^{-1} and relative to the ground-state reactants (^{3}D)Sc$^+$ + H$_2$O asymptote.

Along the both low- and high-spin surfaces and for both Pathways (1) and (2) mechanisms the first step is the ion–molecule complex Sc^+–OH_2 formation. At its ground triplet state this complex has a C_2 symmetry. The experimental value for the energy difference between the ground triplet and excited singlet states is estimated to be 11 kcal mol^{-1} [53]. This experimental finding is in reasonable agreement with the available theoretical [58–60] values of 8.6–20.3 kcal mol^{-1}. From the intermediate Sc^+–OH_2 the reaction along the Pathway (1) proceeds via the transition state TSI and leads to intermediate H–Sc–OH$^+$, which has a singlet spin ground state. Despite numerous attempts we could not locate the triplet of TSI. Pathway (2) proceeds through the transition state TSIII.

Isomerization of H–Sc–OH$^+$ to $(H)_2$-Sc–O$^+$ occurs through the four-center transition state TSII. At the low-spin state of TSII, the calculated H–H distance is quite long, while at the triplet state of this transition state the H–H distance is close to that in a H_2 molecule. For the molecular hydrogen complex $(H_2)ScO^+$ we have calculated several isomers, among which the most stable is a planar structure with an Sc–O bond length close to that in the free ScO$^+$.

Pathway (2), which proceeds via transition state TSIII, has a very high energy barrier, >70 kcal mol^{-1} for both singlet and triplet states. This indicates that the reaction cannot proceed via a 1,1-H_2 elimination mechanism.

Elimination of molecular H_2 from $(H_2)ScO^+$ leads to ScO$^+$, which may have $^3\Delta$ and $^1\Sigma$ states, with the latter lying over 60 kcal mol^{-1} lower in energy. The experimental value of the singlet and triplet energy splitting for ScO$^+$ is 79.6 kcal mol^{-1} [53]. The calculated energy of the reaction $Sc^+ + H_2O \rightarrow H_2 + ScO^+$ is -39.1 kcal mol^{-1}.

Thus, above-presented data indicate that, at the first step, reaction of triplet state Sc$^+$ with water leads to the triplet state ion–molecule complex $Sc(OH_2)^+$ ($^3A_2$). At the next step, intersystem crossing between the singlet state and triplet states occurs and the reaction proceeds via a singlet TSI to give a singlet state intermediate H–Sc–OH$^+$($^1A_1$). This process may not be efficient, but is still energetically accessible at low temperatures. Indeed, experimentally it was observed that this reaction is only 0.7 ± 0.2% efficient at thermal energies [54].

Detailed PES calculations for the reaction of water with the entire series of first-row transition metal ions have been performed by Ugalde and coworkers [59, 61–64]. They have reported similar PESs for all first-row transition metal atoms. Based on these studies the following conclusions were made [64]:

- The key intermediate H–M–OH$^+$ was found to be stable for all the studied reactions except that for M = Cu$^+$. The ground state of this intermediate is a low-spin state, indicating that for the early transition metals, Sc$^+$, Ti$^+$, V$^+$, Cr$^+$ and Mn$^+$, the formation of H–M–OH$^+$ requires spin-flip from the high- and low-spin states. For the late transition metals, Co$^+$, Ni$^+$ and Cu$^+$, the spin change occurs for the step leading to formation of (H_2)M–O$^+$. Fe$^+$ is significantly different from the entire series.

- The calculated barrier at the transition state TSI increases along the row, with the exception of Cr, which shows a unique large barrier.
- The relative energy of the barrier associated with the transition state TSII (for the second hydrogen shift) also increases from the left to the right.
- The reaction is exothermic for M = Sc to V, decreasing from Sc to V. It is endothermic for M = Cr to Cu, with its endothermicity increasing through the series.

Particularly intriguing is the case for M = Fe. Extensive computational studies conducted by Shaik and coworkers [19, 56, 65, 66] on the reverse reaction, $FeO^+ + H_2 \rightarrow Fe^+ + H_2O$, have emphasized the importance of spin–orbit coupling for this reaction. Similarly, Armentrout et al. [67] have shown that the ^4F excited state of Fe^+ is more reactive than its ground ^6D state by a factor of 200.

15.3.2
Insertion of First-row Transition Metal Ions into the N–H bond of Ammonia

In this section we discuss the mechanism of the reaction of first-row transition metal cations with ammonia. Experimental [68–70] investigations indicate that the interaction of early and middle single-charged transition metal ions with ammonia produces mainly the dehydrogenation product, MNH^+. In line with the related system M^+/H_2O, it is assumed that in the first stage of these reactions the reactants form a stable ion–molecule complex, MNH_3^+. Then, the reaction proceeds via oxidative addition, i.e., insertion of M^+ into the N–H bond, to form the intermediate $H–M^+–NH_2$. From this hydrido-intermediate it is hypothesized that the reaction proceeds to yield a molecular hydrogen complex, $(H_2)M^+–NH$, after passing through a four-centered transition state. Elimination of the H_2 molecule leads to the dehydrogenation products. The other possible reaction products of the reaction $M^+ + NH_3$ can be MNH_2^+ and MH^+. For all the first-row transition metal cations, the formation of MNH_2^+ and MH^+ is reported to be endothermic. For M = Co^+, Ni^+, and Cu^+ [68], all products observed are spin-allowed. However, all three metal ions form MH^+ and MNH_2^+, but no MNH^+.

Notably, the $M^+–NH_3$ bond dissociation energies for all first row transition metals have been determined by using collision-induced experiments [71].

On the theoretical side, a series of DFT studies on the reaction of first-row transition metals with NH_3 has been published by the Calabria group [58, 72–76]. MP4(SDTQ) investigation on the insertion of Ti^+ into an N–H bond of NH_3 was also reported [77]. Taketsugu and Gordon have performed *ab initio* studies of the reaction mechanism of Co^+ with NH_3 at the multireference configuration interaction and multireference many-body perturbation theory levels [78]. Vanquickenborne et al. have studied the same reaction at the CASSCF and CASPT2 levels of theory [79]. The reactivity of Ni^+ and Cu^+ with ammonia was studied by Hirao et al. at the both CI and DFT level of theory [80]. Finally, theoretical studies of the M^+–NH_3 binding energy have been performed by Langhoff et al. using the modified coupled-pair functional (MCPF) approach [81].

Recently the reaction of Fe^+, at both its ground ^6D and first excited ^4F states, with ammonia was experimentally and theoretically investigated by Armentrout and coworkers [82].

Computed potential energy surfaces of the $Mn^+ + NH_3 \rightarrow MnNH^+ + H_2$ reaction are reported in Figure 15.2.

As in the case of water (Scheme 15.2), the first step of the reaction of Mn^+ with ammonia (Figure 15.2) is the formation of the ion–molecule complex (Mn–NH_3^+). Stabilization energies with respect to the reactants are very similar at the three employed levels of theory. Comparison shows that our B3LYP/DZVP (37.2 kcal mol^{-1}), B3LYP//B3LYP (37.2 kcal mol^{-1}) and CCSD(T)//B3LYP (38.0 kcal mol^{-1}) results compare very well with experimental (39.0 ± 1.9 kcal mol^{-1}) [70] and with previous MCPF/[8s,6p,4d,1f] (38.4 kcal mol^{-1}) data [81]. In the second step of the reaction, Mn^+ insertion into the N–H bond occurs through a transition state TSI, which leads to the intermediate H–Mn^+–NH_2. Because the high-spin state of TSI lies higher in energy than the low-spin state, it is evident that spin crossing occurs near TSI. From the intermediate (H–Mn^+–NH_2) it is hypothesized that the reaction proceeds to yield the molecular hydrogen complex (H_2)Mn^+–NH after passing through TSII, a four-center transition state. Calculated barriers at TSII are very large. The last step of the reaction, elimination of H_2 occurs without an energy barrier. Inspection of Figure 15.2 reveals that the major products of the reaction could be the ion–molecule complex and the H–Mn^+–NH_2 intermediate.

From a comparison of the B3LYP- and CCSD(T)-calculated potential energy surfaces it appears that these two methods provide similar results. The most relevant difference appears to be the energy gap between quintet and septet states of the H–Mn^+–NH_2 intermediate that are almost degenerate at the CCSD(T)//B3LYP level, whereas the gap is 12 kcal mol^{-1} at the B3LYP/DZVP level. This conclusion is slightly different from that made by the Calabria [58, 72–76] and Donostia [59, 61–64] groups upon investigating the reaction of first-row transition metal cations with X–H (X = O, N, C) prototypical bonds. Notably, the performance of DFT approaches, and of the hybrid B3LYP functional in particular, in predicting transition states geometries and barrier heights has been the subject of extensive theoretical investigations [83–87] and the general conclusion drawn for the examined classes of reactions is that Density Functional Theory based methods tend to underestimate activation barriers.

On the basis of our results presented above, some general trends can be outlined (Figure 15.3).

Data presented in Figure 15.3 have been obtained at B3LYP/DZVP level, except that for Fe, where the B3LYP/DZVP$_{opt}$ approach was used. The use of this recontracted basis set for iron is necessary owing to well-known problems, extensively discussed in the literature [19, 88–90].

As seen in Figure 15.3, the barrier at TSI increases on going from Sc^+ to Cr^+, then decreases, and then increases again to Cu^+. For $M = Sc^+$ to Fe^+, TSI has a

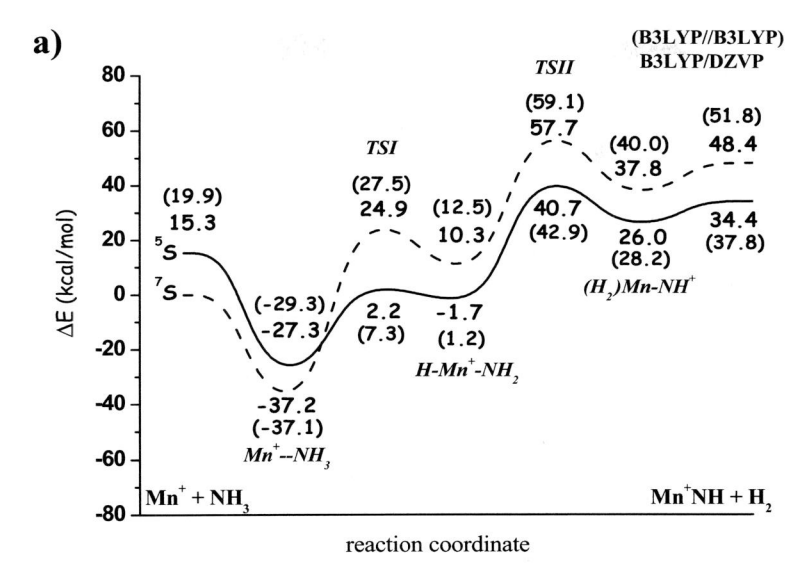

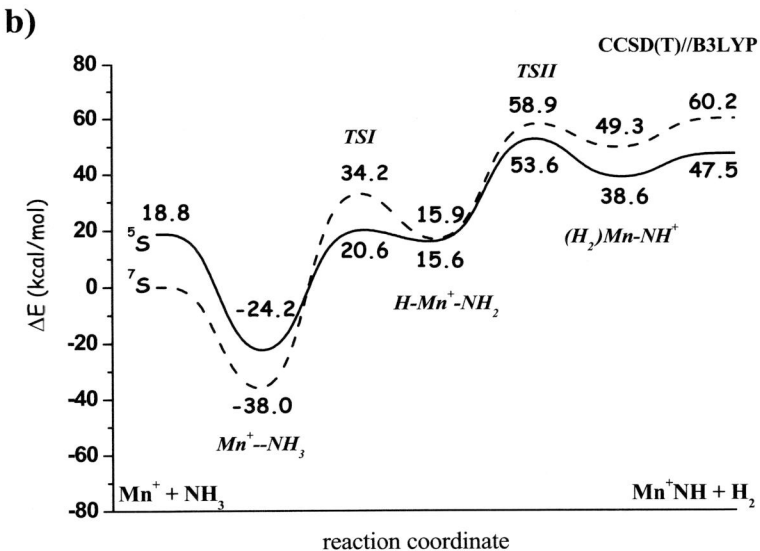

Figure 15.2 Quintet and septet potential energy surfaces for the reaction of Mn^+ with NH_3 computed at (a) B3LYP/DZVP and B3LYP//B3LYP and (b) CCSD(T)//B3LYP levels of theory. Energies are in kcal mol^{-1} and relative to the ground-state reactants (^{7}S)Mn^+ + NH_3 asymptote.

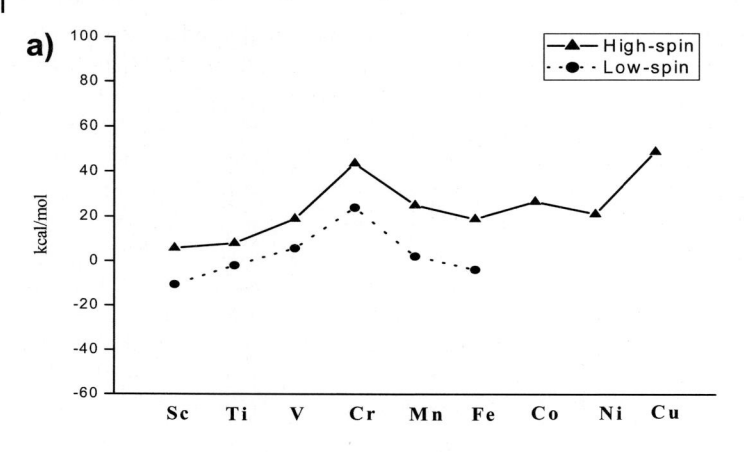

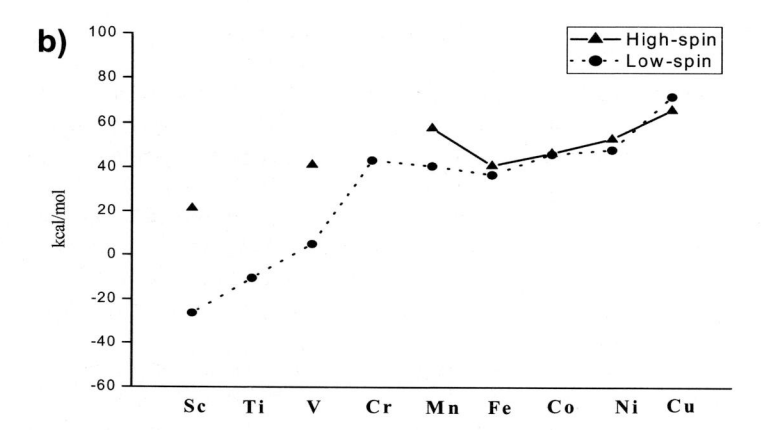

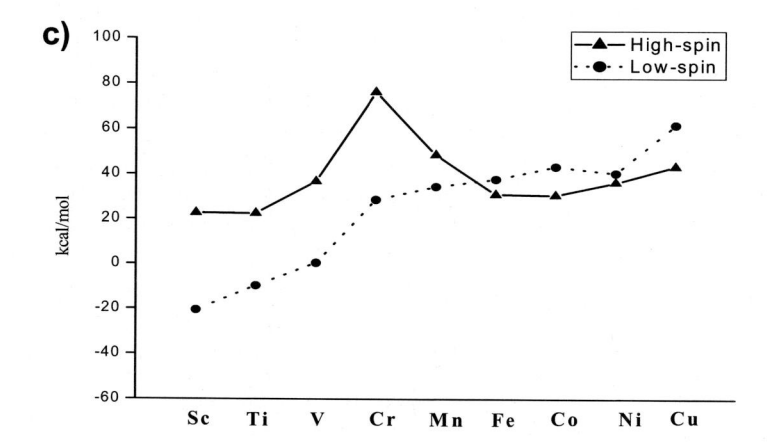

346 | *15 Two State Reactivity Paradigm in Catalysis*

Figure 15.3 (legend see p.)

15.3 Activation of X–H (X = O, N, C) Bonds by First-row Transition Metal Cations

low-spin ground state, which lies lower in energy than the ground state reactants, with the exception of $M=Cr^+$. For the three late transition metals, Co, Ni and Cu, TSI has a high-spin state. From the $H-M^+-NH_2$ produced the reaction is expected to follow through TSII and lead to a $(H_2)M^+-NH$ intermediate. The barrier associated with this process increases upon going from Sc^+ to Cr^+, then slightly decreases for Mn^+ and Fe^+ and then increases again (Figure 15.3b). Finally, elimination of H_2 leads to the dehydrogenation products. The entire process is exothermic only for the three early metals.

For $M=Cr^+$ t-Fe^+ the rate-determining step of the reaction corresponds to the formation of the four-center TSII transition state. It worth noting that, analogously to the energy profile obtained for the insertion reaction of Fe^+ into the O–H bond of water, also for the Fe^+/NH_3 system two spin-crossings should occur to obtain dehydrogenation products.

15.3.3
Bond Activation of CH_4 by bare First-row Transition Metal Cations

C–H bond activation of methane and alkanes has been a subject of considerable interest in transition metal catalytic chemistry [40, 47, 48, 91–98] and, particularly, methane activation by bare transition metal cations has been the focus of several investigations carried out in the gas phase [95b, 99]. These studies show that several electronically excited ions (M^+) dehydrogenate methane and produce metal-carbene complexes, MCH_2^+ [100, 101]. These studies show that excited Sc^+, Ti^+, V^+ and Cr^+ both dehydrogenate and form MH^+ and MCH_3^+ fragments. The PES for methane activation by Cr^+, which will be discussed as an example of the process under investigation, is sketched in Figure 15.4(a).

Analogous to the reaction of transition metal cations with water and ammonia, experimental studies suggest again that the mechanism of oxidative addition of a C–H bond of molecular methane is operative to give an intermediate H–M^+–CH_3, which has a low-spin ground state. Similarly, the first step of the reaction is formation of an ion–molecule complex, which later leads to a hydrido-intermediate via the TSI transition state. Subsequently, the second C–H bond activates (at transition state TSII) to produce an ion–molecule complex, $(H_2)M^+$-CH_2. Finally, dehydrogenation products form from the second insertion intermediate without an energy barrier. In all the considered cases, except Mn, attempts to localize the high-spin transition state TSI were unsuccessful. Our results indicate that high- to low-spin crossing occurs for all metals, except for iron, before

Figure 15.3 Evolution across the first-row of the low- and high-spin relative energies of (a) TS1, (b) TSII and (c) dehydrogenation reaction products for the $M^+ + NH_3$ reaction. Except for Fe^+, where B3LYP/DZVP$_{opt}$ results have been used, B3LYP/DZVP predicted energies are shown. Energies are in kcal mol^{-1} and relative to the ground-state reactants asymptote.

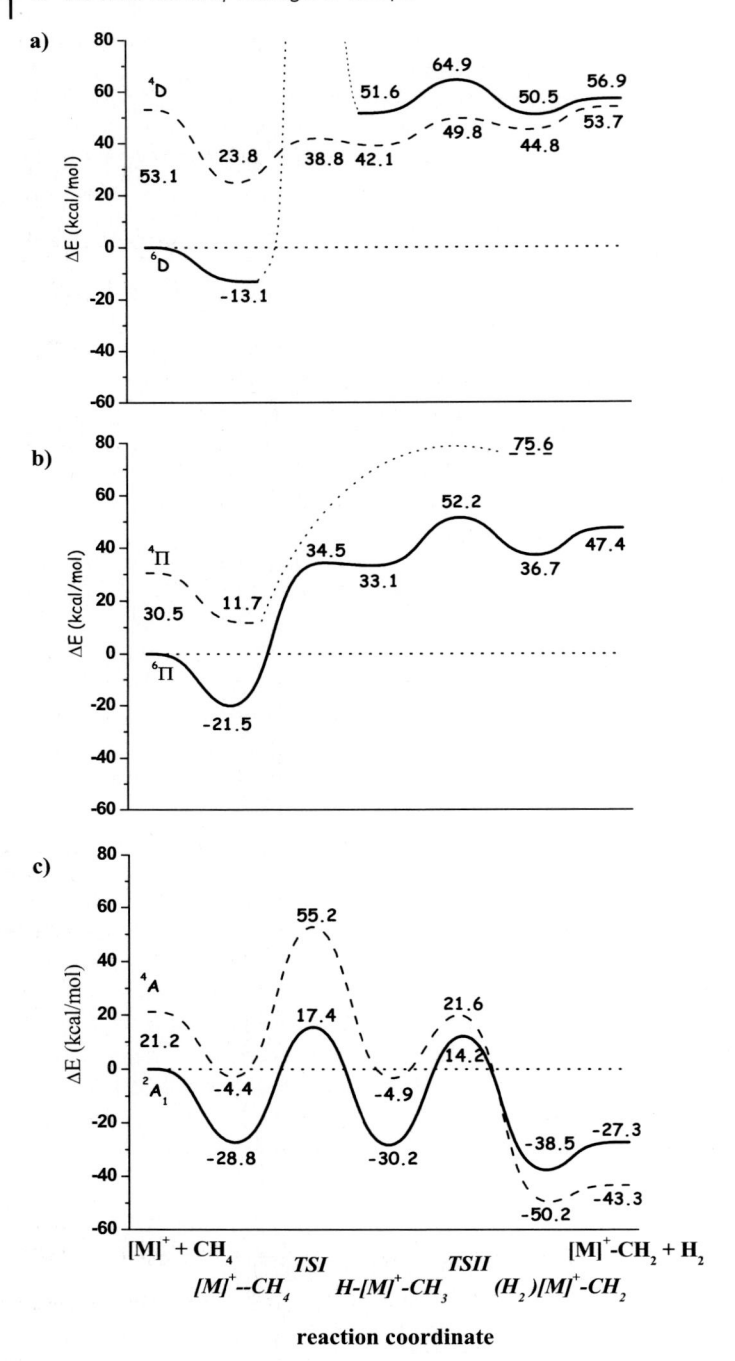

348 | *15 Two State Reactivity Paradigm in Catalysis*

Figure 15.4 (legend see p.)

the TSI transition state. In disagreement with the previous assignment of a sextet state to the FeCH$_4^+$ complex [102], our computations show that the ground state of this complex is a quartet spin state with a binding energy that is consistent with the measured value [103] and other high-level computational findings [104]. Thus, for M = Fe. the spin change takes place quite likely before formation of the encounter ion–molecule complex FeCH$_4^+$.

The picture that emerges from the refined computational works [58, 72–74, 76, 105–108] is in perfect agreement with experimental findings. A long-lived insertion intermediate, H–M$^+$–CH$_3$, is formed for M = Sc, Ti, V, Cr at its low-spin state. From this intermediate, the reaction proceeds via both dehydrogenation and the MH$^+$ and MCH$_3^+$ fragment formation mechanisms. However, for M = Fe$^+$ the cation does not dehydrogenate methane.

Early studies of selective σ bond activation of alkanes and, especially, methane focused mainly on the reactivity of bare metal ions. More recently, numerous papers have reported on the reaction of unsaturated L$_n$M$^+$ systems [44, 51, 53, 57, 67, 109–123]. We have examined the reaction of VO$^+$, FeO$^+$ and CrO$_2^+$ with methane [124, 125]. We have confirmed [126, 127] that reactivity of the bare metal cations and their ligated counterparts can be very different. As an example, we present the calculated potential energy surfaces of the reaction of Cr$^+$, CrO$^+$ and CrO$_2^+$ with methane in Figure 15.4.

Experimentally, Kang and Beauchamp have carried out a detailed examination of the reactions of bare CrO$^+$ with alkanes and alkenes, including methane [51]. Schröder, Schwarz and coworkers have systematically investigated the gas-phase reactivity of first-row transition metal monoxide cations with methane [57]. In particular, it was shown that CrO$^+$ can oxidize alkanes larger than methane. Reactivity of the high-valent Cr(v) dioxide cation CrO$_2^+$ with hydrocarbons has been studied by Fiedler et al. [121] by means of ion-cyclotron resonance and sector-field mass spectrometry. In particular, considering the oxidation of methane, they have shown that CrO$_2^+$ is quite reactive.

Inspection of the PESs shown in Figure 15.4 for M = Cr$^+$, CrO$^+$, and CrO$_2^+$ reveals that the mechanism, through which the molecular hydrogen elimination reaction proceeds, involves the same key features. The first step is formation of the ion–molecule complex [M]$^+$–CH$_4$. From this encounter complex the reaction proceeds to give the first C–H bond insertion intermediate H–[M]$^+$–CH$_3$, through transition state TSI (Figure 15.5). The next step along the path consists of the second C–H bond activation at transition state TSII (Figure 15.5). The resultant complex, (H$_2$)[M]$^+$–CH$_2$, represents the direct precursor of molecular hydrogen elimination.

Figure 15.4 B3LYP/DZVP potential energy surfaces for methane activation leading to molecular hydrogen elimination by (a) sextet and quartet states of Cr$^+$, (b) sextet and quartet states of CrO$^+$ and (c) doublet and quartet states of CrO$_2^+$. Energies are in kcal mol^{-1} and relative to the ground-state reactants.

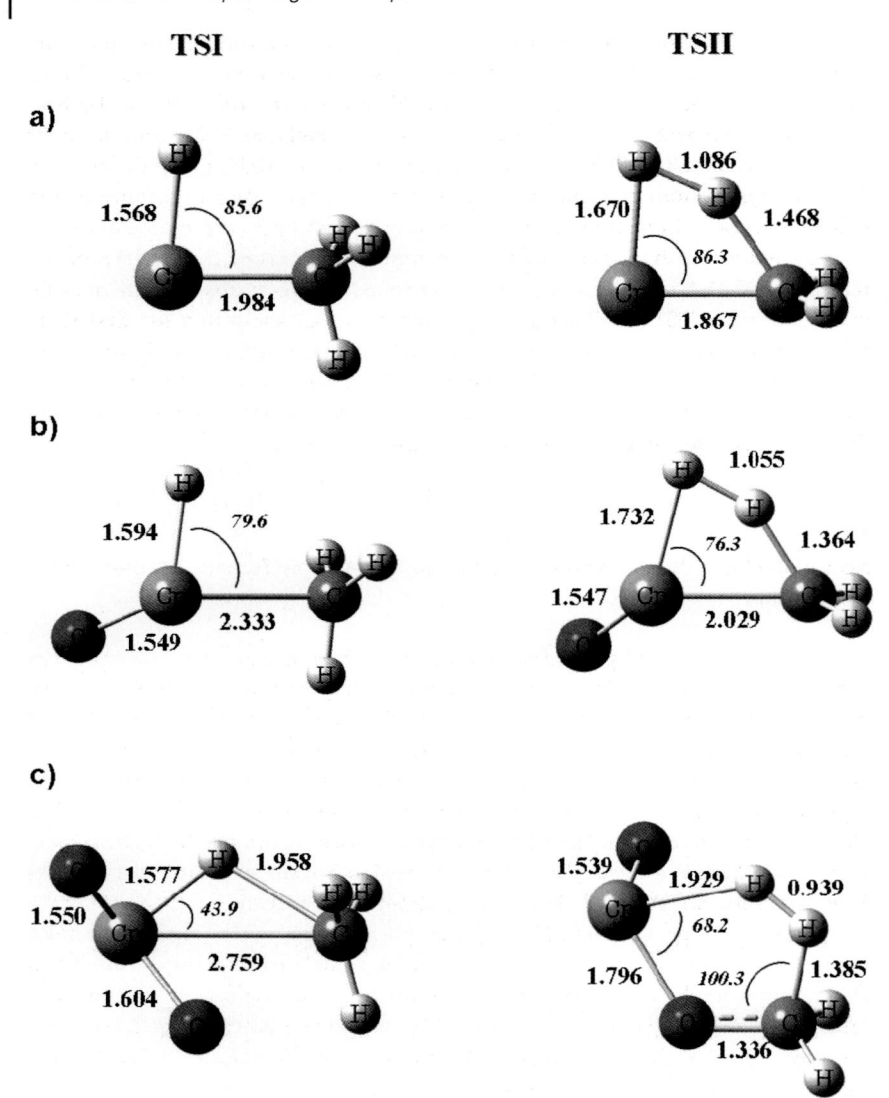

Figure 15.5 B3LYP/DZVP geometrical parameters of the ground state TSI and TSII transition states localized along the PESs for the reaction of methane with (a) Cr^+, (b) CrO^+ and (c) CrO_2^+. Bond lengths are in Å and angles are in degrees.

These results show that the reaction of bare and oxide cations with methane forms only the $[M]^+-CH_4$ ion–molecule complex, at low kinetic energies. Thus, the addition of the oxo ligand to Cr^+ does not change the reactivity of the bare cation with respect to the dehydrogenation of methane. The formation of $(H_2)CrO^+-CH_2$ rather than $(H_2)Cr^+-(OCH_2)$ from the initial complex indicates that the oxygen atom does not participate in the reaction. As a consequence, the relative energies of the associated barriers for Cr^+ and CrO^+ are comparable.

The situation is drastically different for the dioxide cation. On the PES of the reaction of CrO_2^+ with methane all intermediates are accessible and stable. The transition states TSI and TSII given in Figure 15.5(c) show that the first C–H activation involves a hydrogen shift to the metal center combined with the simultaneous Me insertion it to the Cr–O bond. Thus, the second hydrogen is transferred from the formed methoxy group to the metal. The reaction takes place on the excited quartet state surface to avoid seam of crossing.

15.4
Activation of C–C Bond: Cyclotrimerization of Acetylene by Second-row Transition Metal Atoms

Cyclization of alkynes has received a great deal of attention in the last three decades, since it is a key step in the synthesis of polycyclic and heterocyclic compounds [128]. The simplest example of these transformations is the thermal cyclotrimerization of acetylene into benzene, which is an exothermic reaction (about 140 kcal mol^{-1}) [129], but possesses an unexpectedly high activation energy (60–80 kcal mol^{-1}) [130]. Thus, this process requires a catalyst [131–133]. Numerous theoretical and experimental studies, covering many structural, mechanistic, and kinetic aspects of this reaction, have appeared in the literature [134–159]. The great majority of them include group VIII transition metals, with special emphasis on palladium [133, 136]. The polymerization of C_2H_2 has been also studied on small Pd_n ($1 \leq n \leq 30$) clusters supported on thin MgO(100) films [152]. Interestingly, it has been reported that already a single Pd atom adsorbed on MgO is enough for the production of benzene from acetylene at 300 K. Furthermore, theoretical calculations [154] indicate that although a free Pd atom is inert for this reaction, once adsorbed on the surface of MgO it becomes active. Lately, a comparative study of the reactivity of Ag, Rh and Pd, both bare and MgO(001)-supported, was reported [154, 156].

It is particularly interesting to analyze the trends in reactivity of bare atoms. Therefore, we have recently analyzed the acetylene cyclotrimerization reaction mediated by the less studied left-hand side bare transition metal atoms of the 4d series, Y, Zr, Nb and Mo [159]. It has been determined that the cyclotrimerization (Scheme 15.3) occurs through (a) coordination of the first acetylene molecule to the transition metal center; (b) coordination of the second acetylene molecule to the transition metal center; (c) formation of a metallocycle, MC_4H_4; (d) coordination of the third acetylene molecule to the transition metal center to form a (C$_2$H$_2$)-MC$_4$H$_4$ complex; (e) formation of a MC$_6$H$_6$ metallocycle; (f) formation of a final metal–benzene complex followed by benzene elimination.

The overall M + 3C$_2$H$_2$ → MC$_6$H$_6$ process was found to be thermodynamically highly favorable. For Y, the reaction evolves along the low-spin doublet surface, and the formation of all the intermediates along the path takes place with an energy gain. The barrier heights are low, with benzene elimination being the rate-determining step. For M = Zr, multiple crossings between the singlet and triplet spin surfaces occur even if reactants and products have the same spin state. Owing to the high stability of the bis-acetylene complex, the energy barrier associated with the formation of the metallocycle appears to be the rate-determining step. For M = Nb, we have studied three different spin states; the sextet spin state, which is the ground state of the bare Nb atom, and the low-lying quartet and doublet states. All minima involved in this reaction are highly stable; therefore, the calculated barrier heights are quite high, with the formation of the metallacyclopentadiene being the rate-determining step. For M = Mo, the reaction

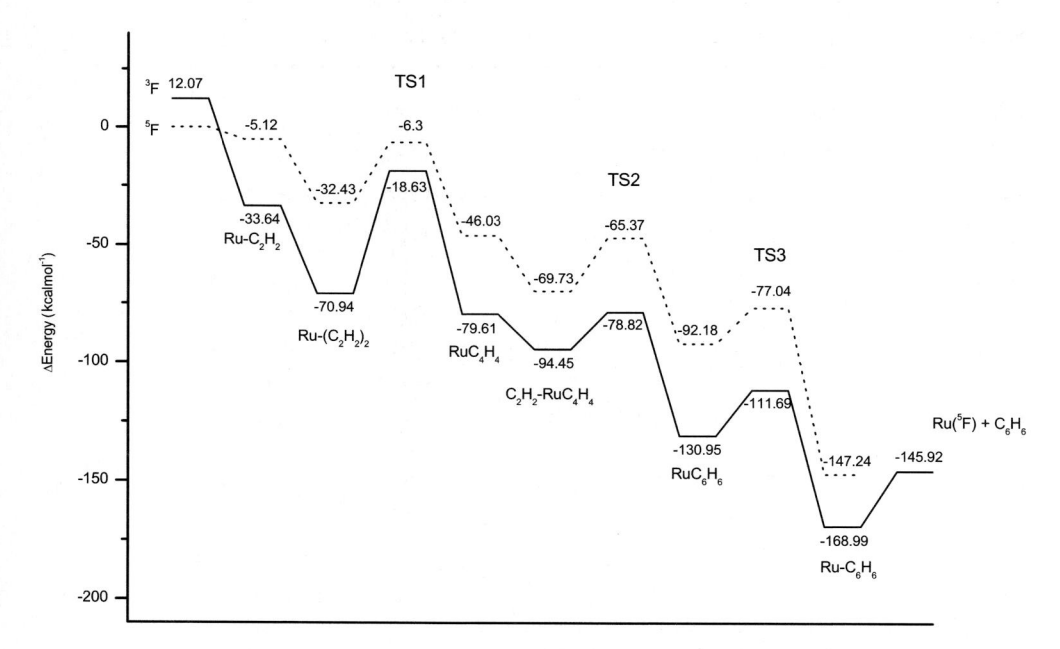

Figure 15.6 B3LYP/LANL2DZ triplet and quintet PESs for the reaction Ru + 3C$_2$H$_2$. Energies are in kcal mol^{-1} and relative to the ground-state reactants.

15.4 Activation of C–C Bond: Cyclotrimerization of Acetylene by Second-row Transition Metal Atoms

evolves completely along the excited quintet spin state of the Mo atom; only the final Mo-benzene adduct possesses the same multiplicity as the reactants.

For M = Ru, we have studied two different spin states, namely the quintet (ground state of the bare atom) and the triplet excited state. Figure 15.6 gives the triplet and quintet PESs for the complete cyclotrimerization reaction calculated at the B3LYP/LANL2DZ level. The most relevant geometrical parameters of ground-state minima and transition states are reported in Table 15.1, according to the notation adopted in Figure 15.7.

Calculations show that the energetic gap between the first ^3F ($4d^75s^1$) excited state and the ground ^5F ($4d^65s^2$) state of Ru is 12.07 kcal mol^{-1}, which is in a good agreement with the experimental value of 18.71 kcal mol^{-1} [160].

The first step of the reaction is the formation of Ru-C$_2$H$_2$ complex, which is a 33.64 kcal mol^{-1} stable relative to the reactants. Coordination of the second acetylene molecule is a 37.3 kcal mol^{-1} exothermic. The formed RuC$_4$H$_4$ intermediate has a planar structure and is more stable than the Ru(C$_2$H$_2$)$_2$ complex by only 8.67 kcal mol^{-1}. The barrier for the formation of this structure (52.31 kcal mol^{-1}) is very high.

Table 15.1 Selected bond lengths (Å) and angles (°), corresponding to the ground spin states of all the stationary points involved in the reaction Ru + 3C$_2$H$_2$ (see Figures 15.6 and 15.7).

Structure–State[a]	a_1, a_2, a_3, a_4 (Å)	b_1, b_2, b_3, b_4, b_5 (Å)	$\alpha_1, \alpha_2, \alpha_3, \alpha_4$ (°)	$\beta_1, \beta_2, \beta_3$ (°)
I–$^3B_1$	1.986	1.295	146.0	–
II–$^3B_3$	2.064, 2.064	1.274, 1.274	149.1, 149.0	−136.9
TS1–^3A	2.126, 2.068	1.314, 1.833	135.9, 138.2	−136.2, −160.9
III–$^3A_2$	2.836, 1.975	1.370, 1.438	121.8, 123.3	180.0, 180.0
IV–^3A	2.910, 2.087, 2.075	1.342, 1.479, 1.266	123.0, 123.1, 152.7, 152.9	−180.0, 180.0, −89.2
TS2–^3A	2.041, 2.002, 2.279, 1.955	1.404, 1.402, 1.399, 2.098	120.7, 122.4, 147.1, 143.4	−169.7, −166.4, −7.5
V–^3A	1.905	1.377, 1.416, 1.384	116.8, 115.4, 113.4	179.9, 180.0, 0.0
TS3–^3A	1.988, 2.984, 3.481	1.377, 1.416, 1.383	117.5, 117.8, 117.2	149.7, 130.5, −19.8
VI–^3A	2.274	1.419	120.0	131.4, 102.2, 0.0

a) Ground spin state structure, a_i: distance M–C as indicated in Figures 15.1, b_i: C–C distances as indicated in Figure 15.7, α_i: C–C–H angles as indicated in Figure 15.1, β_i: selected dihedral angles (β_1: 1234, β_2: 1235, β_3: 2467), following the numeration indicated in Figure 15.7.

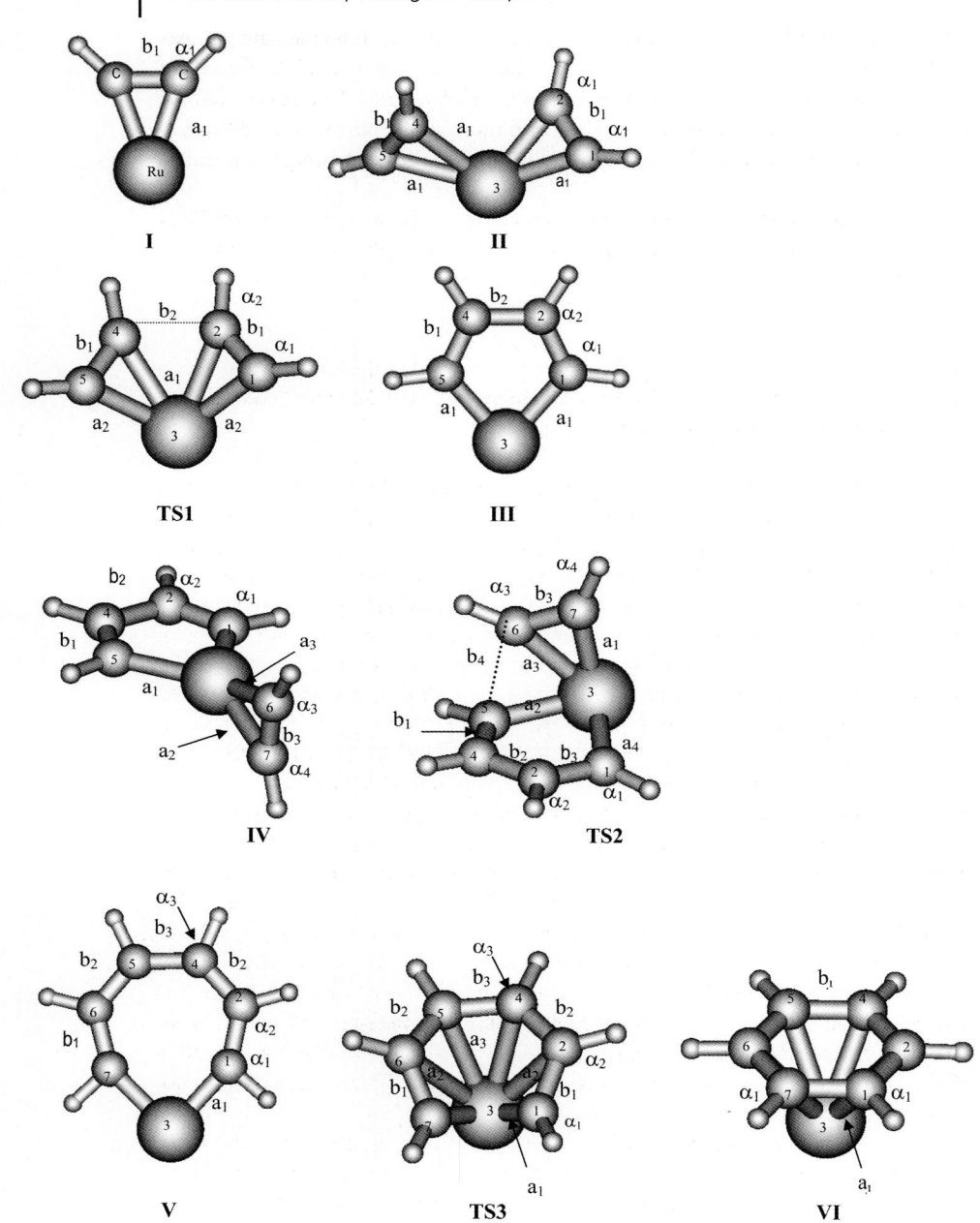

354 | *15 Two State Reactivity Paradigm in Catalysis*

Figure 15.7 Schematic representation of minima and transition state structures involved in the reaction Ru + 3C$_2$H$_2$, and corresponding notation of bond lengths, bond angles and dihedral angles used in Table 15.1.

Coordination of third acetylene molecule to RuC_4H_4 intermediate is exothermic by 15 kcal mol^{-1}. Formation of the MC_6H_6 intermediate from the $(C_2H_2)RuC_4H_4$ intermediate is exothermic by 36.5 kcal mol^{-1}. The energy barrier associated with this process is about 16 kcal mol^{-1}. The formation of the last M-C_6H_6 complex occurs after the system surpasses a barrier height of almost 20 kcal mol^{-1}.

Thus, the formation of the metallopentacycle RuC_4H_4, with a barrier of 52.31 kcal mol^{-1}, is the rate-determining step of the entire $Ru + 3C_2H_2 \rightarrow RuC_6H_6$ reaction.

15.5
Use of the Electron Localization Function to Characterize the Bonding Evolution in Reactions involving Transition Metals

Recently, two topological methodologies have begun to be extensively used to characterize bonding properties: they are the AIM (Atoms in Molecules) approach of Bader [161] and the ELF (Electron Localization Function) approach of Becke and Edgecombe [14], which extensively developed by Silvi and Savin [13, 162]. A wide range of applications of ELF analysis has appeared, including the bonding characterization of transition metal containing systems [75, 124, 163]. The use of catastrophe theory has been proposed [164] along with the ELF formalism to characterize chemical reactions [165]. This type of analysis is called as the Bonding Evolution Theory (BET) [165d].

The ELF, $\rho(r)$, can be interpreted as a measure of the electron localization in atomic and molecular systems, namely, as the conditional probability of finding two electrons with the same spin around a reference point [13]. Analysis of the ELF gradient field provides a mathematical model enabling the partition of the molecular position space in basins of attractors, which present in principle as one to one correspondence with chemical local objects such as bonds and lone pairs. These basins are either core basins, usually labeled C(A), or valence basins, V(A), belonging to the outermost shell and characterized by the number of core basins with which it shares a common boundary, which is called the synaptic order, σ. The number of basins is called the morphic number (μ). In this representation the monosynaptic basins correspond to non-bonded pairs of the usual Lewis representation, whereas the di- and polysynaptic basins are related to bonds.

BET classifies the elementary chemical processes according to the variation of either the number of basins, or the synaptic order of, at least, one basin. There are, accordingly, three types of chemical processes, which correspond to $\Delta\mu > 0$ (plyomorphic process, e.g., a covalent bond breaking), $\Delta\mu < 0$ (miomorphic process, e.g., covalent bond formation) and $\Delta\mu = 0$, $\Delta\sigma = 0$ (tautomorphic differosynaptic process, e.g., breaking of a dative bond). Each structure is only possible for values of the control parameters belonging to definite ranges called structural stability domains. Within a domain of structural stability, the topology of the ELF gradient field is not altered by the variation of the control space parameters, which belong to a definite range. Along the reactions there exist several domains of structural stability, and between two successive domains. Only three elementary catastrophes have been recognized so far in chemical reactions: the fold, cusp and elliptic umbilic catastrophe. The fold catastrophe transforms a wandering point, namely, a point that is not a critical point, into two critical points of different parity. The cusp catastrophe transforms a critical point of a given parity into two critical points of the same parity and one of the opposite parity. Finally, the elliptic umbilic catastrophe changes the index of one critical point by 2.

To illustrate the application of this methodology to the description of the bonding evolution along a chemical reaction, in the next section we describe the similarities and differences of the reaction of Mn^+ (at its ^5S and ^7S states) with three different isoelectronic hydrides, which differ between themselves by the number of lone pairs, i.e., CH_4 (no lone pairs), NH_3 (one lone pair) and H_2O (two lone pairs).

ELF calculations have been carried out with the *TopMod Package* developed at the Laboratoire de Chimie Théorique de l'Université Pierre et Marie Curie [166]. Iso-surfaces have been visualized with the public domain scientific visualization and animation program for high-performance graphic workstations named Scian [167].

15.5.1
Reaction of Mn^+ (in its ^7S and ^5S States) with H_2O, NH_3 and CH_4 Molecules

A detailed description of the reaction mechanisms for the dehydrogenation of H_2O, NH_3 and CH_4 by transition metal cations was presented in Section 15.3. Figure 15.8 shows the calculated reaction pathways of the reaction of Mn^+ with CH_4, NH_3 and H_2O. For all the reactions the overall energetic profiles are similar, showing that high- and low-spin state PESs cross once at the entrance channel, just after the formation of the first ion–molecule complex. Therefore, the reaction starts with the formation of the ion–molecule complex (Mn^+-XH_n) at the septet ground state of the bare cation.

First, we describe the topological changes that take place at the first stage of the dehydrogenation process, namely, from the Mn^+ insertion into the X–H bond (X = C, N, O) to the formation of the first reaction intermediate, $H-Mn^+-XH_{n-1}$. Figure 15.9 shows the localization domains corresponding to key minima and transition states for the reaction with CH_4.

In Figure 15.9 we also include the topological structures of the first minimum, Mn^+-XH_n, at both spin states. The rest of the figures correspond to the lowest quintet spin states. The valence basins present in the initial complex are four $V(C,H)$ basins, which represent four covalent C–H bonds, and a monosynaptic $V(Mn)$. This indicates that the structure can be considered as formed by two fragments. Interaction between the fragments is, therefore, electrostatic in nature. The localization of the monosynaptic valence basin, $V(Mn)$, is quite different for different spin states. The relative locations of the $V(Mn)$ and $V(C,H)$

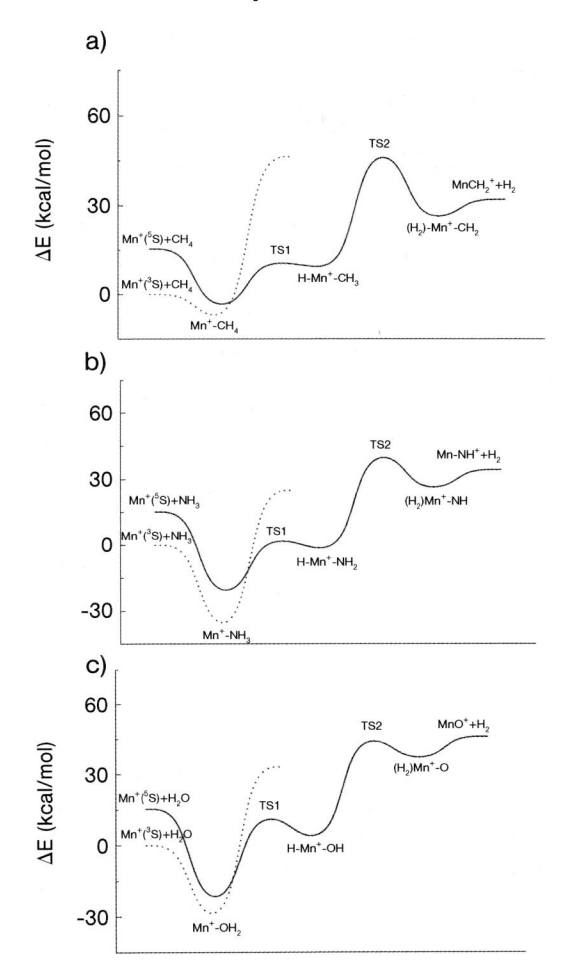

Figure 15.8 Potential energy profiles for the reaction of Mn$^+$ with (a) CH$_4$, (b) NH$_3$ and (c) H$_2$O. Relative energies are in kcal mol^{-1} and calculated with respect to the ground-state reactants asymptote.

attractors in the septet spin complex makes charge transfer from V(Mn) to the CH$_4$ moiety almost impossible. This explains the energy difference between the first minimum and the first transition state, TS1, in the septet spin state (57.33 kcal mol^{-1} at B3LYP/TZVP level of theory, see Ref. [73] for details). Conversely, in the quintet spin state, V(Mn) is a torus (structurally unstable), which promotes the transference of charge from this basin towards the ligand. As a consequence of this charge transfer, V(Mn) disappears through a cusp catastrophe (miomorphic process). At the same time, a hyperbolic umbilic catastrophe promotes the formation of the V(Mn,C) basin. The first transition state, TS1, is

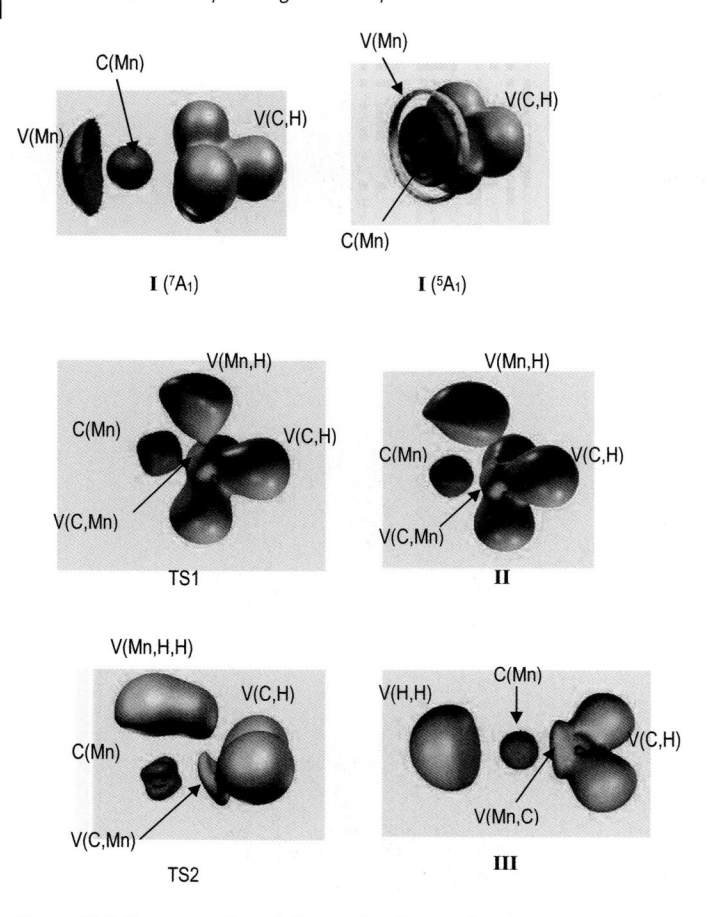

Figure 15.9 Representation of electron localization function (ELF) localization domains for all the key minima involved in the $Mn^+ + CH_4$ reaction path.

therefore characterized by the presence of a disynaptic, V(Mn,H) basin, with a basin population of 1.57e, which is an indication of the formation of a Mn–H bond. At the same time, the absence of one of the V(C,H) basins is an indication of the breaking of a C–H bond, and the presence of a V(Mn,C) disynaptic valence basin, which was absent in the initial minimum, indicates the formation of a Mn–C bond. Continuing on the reaction path, the formation of the first insertion intermediate, $H–Mn^+–XH_{n-1}$, does not involve topological changes but only slight modifications of the basin electron populations.

For NH_3 (H_2O), the mechanism is simplified by the presence of the lone pair (the two lone pairs) of the ligand, which directly forms a disynaptic V(Mn,N) [V(Mn,O), in the case of H_2O] basin in the initial complex. Therefore, a miomorphic process is enough to prepare the complex in the topology of TS1. Figure

15.5 Use of the Electron Localization Function to Characterize the Bonding Evolution

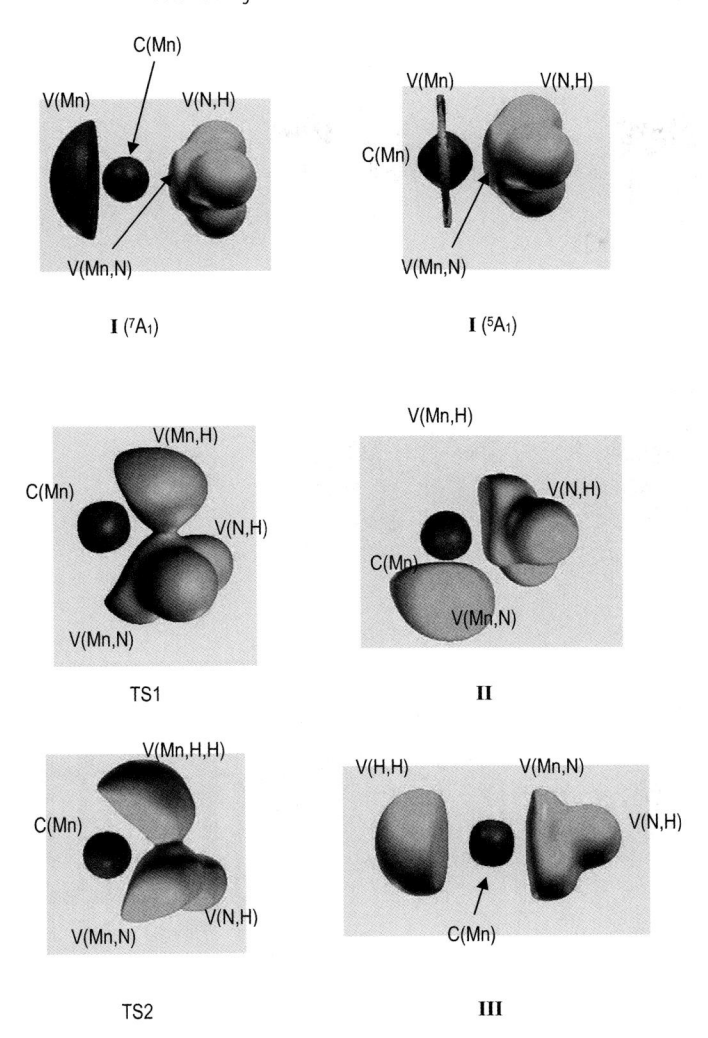

Figure 15.10 Representation of ELF localization domains for all the key minima involved in the Mn$^+$ + NH$_3$ reaction path.

15.10 displays the localization domains corresponding to the species involved in the reaction with ammonia, whereas Figure 15.11 shows the corresponding reaction with a water molecule.

In the second stage of the reaction; namely from the insertion intermediate to the molecular hydrogen complex (H$_2$)Mn$^+$–XH$_{n-2}$, further topological changes take place. The second transition state, TSII, is formed through a cusp catastrophe, which involves the attractors of the V(Mn,H) and of one V(X,H) basin (X = O, N, C), together with the index 1 saddle point lying in the separatrix of

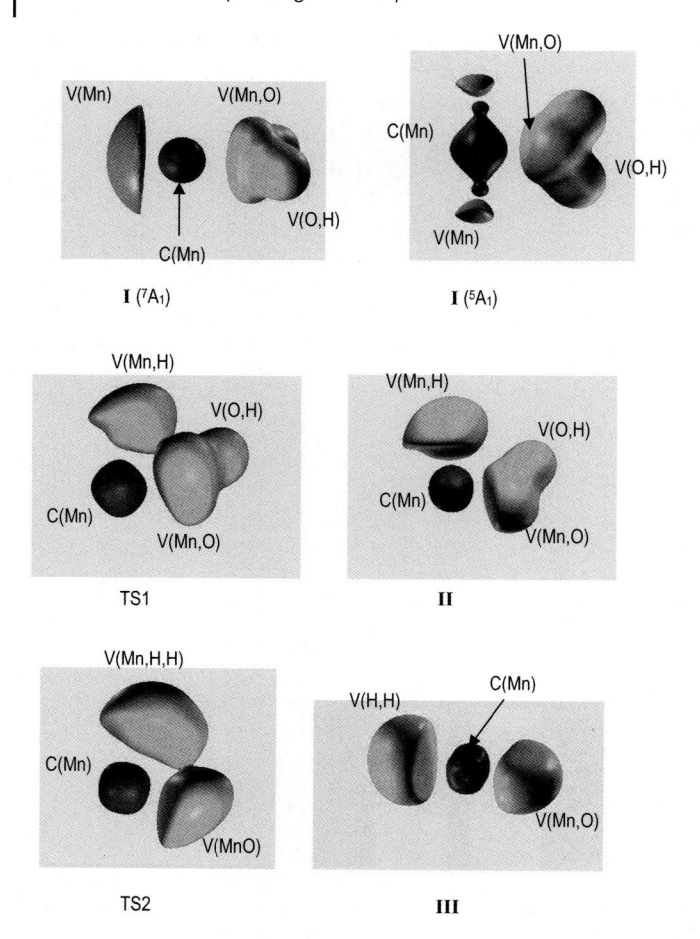

Figure 15.11 Representation of ELF localization domains for all the key minima involved in the $Mn^+ + H_2O$ reaction path.

the two former basins. This process can be identified as miomorphic. The resulting basin is a trisynaptic one V(Mn,H,H), which corresponds to the condensation of two covalent bonds into a three-center bond. The basin population of this basin never exceed 2.5 electrons; the actual values 2.32 (CH_4), 2.15 (NH_3) and 2.09 (H_2O) are correlated with the number of lone pairs of the initial complex. Further evolution of the location of the valence basins is driven mostly by Pauli repulsion between the V(Mn,H,H) and V (Mn,X) basins. The last insertion intermediate is characterized by the presence of a V(H,H) disynaptic basin, with an electron population very close to two electrons, indicating that the H_2 molecule is already formed in this structure.

Topology analysis of the activation of three different binary hydrides by Mn^+ enables us, therefore, to identify two different bonding mechanisms, depending on the presence of lone pairs in the ligand. In all cases it is verified that the formation of the first ion–molecule complex in its quintet spin state promotes the charge transfer from a V(Mn) basin due to its particular spatial distribution. For the reaction with methane the formation of the first intermediate of the reaction is a consequence of the vanishing of the monosynaptic valence basin, V(Mn), followed by the formation of the Mn–C valence basin, V(Mn,C). With lone pair-containing molecules, the first part of the reaction is simpler due to the presence of a dative bond in the initial complex, V(Mn,N) and V(Mn,O) respectively. After formation of the first reaction intermediate, all three reactions are equivalent from a bonding evolution viewpoint, since the presence of a trisynaptic basin, which corresponds to the condensation of two covalent bonds into a three-center bond, is verified in all three cases.

References

1. D. Schroder, S. Shaik, H. Schwarz, *Acc. Chem. Res.* 2000, **33**, 139.
2. P. B. Armentrout, J. L. Beauchamp, *Acc. Chem. Res.* 1989, **22**, 315.
3. P. B. Armentrout, *Science* 1991, **251**, 175.
4. D. R. Yarkony, *Int. Rev. Phys. Chem.* 1992, **11**, 195.
5. D. R. Yarkony, *J. Phys. Chem.* 1996, **100**, 18612.
6. J. N. Harvey, *Computational Organometallic Chemistry*, ed. T. R. Cundari, Marcel Dekker Inc., New York, 2001, p. 291.
7. J. N. Harvey, R. Poli, K. M. Smith, *Coord. Chem. Rev.* 2003, **238/239**, 347.
8. R. Poli, J. N. Harvey, *Chem. Soc. Rev.* 2003, **32**, 1.
9. S. Shaik, S. Cohen, S. P. de Visser, P. K. Sharma, D. Kumar, S. Kozuch, F. Ogliaro, D. Danovich, *Eur. J. Inorg. Chem.* 2004, 207.
10. J. M. Mercero, J. M. Matxain, X. Lopez, A. Largo, L. A. Eriksson, J. M. Ugalde, *Int. J. Mass Spectrom.* 2005, **240**, 37.
11. H. Schwarz, *Int. J. Mass Spectrom.* 2004, **237**, 75.
12. (a) E. R. Davidson (Guest Editor), Computational Transition Metal Chemistry, in *Chem. Rev.*, 2000, 100, issue 2. (b) T. Ziegler, *Chem. Rev.* 1991, **91**, 651.
13. B. Silvi, A. Savin, *Nature* 1994, **371**, 683.
14. A. D. Becke, K. E. Edgecombe, *J. Chem. Phys.* 1990, **92**, 5397.
15. A. D. Becke, *J. Chem. Phys.* 1993, **98**, 1372.
16. A. D. Becke, *J. Chem. Phys.* 1993, **98**, 5648.
17. P. J. Stephens, J. F. Devlin, C. F. Chabalowski, M. J. Frisch, *J. Phys. Chem.* 1994, **98**, 11623.
18. C. W. Bauschlicher, Jr., A. Ricca, H. Partridge, S. R. Langhoff, *Recent Advances in Density Functional Theory*, ed. D. P. Chong, World Scientific Publishing Co., Singapore, 1997, Part II, pp. 165–227.
19. M. Filatov, S. Shaik, *J. Phys. Chem. A* 1998, **102**, 3835.
20. M. Sodupe, V. Branchadell, M. Rosi, C. W. Bauschlicher, Jr., *J. Phys. Chem. A* 1997, **101**, 7854.
21. P. E. M. Siegbahn, Electronic structure calculations for molecules containing transition metals, *Adv. Chem. Phys.* 1996, **43**, 333–367.
22. S. S. Yi, M. R. A. Blomberg, P. E. M. Siegbahn, M. Weisshaar, *J. Phys. Chem.* 1998, **102**, 395.
23. M. Aschi, M. Brönstrup, M. Diefenbach, J. N. Harvey, D. Schröder, H. Schwarz, *Angew. Chem. Int. Ed.* 1998, **37**, 829.
24. A. D. Becke, *Phys. Rev.* 1988, **A38**, 3098.
25. C. Lee, W. Yang, R. G. Parr, *Phys. Rev.* 1988, **B37**, 785.
26. S. H. Vosko, L. Wilk, M. Nusair, *Can. J. Phys.* 1980, **58**, 1200.

27 J. Andzelm, E. Radzio, D. R. Salahub, *J. Comput. Chem.* 1985, **6**, 520.
28 N. Goudbout, D. R. Salahub, J. Andzelm, E. Wimmer, *Can. J. Chem.* 1992, **70**, 560.
29 *Gaussian 03, Revision C.02*, M. J. Frisch, G. W. Trucks, H. B. Schlegel, G. E. Scuseria, M. A. Robb, J. R. Cheeseman, J. A. Montgomery, Jr., T. Vreven, K. N. Kudin, J. C. Burant, J. M. Millam, S. S. Iyengar, J. Tomasi, V. Barone, B. Mennucci, M. Cossi, G. Scalmani, N. Rega, G. A. Petersson, H. Nakatsuji, M. Hada, M. Ehara, K. Toyota, R. Fukuda, J. Hasegawa, M. Ishida, T. Nakajima, Y. Honda, O. Kitao, H. Nakai, M. Klene, X. Li, J. E. Knox, H. P. Hratchian, J. B. Cross, V. Bakken, C. Adamo, J. Jaramillo, R. Gomperts, R. E. Stratmann, O. Yazyev, A. J. Austin, R. Cammi, C. Pomelli, J. W. Ochterski, P. Y. Ayala, K. Morokuma, G. A. Voth, P. Salvador, J. J. Dannenberg, V. G. Zakrzewski, S. Dapprich, A. D. Daniels, M. C. Strain, O. Farkas, D. K. Malick, A. D. Rabuck, K. Raghavachari, J. B. Foresman, J. V. Ortiz, Q. Cui, A. G. Baboul, S. Clifford, J. Cioslowski, B. B. Stefanov, G. Liu, A. Liashenko, P. Piskorz, I. Komaromi, R. L. Martin, D. J. Fox, T. Keith, M. A. Al-Laham, C. Y. Peng, A. Nanayakkara, M. Challacombe, P. M. W. Gill, B. Johnson, W. Chen, M. W. Wong, C. Gonzalez, J. A. Pople, Gaussian, Inc., Wallingford CT, 2004.
30 A. Schäfer, A. Hurbert, R. Ahlrichs, *J. Chem. Phys.* 1994, **100**, 5829.
31 A. J. Wachters, *J. Chem. Phys.* 1970, **52**, 1033.
32 P. J. Hay, *J. Chem. Phys.* 1971, **66**, 4377.
33 K. Raghavachari, G. W. Trucks, *J. Chem. Phys.* 1989, **91**, 1062.
34 M. C. Holthausen, W. Koch, *J. Am. Chem. Soc.* 1996, **118**, 9932.
35 S. Chiodo, N. Russo, E. Sicilia, *J. Comput. Chem.* 2005, **26**, 175.
36 (a) T. J. Lee, J. E. Rice, G. E. Scuseria, H. F. Schaefer III, *Theor. Chem. Acc.*, 1989, **75**, 81. (b) T. J. Lee, P. R. Taylor, *Int. J. Quantum Chem.* 1989, **23S**, 199.
37 J. P. Hay, W. R. Wadt, *J. Chem. Phys.* 1985, **82**, 274.
38 (a) C. Gonzales, H. B. Schlegel, *J. Chem. Phys.* 1989, **90**, 2154. (b) C. Gonzales, H. B. Schlegel, *J. Phys. Chem.* 1990, **94**, 5523.
39 J. Allison, R. B. Freas, D. P. Ridge, *J. Am. Chem. Soc.* 1979, **101**, 1332.
40 *Gas-Phase Inorganic Chemistry*, ed. D. H. Russel, Plenum, New York, 1989, p. 412.
41 P. B. Armentrout, J. L. Beauchamp, *Acc. Chem. Res.* 1993, **26**, 213.
42 P. B. Armentrout, in *Selective Hydrocarbons Activation: Principles and Progress*, eds. J. A. Davies, P. L. Watson, A. Greenberg, J. F. Liebman, VCH, New York, 1990.
43 P. B. Armentrout, *Annu. Rev. Phys. Chem.* 1990, **41**, 313.
44 K. Eller, H. Schwarz, *Chem. Rev.* 1991, **91**, 1121.
45 (a) J. C. Weisshaar, *Adv. Chem. Phys.* 1992, **82**, 213. (b) J. C. Weisshaar, *Acc. Chem. Res.* 1993, **26**, 213.
46 P. B. Armentrout, B. L. Kickel, in *Organometallic Ion Chemistry*, ed. B. S. Freiser, Kluwer, Dordrecht, 1996.
47 P. B. Armentrout, in *Topics in Organometallic Chemistry*, eds. J. M. Brown, P. Hofmann, Springer-Verlag, Berlin, 1999.
48 R. H. Crabtree, *The Organometallic Chemistry of the Transition Metals*, 2nd edn., John Wiley & Sons, New York, 1994.
49 G. A. Somorjai, *Introduction to Surface Chemistry and Catalysis*, John Wiley & Sons, New York, 1994.
50 H. Kang, J. L. Beauchamp, *J. Am. Chem. Soc.* 1986, **108**, 5663.
51 H. Kang, J. L. Beauchamp, *J. Am. Chem. Soc.* 1986, **108**, 7502.
52 B. C. Guo, K. P. Gerns, A. W. Castleman, Jr., *J. Phys. Chem.* 1992, **96**, 4879.
53 D. E. Clemmer, N. Aritsov, P. B. Armentrout, *J. Phys. Chem.* 1993, **97**, 544.
54 Y.-M. Chen, D. E. Clemmer, P. B. Armentrout, *J. Phys. Chem.* 1994, **98**, 11490.
55 D. E. Clemmer, Y.-M. Chen, P. B. Armentrout, *J. Phys. Chem.* 1994, **98**, 7538.
56 D. Schröder, A. Fiedler, M. F. Ryan, H. Schwarz, *J. Phys. Chem.* 1994, **98**, 68.
57 D. Schröder, H. Schwarz, *Angew. Chem. Int. Ed. Engl.* 1995, **34**, 1973.
58 N. Russo, E. Sicilia, *J. Am. Chem. Soc.* 2001, **123**, 2588.

59 A. Irigoras, J. E. Fowler, J. M. Ugalde, *J. Am. Chem. Soc.* 1999, **121**, 574.
60 Y. Song, *J. Mol. Struct. (THEOCHEM)* 1995, **357**, 147.
61 A. Irigoras, J. M. Ugalde, X. Lopez, C. Sarasola, *Can. J. Chem.* 1996, **74**, 1824–1829.
62 A. Irigoras, J. E. Fowler, J. M. Ugalde, *J. Phys. Chem. A* 1998, **102**, 293, 2252.
63 A. Irigoras, J. E. Fowler, J. M. Ugalde, *J. Am. Chem. Soc.* 1999, **121**, 8549.
64 A. Irigoras, O. Elizalde, I. Silanes, J. E. Fowler, J. M. Ugalde, *J. Am. Chem. Soc.* 2000, **122**, 114.
65 S. Shaik, D. Danovich, A. Fiedler, D. Schröder, H. Schwarz, *Helv. Chim. Acta* 1995, **78**, 1393.
66 D. Danovich, S. Shaik, *J. Am. Chem. Soc.* 1997, **119**, 1773.
67 D. E. Clemmer, Y.-M. Chen, F. A. Khan, P. B. Armentrout, *J. Phys. Chem.* 1994, **98**, 6522.
68 D. E. Clemmer, P. B. Armentrout, *J. Phys. Chem.* 1991, **95**, 3084 and references therein.
69 D. E. Clemmer, L. S. Sunderlin, P. B. Armentrout, *J. Phys. Chem.* 1990, **94**, 208.
70 D. E. Clemmer, L. S. Sunderlin, P. B. Armentrout, *J. Phys. Chem.* 1990, **94**, 3008.
71 W. Derek, P. B. Armentrout, *J. Am. Chem. Soc.* 1998, **120**, 3176.
72 E. Sicilia, N. Russo, *J. Am. Chem. Soc.* 2002, **124**, 1471.
73 M. C. Michelini, E. Sicilia, N. Russo, *J. Phys. Chem. A* 2002, **106**, 8937.
74 M. C. Michelini, E. Sicilia, N. Russo, *J. Phys. Chem. A* 2003, **107**, 4862.
75 M. C. Michelini, E. Sicilia, N. Russo, *Inorg. Chem.* 2004 **43**, 4944.
76 S. Chiodo, O. Kondakova, A. Irigoras, M. C. Michelini, N. Russo, E. Sicilia, J. M. Ugalde, *J. Phys. Chem. A* 2004, **108**, 1069.
77 W. Chaojie, Y. Song, *Int. J. Quantum Chem.* 1999, **75**, 47.
78 T. Taketsugu, M. S. Gordon, *J. Chem. Phys.* 1997, **106**, 8504.
79 M. Hendrickx, M. Ceulemans, K. Gong, L. Vanquickenborne, *J. Phys. Chem. A* 1997, **101**, 2465.
80 Y. Nakao, T. Taketsugu, K. Hirao, *J. Chem. Phys.* 1999, **110**, 10863.
81 S. R. Langhoff, C. W. Bauschlicher, H. Partridge, M. Sodupe, *J. Phys. Chem.* 1991, **95**, 10677.
82 R. Liyanage1, P. B. Armentrout, *Int. J. Mass Spectrom.* 2005, **241**, 243.
83 J. L. Durant, *Chem Phys. Lett.* 1996, **256**, 595.
84 B. S. Jursic, *Chem Phys. Lett.* 1997, **264**, 113.
85 H. Basch, S. Hoz, *J. Chem. Phys.* 1997, **101**, 4416.
86 B. J. Lynch, P. L. Fast, M. Harris, D. G. Truhlar, *J. Phys. Chem. A* 2000, **104**, 4811.
87 B. J. Lynch, P. L. Fast, M. Harris, D. G. Truhlar, *J. Phys. Chem. A* 2001, **105**, 2936.
88 A. Ricca, C. W. Bauschlicher, Jr., *Theor. Chim. Acta* 1995, **92**, 123.
89 A. Ricca, C. W. Bauschlicher, Jr., *J. Phys. Chem.* 1995, **99**, 9003.
90 M. C. Holthausen, A. Fiedler, H. Schwarz, W. Koch, *J. Phys. Chem.* 1996, **100**, 6236.
91 A. E. Shilov, *The Activation of Saturated Hydrocarbons by Transition Metal Complexes*, Reidel, Dortrecht, 1984.
92 (a) R. G. Bergman, *Science* 1984, **223**, 902. (b) B. A. Arndtsen, R. G. Bergman, T. A. Mobley, T. H. Peterson, *Acc. Chem. Res.* 1995, **28**, 154.
93 P. B. Armentrout, *Selective Hydrocarbons Activation: Principles and Progress*, eds. J. A. Davies, P. L. Watson, A. Greenberg, J. F. Liebman, VCH, New York, 1990.
94 (a) R. H. Crabtree, *Chem. Rev.* 1985, **85**, 245. (b) R. H. Crabtree, *Chem. Rev.* 1995, **95**, 987.
95 (a) J. C. Weisshaar, *Adv. Chem. Phys.* 1992, **82**, 213. (b) J. C. Weisshaar, *Acc. Chem. Res.* 1993, **26**, 213.
96 J. H. Lunsford, *Angew. Chem., Int. Ed. Engl.* 1995, **34**, 970.
97 J. J. Schneider, *Angew. Chem., Int. Ed. Engl.* 1996, **35**, 1068.
98 C. Hall, R. N. Perutz, *Chem. Rev.* 1996, **96**, 3125.
99 (a) D. Schröder, H. Schwarz, *Angew. Chem., Int. Ed. Engl.* 1990, **29**, 1433. (b) H. Schwarz, D. Schröder, *Pure Appl.*

100 R. Tonkyn, M. Ronan, J. C. Weisshaar, *J. Phys. Chem.* 1988, **92**, 92.
101 (a) L. F. Halle, P. B. Armentrout, J. L. Beauchamp, *J. Am. Chem. Soc.* 1981, **103**, 962. (b) N. Aristov, P. B. Armentrout, *J. Phys. Chem.* 1987, **91**, 6178. (c) R. H. Shultz, J. L. Elkind, P. B. Armentrout, *J. Am. Chem. Soc.* 1988, **110**, 411. (d) L. S. Sunderlin, P. B. Armentrout, *J. Phys. Chem.* 1988, **92**, 1209. (e) R. Geoorgiadis, P. B. Armentrout, *J. Phys. Chem.* 1988, **92**, 7060. (f) P. B. Armentrout, *J. Am. Chem. Soc.* 1990, **112**, 5663. (g) P. A. M. van Koppen, J. Brodbelt-Lustig, M. T. Bowers, D. V. Dearden, J. L. Beauchamp, E. R. Fisher, P. B. Armentrout, *J. Am. Chem. Soc.* 1991, **113**, 2359. (h) Y. Chen, P. B. Armentrout, *J. Phys. Chem.* 1995, **99**, 10775. (i) C. L. Haynes, Y. Chen, P. B. Armentrout, *J. Phys. Chem.* 1995, **99**, 9110.
102 D. G. Musaev, K. Morokuma, *J. Chem. Phys.* 1994, **101**, 10697.
103 R. H. Schultz, P. B. Armentrout, *J. Phys. Chem.* 1993, **97**, 596.
104 M. Hendrickx, K. Gong, L. Vanquickenborne, *J. Chem. Phys.* 1997, **107**, 6299.
105 D. G. Musaev, K. Morokuma, N. Koga, K. A. Nquyen, M. S. Gordon, T. R. Cundari, *J. Phys. Chem.* 1993, **97**, 11435.
106 D. G. Musaev, K. Morokuma, *J. Chem. Phys.* 1994, **101**, 10697.
107 D. G. Musaev, K. Morokuma, *J. Phys. Chem.* 1996, **100**, 11600.
108 Y. G. Abashkin, S. K. Burt, N. Russo *J. Phys. Chem. A* 1997, **101**, 8085.
109 R. B. Freas, D. P. Ridge, *J. Am. Chem. Soc.* 1980, **102**, 7129.
110 S. K. Huang, J. Allison, *Organometallics* 1983, **2**, 833.
111 D. B. Jacobson, B. S. Freiser, *J. Am. Chem. Soc.* 1984, **106**, 3891.
112 T. C. Jackson, D. B. Jacobson, B. S. Freiser, *J. Am. Chem. Soc.* 1984, **106**, 1252.
113 R. M. Pope, S. W. Buckner, *Org. Mass. Spectrom.* 1993, **28**, 1616.
114 Y.-M. Chen, D. E. Clemmer, P. B. Armentrout, *J. Am. Chem. Soc.* 1994, **116**, 7815.
115 J. E. Bushnell, P. R. Kemper, P. Maitre, M. T. Bowers, *J. Am. Chem. Soc.* 1994, **116**, 9710.
116 A. Fiedler, D. Schröder, S. Shaik, H. Schwarz, *J. Am. Chem. Soc.* 1994, **116**, 10734.
117 M. F. Ryan, A. Fiedler, D. Schröder, H. Schwarz, *Organometallics* 1994, **13**, 4072.
118 D. Schröder, A. Fiedler, J. Schwarz, H. Schwarz, *Inorg. Chem.* 1994, **33**, 5094.
119 P. A. M. van Koppen, P. R. Kemper, J. E. Bushnell, M. T. Bowers, *J. Am. Chem. Soc.* 1995, **117**, 2098.
120 B. L. Tjelta, P. B. Armentrout, *J. Am. Chem. Soc.* 1995, **117**, 5531.
121 A. Fiedler, I. Kretzschmar, D. Schröder, H. Schwarz, *J. Am. Chem. Soc.* 1996, **118**, 9941.
122 I. Kretzschmar, A. Fiedler, J. N. Harvey, D. Schröder, H. Schwarz, *J. Phys. Chem. A* 1997, **101**, 6252.
123 M. Engeser, M. Schlangen, D. Schröder, H. Schwarz, T. Yumura, K. Yoshizawa, *Organometallics* 2003, **22**, 3933.
124 S. Chiodo, O. Kondakova, M. C. Michelini, N. Russo, E. Sicilia, *Inorg. Chem.* 2003, **42**, 8773.
125 I. Rivalta, N. Russo, E. Sicilia, *J. Comput. Chem.* 2006, **27**, 174.
126 K. K. Irikura, J. L. Beauchamp, *J. Am. Chem. Soc.* 1989, **111**, 75.
127 C. J. Cassady, S. W. McElvany, *Organometallics* 1992, **11**, 2367.
128 N. E. Schore, *Chem. Rev.* 1988, **88**, 1081.
129 S. W. Benson, *Thermochemical Kinetics* Wiley, New York, 1968.
130 (a) K. N. Houk, R. W. Gandour, R. W. Strozier, N. G. Rondan, L. A. Paquette, *J. Am. Chem. Soc.* 1979, **101**, 6797. (b) R. D. Bach, G. J. Wolber, H. B. Schlegel, *J. Am. Chem. Soc.* 1985, **107**, 2837.
131 W. T. Tysoe, G. L. Nyberg, R. M. Lambert, *J. Chem. Soc., Chem. Commun.* 1983, 623.
132 W. S. Sesselmann, B. Woratschek, G. Ertl, J. Kuppers, H. Haberland, *Surf. Sci.* 1983, **130**, 245.
133 T. M. Gentle, E. L. Muetterties, *J. Phys. Chem.* 1983, **87**, 2469.
134 P. M. Holmblad, D. R. Rainer, D. W. Goodman, *J. Phys. Chem. B* 1997, **101**, 8883.

135 I. M. Abdelrehim, K. Pelhos, T. E. Madey, J. Eng, J. G. Chen, *J. Mol. Catal. A* 1998, **131**, 107.
136 T. G. Rucker, M. A. Logan, T. M. Gentle, E. L. Muetterties, G. A. Somorjai, *J. Phys. Chem.* 1986, **90**, 2703.
137 N. R. Avery, *J. Am. Chem. Soc.* 1985, **107**, 6711.
138 J. R. Lomas, M. S. Baddeley, R. M. Tikhov, R. M. Lambert, *Langmuir* 1995, **11**, 3048.
139 C. H. Patterson, R. M. Lambert, *J. Am. Chem. Soc.* 1988, **110**, 6871.
140 C. H. Patterson, R. M. Lambert, *J. Phys. Chem.* 1988, **92**, 1266.
141 X.-Y. Zhu, J. M. White, *Surf. Sci.* 1989, **214**, 240.
142 C. M. Mate, C.-T. Kao, B. E. Bent, G. A. Somorjai, *Surf. Sci.* 1988, **197**, 183.
143 R. M. Ormerod, R. M. Lambert, *J. Phys. Chem.* 1992, **96**, 8111.
144 R. M. Ormerod, R. M. Lambert, H. Hoffmann, F. Zaera, J. M. Yao, D. K. Saldin, L. P. Wang, D. W. Bennet, W. T. Tysoe, *Surf. Sci.* 1993, **295**, 277.
145 J. A. Gates, L. L. Desmodel, *J. Chem. Phys.* 1982, **76**, 4281.
146 G. Pacchioni, R. M. Lambert, *Surf. Sci.* 1994, **304**, 208.
147 H. Hoffmann, F. Zaera, R. M. Ormerod, R. M. Lambert, J. M. Yao, D. K. Saldin, L. P. Wang, D. W. Bennett, W. T. Tysoe, *Surf. Sci.* 1992, **268**, 1.
148 R. M. Ormerod, C. J. Baddeley, R. M. Lambert, *Surf. Sci. Lett.* 1991, **259**, L709.
149 C. J. Baddeley, R. M. Ormerod, A. W. Stephenson, R. M. Lambert, *J. Phys. Chem.* 1995, **99**, 5146.
150 C. J. Baddeley, M. Tikhov, C. Hardacre, J. R. Lomas, R. M. Lambert, *J. Phys. Chem.* 1996, **100**, 2189.
151 R. M. Ormerod, R. M. Lambert, *J. Chem. Soc., Chem. Commun.* 1990, 1421.
152 S. Abbet, A. Sanchez, U. Heiz, W.-D. Schneider, A. M. Ferrari, G. Pacchioni, N. J. Rösch, *J. Am. Chem. Soc.* 2000, **122**, 3453.
153 S. Abbet, A. Sanchez, U. Heiz, W. D. Schneider, *J. Catal.* 2001, **198**, 122.
154 A. M. Ferrari, L. Giordano, G. Pacchioni, S. Abbet, U. Heiz, *J. Phys. Chem. B* 2002, **106**, 3173.
155 A. S. Wörz, K. Judai, S. Abbet, J. M. Antonietti, U. Heiz, A. Del Vitto, L. Giordano, G. Pacchioni, *Chem. Phys. Lett.* 2004, **399**, 266.
156 K. Judai, A. S. Wörz, S. Abbet, J. M. Antonietti, U. Heiz, A. Del Vitto, L. Giordano, G. Pacchioni, *Phys. Chem. Chem. Phys.* 2005, **7**, 955.
157 K. Judai, S. Abbet, A. S. Wörz, A. M. Ferrari, L. Giordano, G. Pacchioni, U. Heiz, *J. Mol. Catal. A* 2003, **199**, 103.
158 S. Chrétien, D. R. Salahub, *J. Chem. Phys.* 2003, **119**, 12291.
159 M. Martinez, M. C. Michelini, I. Rivalta, N. Russo, E. Sicilia, *Inorg. Chem.* 2005, **44**, 9807.
160 C. E. Moore, *Atomic Energy Levels*, Natl. Bur. Stand. (U.S.) Circ. 467. Vol. III (1958) reprinted as Natl. Stand. Ref. Data Ser., Natl. Bur. Stand. (U.S.) 35 (1971).
161 R. F. Bader, *Atoms in Molecules. A Quantum Theory*, Clarendon, Oxford, 1990.
162 (a) A. Savin, O. Jespen, J. Flad, O. K. Andersen, H. Preuss, H. G. von Schnering, *Angew. Chem., Int. Ed. Engl.* 1992, **31**, 187. (b) M. Kohout, A. Savin, *Int. J. Quantum Chem.* 1996, **60**, 875. (c) U. Häussermann, S. Wengert, R. Nesper, *Angew. Chem., Int. Ed. Engl.* 1994, **33**, 2073. (d) A. Savin, B. Silvi, F. Colonna, *Can. J. Chem.* 1996, **74**, 1088. (e) S. Noury, F. Colonna, A. Savin, B. Silvi, *J. Mol. Struct.* 1998, **450**, 59. (f) A. Savin, R. Nesper, S. Wengert, T. Fässler, *Angew. Chem., Int. Ed. Engl.* 1997, **36**, 1809.
163 See for instance (a) M. Calatayud, B. Silvi, J. Andrés, A. Beltrán, *Chem. Phys. Lett.* 2001, **333**, 493. (b) J. Pilme, B. Silvi, M. E. Alikhani, *J. Phys. Chem. A* 2003, **107**, 4506. (c) M. C. Michelini, E. Sicilia, N. Russo, M. E. Alikhani, B. Silvi, *J. Phys. Chem. A* 2003, **107**, 4862. (d) S. Chiodo, O. Kondakova, M. C. Michelini, N. Russo, E. Sicilia, *J. Phys. Chem. A* 2004, **108**, 1069. (e) M. C. Michelini, N. Russo, M. E. Alikhani, B. Silvi, *J. Comput. Chem.* 2004, **25**, 1647. (f) M. C. Michelini, N. Russo, M. E. Alikhani, B. Silvi, *J. Comput. Chem.* 2005, **26**, 1284. (g) J. Pilme, B. Silvi,

M. E. Alikhani, *J. Phys. Chem. A* 2005, **109**, 10028.
164 R. Thom, *Stabilité Structurelle et Morphogénèse*, Interdictions, Paris, 1972.
165 (a) F. Fuster, A. Savin, B. Silvi, *J. Phys. Chem. A* 2000, **104**, 852. (b) F. Fuster, B. Silvi, *Chem. Phys.* 2000, **252**, 279. (c) F. Fuster, A. Savin, B. Silvi, *J. Comput. Chem.* 2000, **21**, 509. (d) X. Krokidis, S. Noury, B. Silvi, *J. Phys. Chem. A* 1997, **101**, 7277.
166 (a) S. Noury, X. Krokidis, F. Fuster, B. Silvi, *TopMod Package*, Paris, 1997. (b) S. Noury, X. Krokidis, F. Fuster, B. Silvi, *Computers Chem.* 1999, **23**, 597.
167 E. Pepke, J. Muray, J. Lyons, *Scian (Supercomputer Computations Res. Inst.)*, Florida State University, Tallahassee, FL, 1993.

Subject Index

a

acetonitrile 299–300
acetylene, cyclotrimerization 351–355
acrylate-enolate complexes 185–188
acrylates
– complex formation 183–184
– coupling reactions 188
– group transfer polymerization 181–214
activation energy 155, 136, 209
– C–H activation 268
– computed 322
– methane 274
– 2-methylbutane 333
– oxidative dehydrogenation 238
addition reaction, oxidative 112–113
α-agostic interaction 158
β-agostic interaction modes 168
β-agostic precursors, equilibration 160
β-agostic species
– olefin propagation 149–150
– structures 156
alcohols 276–281
– hydrogen transfer 279
– oxidation 241
– solid-state catalysts 231
alkaline-earth metal catalysts 292–293
alkanes
– C–H activation 238–241
– C–H bond activation 276
– hydroxylation 331–334
– oxidative dehydrogenation 235–238
– solid-state catalysts 231
alkenes
– early transition metal complexes 294–296
– epoxidation 223
– hydrosilylation 285
– late transition metal complexes 309–312
– organolanthanide complexes 290
alkyl–alkyl coupling 133

alkynes
– hydrosilylation 285, 287
– hydrosilylation, Wu–Trost mechanism 305
– intermolecular hydrosilylation 304
– late transition metal complexes 303–309
aminoacid ligands 27
ammonia, bond insertion 343–347
apparent catalyst productivity 173
asymmetric ring-closing metathesis 247
asymmetrical R–R′ coupling 137–138
axial ligands
– B_{12} chemistry 38–41
– chlorophylls 50–51
– cytochromes 42–44
– F430 41–42
– globins 42
– heme enzymes 44–47, 52
– His ligand 47–50
– negative 47
– neutral 52
– reduction potentials 51
– tuning of tetrapyrrole cofactors 38–51

b

B_{12} chemistry 38–41
B3LYP 134, 217
– ammonia reactions 344
– di-zirconium complexes 105
– di-zirconocene-N_2 96
– β-elimination 281
– hydrosilylation 298
– olefin metathesis 247
– olefin propagation 152
– rhenium(v)-di-oxo catalyst 302
– Shilov reaction 266
– tetrapyrrole cofactors 29
– two state reactivity paradigm 339
backbiting intermediates, interatomic
    distances 201

backbiting reaction 200–201
– mechanism 199
– profiles 210
– side reactions 198
bacteriochlorin 33
$B(C_6F_5)$ 294
benzene, C–H activation 265–284
bimetallic mechanism 181
bimolecular deactivation reaction 174
block orbitals 319
bond activation 231–245, 270–281, 347–351
– heterolytic 265–284
– methane 270–276
– reactivity paradigm 337–361
σ-bond activation
– heterolytic 265–284
– oxidative addition 112
bond addition 133, 265
bond cleavage 14
– barrier 66
bond distances
– changes 323–324
– epoxidation 225
– epoxidation, counter cation effect 227
bond formation 131
bond insertion 343–347
bond lengths 92–93, 200
– C_2H_2 polymerization 353
– Mg-ligand 50
σ-bond metathesis 279
bonding evolution 355–360
– catastrophe theory 355
bovine erythrocyte GPx 6
BP86 247, 261
$(Bpym)Cl_2$ 273
branching ratios 323
broken symmetry potential 239
broken symmetry solution 234
Buchwald–Hartwig reaction 109
butadiene elimination 143

c

C–C bond activation 351–355
– barriers 269
– intermediate formation 352
– metal atoms 351–355
– reactivity paradigm 337–361
C–C bond formation 131
– dissociative mechanism 142–143
– intermediate III 133
– ligand effect 142
– metal effect 141
– solvent effect 144
– transition metal catalyzed 131–145
– unsaturated ligands 133–145
– unsaturated organic molecules 144
C=C bond synthesis 245
C–C coupling reaction 139
– alkynyl groups 141
– general scheme 134
– profiles 211
C=C double bond, coordination 258
C–F bond breaking 275
C–H activation, population changes 269
C–H bond activation 231–245, 268, 270–274
– $(Bpym)Cl_2$ 273
– methane 270–276
– methane hydroxylation 324
– $(NH_3)_2PtCl_2$ 272
– species 232–233
C_2H_2 351
C_2H_2 polymerization
– ground spin states 353
– rate-determining step 352
$[(\eta^5-C_5H_5)_2Zr]_2[M_2,\eta^2,\eta^2-(NH)_2)]$ 88–89
$[(\eta^5-C_5Me_nH_{5-n})_2M]_2(\mu_2,\eta^2,\eta^2-N_2)$ complexes 95–101
β-carbon
– complex formation 183–184
– coupling reactions 188
carbon-carbon see C–C
carbon dioxide, hydrogenation 277–278
carbonyl compounds
– hydrosilylation reactions 285, 291–292
– reductive elimination 144
catalases 45–46
catalysis
– gas-phase 337–366
– homogeneous 131–145
– ion-pairing 215–230
– two state reactivity paradigm 337–361
catalyst modifications
– enolate-polymer model complexes 206
– ring-opening reaction profile 212
catalyst productivity 173
catalysts 231–244, 270
– C–H bond activation 273–274
– chirality 249
– polymerization process 151
– transition metal oxide 235–238
catalytic cycles
– cross-coupling reactions 110
– GPx 8
– heme proteins 45
– metalloenzymes 58
– molybdenum Co dehydrogenase 78
– NiFe-hydrogenase 75

– NO reduction   71
catalytic functions
– GPx   3–5, 8–12, 16–23
catalyzation   215–230
– ethylene epoxidation   228–232
cationic zirconocene metallacycle   192–193
cations
– CH_4 bond activation   347–351
– X–H (X = O, N, C) bonds activation   339–340
CCSD (T) calculations   339
CH_4
– bond activation   347–351
– Mn^+ reactions   356–360
chain-end control
– olefin propagation   162
– ring-opeing reactions   190
chain-transfer reactions   164, 205
– group transfer polymerization   204–208
– β-hydride elimination   165
– internal barriers   206
chain-transfer-to-ethylene   168
chain-transfer-to-monomer, energies   167
Chalk–Harrod mechanism   286–289
– redox couples   288
change of charge   57
chirality   256
– catalyst   249
chlorin rings   30
chlorophyll   28, 52
– absorption spectra   50
– tetrapyrrole cofactors   50–51
cis-trans isomerization   127
cisplatin   270
Claisen condensation   198, 200
cleavage of N–O bond, transition state   70
closed shell ions   242
closed shell V^v system   232
Co–C bond   36
cobalamin   28
– bond stability   36
coenzyme B12   39, 52
coenzyme F430   52
complex connection pathway
– monophosphine systems   121
– transmetalation   118
complexes
– σ-bond activation   265–284
– C–H bond activation   270–273
– conformations   185–188
– formation   184–185
– K, Ca, Sr and B   292–294
– Mo, W and Re   298–303
– $R_2M(PH_3)_2$ 134–136
– $RR'M\,(PH_3)_2$ 137–139
– Ru and Os   303–312
– Zr and Ta   294–296
computational models
– GPx   5–6
– group transfer polymerization   182–211
– olefin epoxidation   217–218
coordinated complexes
– end on   92
– side-on   91
coordination intermediates
– Mo-based catalysts   257
– olefin metathesis   250
– ring-closing metathesis   255
corrin rings   30, 52
Cossee–Arlman mechanism   149
counter cation effect   222
– olefin epoxidation   221, 226–228
counteranion, reaction rates   170
counterion   152–153
coupling reactions
– calculated energetic   139
– energy profiles   189
– transition states   188–190
Cp_2Zr complex, hydrosilylation   296
Cr^+, geometrical parameters   350
cross-coupling reactions
– mechanisms   110
– palladium-catalyzed   109–129
– Stille reaction   121
crossing seam   325–326
crystal structures
– cytochrome c peroxidase   49
– GPx   1
cusp catastrophe, bonding evolution   356–357
CxH activation, Gibbs free energy   239
cyano groups, reductive elimination   144
cyclic pathways, Stille reaction   122
cyclotrimerization   339, 351–355
cytochrome C oxidase
– binuclear center   65
– bond strength   65
– energy diagrams   67–69
– metalloenzymes   64–70
– N-side   64
– P-side   64
cytochrome C peroxidase, crystal structure   49
cytochrome-like electron transfer, tetrapyrrole cofactors   35–36
cytochrome models   43
cytochrome P-450, catalytic cycles   45
cytochrome P450   220

cytochromes, tetrapyrrole cofactors 42–44
cytosolic (GPx-1) 1

d

d-electrons, metal-oxo bond 319
dehydrogenation 235–238
density functional theory (DFT) 29, 181–214, 337–366
– methane hydroxylation 317–336
– olefin epoxidation 215–230
– olefin metathesis 245–264
50-deoxyadenosylcobalamin (AdoCbl) 36
Desulfomicrobium bacalum, X-ray structure 72–73
desymmetrization, Ru-based catalysts 256
DFT *see* density functional theory
di-V-substituted-γ-Keggin [β-1,2-H$_2$SiV$_2$W$_{10}$O$_{40}$]$^{4-}$ polyoxometalates 215–230
di-V-substituted γ-Keggin POM 223
di-zirconium complexes, stabilization energy 103
di-zirconocene-N$_2$
– dinitrogen hydrogenation 87–91
– energetics 94
– energy gap 96
– reactivity 91
diastereomeric zirconocene metallacycles 196
diazenido complex 88
Dieckmann condensation, side reactions 198
dienes
– ring-closing 258
– ring-closing metathesis 246
dihedral angle, olefin metathesis 248–249
dihydrogen molecules 276–281
dimeric vanadium species, solid-state catalysts 232
dinitrogen, early transition metal complexes 296–298
dinitrogen hydrogenation
– [(η^5-C$_5$Me$_n$H$_{5-n}$)$_2$M]$_2$(μ_2,η^2,η^2-N$_2$) complexes (M = Ti, Zr, or Hf) 95–101
– computational insights 83–105
– coordinated molecules 87–91, 103–105
– di-zirconocene-N$_2$ complexes 87–91
– dizirconium-dinitrogen complexes with bis(amidophosphine) (P2N2) and cyclopentadienyl (CP) ligands 101–103
– N$_2$ coordination modes 91–95
– Zr centers 101–103
dinuclear complexes 104
– hydrogenation 94

dioxide cation, Cr$^+$ 351
dioxygen evolution, energy diagram 62–63
dioxygen reduction, energy diagram 67–69
dissociative pathways
– oxidative addition 112–113
– Stille reaction 123
distal side, tetrapyrrole cofactors 38
ditantalum dinitrogen complex 297
dizirconium-dinitrogen complexes 296
– reactivity 91
– with bis(amidophosphine) (P2N2) and cyclopentadienyl (CP) ligands 101–103
double spin crossing 337

e

E–Se–NOO complex, glutathione (GSH) 22
early transition metal systems 274–276
– polymerization process 151
– Zr and Ta 294–296
effective spin–orbit operator, methane hydroxylation 326
eight-membered metallacycles
– backbiting reaction 200–201
– proton transfer 201
electron affinity (EA), photosystem II 59
electron localization
– bonding evolution 355
– function 355–360
electron uptakes 68
electron withdrawing 202
electronic configurations, transition-metal oxides 319
electronic coupling element 35
electronically excited ions 347
electrophile addition, C–C bond formation 131
electrophilic attack 269
β-elimination 281
elimination reaction
– reductive 125, 131–133
elliptic umbilic catastrophe, bonding evolution 356–357
enantiofaces
– Mo-based catalysts 258
– olefin metathesis 249
enantiomorphic site control, ring-opening reactions 190
enantioselectivity 247, 253, 256
end-on
– coordinated complexes 92
– η^1 isomers 99
energetics
– C$_2$H$_4$ epoxidation 223
– di-zirconocene-N$_2$ 94

Subject Index

enolate α-carbon
– complex formation   183–164
– coupling reactions   188
– proton transfer   201
enolate complexes
– conformations   185–188
– formation   184–185
enolate-polymer model complexes, catalyst modifications   206
epoxidation   215–230
– elementary processes   224
– mechanism   225–226
– model   218
Erker's view, polymer chain propagation   152
ethane, reductive elimination   134
ethylene   220, 228–232
– intrinsic barriers   172
– polymerization   163
– reaction rates   173
extracellular (GPx-3)   1
Eyring equation   173

f

f-block metallocenes, group transfer polymerization   181–214
F430   28
– tetrapyrrole cofactors   41–42
FeO^+ 318
first-row transition metal cations   339–340
– CH_4 bond activation   347–351
first-row transition metal ions
– insertion   343–347
– reactions   340–343
first row transition metal oxides, methane hydroxylation   317–336
fold catastrophe, bonding evolution   356–357
formaldehyde
– C–H bond activation   233–235
– hydrogen transfer   280
– intrinsic barriers   234
– reaction energy   235
Frank–Caro process   84
frontier molecular orbitals   98
– metallacyclopropene complexes   308
frontside attack, ring-opeing reactions   190
Fryzuk complex   87
Fujiwara–Moritani reaction   267–270

g

gas phase   144, 232–233
– C–H activation   238–241
– catalysis   337–366
– reactions   317
– Ru-based catalysts   251

gas-phase calculations, oxidative addition   112
gastrointestinal tract (GPx-2)   1
Gaussian 03 program
– DFT   217
– GPx   5
– olefin metathesis   248
– olefin propagation   152
general mechanistic pathways   286–292
generalized gradient approximation (GGA) functionals   338
GGA   see generalized gradient approximation
Glaser–Tilley mechanism   311
globins, tetrapyrrole cofactors   42
glutamate mutase, optimized structures   40
γ-glutamylcysteinylglycine   6
glutathione (GSH)   22
– GPx catalytic activity   3
GPx   see selenoprotein glutathione peroxidase
group IV metallocenes, homogeneous   149–176
group transfer polymerization   181–214
– computational details   182–211
– interatomic distances   187
– states   188–190
Grubb's catalysts, olefin metathesis   245–264
GTP   see group transfer polymerization

h

H-atom abstraction   329–331
– methane hydroxylation   322
H–H bond activation   276–279
H_2O molecules, Mn^+ reactions   356–360
H_2O_2-based olefin epoxidation   215–230
$[\gamma\text{-}1,2\text{-}H_2SiV_2W_{10}O_{40}]^{4-}$ polyoxometalates   215–230
Haber–Bosch process   84
halide replacement   119
Heck reaction   109
heme enzymes, tetrapyrrole cofactors   44–47
heme ligand   52
heme peroxidases   45–46
heme proteins   44
heterocoupling, potential energy surfaces   139
heterolytic cleavage   74
heterolytic H–H bond activation   277–279
heterolytic σ-bond activation
– characteristic features   266–270
– hydrogen transfer reactions   279–281
high spin state   32
highest occupied molecular orbital (HOMO)   85
His ligands, tetrapyrrole cofactors   47–50

homocoupling, versus heterocoupling pathways 140
homogeneous catalysis 131–145
horseradish peroxidase, amino acids arrangement 48
hybrid B3LYP functional, ammonia reactions 344
hybrid QM/MM approach, GPx 5
hybrid quantum-mechanical (QM) models 183
β-hydride elimination
– olefin polymerization 164–165
– reaction energy 158
– stability 165
hydrocorphin rings 30, 52
hydrogen abstraction 236, 280
hydrogen peroxide 215–230
– ethylene epoxidation 228–232
– GPx catalytic activity 3
hydrogen peroxide coordination, GPx 12–14
hydrogen reduction, energies 242
hydrogen transfer reactions 279–281
hydrogenation
– carbon dioxide 277
– di-zirconocene-N_2 94
– N_2 molecule 104
hydrogenolysis reactions 169
hydrometallation 286, 306
– molecular orbitals 287
hydroperoxy mechanism, epoxidation 225
hydrosilylation 285–316
– alkali and alkaline-earth metal complexes 293
– barriers 303
– bonding 313
– carbonyl compounds. 292
– ditantalum dinitrogen complex 297
– dizirconium dinitrogen complex 296
– energy barriers 303
– ionic 299
– mechanism 298
– metal-oxo catalysts 301
– olefin-first mechanism 295
– organolanthanide-catalyzed 291
– oxidative addition 286
– reductive elimination 286
– relative free energy 312
– tungsten-silylene catalyzed 299–300
hydroxo intermediate, methane hydroxylation 324
hydroxyl mechanism, ethylene epoxidation 228–230
hydroxylation 317–336
– barriers 332

– energy profile 320–324
– regioselectivity 331–334
– spin-orbit coupling (SOC) 326–329

i

imido complexes 276
impurities, homocoupling 139
indirect site-control, olefin propagation 163
inner-sphere reorganization energy
– cytochrome models 43
– tetrapyrrole cofactors 35–36
intermediate spin state 32
intermediates
– GPx structures 9–11
– Mo-based catalysts 258
– ONIOM structures 13
– relative energies 37
– ring-closing metathesis 255
– ring-opening, interatomic distances 194
– silicon-containing 285–316
intrinsic reaction coordinate (IRC) 219
– analysis 324–326
– methane hydroxylation 317–336
iridium complexes, C–C coupling reaction 139
isobacteriochlorin 33
isoelectronic complexes 186
isomerization 21, 126–127, 289
– barriers 94, 127
– water molecule reactions 342

k

kinetic isotope effect (KIE) 329–331
– methane hydroxylation 317–336
– values 330

l

Lacunary $[\gamma\text{-}(SiO_4)W_{10}O_{32}H_4]^{4-}$ polyoxometalates 215–230
lanthanocenes 274
late transition metal catalysts, C–H bond activation 273–274
late transition metal complexes, Ru and Os 303–312
ligand association 143
ligand environment 105
ligand structure 27
ligand substitution, monophosphine systems 121
ligands, axial *see* axial ligands
localization domains, Mn^+ reactions 358
localization function, bonding evolution 355–360
low spin state 32

lowest unoccupied molecular orbital (LUMO) 83–85

m

M$^+$-NH3 bond, dissociation energy 343
main group metal complexes, K, Ca, Sr and B 292–294
Marcus equation 35
Markovnikov hydrosilylation, alkenes 303
Mars–van Krevelen mechanism 237
mechanistic pathways
– general 286–292
– hydrosilylation reactions 313
– new discovered 292–312
MeO$^-$ elimination, backbiting reaction 200
metal-carbene complexes 347
metal cations
– CH$_4$ bond activation 347–351
– X–H (X = O, N, C) bonds activation 339–340
metal complexes
– K, Ca, Sr and B 292–294
– Mo, W and Re 298–303
– reactivity 141
– Ru and Os 303–312
– Zr and Ta 294–296
metal compounds, reactivity paradigm 337–361
metal ions 340–343
– bond insertion 343–347
metal oxides
– C–H bond activation 231–245
– catalysts 235–238
– methane hydroxylation 317–336
metal-oxo bonds 319
– energies 242
metal-oxo catalysts, hydrosilylation 301
metal silylene complexes
– alkene hydrolsilylation 309
– structures 311
metallacycles
– backbiting reaction 200–201
– decomposition 202
– group transfer polymerization 198–204
– interatomic distances 194
– olefin metathesis 250
– proton transfer 201
– ring opening 204
metallacyclopropene complexes 308
metallation reaction, tetrapyrrole cofactors 37–38
metallocene-acrylate-enolate complexes, conformations 185–188

metallocene-enolate complex, formation 184–185
metallocene-MA complex, formation 184
metallocenes
– group IV 149–176
– group transfer polymerization 181–214
– ten-membered 201–204
metalloenzymes
– charge 58
– cytochrome C oxidase 64–70
– energy diagrams 59–79
– modeling 57–80
– molybdenum CO dehydrogenase 76–79
– NiFe-hydrogenase 72–76
– nitric oxide reduction 70–72
– photosystem II 59–64
– reaction energy 58
metallopentacycle formation 355
metathesis transition state 85
methane 271
– σ-bond activation 270–276
– dehydrogenate 347
– metathesis 271
methane C–H bond activation 265–284
– energy barriers 274
– ionic radius 274
– potential energy surfaces 348
methane hydroxylation 317–336
– angular momentum operators 327
– energy profile 320–324
– excited state 325
– ground state 325
– isotope effects 329
– pathways 329
– reaction efficiencies 322–323
– spin inversion 318
– spin-orbit coupling (SOC) 326–329
– states 325
methane monooxygenase, soluble 318
methanol
– binding, model energies 233
– hydrogen transfer 280
– oxidation 233–235
– solid-state catalysts 232
3-methyl-2-butanol, energy diagram 332
2-methylbutane
– C–H bond 333
– regioselectivity 331
methylcobalamin (MeCbl) 36
Mg-ligand bond length 50
microsolvation effects, Shilov reaction 266
middle transition metal complexes, Mo, W and Re 298–303
migratory insertion, olefin 153

miomorphic process
– bonding evolution 355
– NH$_3$ (H$_2$O) 358
MM models *see* molecular-mechanical models
Mn$^+$ reactions 344
– potential energy profiles 357
– states 356–360
– topological structures 356, 359–360
Mo-based catalysts 257–260
– geometry 258–259
Mo-catalyzed olefin metathesis 245–264
Mo$^+$ complexes
– atomic spin densities 320
– bond dissociation energy 320
– reactivity 319–320
– species 319–320
modeling, mechanisms for metalloenzymes 57–80
models
– group transfer polymerization 183
– olefin epoxidation 217–218
– olefin metathesis 247–250
– solid-state catalysts 231–244
molecular-mechanical (MM) models 183
molecular orbitals 319
– analysis 85
– frontier 98, 308
– highest occupied 85
– lowest unoccupied 83–85
molecular weight control 175
molybdenum Co dehydrogenase
– barriers 78
– catalytic cycle 78
– metalloenzymes 76–79
– optimized structure 77
mono nuclear early d- and f-block metallocenes 181–214
monomer addition 200
monomeric unit, GPx 6
monomeric vanadium species, solid-state catalysts 232
mononuclear W(VI)-complex 221
monophosphine systems 120–121
– connecting complex 121
monovanadium, olefin epoxidation 216
Morokuma–Kitaura energy decomposition analysis 102
multinuclear transition metal complexes 104

n

N–H bond insertion 343–347
N–N bond lengths 92–93
N–N stretching frequencies 93
N$_2$ coordination modes, dinitrogen hydrogenation 91–95
N$_2$ molecule hydrogenation 104
NH$_3$ molecules, Mn$^+$ reactions 356–360
NH$_3$ (H$_2$O), miomorphic process 358
(NH$_3$)$_2$PtCl$_2$, C–H bond activation 272
NiFe-hydrogenase
– metalloenzymes 72–76
– reaction rate 75
nitration pathways, GPx 20–23
nitric oxide reduction, metalloenzymes 70–72
nitrogen
– natural population atomic charges 97
– reduced 83
nitrogenase 83
NO reduction, catalytic cycle 71
nomenclature, olefin metathesis 250–251
non-radical mechanism
– isotope effect 330
– methane hydroxylation 318
nucleophile attack 133

o

O–O bond 61
– cleavage 14
– cytochrome C oxidase 65
ODH *see* oxidative dehydrogenation
OH addition, methanol oxidation 233
olefin-alkyl complex 153
olefin binding 158
– reaction rates 170
olefin epoxidation 215–230
– computational details 217–218
– mechanism 218, 223
– reaction mechanism 224–225
olefin-first mechanism, hydrosilylation 295
olefin insertion barriers 172
olefin metathesis 245–264
– nomenclature 250–251
– side-coordinated structures 250
olefin polymerization
– absolute rates of reactions 170–172
– chain propagation 152–163
– chain-transfer-to-hydrogen 168–169
– chain-transfer-to-monomer 166–167
– computational details 152
– homogeneous group IV metallocenes 149–176
– β-hydride elimination 164–165
– kinetics 173–175
olefin propagation
– free energy 154

– interactions   162
– kinetics   154
– mechanism   150
– misinsertion   160
Oligotropha carboxidovarans, molybdenum Co dehydrogenase   76
ONIOM structures, GPx   13
organic electrophile addition   131
organoboronate species   114
organolanthanide complexes, hydrosilylation   290
osmium-silylene complexes, alkene hydrolsilylation   310
oxidation pathway, GPx   16–20
oxidative addition   112–113, 265–284, 306
– ammonia reactions   343
– hydrosilylation   286
– methane   271
– non-polar σ-bonds   112
– water molecule reactions   340
oxidative dehydrogenation   231, 235–238
– C–H bond activation   233–235
– energy barriers   234
– propane   237
oxygen-evolving complex (OEC)
– photosystem II   59
– structure   60
oxygen radical sites   240
oxyheme   42

p

palladium(II) diphosphine   120
palladium-catalyzed cross-coupling reactions   109–129
– oxidative addition   112–113
– reductive elimination   125
– transmetalation   114–124
peroxidase activity
– GPx   3, 8–12
peroxidase mechanism, GPx   3
peroxo species   218
peroxynitrite   15
– concerted oxidation mechanism   17
peroxynitrite/peroxynitrous acid   4
– coordination   17
peroxynitrous acid   15
– concerted oxidation mechanism   18
– stepwise oxidation mechanism   18
PES   *see* potential energy surfaces
bis(phosphine) systems   118–119
phospholipid hydroperoxide (GPx-4)r   1
photosystem II
– metalloenzymes   59–64
– reaction center   59

platinum catalyzation, hydrosilylation   287
platinum redox couples, Chalk–Harrod mechanism   288
plyomorphic process, bonding evolution   355
polymer chain propagation   152
polymerization   151
– acrylates   181–214
– mechanism   184–195
– olefin   149–176
polymerization barrier   153
polymers
– chain transfer reactions   204–207
– syndio-enriched   190
polyoxometalates   215–230, 228–232
– catalysts   215–225
porphyrin   28, 30, 52
– model   49
potential energy surfaces (PES)   338
– alkene hydrolsilylation   310
– ammonia reactions   344
– homocoupling   140
– β-hydride elimination step   164
– methane activation   348
– methane hydroxylation   321
– Mn^+ reactions   345
– olefin propagation   159–160
– Sc^+ 341
propane
– oxidative dehydrogenation   237
– solid-state catalysts   232
propyl radical, oxidative dehydrogenation   237
propylene polymerization   163
proton affinity (PA)
– molybdenum Co dehydrogenase   78
– photosystem II   59, 62
proton transfer   199, 201, 204
– epoxidation   224–225
– pathways   74
proton uptakes   68
PSII   *see* photosystem II
Pt(II) complexes, C–H bond activation   270–273
Pt(Bpym)Cl$_2$ 270

q

QM/MM calculations   209
– olefin metathesis   245–264
quantum-mechanical models   183

r

$R_2M(PH_3)_2$ complexes   134–136
radical cations, very reactive   242

reaction profiles
– backbiting 210
– C–C coupling 211
– formaldehyde formation 234
– polymer chain transfer reaction 205–206
– ring-opening 212
reactions
– backbiting 200–201
– first-row transition metal ions 340–343
– Fujiwara–Moritani 267–270
– hydrosilylation 285–316
– involving transition metals 355–360
– Mn^+ 356–360
– ring-opening 190–195
– Shilov 266–267
reactivity paradigm 337–361
recombination transition state 324
redox couples, Chalk–Harrod mechanism 288
reductase activity
– GPx 3–5, 16–23
reductase mechanism, GPx 4
reduction potentials
reductive elimination 125
– asymmetrical RR0M $(PH_3)_2$ complexes 137–139
– C–C bond formation 131–132
– carbon-carbon bond formation 131–133
– carbon dioxide 277
– Chalk–Harrod mechanism 289
– homocoupling versus heterocoupling pathways 140
– hydrosilylation 286
– symmetrical $R_2M(PH_3)_2$ complexes 134–136
– unsaturated organic molecules 144
regioselectivity 290, 331–334
– hydrosilylation 293
reorganization energy see inner-sphere reorganization energy
reversible coordination 150
rhenacyclopropene 307
rhenium(v)-di-oxo catalyst 301–302
rhodium catalyst 288
ring-closing, reaction profile 258
ring-closing metathesis 246
– energies 251
– geometric parameters 254
ring-opening reactions 190–195
– backside attack 190
– energy profiles 191
– geometric rearrangement 192
– metallacycle 204
– reaction energy 195

– reaction profile 212
RR′M $(PH_3)_2$ complexes 137–139
Ru-based catalysts 251–257
– geometry 252–253
– isomers 251
Ru-catalyzed olefin metathesis 245–264
ruthenacyclopropene, X-ray structures 307
ruthenium-silylene complexes 310

s
S–H bond, GPx 10
samarocene enolates
– initiation energies 185
– MA complexes 204
samarocene species, energies 209
samarocene systems 181
– interatomic distances 187
– relative energies 188
sandwich complexes 95
Sc^+, potential energy surface 341
Schrock's catalysts, olefin metathesis 245–264
second-row transition metal atoms, C–C bond activation 351–355
second-sphere hydrogen bonds 44
selenenic acid, [E–Se–OH] 14–16
selenocysteine
– active state 12
– residue 6
selenol, nitration 20
selenolate anion, formation 14
selenoprotein glutathione peroxidase (GPx) 1–24
– active site 2
– calculated barrier 9
– catalytic functions 3–5, 8–12, 16–23
– computational models 5–6
– concerted oxidation mechanism 18
– elementary reaction 10
– energy diagrams 11, 15
– formation of selenenic acid [E–Se–OH] 14–16
– hydrogen peroxide coordination 12–14
– nitration pathways 20–23
– oxidation pathway 17–20
– peroxidase activity 3, 8–12
– peroxynitrite/peroxynitrous acid coordination 17
– reductase activity 3–5, 16–23
– refinement of the active site 6–8
– surrounding protein effect 12–16
Shilov reaction 266–270
Si-H bond 285
side-on coordinated complexes 91

side-on η^2 isomers   99
side reactions   182
– condensation   198
– equivalent structures   207
– group transfer polymerization   198–204, 207–211
– minimization   207
silanes, hydrosilylation   286
silica surfaces, vanadium oxides   232
silicon-containing intermediates   285–316
silsesquioxane structures   232
silylium ion   298
silylmetallation   294, 305
– molecular orbitals   287
single-point energy calculations   339
$[\gamma\text{-}(SiO_4)W_{10}O_{32}H_4]^{4-}$ 218–223
Sm-catalyzed processes   192
Sm-eight-membered metallacycles
– backbiting reaction   200–201
– proton transfer   201
SOC   *see* spin-orbit coupling
solid-state catalysts   231–244
solvent effects   31, 144
– counter cation effect   228
– Shilov reaction   266
– $[\gamma\text{-}(SiO_4)W_{10}O_{32}H_4]^{4-}$ 219
Sonogashira reaction   109
spin crossover, two state reactivity paradigm   338
spin-orbit coupling (SOC), methane hydroxylation   317–336
state-selective reactivity   337
stepwise oxidation mechanism
– barriers   19
– GPx   18
stereochemistry, hydrosilylation reactions   289–291
stereoregularity   195–204
– control   192
– kinetic scheme   195–204
– rotational movement   195
steric repulsion   195
Stille reaction   122–124
stretching frequencies   93
surface species, C–H bond activation   232–233
surrounding protein effect, GPx   12–16
Suzuki–Miyaura reaction   114–121
– bis(phosphine) systems   118–119
– cross-coupling reactions   109
– monophosphine systems   120–121
symmetrical R–R coupling   135
syndio-enriched polymers   190
syndiotacticities, GTP   197

t
ten-membered metallocenes   201–204
tert-butyl hydroperoxide (TBHP)   216
tetra-n-butylammonium ion, counter cation effect   226
tetracoordinated metallocene-enolate-MA complex, symmetry   185
tetradentate ligands   27
tetrahedral intermediate, eight-membered metallacycles   203
tetrahydrofuran (THF), group transfer polymerization   184
tetramethylammonium ion   218
3,4,3′,4′-tetramethylated *ansa* system   207
tetrapyrrole cofactors   27–53
– B_{12} chemistry   38–41
– bond dissociation   40
– cavity size and flexibility   33–35
– chlorophylls   50–51
– comparison of intrinsic chemical properties   31–38
– cytochrome-like electron transfer   35–36
– cytochromes   42–44
– F430   41–42
– globins   42
– heme enzymes   44–47
– His ligand   47–50
– hole size   33
– lower axial ligands   38–42
– metallation reaction   37–38
– potential energy curves   34
– proximal side   38
– reaction energy   37
– spin states   32–33
– stability of metal–carbon bond   36–37
– tuning by axial ligands   38–51
Ti complexes   95
Toste mechanism, hydrosilylation   302
transition metal catalysts   149
– C–C bond activation   131–145, 351–355
– C–H bond activation   273–274
transition metal cations
– CH_4 bond activation   347–351
– X–H (X = O, N, C) bonds activation   339–340
transition metal center, side-on η^2 coordination   104
transition metal complexes   274–276, 337–361
– σ-bond activation   265–284
– electron localization function   355–360
– hydrosilylation   292
– Mo, W and Re   298–303
– Ru and Os   303–312
– Zr and Ta   294–296

transition metal ions 340–343
- bond insertion 343–347
transition metal oxides
- C–H bond activation 231–245
- catalysts 235–238
- methane hydroxylation 317–336
transition metal oxo bonds 241
transition metal–dinitrogen complexes 86
transition states 135, 153, 157, 208, 255–256
- C_2H_2 polymerization 354
- cleavage of N–O bond 70
- counter-cations 222
- GPx structures 9–11
- group transfer polymerization 188–190
- intramolecular hydrosilylation 306
- isomerization 127
- kinetic isotope effect 329
- Mo-based catalysts 260
- olefin metathesis 250
- olefin propagation 159
- ONIOM structures 13
- ring-opening reactions 191, 194
- Shilov reaction 266
- side reactions 207
- $[\gamma\text{-}(SiO_4)W_{10}O_{32}H_4]^{4-}$ 219
- $[\gamma\text{-}(SiO_4)W_{10}O_{28}(OH)_3(OOH)(H_2O)]^{4-}$ 220
- Stille reaction 123
- water molecule reactions 342
transmetalation 114–124
- base 117–118
- cis alternative 118
- cross-coupling reactions 111
- energy profiles 114
tripletelectronic states, frontier orbitals 98
trivanadium, olefin epoxidation 216
tungsten catalyst, hydrosilylation 298
tungsten-silylene complex 299–300
turnover-limiting step 154
two state reactivity paradigm 337–361

u

unsaturated ligands 133–145
unsaturated organic molecules 144

v

vanadium oxide 231–244
- reducibility 235
Vin–Ph reductive elimination 138
Vin–Vin reductive elimination 137
VinaVin coupling, different ligands 142
vinyl bromide, cross-coupling reactions 111
vinyl groups, reductive elimination 134
vinylboronic acid, cross-coupling reactions 111
vinylsilanes, formation 288

w

W-hydroperoxy (W–OOH) species, formation 218–220
water-derived ligands, photosystem II 59
water molecule 340–343
- bimolecular reactions 339
- reaction barriers 343
Wu–Trost mechanism, alkyne hydrosilylation 305

x

X–H (X = O, N, C) bonds activation 339–340
- reactivity paradigm 337–361

z

zirconocene enolate-MA complexes, formation energies 204
zirconocene metallacycles, diastereomeric 196
zirconocene systems 181
- energies 209
- interatomic distances 187
Zr-catalyzed processes, stereoregularity control 192
Zr centers, dinitrogen hydrogenation 101–103
Zr-eight-membered metallacycles
- backbiting reaction 200–201
- proton transfer 201